Handbook of Engineering Hydrology

Handbook of Engineering Hydrology

Editor:

Dr. Tanjina Nur, Ph.D.

ArclerPress

www.arclerpress.com

Handbook of Engineering Hydrology

Editor: Dr. Tanjina Nur, Ph.D.

708 3rd Avenue, 6th Floor
New York
NY 10017
United States of America
www.arclerpress.com

ISBN: 978-1-68094-379-5

Library of Congress Control Number: 2016950961

About the Editor

Dr. Tanjina Nur, Ph.D.

Tanjina finished her PhD in Civil and Environmental Engineering in 2014 from University of Technology Sydney (UTS). Now she is working as Post-Doctoral Researcher in the Centre for Technology in Water and Wastewater (CTWW) and published about eight International journal papers with 80 citations. Her research interest is wastewater treatment technology using adsorption process.

Contents

List of Contributors

Claudia Canedo
Department of Water Resources Engineering, Lund University
Institute of Hydraulic & Hydrology, Higher University of San Andres, Bolivia

Ramiro Pillco Zolá
Institute of Hydraulic & Hydrology, Higher University of San Andres, Bolivia

Ronny Berndtsson
Department of Water Resources Engineering, Lund University
Center for Middle Eastern Studies, Lund University

Mario Lefebvre
Department of Mathematics and Industrial Engineering, École Polytechnique, Station Centre-ville, Montreal, QC, Canada H3C 3A7

Fatima Bensalma
Department of Mathematics and Industrial Engineering, École Polytechnique, Det Box 6079, Station Centre-ville, Montreal, QC, Canada H3C 3A7

Kara A. Smith
Department of Marine, Earth and Atmospheric Sciences, North Carolina State University, Raleigh, NC 27695, USA

Fredrick H. M. Semazzi
Department of Marine, Earth and Atmospheric Sciences, North Carolina State University, Raleigh, NC 27695, USA

Dong Jiang
Institute of Geographical Sciences and Natural Resources Research, Chinese Academy of Sciences, Beijing 100101, China

Jianhua Wang
State Key Laboratory of Simulation and Regulation of Water Cycle in River Basin, Department of Water Resources, China Institute of Hydropower & Water Resources Research, Beijing 100038, China

Yaohuan Huang
Institute of Geographical Sciences and Natural Resources Research, Chinese Academy of Sciences, Beijing 100101, China

Kang Zhou
Computer Network Information Center, Chinese Academy of Sciences, Beijing 100190, China

Xiangyi Ding
State Key Laboratory of Simulation and Regulation of Water Cycle in River Basin, Department of Water Resources, China Institute of Hydropower & Water Resources Research, Beijing 100038, China

Jingying Fu
Institute of Geographical Sciences and Natural Resources Research, Chinese Academy of Sciences, Beijing 100101, China

Xiaohua Wei
Earth and Environmental Science Department, University of British Columbia (Okanagan), 3333 University Way, Kelowna BC V1V 1V7, Canada

Wenfei Liu
College of Landscape and Art, Jiangxi Agricultural University, No.1101, Zhi Min Road, Nanchang 330045, China;
Institute of Ecology and Environmental Science, Nanchang Institute of Technology, No. 289, Tianxiang Road, Nanchang 330099, China

Peicong Zhou
College of Landscape and Art, Jiangxi Agricultural University, No.1101, Zhi Min Road, Nanchang 330045, China;

Gonghuan Fang
State Key Laboratory of Desert and Oasis Ecology, Xinjiang Institute of Ecology and Geography, Chinese Academy of Sciences, Urumqi 830011, China
University of Chinese Academy of Sciences, Beijing 100049, China
Department of Geography, Ghent University, 9000 Ghent, Belgium

Jing Yang
State Key Laboratory of Desert and Oasis Ecology, Xinjiang Institute of Ecology and Geography, Chinese Academy of Sciences, Urumqi 830011, China
National Institute of Water and Atmospheric Research, Christchurch 8011, New Zealand

Yaning Chen
State Key Laboratory of Desert and Oasis Ecology, Xinjiang Institute of Ecology and Geography, Chinese Academy of Sciences, Urumqi 830011, China

Shuhua Zhang
State Key Laboratory of Desert and Oasis Ecology, Xinjiang Institute of Ecology and Geography, Chinese Academy of Sciences, Urumqi 830011, China
University of Chinese Academy of Sciences, Beijing 100049, China

Haijun Deng
State Key Laboratory of Desert and Oasis Ecology, Xinjiang Institute of Ecology and Geography, Chinese Academy of Sciences, Urumqi 830011, China
University of Chinese Academy of Sciences, Beijing 100049, China

Haimeng Liu
University of Chinese Academy of Sciences, Beijing 100049, China
Institute of Geographic Sciences and Natural Resources Research, Chinese Academy of Sciences, Beijing 100101, China

Philippe De Maeyer
Department of Geography, Ghent University, 9000 Ghent, Belgium

Johanna Springer
Department of Geography, Ludwig-Maximilians-Universitaet Muenchen (LMU), Luisenstr. 37, D-80333 Munich, Germany

Ralf Ludwig
Department of Geography, Ludwig-Maximilians-Universitaet Muenchen (LMU), Luisenstr. 37, D-80333 Munich, Germany

Stefan W. Kienzle
Department of Geography, University of Lethbridge, Alberta Water and Environmental Science Building, 4401 University Drive, Lethbridge, AB T1K-3M4, Canada
Applied Behavioral Ecology and Ecosystems Research Unit, University of South Africa, PO Box 392, Florida, 1710 Pretoria, South Africa

Feng Wu
State Key Joint Laboratory of Environmental Simulation and Pollution Control, School of Environment, Beijing Normal University, Beijing 100875, China

Jinyan Zhan
State Key Joint Laboratory of Environmental Simulation and Pollution Control, School of Environment, Beijing Normal University, Beijing 100875, China

Hongbo Su
Key Laboratory of Water Cycle & Related Land Surface Processes, Institute of Geographic Sciences and Natural Resources Research, CAS, Beijing 100101, China

Haiming Yan
State Key Joint Laboratory of Environmental Simulation and Pollution Control, School of Environment, Beijing Normal University, Beijing 100875, China

Enjun Ma
School of Mathematics and Physics, China University of Geosciences, Wuhan 430074, China

Zhixiang Lu
Key Laboratory of Ecohydrology of Inland River Basin, Cold and Arid Regions Environmental and Engineering Research Institute, Chinese Academy of Science, Lanzhou 730000, China
University of the Chinese Academy of Science, Beijing 100049, China

Songbing Zou
Key Laboratory of Ecohydrology of Inland River Basin, Cold and Arid Regions Environmental and Engineering Research Institute, Chinese Academy of Science, Lanzhou 730000, China

Zuodong Qin
Institute of Loess Plateau, Shanxi University, Taiyuan 030006, China

Yonggang Yang
Institute of Loess Plateau, Shanxi University, Taiyuan 030006, China

Honglang Xiao
Key Laboratory of Ecohydrology of Inland River Basin, Cold and Arid Regions Environmental and Engineering Research Institute, Chinese Academy of Science, Lanzhou 730000, China

Yongping Wei
The Australia China Centre on Water Resources Research, University of Melbourne, Parkville, VIC 3010, Australia

Kai Zhang
School of Soil and Water Conservation, Beijing Forestry University, Beijing 100083, China

Jiali Xie
Key Laboratory of Desert and Desertification, Cold and Arid Regions Environmental and Engineering Research

Tao Zhang
College of Hydrology and Water Resources, Hohai University, No. 1 Xikang Road, Nanjing 210098, Jiangsu Province, China

Xingnan Zhang
College of Hydrology and Water Resources, Hohai University, No. 1 Xikang Road, Nanjing 210098, Jiangsu Province, China
National Engineering Research Center of Water Resources Efficient Utilization and Engineering Safety, Hohai University, No. 1 Xikang Road, Nanjing 210098, Jiangsu Province, China

Dazhong Xia
College of Hydrology and Water Resources, Hohai University, No. 1 Xikang Road, Nanjing 210098, Jiangsu Province, China

Yangyang Liu
Changjiang Water Resources Protection Institute, No. 515 Qintai Street, Wuhan 430051, Hubei Province, China

Sandipan Ghosh
Department of Geography, The University of Burdwan, Barddhaman, West Bengal 713104, India

Biswaranjan Mistri
Department of Geography, The University of Burdwan, Barddhaman, West Bengal 713104, India

Christian Massari
Research Institute for Geo-Hydrological Protection, National Research Council CNR, Via di Madonna Alta 126, 06128 Perugia, Italy

Luca Brocca
Research Institute for Geo-Hydrological Protection, National Research Council CNR, Via di Madonna Alta 126, 06128 Perugia, Italy

Luca Ciabatta
Research Institute for Geo-Hydrological Protection, National Research Council CNR, Via di Madonna Alta 126, 06128 Perugia, Italy

Tommaso Moramarco
Research Institute for Geo-Hydrological Protection, National Research Council CNR, Via di Madonna Alta 126, 06128 Perugia, Italy

Simone Gabellani
International Centre on Environmental Monitoring (CIMA) Research Foundation, Via A. Magliotto, 2-17100 Savona, Italy

Clement Albergel
European Centre for Medium-Range Weather Forecasts (ECMWF), Shinfield Park, RG2 9AX Reading, UK

Patricia De Rosnay
European Centre for Medium-Range Weather Forecasts (ECMWF), Shinfield Park, RG2 9AX Reading, UK

Silvia Puca
Italian Civil Protection Department, Via Vitorchiano 2, 00189 Rome, Italy

Wolfgang Wagner
Department of Geodesy and Geoinformation, Vienna University of Technology, 1040 Vienna, Austria

Valentin Heimhuber
Hydrology and River Basin Management, Technical University of Munich (TUM), Arcisstrasse 21, Munich 80333, Germany

Johann-Christian Hannemann
Landscape Architecture and Public Space, Technical University of Munich (TUM), Emil-Ramann-Straße 6, Freising 85350, Germany

Wolfgang Rieger
Hydrology and River Basin Management, Technical University of Munich (TUM), Arcisstrasse 21, Munich 80333, Germany

Eric Pohl
Remote Sensing Group, Institute of Geology, Technische Universität Bergakademie Freiberg, B.-von-Cotta-Str. 2, Freiberg D-09599, Germany

Richard Gloaguen
Remote Sensing Group, Institute of Geology, Technische Universität Bergakademie Freiberg, B.-von-Cotta-Str. 2, Freiberg D-09599, Germany
Remote Sensing Group, Helmholtz Institute Freiberg for Resource Technology, Freiberg D-09599, Germany

Ralf Seiler
Remote Sensing Group, Institute of Geology, Technische Universität Bergakademie Freiberg, B.-von-Cotta-Str. 2, Freiberg D-09599, Germany

L. Seoane
CNRS/UMR8630, Observatoire de Paris, 61 avenue de l'Observatoire, 75014 Paris, France

J. Nastula
Space Research Center, Polish Academy of Sciences, Bartycka 18 A, 00-716 Warsaw, Poland

C. Bizouard
CNRS/UMR8630, Observatoire de Paris, 61 avenue de l'Observatoire, 75014 Paris, France

D. Gambis
CNRS/UMR8630, Observatoire de Paris, 61 avenue de l'Observatoire, 75014 Paris, France

Yong Niu
College of Hydrology and Water Resources, Hohai University, Nanjing 210098, China
Research Center of Lake Environment, Chinese Research Academy of Environmental Science, Beijing 100012, China

Wei Jiao
School of Environment, Beijing Normal University, Beijing 100875, China

Hui Yu
Research Center of Lake Environment, Chinese Research Academy of Environmental Science, Beijing 100012, China

Yuan Niu
Research Center of Lake Environment, Chinese Research Academy of Environmental Science, Beijing 100012, China

Yong Pang
College of Hydrology and Water Resources, Hohai University, Nanjing 210098, China

Xiangyang Xu
College of Hydrology and Water Resources, Hohai University, Nanjing 210098, China

Xiaochun Guo
Research Center of Lake Environment, Chinese Research Academy of Environmental Science, Beijing 100012, China

Susan M. Crooks
Centre for Ecology & Hydrology, Crowmarsh Gifford, Wallingford OX10 8BB, UK

Alison L. Kay
Centre for Ecology & Hydrology, Crowmarsh Gifford, Wallingford OX10 8BB, UK

Helen N. Davies
Centre for Ecology & Hydrology, Crowmarsh Gifford, Wallingford OX10 8BB, UK

Victoria A. Bell
Centre for Ecology & Hydrology, Crowmarsh Gifford, Wallingford OX10 8BB, UK

Yang Yu
Chair of Hydrology and River Basin Management, Technical University of Munich (TUM), Arcisstrasse 21, Munich 80333, Germany
Xinjiang Institute of Ecology and Geography, Chinese Academy of Sciences (CAS), No. 818 South Beijing Road, Urumqi 830011, Xinjiang, China

Markus Disse
Chair of Hydrology and River Basin Management, Technical University of Munich (TUM), Arcisstrasse 21, Munich 80333, Germany

Ruide Yu
Xinjiang Institute of Ecology and Geography, Chinese Academy of Sciences (CAS), No. 818 South Beijing Road, Urumqi 830011, Xinjiang, China

Guoan Yu
Chair of Hydrology and River Basin Management, Technical University of Munich (TUM), Arcisstrasse 21, Munich 80333, Germany
Key Laboratory of Water Cycle and Related Land Surface Processes, Institute of Geographic Sciences and Natural Resources Research, Chinese Academy of Sciences, 11A Datun Road, Chaoyang District, Beijing 100101, China

Lingxiao Sun
Xinjiang Institute of Ecology and Geography, Chinese Academy of Sciences (CAS), No. 818 South Beijing Road, Urumqi 830011, Xinjiang, China

Philipp Huttner
Chair of Hydrology and River Basin Management, Technical University of Munich (TUM), Arcisstrasse 21, Munich 80333, Germany

Christian Rumbaur
Chair of Hydrology and River Basin Management, Technical University of Munich (TUM), Arcisstrasse 21, Munich 80333, Germany

Preface

Engineering Hydrology, sometimes called hydrologic engineering, is an engineering specialty that focuses on water resources. Engineering Hydrology is chiefly concerned with the flow and storage of water including urban drainage, water supply, wastewater treatment, river management and coastal protection. Hydrological engineering also focuses on preventing floods and lessening the effects of floods, droughts and other natural disasters. In this book, we try to cover different aspects of Engineering Hydrology. Firstly, we discuss the role of hydrological studies for the development of the TDPS System and an application of filtered renewal processes. Then we present the role of the dominant modes of precipitation variability over Eastern Africa in modulating the hydrology of Lake Victoria and the review of GRACE Data Applications in terrestrial hydrology monitoring. Relative contributions of forest change and climatic variability to hydrology in large watersheds and climate change impact on the hydrology of a typical watershed in the Tianshan Mountains are also discussed.

In the middle part of the book, we focus on impacts of forest fires and climate variability on the hydrology of an Alpine medium sized catchment in the Canadian Rocky Mountains and scenario-based impact assessment of land use/cover and climate changes on watershed hydrology in Heihe river basin of Northwest China. Then hydrologic responses to land Use change in the Upper Fenhe River Watershed and an analysis of land use change dynamics and its impacts on hydrological processes in the Jialing River Basin are also discussed. We also try to present geographic concerns on flood climate and flood hydrology in monsoon-dominated Damodar river basin of Eastern India and the use of H-SAF soil moisture products for Operational Hydrology along with flood risk management in remote and impoverished Areas of Haiti. The last chapters of the book describe remote sensing-based assessment of the variability of winter and summer precipitation in the Pamirs and their effects on hydrology and hazards using harmonic time series analysis. We also try to find out the hydrological excitation of polar motion derived from GRACE gravity field solutions and spatial evaluation of heavy metals concentrations in the surface sediment of Taihu Lake. At the end, we describe National scale rainfall-runoff modelling with demonstration of a hydrological modelling framework and large-Scale hydrological modeling and decision-making for agricultural water consumption and allocation in the main stem tarim river of China.

Editor
Dr. Tanjina Nur, Ph.D.

INTRODUCTION

Hydrological engineering, sometimes called hydrologic engineering, is an engineering specialty that focuses on water resources. There are both undergraduate and graduate degree programs available in this field, leading to potential career options in education, engineering and consultancy. Hydrological engineering, also called water resources engineering, is a civil engineering specialty offered at both the undergraduate and graduate levels. Hydrological engineering is chiefly concerned with the flow and storage of water. Topics commonly covered include urban drainage, water supply, wastewater treatment, river management and coastal protection. Hydrological engineering also focuses on preventing floods and lessening the effects of floods, droughts and other natural disasters.

Scope of Hydrology

- Water is one the most valuable natural resources essential for human and animal life, industry and agriculture.
- It is also used for Power generation, navigation and fisheries.
- Tremendous importance is given to the hydrology all over the world in the development and management of water resources for irrigation, water supply, flood control, water-logging and salinity control, Hydro power and navigation.

It uses hydrologic principles in the solution of engineering problems arising from human exploitation of water resources of the earth. The engineering hydrologist, or water resources engineer, is involved in the planning, analysis, design, construction and operation of projects for the control, utilization and management of water resources.

Hydrologic calculations are estimates because mostly the empirical and approximate methods are used to describe various hydrological processes.

Uses & Applications of Hydrology in Engineering

In hydrology we apply scientific knowledge and mathematical principles to solve water-related problems in society: problems of quantity, quality and availability. Mathematical models of all Hydrological phenomena are made.

They may be concerned with finding water supplies for cities or irrigated farms, or controlling river flooding or soil erosion. Or, they may work in environmental protection: preventing or cleaning up pollution or locating sites for safe disposal of hazardous wastes.

Applications of Engineering Hydrology

- Hydrology is used to find out maximum probable flood at proposed sites e.g. Dams.
- The variation of water production from catchments can be calculated and described by hydrology.
- Engineering hydrology enables us to find out the relationship between a catchments's surface water and groundwater resources
- The expected flood flows over a spillway, at a highway Culvert, or in an urban storm drainage system can be known by this very subject.
- It helps us to know the required reservoir capacity to assure adequate water for irrigation or municipal water supply in droughts condition.
- It tells us what hydrologic hardware (e.g. rain gauges, stream gauges etc) and software (computer models) are needed for real-time flood forecasting
- Used in connection with design and operations of hydraulic structure
- Used in prediction of flood over a spillway, at highway culvert or in urban storm drainage
- Used to assess the reservoir capacity required to assure adequate water for irrigation or municipal water supply during drought
- Hydrology is an indispensable tool in planning and building hydraulic structures.

- Hydrology is used for city water supply design which is based on catchments area, amount of rainfall, dry period, storage capacity, runoff evaporation and transpiration.
- Dam construction, reservoir capacity, spillway capacity, sizes of water supply pipelines and effect of afforest on water supply schemes, all are designed on basis of hydrological equations.
- Determining the water balance of a region.
- Determining the agricultural water balance.
- Designing riparian restoration projects.
- Mitigation and predicting floods, landslides and drought risk.
- Flood forecasting and flood warnings.
- Designing irrigation schemes and managing agricultural productivity.
- Designing dams for water supply or hydroelectric power generation.
- Designing bridges.
- Designing sewers and urban drainage system.
- Predicting geomorphologic changes, such as, erosion or sedimentation.
- Assessing the impact of natural and anthropogenic environmental change.
- Assessing containment transport risk and establishing environmental policy guidelines

Global Hydrological Cycle Processes - The Hydro-Cycle

- The hydrologic cycle describes the continuous re-circulating transport of the waters of the earth, linking atmosphere, land and oceans.
- Water evaporates from the ocean surface, driven by energy from the Sun, and joins the atmosphere, moving inland as clouds. Once inland, atmospheric conditions act to condense and precipitate water onto the land surface, where, driven by gravitational forces, it returns to the ocean through river and streams.
- The process is quite complex, containing many sub-cycles.
- Engineering Hydrology takes a quantitative view of the hydrologic cycle.
- The quantification of the hydrologic cycle which is an open system can be represented by a mass balance equation, where inputs minus outputs are equal to the change in storage.
- It is a basic Hydrologic Principle or equation that may be applied either on global or regional scale

The water holding elements of the hydrological cycle are Hydrological Cycle:

- Atmosphere
- Vegetation
- Snow packs
- Land surface
- Soil
- Streams, lakes and rivers
- Aquifers
- Oceans

Hydrological Cycle also known as Water Cycle is the study of continuous movement of water on, above and below the earth surface.

The main processes involved in hydrological cycle are:

- Evaporation
- Condensation
- Precipitation
- Interception
- Infiltration
- Percolation
- Transpiration
- Runoff
- Storage

The water cycle begins with the evaporation of water from oceans and other water bodies. The resulting vapors are transported by moving air and under proper conditions, the vapor are condensed to form clouds, which in turn results in precipitation.

The precipitation which falls upon land is dispersed in several ways. The greater part is temporarily retained in the soil near where it falls and is ultimately returned to the atmosphere by evaporation and transpiration by plants. A portion of the water flows over surface soil to stream channels, while other penetrates into the ground to become part of the ground water. Under the influence of gravity, both surface and underground water move towards lower elevations and may eventually discharge into the oceans.

This Hydrologic Cycle recycles the earth's valuable water supply. In other words, the water keeps getting reused over and over.

Hydrological Systems

A hydrologic system is as a structure or volume in space, surrounded by a boundary, that accepts water and other inputs, operates on them internally, and produces them as outputs. The structure (for surface or subsurface flow) or volume in space (for atmospheric moisture flow) is the totality of the flow paths through which the water may pass from the point it enters the system to the point it leaves. The boundary is a continuous surface defined in three dimensions enclosing the volume or structure. A working medium enters the system as input, interacts with the structure and other media, and leaves as output. Physical, chemical and biological processes operate on the working media within the system; the most common working media involved in hydrologic analysis are water, air and heat energy. The global hydrologic cycle can be represented as a system containing three subsystems: the atmospheric water system, the surface water system, and the subsurface water system.

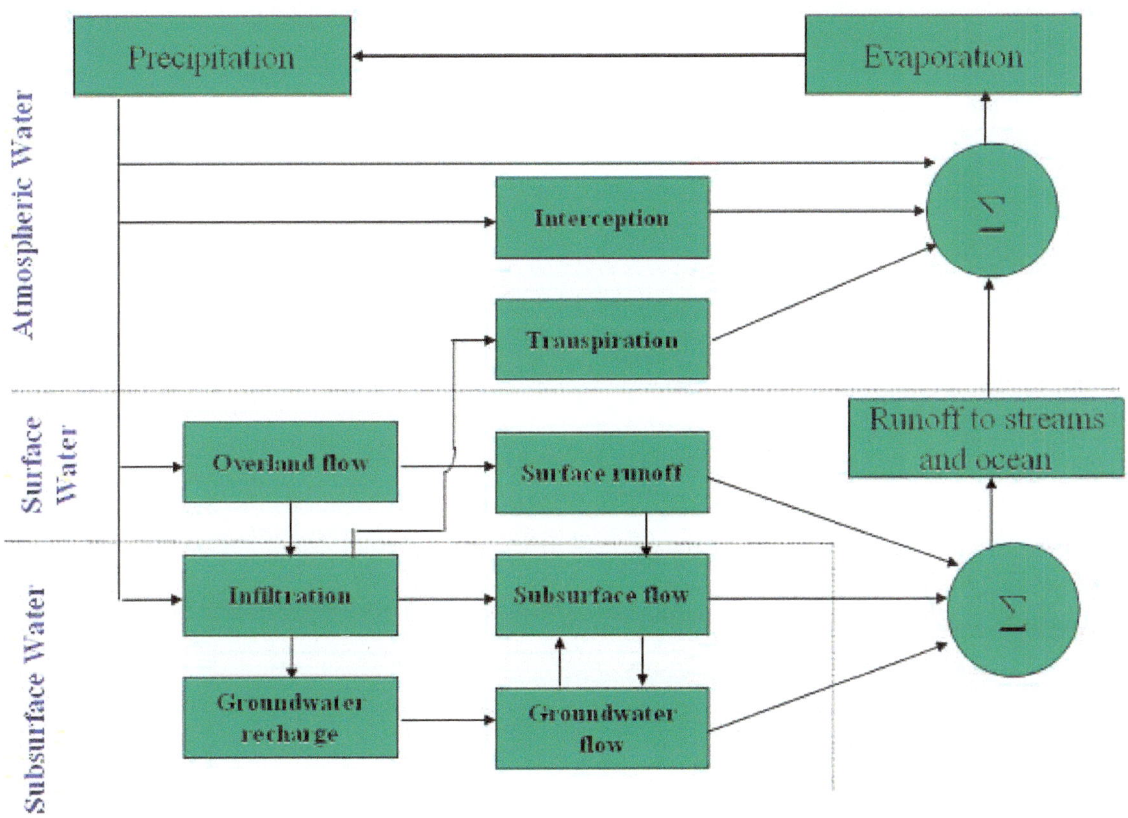

Catchment and Basin

- A Catchment is a portion of the earth's surface that collects runoff and concentrates it at its furthest downstream point, referred to as the catchment outlet.
- The runoff concentrated by a catchment flows either into a larger catchment or into the ocean.
- The place where a stream enters a larger stream or body of water is referred to as the mouth.
- The terms watershed and basin are commonly used to refer to catchments. Generally, watershed is used to describe a small catchment (stream watershed), whereas basin is reserved for large catchments (river basins).

Watershed and Stream order

- The watershed or basin is defined by the surrounding topography, the perimeter of which is called a divide. It is the highest elevation surrounding the watershed. All of the water that falls on the inside of the divided has the potential to be shed into the streams of the basin encompassed by the divide. Water falling outside of the divide is shed to another basin.

- The water flowing in streams is called stream flow

Horton suggested a classification of stream order as a measure of the amount of branching within a basin. A first order stream is a small, un-branched tributary. A second order stream has only first order tributaries. A third order stream has only first and second order tributaries and so on. When a channel of lower order joins a channel of higher order, the channel downstream retains the higher of the two orders.

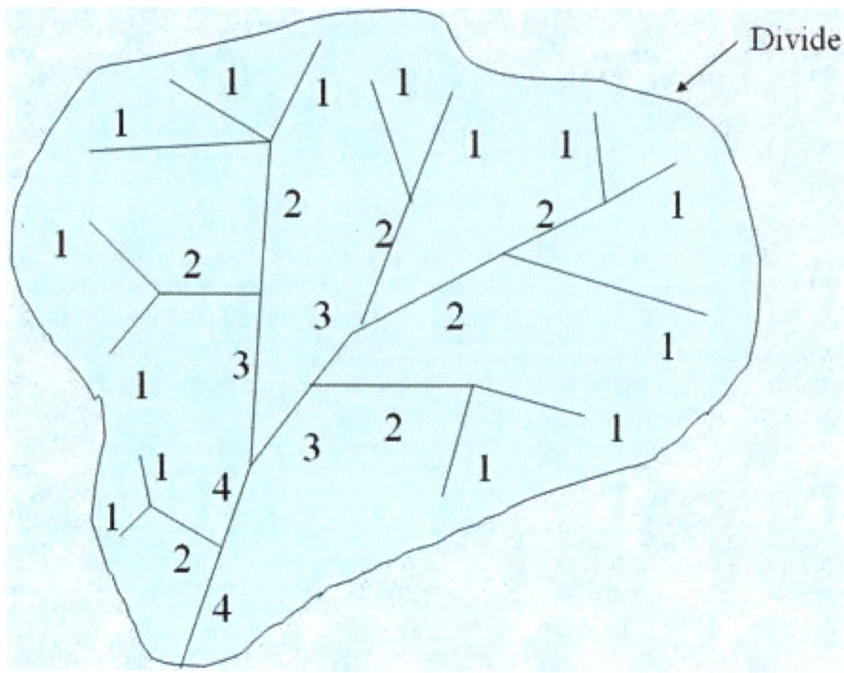

Water Balance Problem

In a given year, a catchment with an area of 2500 km^2 received 1.3 m of precipitation. The average rate of flow measured in a river draining the catchment was 30 m3s^{-1}.

1. How much total river runoff occurred in the year (in m3)?
2. What is the runoff coefficient?
3. How much water is lost due to the combined effects of evaporation, transpiration, and infiltration? (Expressed in m).

Solution

Total runoff volume

= number of seconds in a year ´ average flow rate

= 31 536 000 ´ 30

= 9.4608´108 m^3

Runoff coefficient

= Runoff volume/ precipitation volume

= (9.4608´108) / (1.3 ´ 2500 ´ 106)

= 0.29 (29 %)

The water balance equation can be arranged to produce:

ET+F= P - R - ΔS

So,

ET + F = 3.25´109 - 9.4608´108

= 2.30392´109 m^3

= (2.30392´109) / (2500´106)

= 0.92 m

Precipitation

What is Precipitation? Defined:

Water falling in solid or liquid form e.g. rain, snow, and hail.

Uses of Precipitation Data

- Runoff estimation analysis
- Groundwater recharge analysis
- Water balance studies of catchments
- Flood analysis for design of hydraulic structures
- Real-time flood forecasting
- low flow studies

Factors Affecting Precipitation Formation

Following are the factors mainly responsible for the formation of precipitation

- A lifting mechanism to produce cooling of the air.
- A mechanism to produce condensation of water vapors and formation of cloud droplets.
- A mechanism to produce growth of cloud droplets to size capable of falling to the ground against the lifting force of air.
- When air ascends from near the surface to upper
- Levels in the atmosphere it cools.
- Cooling lowers the capacity of a given volume of air to hold a certain amount of water vapor.
- As a result super saturation occurs and the excess moisture over saturation condenses through the cooling process.

Mechanism of Cooling

This is the only mechanism capable of producing the degree and rate of cooling needed to account for heavy rainfall.

Condensation of Water Vapor

- Condensation of water into cloud droplets takes place on hygroscopic nuclei which are small particles having an affinity for water.
- The source of these condensation nuclei are the particles of sea salt or products of combustion of certain sulfurous and nitrous acid and carbon dioxide.
- There are always sufficient nuclei present in the atmosphere.
- Growth of droplets is required if the liquid water present in the cloud is to reach the ground. The two processes regarded as most effective for droplet growth are:
- Coalescence of droplets through collision due to difference in speed of motion between larger and smaller droplets.
- Co-existence of ice crystals and water droplets.
- Co-existence effect generally happens in the temperature range from 100 to 20 °F

CHAPTER 1

ROLE OF HYDROLOGICAL STUDIES FOR THE DEVELOPMENT OF THE TDPS SYSTEM

Claudia Canedo, Ramiro Pillco Zolá and Ronny Berndtsson

Keywords: Water resources, climate, hydrology, climate change, TDPS system, Altiplano

INTRODUCTION

The South American Altiplano is one of the largest high plateaus in the world. The highland is faced with extreme climatic variability that determines the region´s poverty and harsh living conditions [1]. The area covers about 192,000 km^2, and presents a north–south gradient variation with an annual precipitation decreasing from 1100 mm to 200 mm [2] and about 1800 mm average annual potential evaporation. The Altiplano contains two large lakes, Lake Titicaca in the northern part and Lake Poopó in the central part. The two lakes are connected through the Desaguadero River and salt content in the water is gradually increasing from north to south, with Lake Poopó's extremely salty water. This is due to increasing upstream irrigation leading to a more salty return flow and lower precipitation occurrence in the southern part of the system. The total water storage is about 936 km^3. The water in Lake Titicaca represents about 99% of the total surface water in the Altiplano, meaning that this lake supplies a major part of water and humidity into the southern arid region. Together, the semiarid region around Lake Titicaca, the arid region around Lake Poopó, and the saltpans constitute a large watershed, called the Lake Titicaca, Desaguadero River, Lake Poopó, and Coipasa Salt Flat System (TDPS hydrologic system; Figure 1).

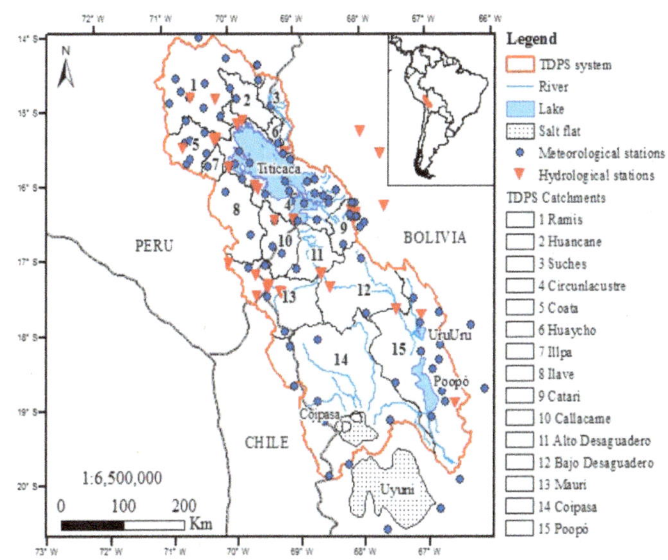

Figure 1: Catchments and hydroclimatological control points of the Lake Titicaca, Desaguadero River, Lake Poopó, and Coipasa Salt Flat System (TDPS). In total, there are 91 weather stations (43 in Bolivia, 37 in Peru, and 11 in Chile) in the TDPS system (one station per 1600 km^2).

The TDPS system has an average elevation of about 4000 m above mean sea level (amsl). It extends from the Amazon and La Plata basins to the east and the Atacama Desert, influenced by the southeast Pacific anticyclone to the west, along the central portion of the Andes cordillera [3,4]. The Andes cordillera, the longest mountain chain in the world, runs through the western part of the continent (between latitudes ~10° N and 53° S). In contrast to its length, it is relatively narrow (~200 km). However, at its widest point it is divided into two mountain ranges surrounding the Altiplano. The

Citation: Canedo, C.; Pillco Zolá, R.; Berndtsson, R. Role of Hydrological Studies for the Development of the TDPS System. Water 2016, 8, 144. doi:10.3390/w8040144.

highest peak in the Bolivian Altiplano is the Sajama volcano which is 6542 m amsl (18°06' S, 68°53' W) [5]. The TDPS system covers an area of approximately 144,590 km^2, with 34% in Peru, 61% in Bolivia, and 5% in Chile (Figure 1). The Desaguadero River is Titicaca's only outlet that flows into Lake Poopó. Sporadic overflow from Lake Poopó in turn ends up in the Coipasa Saltpan (e.g., [6]). The world's largest salt desert, the Salar de Uyuni, is located south of the TDPS system [7].

The total population of the TDPS system was about 2.5 million in 2003 (48% Peruvians and 52% Bolivians) [8]. In 2009, the population had increased to more than 3 million [9]. The principal economic activities in the TDPS system are agriculture, industry, and trade [10]. The common agricultural production is focused on fishing, cattle, and crops, such as quinoa and potato [8]. Generally, the agricultural productivity in the Altiplano is low due to the high susceptibility to climate, poor soils, and manual labor. The high evaporative demand and the erratic distribution of rainfall increase the agricultural vulnerability. In addition, the precipitation dependency of rain-fed agriculture and the high frost risk limit the farming production to the moist austral summer [11].

Important constraints to agricultural development in the Altiplano are the lacking data on climatic variability and related water resources availability. Lacking data are also hampering the scientific publications for the region and restrict the accessibility of climatological and hydrological information. The situation is not improved by the lack of collaboration between scientists, local populations, and governmental and non-governmental organizations [12,13]. Most of the research has been carried out for the Lake Titicaca and Desaguadero River basins. The spatially-uneven meteorological and hydrological data collection restricts the water resources assessment for the rest of the TDPS system and for sub-basin scales. In spite of the scarce data situation and little knowledge on rainfall-runoff relationships for the arid catchments along the Altiplano, with common problems of flash floods, rivers are now being diverted and dams are being built at a large scale. The Bolivian government plan for 2015–2020 includes building dams along intermittent rivers. With the present lack of knowledge of the area's hydrology, the ecological and environmental consequences of such large-scale infrastructure are difficult to foresee.

In view of the above, it is clear that the Altiplano poses specific intricacies regarding topography and climatic variability. This, in combination with significant data scarcity, makes it a difficult area to model hydrologically. Henceforth, the objective of this study is to produce a state of the art review of the research with respect to climate and water resources in the TDPS hydrological system, define important research gaps in hydrological data information of the region, and the key role that these have for the sustainable development in view of climate change and population increase.

LIVING CONDITIONS AND ENVIRONMENT

The Central Andes are mainly inhabited by people of indigenous origin. These communities have developed cultures and technologies adapted to the harsh climate conditions in the region [14]. An example of adaptive agricultural technology is the agrodiversity system. This means that agricultural practices are managed over diverse ecological zones and the social organization relates to this diversification. The prosperous Tiwanaku civilization (300 BC–AD 300) developed agrodiversity to successfully deal with three principal problems in the Altiplano, namely; poor soil, aridity, and frost. The Tiwanakus thus created micro-climates for a sustainable agriculture by use of raised terraced fields, called "camellones". The camellones or canals brought water and nutrients to the fields and acted as heat storages during the night [15]. Nevertheless, at present, poverty is the major barrier to overcome in order to improve living conditions. The limited availability of local resources and services, like water and sanitation access (about 20%), compel the rural population to migrate to cities [8].

The traditional quinoa cultivation extends along the high Andes, and it is one of the most important crops of the region. The quinoa production in Bolivia increased from 34,000 ton in 2009 to 84,000 ton in 2014, with an export rise from 43,000 to 196,000 USD, respectively [16]. This increase improved farmers' income, and increased national and international trade. The demand was especially pronounced in 2013, named the International Year of Quinoa by the United Nations. In 2013, Bolivia and Peru supplied over 95% of the global quinoa production. The quinoa price fell from 6,600 USD/ton in 2013 to 4,500 USD/ton in 2014, due to the increase in production, especially in Peru, that surpassed Bolivia as the largest exporter of quinoa in the world [17].

CLIMATE VARIABILITY

Two factors play a main role for the extreme climate of the Altiplano. The first is the Bolivian High that acts as a buffer against moist air inflow to the region. The second is the orographic barrier of the Andes that affects not only the regional climate, but also the world climate [18]. The Altiplano summertime (DJFM) precipitation variability, and its relationship to the atmospheric circulation, were investigated by among others Thibeault *et al.* [19], Lenters and Cook [20], Vuille [5], and Garreaud and Aceituno [21]. Easterly flow transports humid air from the lowlands east of the Altiplano, while westerly flow inhibits moisture transport from the east. This mechanism affects the precipitation variability and relates to changes in the mid- and upper-level wind patterns [19]. Summertime precipitation variability of the Altiplano relates to the strength and position of the upper-level anticyclone called the Bolivian High. This affects the direction of low-level flow east of the Altiplano [20,22]. Northward (southward) displacement of the Bolivian High is associated with low-level southerly (northwesterly) flow and dry, stable (humid, unstable) conditions that are unfavorable (favorable) for convection in the Altiplano [19]. During the austral summer establishment of the upper-level Bolivian High is centered at 17° S and 70° W [23].

The TDPS system within the Altiplano area is a semiarid to arid region, where the seasonal distribution of rain is similar throughout the region. The rainy season occurs generally from December to March, and the dry season from May to August [24]. Annual rainfall can fluctuate 50% from the mean. The largest rainfall in the TDPS System occurs in and around the Lake Titicaca [25]. Outside the Lake Titicaca area, the northern part of the system (surrounding Coata and Ramis Rivers) has an annual precipitation of about 800 to 1400 mm

[24]. However, in the southern part of the TDPS system the annual precipitation decreases to about 200 mm along the southwestern part of the system (Salar de Coipasa). Along the east, precipitation is caused by moisture lifting from the Amazon through the Cordillera.

The Bolivian Altiplano wet season takes place from November to March (Figure 2) [3]. Over 60% of the annual precipitation fall during the summer months (DJF) in association with the South American Monsoon (SAM) and inter-annual variations are large (e.g., [3,26]). During the winter months cool and dry westerly flow is present along the Altiplano. During austral summer, warm, moist, easterly flow spans the central Andes, providing abundant moisture and instability at low levels [5,26,27]. By way of these eastern humidity fluxes, rainfall on the Lake Titicaca Basin is about 550 mm during the rainy season and represents about 70% of the average annual rate (800 mm) [28]. During the austral winter, westerly winds become the predominant flux, transporting dry air from the Pacific Ocean. The average precipitation during the dry eight months over the Lake Titicaca is about 250 mm. The annual precipitation over the Lake Poopó in the northern part of the basin is about 420 mm and, in the southern part, 270 mm [29].

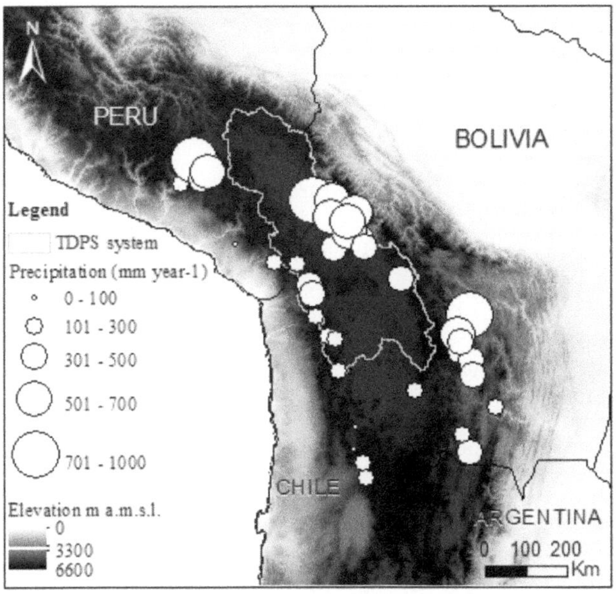

Figure 2: Average annual precipitation over the Altiplano. The symbol size indicates the amount of rainfall (the meteorological data were supplied by SENAMHI Bolivia, SENAMHI Peru, and CAZALAC Chile) (adapted from Aceituno [30] and Garreaud *et al.* [3]).

In the Altiplano, drought and frost are frequent, with consequences in yield reductions, leading to shortages of food production [11]. Vizy and Cook found that rainfall variability over the central Andes is not directly controlled by moisture changes in the Amazon basin [31]. Drought presents a decadal variation related to the Pacific Decadal Oscillation in Bolivia, with a higher frequency in the Altiplano, that is affected principally by the El Niño Southern Oscillation (ENSO) anomalies [32]. Furthermore, due to the increase of evaporation, the Altiplano shows a trend towards drier conditions, and consequently an increase in the magnitude and duration of drought events.

Previous studies have shown a relationship between the Altiplano precipitation anomaly and the Southern Oscillation. Wet (dry) summers relate to cooling (warming) of the tropical troposphere and associated changes in the upper-level zonal flow during La Niña (El Niño) events [3,19,21]. Garreaud *et al.* found that the inter-annual variability in South America is manly influenced by ENSO [33]. The same study determined anomalies during the ENSO phases with an annual decrease (increase) of 1 °C in temperature during La Niña (El Niño) events and higher (lower) precipitation in austral summer over the tropical Andes. Seiler *et al.* investigated impacts of ENSO on temperature and precipitation from 1960 to 2009, with generally warmer temperatures leading to drier conditions with more variable precipitation during the austral summer (DJF) in the Andes [34]. Thibeault *et al.* showed that the La Niña causes dry (wet) anomalies in the central (western) Pacific [19]. López-Moreno *et al.* studied the inter-annual variability of air temperature from 1945 to 2012 over the Altiplano in Peru and Bolivia, concluding that the Southern Oscillation Index (SOI) and the maximum temperature have significant negative (positive) correlations during the austral summer (winter). Minimum temperature and SOI presented lower correlation coefficients [35].

WATER RESOURCES UTILIZATION

The fragile TDPS system is experiencing an increasing water demand from both Bolivia and Peru (Figure 3; [6]). The TDPS surface water represents the most important water source for the entire Altiplano. The possibilities for extended groundwater use are not well known at present. For instance water supply in Oruro city (with more than 500,000 inhabitants [9]) comes mainly from groundwater. Pressure from the population is mainly related to water extraction for irrigation purposes rather than for mining. Equitable distribution between Bolivia and Peru regarding the TDPS system goes back to 1987 when the two countries signed a joint agreement on the Binational General Master Plan for the Development of the Integrated Region of the Lake Basin. In 1988, an agreement to maintain the TDPS as a unit was added. The PELT (Proyecto Especial del Lago Titicaca) project ran from 1989 to1993 and the outcome of this project is still the most important source of hydrological information. At the end of the project the two countries established the Authority of the Lake Titicaca (ALT-Autoridad para el Lago Titicaca; [6]). However, since 1993, especially for the downstream of the Lake Titicaca, not many hydrological studies or increase in hydrological knowledge of the region have been achieved.

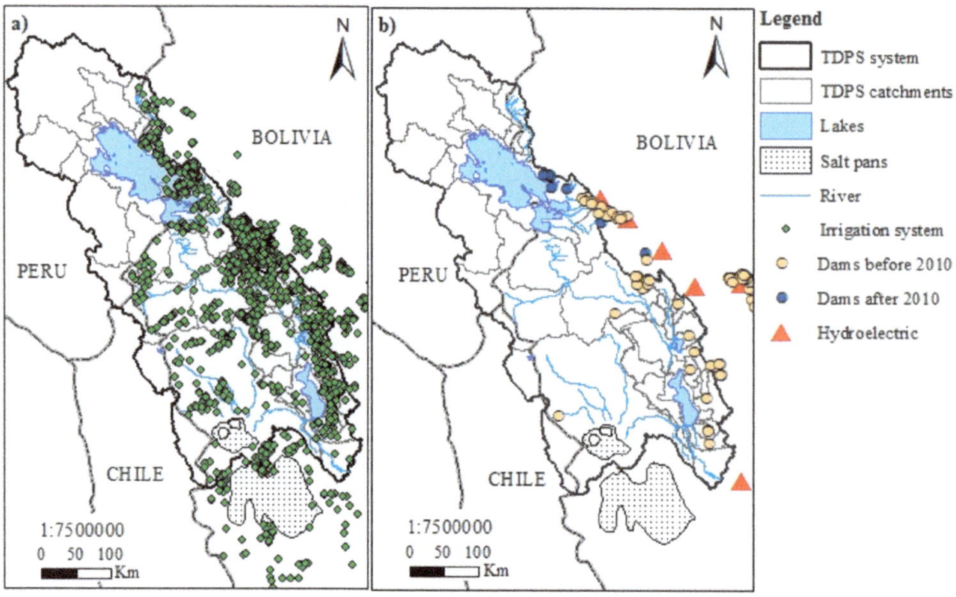

Figure 3: The Bolivian TDPS water use with irrigation systems (**a**) and water reservoirs; (**b**) (adapted from VRHR-MMAyA [36] and PRO-AGRO [37]).

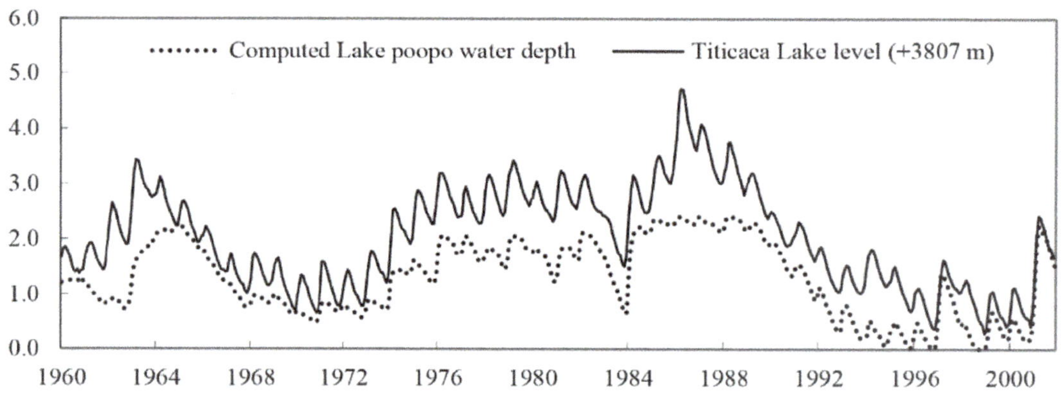

Figure 4: Long-term water level relationship between the Lake Titicaca and the Lake Poopó (adapted from Pillco and Bengtsson [29]).

Table 1: Sub-catchment average discharge in the TDPS system during two periods

Catchment	Discharge Point	Q_a 1965–1989 (m^3/s)	Q_b 1965–2012 (m^3/s)	Q_b/Q_a (–)
Lake Titicaca	Titicaca outflow	51.3	35.0	0.68
Mauri River	Mauri-Calacoto	18.7	15.5	0.82
Desaguadero River Alto	Ulloma	78.5	58.3	0.74
Desaguadero River Bajo	Chuquiña	90.7	68.1	0.75

The PELT project defined a flow rate of 20 m^3/s to be equally shared between the two countries. The flow rate was established from water balance calculations for the upper part of the hydraulic network of TDPS just downstream of the Lake Titicaca [38]. In addition, a hydrological study of the TDPS was delimited to the upstream of the Chuquiña discharge station (located on the Desaguadero, upstream of Lake Poopó) due to the lack of hydrological observations in the southern part of the TDPS. Moreover, the water supply capacity requires an updated analysis, considering possible climate and land use changes in the region.

Lake Titicaca in the upper part of TDPS is the most important reservoir of fresh water for the Altiplano with 56,270 km^2 of drainage area. The mean water level is 3810 m amsl with a corresponding surface area of 8400 km^2 [39]. The lake is fed by ten main tributaries and direct rainfall on the lake mainly during the rainy season between March and April. From the total river inflow, five rivers contribute about 67% of inflow (Ramis, Coata, Ilave, Huancane, and Suches) [40]. The only single outlet from the lake is the Desaguadero River, where discharge appears to have decreased during the last decades according to Table 1. Generally, the maximum discharge occurs in April [28,41]. During April 1986 a water level peak was recorded at 3811.7 m amsl. A minimum of 3807.4 m amsl was recorded in 1996 (Figure 4). The Titicaca drainage area receives the highest annual precipitation of the

TDPS (from 800 to 1400 mm) [42], and the evaporation rate from the lake is close to 1500 mm/year.

From Lake Titicaca via the Desaguadero River water is transported about 300 km to the shallow lakes of Poopó and Uru-Uru. Here, the average areal precipitation is about 350 mm/year [29], concentrated to four months (DJFM). The strong seasonal rainfall defines the intermittent flows in the regional rivers. Pillco and Bengtsson found that the long-term Lake Poopó water level can be predicted using a simple regression equation with the Lake Titicaca water level (Figure 4) [29]. They also found that the Lake Poopó water storage depends on 65% of the Desaguadero River discharge. On the other hand, most of this discharge represents outflow from Lake Titicaca. The rest depends on the drainage area contribution between Titicaca and Poopó. According to Figure 4, the water level fluctuation in the Lake Titicaca is greater as compared to the Lake Poopó in terms of seasonal and annual variation. The highest water level fluctuation in Titicaca occurred from 1980 to 1992. This was not noted in the Lake Poopó, due to rapid outflow to the southern basins. During the last century, only during 1986–1987 the entire TDPS systems, including the salt flat areas, have displayed continuous discharge in all parts of the drainage basin.

Four main contributors define the discharge at the outlet of Desaguadero River. From upstream to downstream, the first is Lake Titicaca discharge (see Figure 5) with the highest outflow in 1986, reaching 380 m^3/s. The second is the Desaguadero-Alto catchment with a drainage area of 3616 km^2. The third most important river contributor is Mauri, with an area of 9803 km^2 and with a mean discharge of 15 m^3/s at its outlet at Mauri-Calacoto. The fourth contributor is the Desaguadero-Bajo catchment upstream the Chuquiña discharge point, with a total area of 10,564 km^2. The mean discharge from these areas is 70 m^3/s. Table 1 shows a comparison of flow using available data and comparing the mean annual discharge of the periods 1965–1989 and 1965–2012. From the relation Q_b/Q_a, it can be seen that the Lake Titicaca contribution to the Desaguadero River has experienced a major change of the average water flow for the period 1965–2012 (35.0 m^3/s) compared to the period 1965–1989 (51.3 m^3/s). This change shows an important need to update the water budget of the TDPS system, in order to redefine the water distribution in the TDPS system.

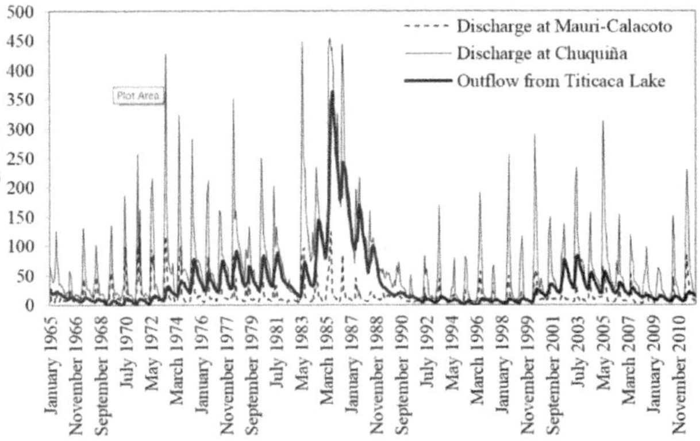

Figure 5: Long-term discharge of the main contributors in the TDPS system (adapted from Autoridad Binacional Lago Titicaca (ALT) and Uría and Molina [43,44]).

The Lake Poopó subsystem comprises two interconnected lakes, the Uru-Uru and the Poopó. Lake Poopó is remarkably shallow with a depth ranging from only 1.30 to 2.37 m. Thus, between wet and dry seasons the surface water area varies from 1000 to 2,500 km^2 [45]. In addition, Pillco and Bengtsson found that the lake area reduces by more than 50% during the dry season as compared to the wet season [45]. The lake has dried out at several occasions. The longest period of such an event occurred from 1994 to 1997. This situation destroyed the lake biomass and strongly affected fishermen's income. Otherwise, the temporal Lake Uru Uru represents on average a surface area of about 280 km^2, a volume of 0.08 km^3, and a depth of 0.8 m when the Lake Poopó surface area is about 3011 km^2, with a volume close to 4 km^3. Conversely, Lake Uru-Uru dries out when Lake Poopó water level is less than 1.62 m.

CLIMATE CHANGE SCENARIOS

In recent years, several studies have tried to decipher climate change in the Andes (e.g., [12,46,47,48,49,50]). Regional Climate Models (e.g., [50]) and Global Climate Models ([46]) were used to predict the future climate in the region. Urrutia and Vuille noted that the climate in tropical South America at the end of the 21st century is likely to be dominated by a significant warming over the entire area, with temperatures increasing between 2 and 7 °C depending on the location and scenario considered [50]. Rabatel *et al.* concluded the precipitation has not displayed a significant and spatially-coherent trend in the tropical Andes since the middle of the 20th century [47]. Bradley *et al.* showed changes in mean annual air temperature projections between 1990–1999 and 2090–2099, using the A2 IPCC Special Report on Emissions Scenario (SRES) considered in AR4 (see [51]), likely to increase by 4 to 5 °C along the tropical Andes, where the maximum temperatures are projected to occur over the high mountains [52].

Seiler *et al.* simulated precipitation and temperature changes over Bolivia using 35 GCMs (23 GCMs corresponding to CMIP3 contributing to AR4, see [51]; and 12 GCMs corresponding to CMIP5 contributing to AR5, see [53]) with IPCC SRESs: B1, A1B, A2, RCP4.5, and RCP8.5 [46]. The study projected an increase in temperature (2.5–5.9 °C) and shortwave radiation (1% to 5%). The changes in annual precipitation in the Andes showed a decrease of 3% for CMIP3 while the CMIP5 GCMs projected an increase by 3%. The monthly precipitation in the Andes may decrease 9% according to CMIP3 GCMs, while CMIP5 forecasts an increase by 20% for the wet season. Furthermore, Escurra *et al.* predicted a considerable decrease of precipitation (−5% to −25%), and evapotranspiration (−5% to −25%) in the south part of the Lake Titicaca Basin and a general trend toward less rainfall along the Altiplano based on CMI from 17 GCMs (SRES A2 projected for 2050) considered in AR4 [12].

Regarding hydrological projections, the same authors studied climate change effects for rural water balances. The study concluded that the climate change projections (17 GCMs for A2 SRES by 2050) show a general decrease in the total annual renewable water in Bolivia. In addition, the projected precipitation and evapotranspiration may decrease significantly (−5% to −10%) in the southern part of the Lake Titicaca. Thus, the study indicated drier future conditions, but not excluding higher probability of flood occurrence. However, Seiler *et al.* projected an increase of precipitation according to CMIP5 (12 GCMs for RCP4.5 and RCP8.5 SRES by 2100; see [53]) in the Andes [46]. As a result, the Altiplano could present wetter conditions and the runoff could also increase. However, the temperature has increased during the last decades over the Altiplano [34,54], and it is projected that this trend will continue in the future [50,51,52]. Thus, in a warmer environment, the water losses through evapotranspiration could also increase. Climate modelling of the TDPS system presents similar results as for the Altiplano. Whereas climate models projections, in general, agree on a temperature increase over the area, there is significant discrepancy in the precipitation projections (e.g., [13,55,56]). Moreover, the climate change analysis at basin scale in mountainous regions presents major challenges due to the complex topography and locally-distinct microclimates influencing environmental factors.

CLIMATIC EFFECTS ON SOCIOECONOMY

The Altiplano vulnerability to climate is reflected by the sensitivity to hazardous events. The recurrent extreme events in the region are droughts, floods, and frosts, with a significant socioeconomic and environmental impact (Table 2; [13]). The Southern Oscillation strongly affects the rainfall pattern over the Altiplano. Generally, El Niño events result in reduced rainfall, while La Niña (LN) shows the opposite effects, causing above-average precipitation. The EN events of 1982/83 and 1997/98 and LN events of 2007/08 affected thousands of people leading to millions of US$ economic losses [56]. Droughts and floods have significant environmental, social, and economic impact on the TDPS system. In the 1990s, droughts caused economic losses estimated at US$ 216 million, while the great flood in 1986–1987 inundated 46,000 ha of the shores of Lake Titicaca and caused losses totaling US$ 125 million throughout the TDPS system. Both were due to imbalances caused by the decreasing storage capacity of the basins as a result of poor land use and inappropriate location of productive activities and infrastructure [6]. The EN phenomenon is important for the Lake Poopó water levels. In 1983 the lake water level presented a fast decline, as well as from 1992 to 1996 [29].

Table 2: Socio-economic impacts of extreme events over the TDPS system in the Altiplano (after Mamani-Salinas [10] * and UNDP 2011 **)

Period	Extreme Event	Social Impact	Economic Impact Million US$
1982–1983	El Niño event	Agricultural and livestock damages	128 *
1989–1990	Droughts	Agricultural damages	88 *
1986–1987	La Niña event	Agricultural and infrastructure damages	125 *
1997–1998	El Niño event	135,000 people affected	515 **
2006–2007	El Niño event	562,000 people affected	443 **
2007–2008	La Niña event	Agricultural damages	273 **
2009–2010	El Niño event	459,000 people affected	236 **

During the last decades the occurrence of extreme events in Bolivia is becoming more frequent. In 2006/2007 and 2009/2010 the EN episodes affected 562,000 and 459,000 people, with economic losses of US$ 443 million and US$ 236 million, respectively, and LN event in 2007/2008 caused US$ 273 million of economic losses in the country [56].

WATER RESOURCES MANAGEMENT

In the Altiplano, an increasing population means a rising demand for the water supply for drinking water, water used in agriculture, and power generation [52]. In addition, the under-developed water infrastructure (leaking pipes and illegal water distribution pipes) and low storage capacity reduce water availability [57]. Furthermore, water sustainability and management in arid regions face new challenges, caused by climate change and glacier recession. The glacial meltwater in the Andes provides river flow during the dry season [58]. Surrounding the TDPS system, the Occidental and Oriental Cordilleras represent numerous glaciers at an altitude higher than 5000 m amsl, (e.g., Palomani Grande (5330 m), Illampu (6383 m), Huayna Potosí (6094 m), and Illimani (6490 m) [25]. The glacier meltwater is generally used in agriculture, water consumption, and electricity generation [57]. Glaciers in the Andes have been retreating at an increasing rate since the late 1970s. The retreat of tropical glaciers in the Andes accelerated in the late 1970s, in the mid-1990s, and in the early 2000s [47]. The magnitude of glacier mass loss is directly related to the size and elevation of the glacier, and those located below 5400 m amsl are the most vulnerable to loss of water [47].

In the TDPS system, the demand for drinking water, irrigation, mining, industry, livestock, and others is higher than the available flow [6]. Lake Titicaca's main tributary is the Ramis River and the main outflow is the Desaguadero River [8]. The inflow is mostly used to maintain the lake water level, itself, considering the high water volume loss through evaporation, reaching an annual mean of 1450 mm at Lake Titicaca and 1900 mm along the south of the TDPS system [6]. In addition, the Desaguadero River shows high salinity and low water quality [59]. The pollution of the TDPS system's water resources is caused by wastewater and solid waste from Puno (to Lake Titicaca), Oruro (to Lake Uru Uru and Poopo), Juliaca (to Coata River), and El Alto (to Seco River) [24]. This organic and bacteriological contamination causes eutrophication in the Puno bay. Wastewater and solid waste are frequently discharged into water bodies without treatment [25]. Heavy metals from mining residues generate chemical pollution in Lake Poopó and Uru Uru, part of the Desaguadero River, and Coata River [24]. Therefore, the water of the rivers is used for irrigation, but is not recommended for domestic consumption. In consequence, the population extracts groundwater and spring water to cover the water demand especially during the dry season. Inventories identified 900 wells throughout the TDPS system. Generally the well depth is less than 10 m, and the deepest is 110 m [6].

Adaptation measures to climate change and variability involve water supply conservation and protection, improving water usage in agriculture, and implementation of water reservoirs [52]. The physical adaptation should improve existing systems for management and distribution, with a designed technology that could be replicated in the Andean communities and self-run by local organizations [57].

RESEARCH NEEDS

The lack of hydrological and climatological long-term data is the largest barrier for the research of the Altiplano basins. The records of the hydro-meteorological stations show large gaps. In addition, these stations are unevenly distributed, with higher station density around the Lake Titicaca and lower station density in the southern part of the TDPS system. Scientific publications mainly deal with the subject of large-scale climatology in the Altiplano and hydrological studies are concentrated the Lake Titicaca region in the northern part of the Altiplano. There is an important need to develop hydroclimatological studies related to water availability, management, quality, and supply in order to provide this information to policy-makers and local stakeholders. Improving the interaction and communication between all stakeholders could be possible way to achieve a better water management and information exchange. In consequence, the improvement of the hydrological and climatological knowledge could be achieved through teamwork, data-sharing, knowledge development, and communication between researchers, policy-makers, and local people.

To improve the understanding of hydrology and climate over the Altiplano basins, research investment is required across a broad range of disciplines, reducing uncertainties related to the physical and human-induced processes that are affected by the hydro-climatic system. This review demonstrates the continued need for high-quality data and observations, analysis of Earth system processes and changes, and modeling that increases hydrological understanding and climate change projections at different scales. Hydro-climatic research is also essential to improve the understanding of water use and land use, as well as impacts of climate change. Assessing the hydrology and climate requires understanding of the roles of the hydrologic cycle, water resources, environmental watershed sustainability, climate variability, and change that are driven by the hydro-climatic system.

Through research investments, it would be possible to install hydrological and climatological observation systems and collect much needed data. Other accessible source of information are remote sensing and reanalysis data, for hydro-climatological analysis and monitoring network. During previous years, remote sensing and GIS techniques have been used in hydrological assessment issues, such as water resource management, water quality, and hydrological modelling (e.g., [60,61,62]). In addition, reanalysis data products could generate data sets, as the Tropical Rainfall Measuring Mission (TRMM) Multi-satellite Precipitation Analysis (TMPA) that produced almost two decades of precipitation data.

In addition, regionalization studies could develop hydrological models for estimating continuous streamflow used in transferring hydrological information from gauged to ungauged basins [63]. On the other hand, groundwater is an important source of water supply in the highlands. As the demand for water increases, a better knowledge of groundwater resources is needed to manage and develop sustainable water management. In the TDPS system, fresh water is mainly supplied by aquifers (e.g., Oruro city). However, there is an important need to understand the groundwater characteristics, such as water quality and aquifer recharge. In addition, the wetlands, known as "bofedales" in the Andes, are also an important water reservoir that act as natural filters of water, improving its quality. The study of the hydro-climatological impact on these reservoirs also requires scientific research.

SUMMARY AND CONCLUSIONS

The TDPS is highly sensitive and vulnerable to climate variability and change impacts for its water availability. Changes in temperature and precipitation will significantly affect the hydrological cycle in the Andes. The climate change adaptation, however, faces political and financial barriers. Long-term solutions to regional water challenges require inter-disciplinary approaches between scientists and local stakeholders (e.g., [57]). Climate variability and change impacts increase social and economic inequalities and affect people´s exposure to climatic hazards. There is evidence of recurrent droughts and floods that negatively affect important economic activities. There are already now negative impacts of rising temperature, increased variability, and weather extremes on crops and livestock of indigenous people in the Altiplano [64].

The lack of observations and monitoring of climate systems and water supplies over the Andean region limit the effectiveness of water resources management [57]. Additionally, the coarse spatial resolution of the present generation of GCMs limits the validity of climate model results in the Andes [12]. Scarce data with high spatial and temporal variability represent a challenge to correctly simulate the hydrology of the Altiplano [65]. Moreover, this simulations are concentrated in the northern part of the system. Considering this, the water distribution between Peru and Bolivia requires a water balance update, including the anthropogenic and climate change impacts. It is necessary to include the entire TDPS area for future hydrological studies, taking into account that the southern part has often been excluded.

Consequently, there is an important need to improve the understanding of water resources in the current and future climate over the TDPS system and to provide better information that can be used in improved decision-making. For this propose, improving the observation data network of the TDPS is required. In addition, it is possible to apply new techniques (e.g., remote sensing tools and data reanalysis), for developing hydrological and climatological research in data-scarce regions.

ACKNOWLEDGMENTS

The study was funded by the Swedish International Development Cooperation Agency (SIDA), supported by the San Andres Mayor University in Bolivia and Lund University in Sweden. The authors would like to express their gratitude to the project "Participatory research and development of adaptation plans to climate change in the Bolivian Central Altiplano" for providing the discharge data of the TDPS. In addition, the meteorological and hydrological data were provided by the Servicio Nacional de Meteorologia e Hidrologia (SENAMHI) and Autoridad Binacional Autónoma del Lago Titicaca (ALT).

REFERENCES

1. Winters, C. Impact of climate change on the poor in Bolivia. *Glob. Major.* 2012, *3*, 33–43.
2. Pillco, R.; Uvo, C.B.; Bengtsson, L.; Villegas, R. Precipitation variability and regionalization over the Southern Altiplano, Bolivia. *Int. J. Climatol.* 2007, 149–164.
3. Garreaud, R.; Vuille, M.; Clement, A.C. The climate of the Altiplano: Observed current conditions and mechanisms of past changes. *Palaeogeogr. Palaeoclimatol. Palaeoecol.* 2003, *194*, 5–22.
4. Vuille, M.; Keimig, F. Interannual variability of summertime convective cloudiness and precipitation in the Central Andes derived from ISCCP-B3 Data. *J. Clim.* 2004, *17*, 3334–3348.
5. Vuille, M. Atmospheric circulation over the Bolivian Altiplano during dry and wet periods and extreme phases of the Southern Oscillation. *Int. J. Climatol.* 1999, *19*, 1579–1600.
6. Revollo, M.M. Management issues in the Lake Titicaca and Lake Poopo system: Importance of developing a water budget. *Lakes Reserv. Res. Manag.* 2001, *6*, 225–229.
7. Baker, P.A.; Rigsby, C.A.; Seltzer, G.O.; Fritz, S.C.; Lowenstein, T.K.; Bacher, N.P.; Veliz, C. Tropical climate changes at millennial and orbital timescales on the Bolivian Altiplano. *Nature* 2001, *409*, 698–701. [PubMed]
8. UNESCO-WWAP. *Water for People, Water for Life*; UNESCO Pub.: Paris, France; New York, NY, USA, 2003; Volume 1, p. 576.
9. INE. *Estadísticas Demográficas*; Instituto Nacional de Estadística, Ed.; The Bolivian National Institute of Statistics: La Paz, Bolivia, 2014. (In Spanish).
10. Mamani Salinas, A. First workshop river basin commissions and other joint bodies for transboundary water cooperation: Legal and institutional aspects. In *Authority of Lake Titicaca (ALT)—A Bilateral Cooperation Experience*; Autoridad Binacional del Lago Titicaca: Geneva, Switzerland, 2013.
11. Garcia, M.; Raes, D.; Jacobsen, S.E.; Michel, T. Agroclimatic constraints for rainfed agriculture in the Bolivian Altiplano. *Arid Environ.* 2007, *71*, 109–121.
12. Escurra, J.J.; Vazquez, V.; Cestti, R.; De Nys, E.; Srinivasan, R. Climate change impact on countrywide water balance in Bolivia. *Reg. Environ. Chang.* 2014, *14*, 727–742.
13. Valdivia, C.; Thibeault, J.; Gilles, J.L.; García, M.; Seth, A. Climate trends and projections for the Andean Altiplano and strategies for adaptation. *Adv. Geosci.* 2013, *33*, 69–77.
14. Vidaurre de la Riva, M.; Lindner, A.; Pretzsch, J. Assessing adaptation—Climate change and indigenous livelihood in the Andes of Bolivia. *J. Agric. Rural Dev. Trop. Subtrop.* 2013, *114*, 109–122.
15. UNESCO-WHC. Tiwanaku: Spiritual and Political Centre of the Tiwanaku Culture. Available online: http://whc.unesco.org/en/list/567 (accessed on 30 November 2000).
16. Instituto Nacional de Estadística. *Estadísticas económicas*; Instituto Nacional de Estadística, Ed.; The Bolivian National Institute of Statistics: La Paz, Bolivia, 2015. (In Spanish).
17. Quispe, A. Perú producirá esta gestión un 41% más de quinua que Bolivia. In *La Razon*; Lazcano, M.: La Paz, Bolivia, 2015.
18. Takahashi, K.; Battisti, D.S. Processes controlling the mean tropical pacific precipitation pattern. part I: The Andes and the eastern Pacific ITCZ. *J. Clim.* 2007, *20*, 3434–3451.
19. Thibeault, J.; Seth, A.; Wang, G.L. Mechanisms of summertime precipitation variability in the Bolivian Altiplano: Present and future. *Int. J. Climatol.* 2012, *32*, 2033–2041.
20. Lenters, J.D.; Cook, K.H. Summertime precipitation variability over South America: Role of the large-scale circulation.*Mon. Weather Rev.* 1999, *127*, 409–431.
21. Garreaud, R.D.; Aceituno, P. Interannual rainfall variability over the South American Altiplano. *J. Clim.* 2001, *14*, 2779–2789.
22. Lenters, J.D.; Cook, K.H. On the origin of the Bolivian high and related circulation features of the South American climate. *J. Atmos. Sci.* 1997, *54*, 656–678.
23. Garreaud, R.D. The Andes climate and weather. *Adv. Geosci.* 2009, *22*, 3–11.
24. UNEP. *Diagnostico Ambiental del Sistema Titicaca-Desaguadero-Poopo-Salar de Coipasa (Sistema TDPS) Bolivia-Perú*; División de Aguas Continentales del Programa de las Naciones Unidas para el Medio Ambiente y Autoridad Binacional del Sistema TDPS: Washington, DC, USA, 1996; p. 192.
25. Comisión multisectorial para la prevención y recuperación ambiental del Lago Titicaca y sus afluentes (CMLT). *Estado de la calidad ambiental de la Cuenca del Lago Titicaca ámbito Peruano*; (CMLT): Puno, Peru, 2014; p. 152.
26. Garreaud, R.D. Multiscale analysis of the summertime precipitation over the central Andes. *Mon. Weather Rev.* 1999,*127*, 901–921.
27. Vuille, M.; Bradley, R.S.; Keimig, F. Interannual climate variability in the Central Andes and its relation to tropical Pacific and Atlantic forcing. *J. Geophys. Res. Atmos.* 2000, *105*, 12447–12460.
28. Delclaux, F.; Coudrain, A.; Condom, T. Evaporation estimation on Lake Titicaca: A synthesis review and modelling.*Hydrol. Process.* 2007, *21*, 1664–1677.
29. Pillco, R.; Bengtsson, L. Long-term and extreme water level variations of the shallow Lake Poopó, Bolivia. *Hydrol. Sci. J.* 2006, *51*, 98–114.
30. Aceituno, P. On the Functioning of the Southern Oscillation in the South American Sector. Part I: Surface Climate.*Mon. Weather Rev.* 1988, *116*, 505–524.
31. Vizy, E.K.; Cook, K.H. Relationship between Amazon and high Andes rainfall. *J. Geophys. Res. Atmos.* 2007, *112*.
32. Vicente-Serrano, S.M.; Chura, O.; López-Moreno, J.I.;

Azorin-Molina, C.; Sanchez-Lorenzo, A.; Aguilar, E.; Moran-Tejeda, E.; Trujillo, F.; Martínez, R.; Nieto, J.J. Spatio-temporal variability of droughts in Bolivia: 1955–2012. *Int. J. Climatol.* 2015, *35*, 3024–3040.

33. Garreaud, R.D.; Vuille, M.; Compagnucci, R.; Marengo, J. Present-day South American climate. *Palaeogeogr. Palaeoclimatol. Palaeoecol.* 2009, *281*, 180–195.

34. Seiler, C.; Hutjes, R.W.A.; Kabat, P. Climate variability and trends in Bolivia. *J. Appl. Meteorol. Clim.* 2013, *52*, 130–146.

35. López-Moreno, J.I.; Morán-Tejeda, E.; Vicente-Serrano, S.M.; Bazo, J.; Azorin-Molina, C.; Revuelto, J.; Sánchez-Lorenzo, A.; Navarro-Serrano, F.; Aguilar, E.; Chura, O. Recent temperature variability and change in the Altiplano of Bolivia and Peru. *Int. J. Climatol.* 2016, *36*, 1773–1796.

36. VRHR-MMAyA. *Inventario Nacional de Sistemas de Riego*; PROAGRO: Cochabamba, Bolivia, 2013; Volume 1, p. 29.

37. PROAGRO. *Inventario Nacional de Presas*; PROAGRO: Cochabamba, Bolivia, 2010; Volume 1, p. 406.

38. INTECSA-AIC-CNR. *Plan Director Global Binacional de Protección: Prevención de inundaciones y aprovechamiento de los recursos del Lago Titicaca, Rio Desaguadero, Lago Poopó y Lago Salar de Coipasa (Sistema TDPS)*; Autoridad Binacional del Lago Titicaca: La Paz, Bolivia, 1993; p. 80.

39. INTECSA-AIC-CNR. *Plan Director Global Binacional: Modelos matemáticos del Sistema Hídrico TDPS. Balance hídrico del Lago Titicaca, apendices*; Autoridad Binacional del Lago Titicaca: La Paz, Bolivia, 1993; p. 327.

40. Carmouze, J.-P.; Aquize, E. La régulation hydrique du lac Titicaca et l'hydrologie de ses tribitaries. *Hydrobiol. Trop.* 1981, *14*, 311–328.

41. Roche, M.A.; Bourges, J.; Cortes, J.; Mattos, R. Climatology and hydrology of the Lake Titicaca basin. In *Lake Titicaca: A Synthesis of Limnological Knowledge*; Dejoux, C., Iltis, A., Eds.; Kluwer Academic: Dordrecht, The Netherlands, 1992; pp. 63–88.

42. INTECSA-AIC-CNR. *Plan Director Global Binacional: Estudio de climatologia*; Autoridad Binacional del Lago Titicaca: La Paz, Bolivia, 1993; p. 111.

43. Uría, Á.; Molina, J. *Análisis y tratamiento de los datos hidrológicos de las cuencas Mauri-Desaguadero y Lago Titicaca*; Agua Sustentable e Instituto de Hidráulica e Hidrología: La Paz, Bolivia, 2013; p. 86.

44. Uría, Á. *Análisis de la Información Hidrométrica y Régimen Hidrológico en las Cuencas del Río Mauri, Desaguadero y Lago Titicaca*; Universidad Mayor de San Andrés: La Paz, Bolivia, 2014; p. 177.

45. Pillco, R.; Bengtsson, L. Three methods for determining the area-depth relationship of Lake Poopó, a large shallow lake in Bolivia. *Lakes Reserv. Res. Manag.* 2007, *12*, 275–284.

46. Seiler, C.; Hutjes, W.A.R.; Kabat, P. Likely ranges of climate change in Bolivia. *J. Appl. Meteorol. Clim.* 2013, *52*, 1303–1317.

47. Rabatel, A.; Francou, B.; Soruco, A.; Gomez, J.; Caceres, B.; Ceballos, J.L.; Basantes, R.; Vuille, M.; Sicart, J.E.; Huggel, C.; *et al.* Current state of glaciers in the tropical Andes: A multi-century perspective on glacier evolution and climate change. *Cryosphere* 2013, *7*, 81–102.

48. Buytaert, W.; Celleri, R.; Timbe, L. Predicting climate change impacts on water resources in the tropical Andes: Effects of GCM uncertainty. *Geophys. Res. Lett.* 2009, *36*.

49. Buytaert, W.; Vuille, M.; Dewulf, A.; Urrutia, R.; Karmalkar, A.; Celleri, R. Uncertainties in climate change projections and regional downscaling in the tropical Andes: Implications for water resources management. *Hydrol. Earth Syst. Sci.* 2010, *14*, 1247–1258.

50. Urrutia, R.; Vuille, M. Climate change projections for the tropical Andes using a regional climate model: Temperature and precipitation simulations for the end of the 21st century. *J. Geophys. Res. Atmos.* 2009, *114*.

51. IPCC. *Climate Change 2007: The Physical Science Basis. Contribution of Working Group I to the Fourth Assessment Report of the Intergovernmental Panel on Climate Change*; Cambridge University Press: Cambridge, UK; New York, NY, USA, 2007; p. 996.

52. Bradley, R.S.; Vuille, M.; Diaz, H.F.; Vergara, W. Threats to water supplies in the tropical Andes. *Science* 2006, *312*, 1755–1756. [PubMed]

53. IPCC. *Climate Change 2013: The Physical Science Basis. Contribution of Working Group I to the Fifth Assessment Report of the Intergovernmental Panel on Climate Change*; Cambridge University Press: Cambridge, UK; New York, NY, USA, 2013; p. 1535.

54. Vuille, M.; Francou, B.; Wagnon, P.; Juen, I.; Kaser, G.; Mark, B.G.; Bradley, R.S. Climate change and tropical Andean glaciers: Past, present and future. *Earth Sci. Rev.* 2008, *89*, 79–96.

55. Thibeault, J.M.; Seth, A.; Garcia, M. Changing climate in the Bolivian Altiplano: CMIP3 projections for temperature and precipitation extremes. *J. Geophys. Res. Atmos.* 2010, *115*.

56. *UNDP. Tras las huellas del cambio climático en Bolivia: estado del arte del conocimiento sobre adaptación al cambio climático: agua y seguridad alimentaria*; PNUD-Bolivia: La Paz, Bolivia, 2011; p. 144.

57. Rangecroft, S.; Harrison, S.; Anderson, K.; Magrath, J.; Castel, A.P.; Pacheco, P. Climate change and water resources in arid mountains: An example from the Bolivian Andes. *Ambio* 2013, *42*, 852–863. [PubMed]

58. Coudrain, A.; Francou, B.; Kundzewicz, Z.W. Glacier shrinkage in the Andes and consequences for water resources. *Hydrol. Sci.* 2005, *50*, 925–932.

59. Calizaya, A.; Meixner, O.; Bengtsson, L.; Berndtsson, R. Multi-criteria decision analysis (MCDA) for integrated water resources management (IWRM) in the Lake Poopo Basin, Bolivia. *Water. Resour. Manag.* 2010, *24*, 2267–2289.

60. Singh, P.; Gupta, A.; Singh, M. Hydrological inferences from watershed analysis for water resource management

using remote sensing and GIS techniques. *Egypt. J. Remote Sens. Space Sci.* 2014, *17*, 111–121.

61. Ritchie, J.C.; Zimba, P.V.; Everitt, J.H. Remote sensing techniques to assess water quality. *Photogramm. Eng. Remote Sens.* 2003, *69*, 695–704.

62. Boegh, E.; Thorsen, M.; Butts, M.B.; Hansen, S.; Christiansen, J.S.; Abrahamsen, P.; Hasager, C.B.; Jensen, N.O.; van der Keur, P.; Refsgaard, J.C.; *et al.* Incorporating remote sensing data in physically based distributed agro-hydrological modelling. *J. Hydrol.* 2004, *287*, 279–299.

63. Razavi, T.; Coulibaly, P. Streamflow prediction in Ungauged Basins: Review of regionalization methods. *J. Hydrol. Eng.*2013, *18*, 958–975.

64. IPCC. *Climate Change 2014: Impacts, Adaptation, and Vulnerability. Part A: Global and Sectoral Aspects. Contribution of Working Group II to the Fifth Assessment Report of the Intergovernmental Panel on Climate Change*; Cambridge University Press: Cambridge, UK; New York, NY, USA, 2014; p. 1132.

65. Condom, T.; Escobar, M.; Purkey, D.; Pouget, J.C.; Suarez, W.; Ramos, C.; Apaestegui, J.; Tacsi, A.; Gomez, J. Simulating the implications of glaciers' retreat for water management: A case study in the Rio Santa Basin, Peru.*Water Int.* 2012, *37*, 442–459.

CHAPTER 2

AN APPLICATION OF FILTERED RENEWAL PROCESSES IN HYDROLOGY

Mario Lefebvre and Fatima Bensalma

INTRODUCTION

Let $\{(t), t \geq 0\}$ be a Poisson process with rate λ. A filtered Poisson (sometimes called shot noise) process is a continuoustime stochastic process $\{(t), t \geq 0\}$ defined by

$$X(t) = \sum_{n=1}^{N(t)} w(t, \tau_n, Y_n) \quad (X(t) = 0 \text{ if } N(t) = 0), \tag{1}$$

in which the random variables τ_1, τ_2,... denote the arrival times of the events of the Poisson process and Y_1, Y_2,... are assumed to be independent and identically distributed random variables that are also independent of $\{N(t), t \geq 0\}$. The function $(\cdot, \cdot, \cdot)$ is called the response function. In many applications, the response function is chosen of the form

$$w(t, \tau_n, Y_n) = Y_n e^{-(t-\tau_n)/c}, \tag{2}$$

where c is a parameter that must be estimated. It then gives the value at time t of an event of magnitude Y_n of the Poisson process that occurred at time τ_n. Moreover, the random variables Y_1, Y_2,... are generally assumed to be exponentially distributed with parameter μ. With the above response function, the filtered Poisson process behaves as in Figure 1. Actually, this behavior depends on the form of the response function, but not on the distribution of the time between the events. Therefore, the same behavior would be observed in the case when $\{(t), t \geq 0\}$ is a renewal process. Remember that a Poisson process is a particular renewal process. To model the daily flows of rivers given their most recent observed values, conceptual and physical models based on the filtered Poisson process have been widely used successfully for many decades; see, for example, Weiss [1], Kelman [2], Koch [3], and Konecny [4]. Filtered Poisson processes are still being used to model various phenomena in civil engineering; see Yin et al. [5] and Miyamoto et al. [6]. Now, especially in hydrological applications, the form of the response function in (2) is taken for granted rather than being justified by the observations of the variable of interest. Actually, it is generally simply an assumption made to obtain a mathematically tractable model.

Similarly, the actual distribution of the Y_n's is not investigated. However, if one is only interested in forecasting the value of $(t+\delta)$, based on the values of the process up to time t, this does not really cause a problem because the forecast is normally based on the mean of the Y_n's, and the distribution itself is not needed. Finally, and even more importantly, the assumption that $\{(t), t \geq 0\}$ is indeed a Poisson process is not tested. Again, this is a simplifying assumption, because one can then make use of the nice properties of the Poisson process, notably the fact that it has independent and stationary increments. Next, notice that, with the response function in (2), the effect on the value of the process of an event that occurred at

Citation: Mario Lefebvre and Fatima Bensalma, "An Application of Filtered Renewal Processes in Hydrology," International Journal of Engineering Mathematics, vol. 2014, Article ID 593243, 9 pages, 2014. doi:10.1155/2014/593243.

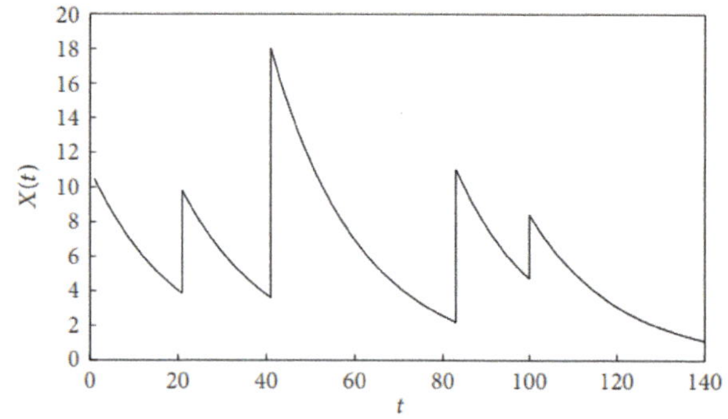

Figure 1: Example of a trajectory of a filtered Poisson process defined by (1) and (2).

time τ_n is maximum at that time instant and $\{X(t), t \geq 0\}$ has discontinuities at the arrival times of the events of the Poisson process. In the case of the application in hydrology that consists in modeling the flow of rivers, if one looks at real hydrographs, one sees that there is a more or less extended period of time during which the flow increases from a minimum to a maximum. Even if river flows data are generally available on a daily basis, one does not observe very sudden increases of the flow, followed by an exponential decrease. Rather, it takes on average two to three days for the flow to reach a maximum, and the decrease is more or less rapid. To obtain a more realistic model for the variations of river flows, by taking into account the periods of flow increase, Lefebvre and Guilbault [7] used the following response function:

$$w(t, \tau_n, Y_n) = Y_n(t - \tau_n)^d e^{-(t-\tau_n)/c}, \tag{3}$$

where d is a positive parameter. With this response function, the river flow begins to increase immediately after time τn, but the maximum is reached after dc time units (i.e., generally days) and then it starts to decrease. Their work was improved by the authors, who provided a method to estimate the parameter d. They found that, for the applications that they considered, d was in the interval (0,1).

With a Poisson process being a continuous-time Markov chain, the time that the process spends in a given state is exponentially distributed, which is a strictly decreasing density function. Let T_1, T_2,... be the times between two successive events, that is, the interarrival times. In the case of river flows, the interarrival time is generally a random variable with a density function that is increasing at first. Hence, the assumption that $\{(t), t \geq 0\}$ is a Poisson process is at least doubtful.

In Lefebvre [8], the author tested the hypothesis that the T_n's are exponentially distributed. He found that, for the Delaware River, it was not accurate. Instead, it was shown that both a gamma and a Rayleigh distribution fitted the data much better. The objective was to forecast the peak flows of the Delaware River, which is a difficult task due to the very weak correlation between the successive peaks. To try to forecast these peak flows, the author used a filtered renewal process as a model for the river flow. An advantage in using a Rayleigh distribution for the interarrival time is that the filtered renewal process can be transformed into a filtered Poisson process by working on a different time scale. Poisson and renewal processes have been used in hydrology to model low flows (see, for instance, Loaiciga and Leipnik [9] and Yagouti et al. [10]) as well as various physical phenomena such as traffic noise (see Marcus [11]). Andersen [12, 13] gave new theoretical results on filtered renewal processes (see also the references therein). The aim of the present paper is to show that we can obtain more accurate short-term forecasts of river flows by making use of filtered renewal processes rather than filtered Poisson processes.

To do so, we will first need to develop a formula to forecast the flow at time $(t + \delta)$, given the history of the process up to time t. Because we cannot appeal to the memoryless property of the exponential distribution, the task of finding an estimator for $X(t+\delta)$is more difficult.The values of the river flow before time t must be taken into account, contrary to the case of filtered Poisson process, for which the forecasts depend only on $X(t)$.

Moreover, because we want to clearly see the improvement obtained by using a more realistic distribution for the interarrival time, we will take the response function defined in (2). Indeed, as mentioned above, filtered Poisson processes having this response function are commonly used in hydrology, and the forecasting formulas are then easy to derive and implement. Therefore, it will be easier to judge the quality of the model that we propose. In the next section, filtered renewal processes will be formally defined and the formulas needed to forecast the river flows will be developed. Then, in Section 3, the results will be applied to forecast the flows of the Delaware and Hudson Rivers. Finally, a few concluding remarks will be made in Section 4.

RIVER FLOWS MODELED AS A FILTERED RENEWAL PROCESS

Let (t) denote the flow of a certain river at time t.We assume that

$$X(t) = \sum_{n=1}^{N(t)} Y_n e^{-(t-\tau_n)/c} \quad (X(t) = 0 \text{ if } N(t) = 0), \tag{4}$$

in which $\{N(t), t \geq 0\}$ is a renewal process. The τ_n's are the arrival times of the events of the process $\{N(t), t \geq 0\}$, and the times $T_n = \tau_n - \tau_{n-1}$ between the successive renewals are independent and identically distributed random variables. Finally, the (positive) random variables Y_1, Y_2,... are independent and identically distributed (and independent of $\{N(t), t \geq 0\}$).

Thus, the process is characterized by the sudden increases of the river flow caused by the events that occur at times τ_1, τ_2,.... A renewal cycle is defined as the time Tn between two consecutive peaks. For the interarrival time, the following distributions (in addition to the exponential distribution) will be considered in Section 3: lognormal, gamma, Weibull, and Rayleigh.

Next, in order to be able to forecast the river flow at time $t+\delta$, we must first estimate the parameter c that appears in the response function. If there are no signals between t and $t+1$, then $(t + 1)$ is given by

$$X(t + 1) = e^{-1/c} X(t). \tag{5}$$

Hence, to estimate the parameter c, we can consider all the days where the flow decreased in the data set and calculate the ratio $(t+1)/(t)$. The point estimate of c is then derived from the arithmetic mean $\overline{R}$ of this ratio:

$$\widehat{c} = -\frac{1}{\ln \overline{R}}. \tag{6}$$

In the case of the random variables Y_n, which constitute a random sample of a parent random variable Y, their exact distribution is not needed. Indeed, only the expected value of Y will be used to forecast the value of $(t + \delta)$. Therefore, we simply have to compute the mean increase of the river flow during the calibration period of the model. We will now derive formulas to estimate the river flow one and two days in advance. These formulas can be used with any distribution for the interarrival time T, which denotes the parent variable of the T_n's.

Forecasting the Flow at Time t+1

We will first try to forecast the river flow at time t+1, given all the information available up to t. To do so, we assume that there will be at most one event in the interval $(t, t + 1]$. At any rate, as mentioned above, in practice flow values are generally available on a daily basis. Therefore, it is not really possible to determine whether more than one event occurred in $(t, t + 1]$. When it does happen, these events can be considered as a single large event. Apart from the case when the interarrival time T has an exponential distribution, we must take the history of the process into account to calculate the probability that there will indeed be an event in $(t, t + 1]$. Let

$$U_i = \tau_{N(t)+i} - \tau_{N(t)+i-1} \quad \text{for } i = 1, 2, \ldots. \tag{7}$$

That is, U_i is the time elapsed between the $(i - 1)$st and the ith event after the most recent one that occurred before or at time t. The random variables T and U_1 are identically distributed. We must compute

$$p_k := P\left[U_1 \leq k + 1 \mid U_1 > k\right], \tag{8}$$

where k is the (integer) number of days since the most recent signal was observed. We have

$$p_k = \frac{\int_k^{k+1} f_{U_1}(u)\, du}{\int_k^{\infty} f_{U_1}(u)\, du}. \tag{9}$$

With the help of a mathematical software, we can obtain numerical values of this probability for any density function f_{U1} and any value of k.

Next, if T is exponentially distributed, it is well known that, given that there is a signal (i.e., an event) between t and t+1, the conditional density of T is uniform on the interval $(t, t + 1]$, so that the signal in question occurred on average at time $t + 1/2$. In the general case, we must compute the conditional density function of U_1 in the interval $(k, k + 1]$, given that $U_1 > k$. If there is indeed a signal between t and t+1, the forecast of the river flow at time t+1 would be

$$Z_k := X(t)\, e^{-1/\widehat{c}} + E[Y] \frac{1}{P[U_1 > k]} \times \int_k^{k+1} e^{-(k+1-u)/\widehat{c}} f_{U_1}(u)\, du. \tag{10}$$

Therefore, the forecast of the river flow at time t+1, given the entire history of the process up to t, is given by

$$\begin{aligned}\widehat{X}(t+1) &= X(t)\, e^{-1/\widehat{c}}(1 - p_k) + Z_k p_k \\ &= X(t)\, e^{-1/\widehat{c}} + p_k E[Y] \frac{1}{P[U_1 > k]} \\ &\quad \times e^{-(k+1)/\widehat{c}} \int_k^{k+1} e^{u/\widehat{c}} f_{U_1}(u)\, du.\end{aligned} \tag{11}$$

Forecasting the Flow at Time t+2

Let $U_{1'} = U_1 \mid \{U_1 > k\}$. To forecast the river flow at time t+2, given all the information available up to time t, we need the distribution of the random variable $S_k := U_{1,} + U_2$, where U_2 is defined in (7). We now assume that the probability that there will be more than two signals in the interval $(t, t + 2]$ is negligible (irrespective of the value of k). The density function of S_k is given by the convolution of the density functions of $U_{1'}$ and U_2:

$$f_{S_k}(s + k) = \int_k^{\infty} f_{U_{1,k}}(u)\, f_{U_2}(s + k - u)\, du. \tag{12}$$

Since U_2 is a nonnegative random variable, we can write that

$$f_{S_k}(s + k) = \int_k^{s+k} f_{U_{1,k}}(u)\, f_{U_2}(s + k - u)\, du \quad \text{for } s > 0. \tag{13}$$

In the case when T has an exponential distribution with parameter λ, we know that $(S_k - k)$ has a gamma distribution with parameters α=2 and λ. In the other cases, we must generally evaluate the above integral numerically.

The value of $(t + 2)$ is given by

$$X(t+2) = \begin{cases} e^{-2/c} X(t) & \text{if } U_{1,k} > k + 2, \\ e^{-2/c} X(t) + Y_1^* e^{-(k+2-U_{1,k})/c} & \text{if } U_{1,k} \leq k + 2 \\ & \text{and } U_2 > 2 - (U_{1,k} - k), \\ e^{-2/c} X(t) + Y_1^* e^{-(k+2-U_{1,k})/c} & \text{if } U_{1,k} + U_2 \leq k + 2 \\ \quad + Y_2^* e^{-(2-U_2)/c} & (\text{and } U_3 > 2 - (U_{1,k} - k) - U_2), \end{cases} \tag{14}$$

where Y_i^* denotes the size of the ith signal in the interval$(t, t+ 2]$. As mentioned above, we assume that

$$P[U_3 > 2 - (U_{1,k} - k) - U_2] \simeq 1. \tag{15}$$

We must first compute

$$p_{1,k} := P\left[U_{1,k} \in (k, k+2]\right]. \tag{16}$$

When U_1 has an exponential distribution with parameter λ,

$$p_{1,k} = P\left[U_1 \in (0, 2]\right] = 1 - e^{-2\lambda}. \tag{17}$$

In general, we have

$$p_{1,k} = \frac{\int_k^{k+2} f_{U_1}(u)\, du}{\int_k^{\infty} f_{U_1}(u)\, du}. \tag{18}$$

Next,

$$\begin{aligned} p_{2,k} &:= P\left[U_{1,k} + U_2 \in (k, k+2]\right] \equiv P\left[S_k \in (k, k+2]\right] \\ &= \int_0^2 f_{S_k}(s+k)\, ds \\ &= \int_0^2 \left[\int_k^{s+k} f_{U_{1,k}}(u)\, f_{U_2}(s+k-u)\, du\right] ds. \end{aligned} \tag{19}$$

Alternatively, using the fact that U_1 and U_2 (≥ 0) are independent random variables, we may write that

$$\begin{aligned} p_{2,k} &:= P\left[U_{1,k} + U_2 \in (k, k+2]\right] \\ &= \int_k^{k+2} P\left[U_{1,k} + U_2 \in (k, k+2] \mid U_{1,k} = u\right] f_{U_{1,k}}(u)\, du \\ &= \int_k^{k+2} P\left[U_2 \in (k-u, k+2-u]\right] f_{U_{1,k}}(u)\, du \\ &= \int_k^{k+2} \left[\int_0^{k+2-u} f_{U_2}(v)\, dv\right] f_{U_{1,k}}(u)\, du. \end{aligned} \tag{20}$$

Finally, by independence,

$$\begin{aligned} &P\left[\{U_{1,k} \leq k+2\} \cap \{U_2 > 2 - (U_{1,k} - k)\}\right] \\ &= \int_k^{k+2} P\left[U_2 \in (k+2-u, \infty)\right] f_{U_{1,k}}(u)\, du \\ &= \int_k^{k+2} \left[\int_{k+2-u}^{\infty} f_{U_2}(v)\, dv\right] f_{U_{1,k}}(u)\, du \\ &= \int_k^{k+2} \left[1 - \int_0^{k+2-u} f_{U_2}(v)\, dv\right] f_{U_{1,k}}(u)\, du \\ &= p_{1,k} - p_{2,k}. \end{aligned} \tag{21}$$

The forecast of the river flow at time $t+2$, given the entire history of the process up to time t, is thus given by

$$\begin{aligned} &\widehat{X}(t+2) \\ &= X(t)\, e^{-2/\tilde{c}} (1 - p_{1,k}) \\ &\quad + \left\{X(t)\, e^{-2/\tilde{c}} + E[Y] \int_k^{k+2} e^{-(k+2-u)/\tilde{c}} f_{U_{1,k}}(u)\, du\right\} \\ &\quad \times (p_{1,k} - p_{2,k}) \\ &\quad + \left\{X(t)\, e^{-2/\tilde{c}} + E[Y] \int_k^{k+2} e^{-(k+2-u)/\tilde{c}} f_{U_{1,k}}(u)\, du\right. \\ &\qquad + E[Y] \int_k^{k+2} \left[\int_0^{k+2-u} e^{-(2-v)/\tilde{c}} f_{U_2}(v)\, dv\right] \\ &\qquad \left. \times f_{U_{1,k}}(u)\, du\right\} p_{2,k} \\ &= X(t)\, e^{-2/\tilde{c}} + \left\{E[Y] \int_k^{k+2} e^{-(k+2-u)/\tilde{c}} f_{U_{1,k}}(u)\, du\right\} p_{1,k} \\ &\quad + \left\{E[Y] \int_k^{k+2} \left[\int_0^{k+2-u} e^{-(2-v)/\tilde{c}} f_{U_2}(v)\, dv\right]\right. \\ &\qquad \left. \times f_{U_{1,k}}(u)\, du\right\} p_{2,k} \\ &= X(t)\, e^{-2/\tilde{c}} + \left\{E[Y]\, e^{-(k+2)/\tilde{c}} \int_k^{k+2} e^{u/\tilde{c}} f_{U_{1,k}}(u)\, du\right\} p_{1,k} \\ &\quad + \left\{E[Y] \int_k^{k+2} \left[e^{-2/\tilde{c}} \int_0^{k+2-u} e^{v/\tilde{c}} f_{U_2}(v)\, dv\right]\right. \\ &\qquad \left. \times f_{U_{1,k}}(u)\, du\right\} p_{2,k}. \end{aligned} \tag{22}$$

FORECASTING THE FLOWS OF THE DELAWARE AND HUDSON RIVERS

To assess the quality of the forecasts obtained with the renewal filtered process, we need to apply the formulas developed in the previous section to real-life data. As in [7], we chose the Delaware and Hudson Rivers, located in the United States, at the Montague (01438500) and the North Creek (01315500) gage stations, respectively. The flow values are available on the Internet at the following address: http://nwis.waterdata.usgs.gov. To calibrate the model, we used the data from October 2008 to September 2009.

Remarks. (i)When one is dealing with real-life data, things are not as simple as the mathematical models suggest. Looking at the actual hydrograph of the Delaware River, we observe a number of weak peaks that occur while the flow is decreasing and for which the increasing period lasts only one day before the flow resumes its decline. Similarly, there are sometimes small increases of the flow value occurring just after a minimum was observed that last only one day. In building the data set, there is always a subjective part. We decided to neglect these weak peaks, thus considering only the peaks that appear quite clearly in the hydrograph.

(ii) Because the river flow is observed on a daily basis, we must discretize the set of possible values taken by the interarrival time T. We computed the number k of days elapsed between consecutive peaks in the data set. In the case of the first observed peak, k is the number of days elapsed since the beginning of the calibration period. Then, k is obtained by subtracting the arrival times of the consecutive peaks. Notice that k is greater than or equal to 1.

We first consider the Delaware River. For this river, we found that the average value $\overline{T}$ of the interarrival times in the data set is equal to 6.8889 days. Moreover, the standard deviation of the observations is given by $s_T = 3.8977$ days. Based on these values, one may at once conclude that T is very unlikely to be exponentially distributed. Indeed, remember that the mean and the standard deviation of an exponential distribution with parameter λ are both equal to $1/\lambda$. Next, the value of the point estimate of the (unitless) parameter c (see (6)) is $\widehat{c} = 9.2592$, and the average value of the magnitude Y of the signals is given by $\overline{Y} = 984.6121$ cubic feet per second. For the distribution of the random variable T, we tested the following models:

(i) an exponential distribution with parameter λ,

(ii) a lognormal distribution with parameters μ and σ,

(iii) a gamma distribution with shape parameter α and scale parameter λ,

(iv) a Weibull distribution with shape parameter λ and scale parameter κ,

(v) a Rayleigh distribution with parameter σ.

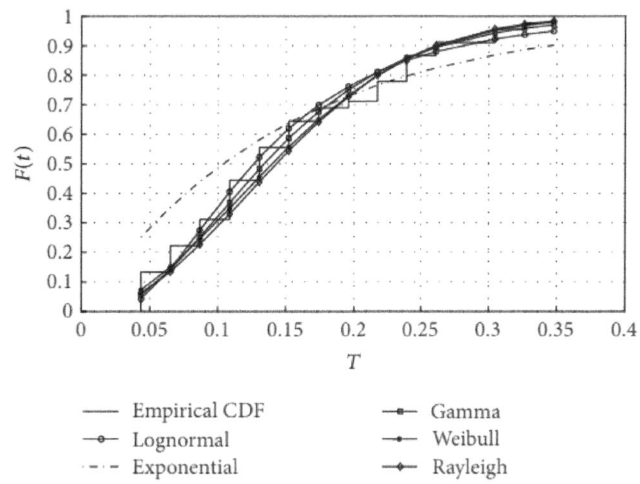

Figure 2: Empirical and fitted distribution functions for the random variable T (Delaware River).

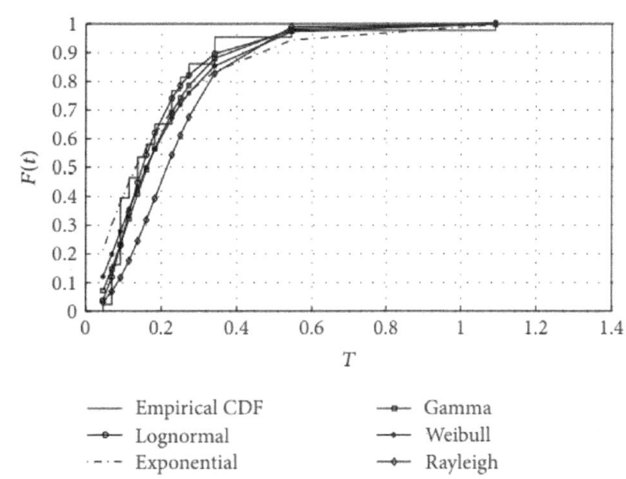

Figure 3: Empirical and fitted distribution functions for the random variable T (Hudson River).

Table 1: Goodness-of-fit tests for the distribution of the random variable T (Delaware River)

Distribution	Exponential	Lognormal	Gamma	Weibull	Rayleigh
Parameters	1/6.8889	1.7574, 0.6187	3.0544, 2.2554	7.7974, 1.8975	5.5817
P-value	0.0344	0.7887	0.7723	0.6927	0.7008

Table 2: Criteria for comparing the forecasting models (Delaware River)

		Filtered renewal process				Filtered Poisson process
		Lognormal	Gamma	Weibull	Rayleigh	Exponential
$X(t+1)$	Mape	12.58%	12.49%	12.55%	12.56%	12.76%
	Nash	0.8288	0.8290	0.8287	0.8287	0.8356
	Pc	0.1674	0.1664	0.1663	0.1669	0.2522
	Corr	0.9188	0.9186	0.9180	0.9188	0.9187
$X(t+2)$	Mape	20.36%	20.33%	20.34%	20.34%	21.83%
	Nash	0.5447	0.5445	0.5445	0.5445	0.5573
	Pc	0.2930	0.2893	0.2869	0.2863	0.3344
	Corr	0.7735	0.7732	0.7731	0.7730	0.7734

Table 1 summarizes the results of the chi-square goodness-of-fit tests for each distribution and gives the respective parameter estimates. We find that, as expected, the exponential distribution is rejected, whereas the other distributions are all accepted, according to the P-values of the tests. The empirical cumulative distribution function as well as the distribution functions of the various models considered above is shown in Figure 2. We clearly see that the exponential distribution does not fit the data nearly as well as the other four models. Now, in the case of the filtered Poisson process (1) with the response function defined in (2), to estimate the river flow at time t+1 we compute the expected value of $X(t + 1)$ given all the past observations.

Because of the memoryless property of the exponential distribution, we find that this conditional expectation only depends on the most recent value of the flow, that is, (t). We obtain (see [7]) that

$$E[X(t+1)X] = (t)e^{-1/c}X(t) + E[X(1)], \tag{23}$$

Where

$$E[X(1)] = \frac{\lambda c}{\mu}\left(1 - e^{-1/c}\right). \tag{24}$$

Similarly, to estimate the river flow at time t+2, we only need to calculate

$$\begin{aligned} E[X(t+2) \mid X(t)] &= E\left[\sum_{n:0 \le \tau_n \le t+2} Y_n e^{-(t+2-\tau_n)/c} \mid X(t)\right] \\ &= e^{-2/c} E\left[\sum_{n:0 \le \tau_n \le t} Y_n e^{-(t-\tau_n)/c} \mid X(t)\right] \\ &\quad + E\left[\sum_{n:t < \tau_n \le t+2} Y_n e^{-(t+2-\tau_n)/c} \mid X(t)\right] \\ &= e^{-2/c} X(t) + E[X(2)], \end{aligned} \tag{25}$$

Where

$$E[X(2)] = \frac{\lambda c}{\mu}\left(1 - e^{-2/c}\right). \tag{26}$$

Moreover, we have that

$$\lim_{t \to \infty} E[X(t)] = \frac{\lambda c}{\mu}, \tag{27}$$

$$\lim_{t \to \infty} \mathrm{Var}[X(t)] = \frac{\lambda c}{\mu^2}, \tag{28}$$

$$\lim_{t \to \infty} \rho_{X(t),X(t+\delta)} = e^{-\delta/c}, \tag{29}$$

where $\rho X(t),X(t+\delta)$ denotes the correlation coefficient of $X(t)$ and $X(t + \delta)$.

By making use of these three equations, we can estimate the parameter λ of the Poisson process, the parameter μ of the assumed exponential distribution of the Y_n's, and the parameter c in the response function. Actually, we do not need point estimates of λ and μ to forecast the flow. From (27), we deduce that we can simply estimate the ratio $\lambda c/\mu$ that appears in both (24) and (26) by the mean value of the observed data when the process is in steady state. Finally, to estimate the parameter c, we first compute the empirical correlation coefficient r_1 between the flow values at times t and t+1 and then we set $\hat{c} = -1/\ln(r_1)$ (see (29)). With the value of r_1 being close to 1, the previous formula is well defined.

Remark. If one is only interested in forecasting the river flow at time $t+\delta$, one may estimate c by computing the empirical value of $\rho X(t),X(t+\delta)$ and making use of (29). We are now ready to compare the forecasts derived from both the filtered Poisson process and the filtered renewal process with the various distributions considered for the interarrival time T. We used the formulas given above to forecast the flow of the Delaware River for the 89-day period from February to April 2010. To assess the accuracy of the forecasts, we consider four criteria commonly used in hydrology: the mean absolute percentage error (Mape), the Nash criterion (Nash), the peak criterion (Pc), and the correlation coefficient (Corr) between the observed and forecasted values. The Nash criterion is based on the mean square error and is used to assess the predictive power of hydrological models. It evaluates the quality of the forecasts from the differences between the expected and observed daily values.

Table 3: Goodness-of-fit tests for the distribution of the random variable T (Hudson River)

Distribution	Exponential	Lognormal	Gamma	Weibull	Rayleigh
Parameters	1/8.3488	1.8754, 0.6625	9.2134, 1.3399	2.1785, 3.8324	7.9978
P-value	0.0013	0.6983	0.3203	0.2532	0.000139

Table 4: Criteria for comparing the forecasting models (Hudson River).

		Filtered renewal process			Filtered Poisson process
		Lognormal	Gamma	Weibull	Exponential
$X(t+1)$	Mape	12.33%	12.14%	12.33%	15.58%
	Nash	0.8240	0.8243	0.8240	0.8360
	Pc	0.2875	0.2895	0.2872	0.4818
	Corr	0.9146	0.9146	0.9146	0.9146
$X(t+2)$	Mape	20.84%	21.00%	21.15%	32.78%
	Nash	0.5841	0.5838	0.5835	0.6222
	Pc	0.4558	0.4533	0.4518	0.5575
	Corr	0.7908	0.7909	0.7910	0.7910

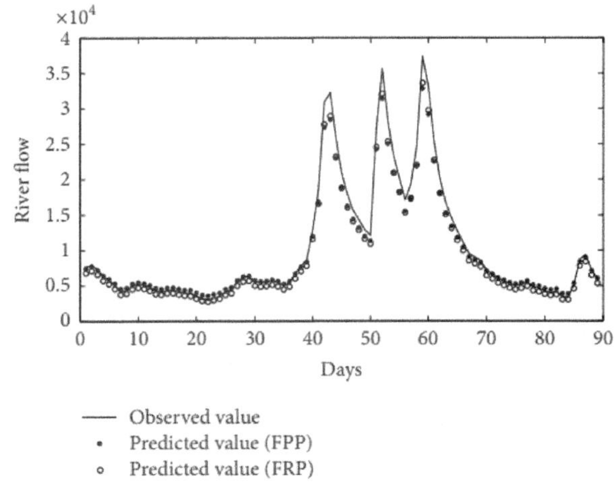

Figure 4: Observed and forecasted values of the flow of the Delaware River at time t+1.

It is equal to 1 in case of a perfect fit. For its part, the peak criterion is used to measure the quality of the forecasts during the critical peak period. The closer to 0 it is, the better the forecasts are. Table 2 displays the results obtained with the two competing models. Looking at these results, we notice that the Mape, Nash, and Corr criteria yield practically the same values for the filtered renewal and filtered Poisson processes. However, the filtered renewal process did much better than the filtered Poisson process when we consider the peak criterion. This criterion is especially important because accurate forecasts are really needed during the peak period.

We now turn to the Hudson River. The value of the point estimate of the constant c for this river is $\hat{c}$ = 9.1172 and the average magnitude of the signals is $\overline{Y}$ = 229.2533, which is much smaller than in the case of the Delaware River. Proceeding as above, we performed chi-square goodness-of-fit tests for the various distributions considered. The results are presented in Table 3 (see also Figure 3). This time, we see that both the exponential and Rayleigh distributions must be rejected, while the lognormal distribution clearly provides the best fit to the data.

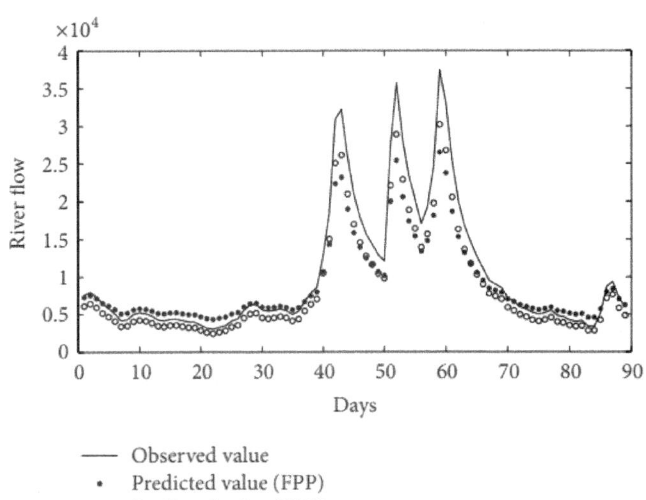

Figure 5: Observed and forecasted values of the flow of the Delaware River at time t+2.

The values of the four criteria used to compare the models are shown in Table 4. We discarded the Rayleigh distribution, because it yielded a very bad fit to the data. As in the case of the Delaware River, the values of the Nash and Corr criteria are quite similar, while the important peak criterion and the Mape criterion are much smaller for the filtered renewal process than for the filtered Poisson process. Finally, we present the observed and forecasted values of the flows of the Delaware and Hudson Rivers at times t+1 and t+2 in Figures 4, 5, 6, and 7, based on the filtered Poisson process (FPP) and the filtered renewal process (FRP) with a lognormal distribution for the interarrival time T. Notice that the filtered renewal process is generally better than the filtered Poisson process at forecasting high flow values, which is our main concern.

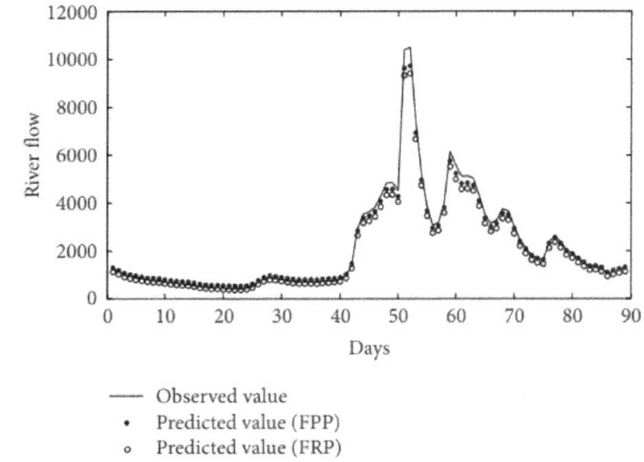

Figure 6: Observed and forecasted values of the flow of the Hudson River at time t+1.

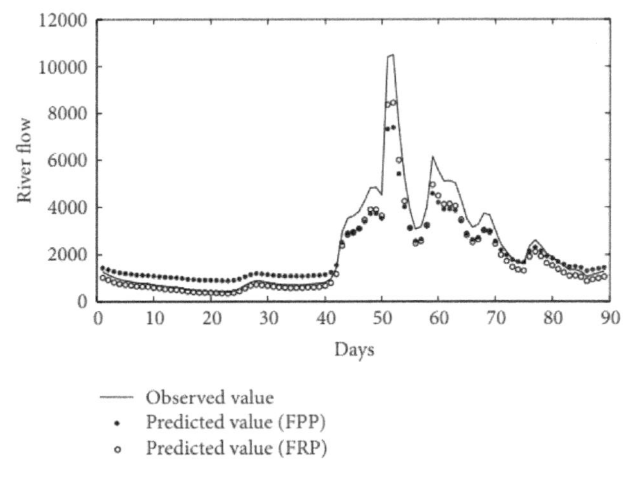

Figure 7: Observed and forecasted values of the flow of the Hudson River at time t+2.

CONCLUSION

In this paper, we were able to significantly improve the shortterm hydrological forecasts produced by filtered Poisson processes by using a more realistic distribution for the interarrival time of the signals, thus working with filtered renewal processes. As we saw in the previous section, the filtered Poisson process, despite its lack of realism, is able to yield reasonable short-term forecasts. One possible explanation is the fact that the forecasting formulas derived from this model are very easy to implement. Also, contrary to the filtered renewal process, it does not require subjective decisions to estimate its various parameters. However, the filtered Poisson process was not able to produce forecasts as accurate as those derived from the filtered renewal process during the important peak period.

In addition to being used to forecast flow values, the models considered in this paper can also give us an estimate of the probability that the river flow will exceed a certain threshold in the next few days. This threshold may be a value that corresponds to a flow above which the risk of flooding is very high. Finally, we could use the response function in (3) to further improve the forecasts. To do so, we would have to be able to estimate the parameter d in that response function and then obtain formulas that generalize the ones in (11) and (22), which is probably not an easy task.

ACKNOWLEDGMENT

The authors want to thank the anonymous reviewers for the constructive comments.

REFERENCES

1. G. Weiss, "Shot noise models for the generation of synthetic streamflow data," Water Resources Research, vol. 13, no. 1, pp. 101–108, 1977. ·
2. J. Kelman, "A stochastic model for daily streamflow," Journal of Hydrology, vol. 47, no. 3-4, pp. 235–249, 1980. ·
3. R. W. Koch, "A stochastic streamflow model based on physical principles," Water Resources Research, vol. 21, no. 4, pp. 545–553, 1985. · ·
4. F. Konecny, "On the shot-noise streamflow model and its applications," Stochastic Hydrology and Hydraulics, vol. 6, no. 4, pp. 289–303, 1992. · ·
5. Y.-J. Yin, Y. Li, and W. M. Bulleit, "Stochastic modeling of snow loads using a filtered Poisson process,"Journal of Cold Regions Engineering, vol. 25, no. 1, pp. 16–36, 2010. · ·
6. H. Miyamoto, J. Morioka, K. Kanda et al., "A stochastic model for tree vegetation dynamics in river courses with interaction by discharge fluctuation impacts," Journal of Japan Society of Civil Engineers B1: Hydraulic Engineering, vol. 67, no. 4, pp. 1405–1410, 2012.
7. M. Lefebvre and J.-L. Guilbault, "Using filtered Poisson processes to model a river flow," Applied Mathematical Modelling, vol. 32, no. 12, pp. 2792–2805, 2008. · ·
8. M. Lefebvre, "A filtered renewal process as a model for a river flow," Mathematical Problems in Engineering, vol. 2005, no. 1, pp. 49–59, 2005. · ·
9. H. A. Loaiciga and R. B. Leipnik, "Stochastic renewal model of low-flow streamflow sequences,"Stochastic Hydrology and Hydraulics, vol. 10, no. 1, pp. 65–85, 1996. ·
10. A. Yagouti, I. Abi-Zeid, T. B. M. J. Ouarda, and B. Bobée, "Revue de processus ponctuels et synthèse de tests statistiques pour le choix d'un type de processus," Journal of Water Science, vol. 14, no. 3, pp. 323–361, 2001. ·
11. A. H. Marcus, "Traffic noise as a filtered Markov renewal process," Journal of Applied Probability, vol. 10, no. 2, pp. 377–386, 1973. ·
12. P. K. Andersen, "Filtered renewal processes with a two-sided impact function," Journal of Applied Probability, vol. 16, no. 4, pp. 813–821, 1979. ·
13. P. K. Andersen, "A note on filtered Markov renewal processes," Journal of Applied Probability, vol. 18, no. 3, pp. 752–756, 1981. ·

CHAPTER 3 THE ROLE OF THE DOMINANT MODES OF PRECIPITATION VARIABILITY IN MODULATING THE HYDROLOGY

Kara A. Smith and Fredrick H. M. Semazzi

INTRODUCTION

Lake Victoria is approximately 67,000 km2 in area but shallow with an average depth of only 40 m. It is territorially administrated by Tanzania, Uganda, and Kenya and is a major source of income in those countries producing $3-4 billion annually. The lake basin provides fresh water and hydroelectric power for over 30 million people who live in the Lake Victoria basin, as well as supporting agriculture, fisheries, trade, and tourism.

The lake is the primary source of the White Nile which flows through Lake Kyoga in Uganda and Lake Albert into Sudan, where it joins with the Blue Nile at Khartoum to form the Nile River, which supports the livelihood of over 300 million people. The climate of the lake basin is dominated by the bimodal signature of the intertropical convergence zone (ITCZ). It is well known that El Nino-Southern Oscillation (ENSO) ˜ and the Indian Ocean Zonal Mode (IOZM) dominate the interannnual climate variability of Eastern Africa [1, 2].

Bowden and Semazzi [2] examined the intraseasonal climate variability over the Greater Horn of Africa during the October–December rainy season using empirical orthogonal function (EOF) analysis of pentad data for 1979–2001. The two dominant modes for that season were found to be a mixture of El Nino-Southern Oscillation-Indian Ocean ˜ Dipole and a decadal mode. Schreck III and Semazzi [3] found that the second most significant October–December regional precipitation EOF mode was significant over this region. The mode is characterized by decadal variability and a dipole rainfall loading pattern. Indeje et al. [4] analyzed the evolution of ENSO throughout the region and found that relatively wet conditions were observed during the March–May and October–December rainfall seasons of the El Nino years. ˜

The spatial distribution of rainfall over Lake Victoria basin is complex, consisting of a wave-like dry-wet-dry-wet climatological rainfall pattern. This climatic pattern, its variability, and associated atmospheric and marine conditions determine the performance of the primary socioeconomic activities across the basin, including agriculture, fisheries, and hydroelectric power generation. The productivity of hydroelectric dams along the Nile River is, to a large extent, determined by the level of Lake Victoria, which is primarily dictated by the rainfall variability over the Lake Victoria basin. Because of this, it is important to determine the modes of variability in precipitation, which dominate the variability in lake levels. Many studies [5–8] have utilized water balance models of Lake Victoria to close the water balance for the lake as well as to examine the sharp increase in levels from 1961 to 1964 and the decline afterwards. Yin and Nicholson [9] used observed lake levels and a water balance model to estimate rainfall over the lake basin. More recently, studies have been done in order to attribute the recent decrease in Lake Victoria to droughts and release from the hydroelectric dams which control the lake outflow [10–12]. While previous studies in this region have examined either the dominant modes of climate variability or the impact of total rainfall on Lake Victoria levels, the primary objective of this study is to estimate the relative contributions of the dominant modes of annual precipitation variability to the modulation of Lake Victoria levels. The

Citation: Kara A. Smith and Fredrick H. M. Semazzi, "The Role of the Dominant Modes of Precipitation Variability in Modulating the Hydrology," Advances in Meteorology, vol. 2014, Article ID 516762, 11 pages, 2014. doi:10.1155/2014/516762.

relationship between precipitation over Lake Victoria basin and lake levels is nonlinear. This relationship should be studied to determine which modes need to be accurately modeled in order to accurately model Lake Victoria levels, which are important to the hydroelectric industry in East Africa.

METHODS

The primary method of investigation is a water balance model in combination with application of EOF analysis of the regional Eastern Africa rainfall using the University of East Anglia Climate Research Unit (CRU) version 3.21 [13] over a region covering 10.67° S–4.25° N and 29.98° E–41.9° E. The water balance model employs rainfall at six discrete locations around the lake to construct Lake Victoria levels. We adopt the EOF approach to isolate the contribution of individual significant EOF modes and to reconstruct the lake levels based on the water balance model. The results are compared with Lake Victoria levels constructed using the total rainfall. The time series from the significant seasonal modes are correlated with NOAA Extended Reconstructed Sea Surface Temperature (ERSST) v3b global sea surface temperatures (SSTs) to investigate the physical mechanisms responsible for each mode.

Empirical Orthogonal Function (EOF) Analysis

Empirical orthogonal function (EOF) analysis [14] was performed on precipitation in Eastern Africa in order to examine the effect of different modes of precipitation variability on rainfall over the region. We test for significance of the modes using Kendall's [15] criterion for distinctly separated EOFs, for sample size N, that the sampling error associated with a given eigenvalue must be smaller than its spacing from the neighboring eigenvalue. We then relate the spatial patterns to known physical properties, such as ENSO and IOZM. EOF analysis is performed on annual precipitation since annual precipitation is the input in the water balance model used in this study. Seasonal modes of variability are also found and compared to the significant annual modes to determine their physical properties. Reconstruction of variables utilizing a systematic inclusion of a subset of eigenmodes which account for most of the data variability has been utilized to filter out statistical noise by reconstructing the data using a reduced number of significant EOF modes. We reconstructed rainfall over the lake basin with only one significant model of variability at a time to examine the contributions of the individual modes to Lake Victoria levels. Since the water balance model used in this study takes in annual precipitation, precipitation is only reconstructed for the annual modes of variability. Our data reconstruction procedure is similar to the approach used in Weickmann and Chervin [16] and Semazzi et al. [17]. Since the EOF analysis was calculated based on the normalized correlation matrix, we first denormalize the data by multiplying the data by the standard deviation for each grid point and then adding the resulting anomalies to the mean. The reconstruction was then performed using the equation

$$q_j(\lambda,\phi) = \bar{q}(\lambda,\phi) + \sum^{M} g_{ij} e_i(\lambda,\phi)\, s(\lambda,\phi), \tag{1}$$

where $\bar{q}(\lambda,\phi)$ is the time mean vector, M is the number of the selected set of EOFs used in the reconstruction, g_{ij} is the principal component at time j for the eigenvector, i, $e_i(\lambda, \phi)$ is the ith eigenvector, and $s_i(\lambda, \phi)$ is the standard deviation. The annual mean for the reconstructed data is identical to the corresponding mean for the reconstructed data, independent of the number of modes used in the reconstruction of the percentage of variability covered by the modes.

Water Balance Model

Tate et al. [8] developed a water balance model based on those used in previous studies [5– 7,18] to estimate the level of Lake Victoria at the end of a given year(L_n) by calculating the change in the lake level during the year (ΔL) and adding it to the projected level for the previous year ($L_{n-1(\text{estimated})}$). The change in lake level is calculated as follows:

$$\Delta L_n = \left[P - E + \frac{Q_{\text{in}} - Q_{\text{out}}}{A}\right]_n, \tag{2}$$

where P and E are precipitation and evaporation over the lake, respectively, Q_{in} is inflow into the lake from tributaries, Q_{out} is outflow from the dams at Jinja, and A is the surface area of Lake Victoria. Evaporation is assumed constant at 1595 mm yr−1. Precipitation over the lake, the dominant source of water for Lake Victoria, is assumed to be a nonlinear function of the average rain gauge precipitation at six stations (Jinja, Entebbe, Kisumu, Musoma, Bukoba, and Mwanza) along the perimeter of the lake (Figure 1). Since station data is not readily available, CRU TS 3.21 gridded precipitation was used to investigate the modes of variability in Eastern Africa and their effects on Lake Victoria levels over the years 1950–2009. The CRU precipitation is interpolated to the six stations using WENO interpolation as described in [18]. The interpolated station precipitation values are then used in place of the station gauge precipitation.

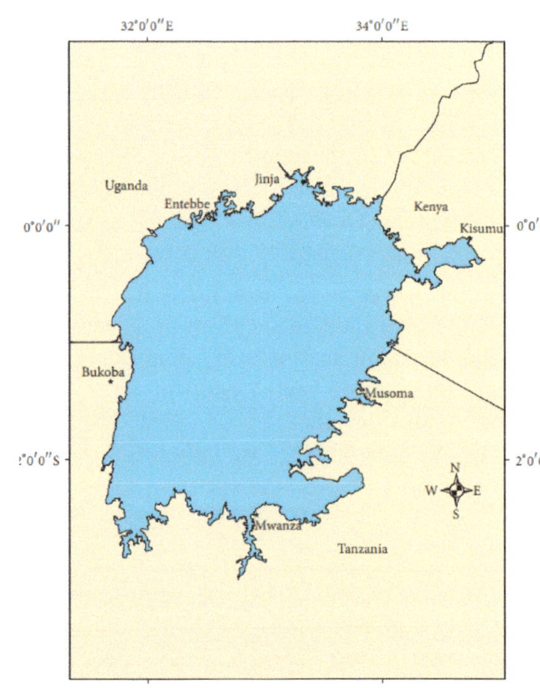

Figure 1: Location of stations used in water balance model.

The tributary inflow, Q_{in}, is solved using the technique of [8], while Q_{out} is found by solving the equation from [19]. Since ΔL_n appears on the left-hand side of the equation, and as part of the equation used to calculate Q_{out}, the equation must be solved iteratively. This procedure is executed using two iterations, setting $\Delta L_n = 0$ on the first iteration. Lake Victoria levels calculated using the water balance model are shown as a black line in Figure 2, while observed lake levels are shown in purple. The Lake Victoria level observations used in this study are based on a data set of lake level observations from January 1949 to May 1998 at a gauge at the main river outlet in Jinja, Uganda, obtained from an archive at the Ministry of Water Resources in Uganda [20]. This data set was extended through the present using TOPEX/POSEIDON and Jason-1 satellite altimetry data [21]. Since the heights obtained by satellite altimetry are an average of all topography within the instrument footprint further averaged in the direction of the satellite motion, these values differ from traditional gauge measurements which are at specific points.

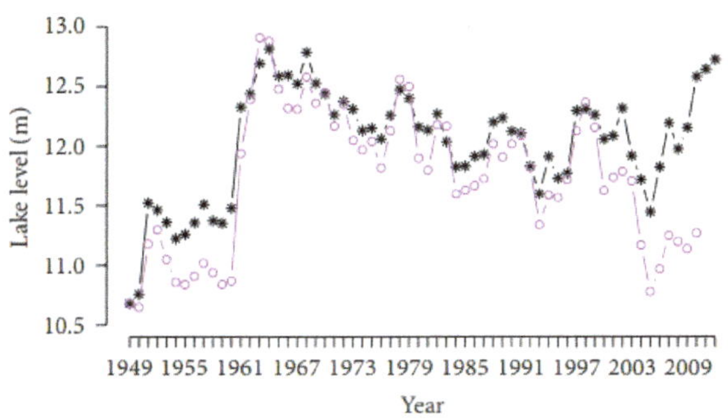

Figure 2: Comparison of Lake Victoria levels modeled using CRU 3.21 (black with asterisks) and observed lake levels (purple with circles).

Therefore, the TOPEX/POSEIDON and Jason-1 altimetry data is supplied as a lake height variation with respect to TOPEX/POSEIDON 10-year mean level. The climatological mean was found by subtracting the satellite data from the observations for the time period the two data sets had in common to determine the difference between the two. From this, a mean was found and added to the satellite anomalies. The mean was adjusted until bias and error were reduced to $-4 * 10^{-4}$ and 0.0585, respectively [20]. The remaining satellite altimetry time series was then added to the resulting mean, creating a combined data set for January 1949 through the present. The first two years of the jump in levels from 1961 to 1964 are captured, while the modeled lake levels are slightly lower than observed in 1963-64. The model does well until the 1980s when it underpredicts lake levels by up to 0.5 meters in 1983 and then predicts higher than observed levels in the late 1990s through 2005, overpredicting by as much as 0.73 m in 2005, which is most likely related to the overrelease from Nalubaale and Kiira dams found by Kull [11] and expanded upon in [12]

RESULTS

Annual Rainfall

The first mode of annual variability accounts for 37.7 percent of variability. The loading (Figure 3(a)) is mostly positive with a time series showing a high amount of variability but a general drying trend after 1961 (Figure 3(b)). The second mode of annual precipitation variability, which accounts for 9 percent of variability, has a positive loading surrounding the lake and to the north into Uganda and Kenya, with a negative loading to the south (Figure 3(c)). The time series shows a decadal oscillation with the largest variation in the late 1970s and early 1980s (Figure 3(d)). The third mode of variability is also significant and accounts for just over 7 percent of the variability. The loading has a positive pattern to the north and south of the lake basin, with a negative loading to the east (Figure 3(e)). Over the lake, there is close to neutral negative loading (Figure3(f)).

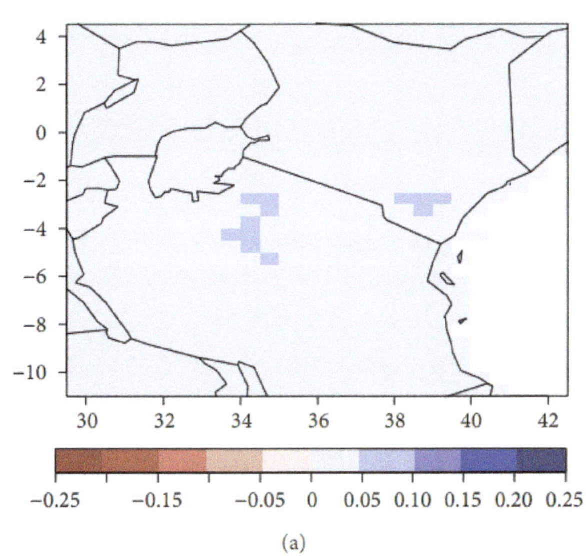

(a)

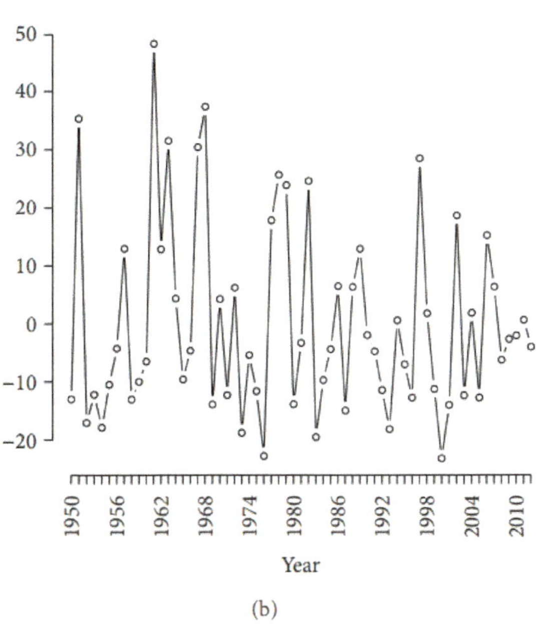

(b)

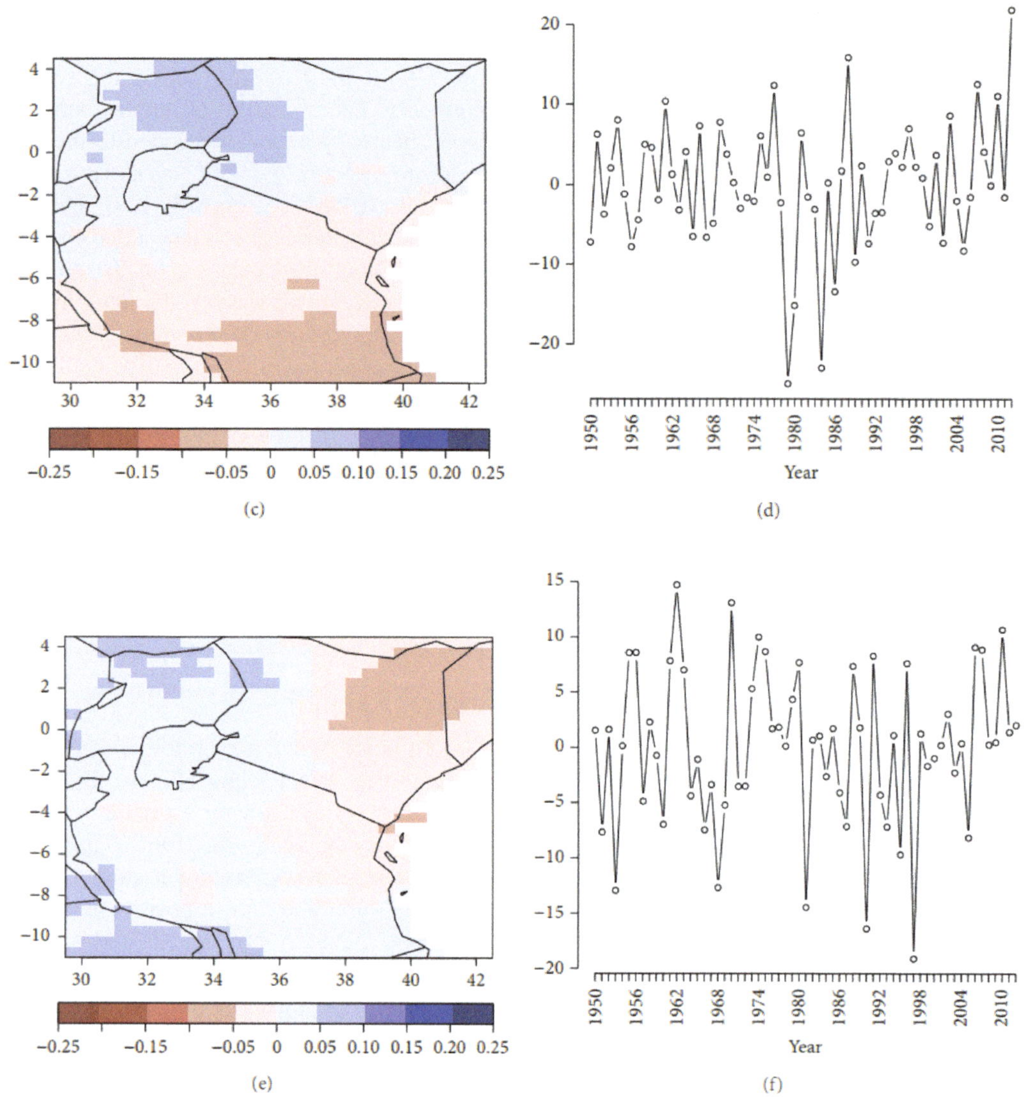

Figure 3: EOF loadings and time series for first ((a), (b)), second ((c), (d)), and third ((e), (f)) modes of variability for annual precipitation.

Lake levels calculated using precipitation reconstructed from individual significant modes of annual precipitation variability, compared to the level obtained using the total annual precipitation, are shown in Figure 4. The precipitation reconstructed from the first significant mode of variability accounts for the majority of variability in Lake Victoria levels, while the effects of the second and third modes on the variability of lake levels are negligible. The increase in modeled levels from 1949 to 1963 using each of the filtered precipitation time series occurs as the model adjusts to the increase in average annual precipitation during that period.

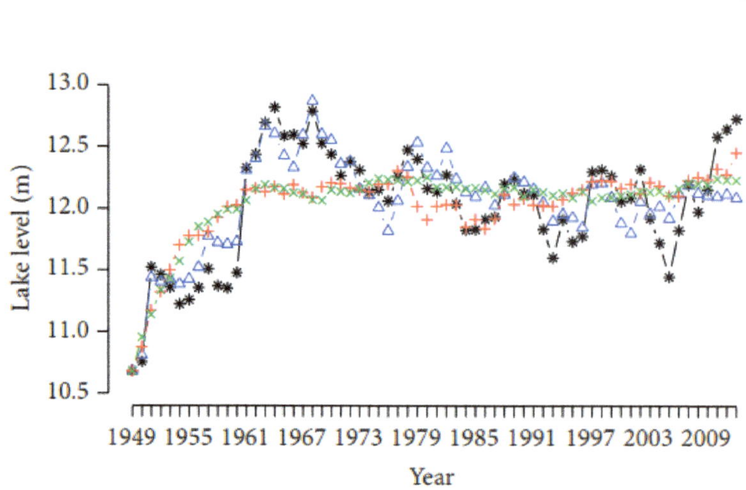

Figure 4: Results of using individual modes of annual precipitation variability in the water balance model for Lake Victoria. Levels modeled using the unfiltered CRU 3.21 dataset are in black (asterisks); levels modeled using precipitation accounted for by the first mode of annual

variability are in blue (triangles); levels modeled using precipitation accounted for by the second mode of variability are in red (crosses), while levels modeled using precipitation accounted for by the third mode of variability are in green (×'s).

Seasonal Variability

In order to understand the physical meaning behind the annual modes of variability, we examine the significant modes of variability for the short rains of October through December (OND) and the long rains of March through May (MAM). We relate these seasonal modes to the annual modes used above.

Short Rains (OND Season)

The first three modes of variability for the short rains, OND season, are significant and account for just over 71% of the variability for this season. We will examine these modes in detail.

The first mode of OND station precipitation variability has a loading which is entirely positive (Figure 5(a)). The time series we obtain for the first mode of variability of OND station precipitation reproduces the time series found by [3] for 1961–90. This first mode explains 57.4% of the variability. Schreck III and Semazzi [3] related their EOF1 to ENSO by correlating their time series with the first significant EOF mode from global CMAP precipitation. They found that the first mode from the EOF analysis on regional precipitation corresponds to a combination of the first global EOF mode representing ENSO and the second global mode representing IOZM. The major warm ENSO events for the period correspond to the high peaks of the time series. In particular, the ENSO of 1982 is distinctly conspicuous. When the first mode from OND precipitation is correlated with SSTs for 1950–2012, an ENSO pattern emerges (Figure 6). IOZM often occurs simultaneously with ENSO; however, the difference between the IO ZM positive and negative years without ENSO for this mode is not statistically significant. This mode has a high correlation (0.77) with the first annual mode. Therefore, the first mode of OND variability and the first annual mode are considered to be ENSO.

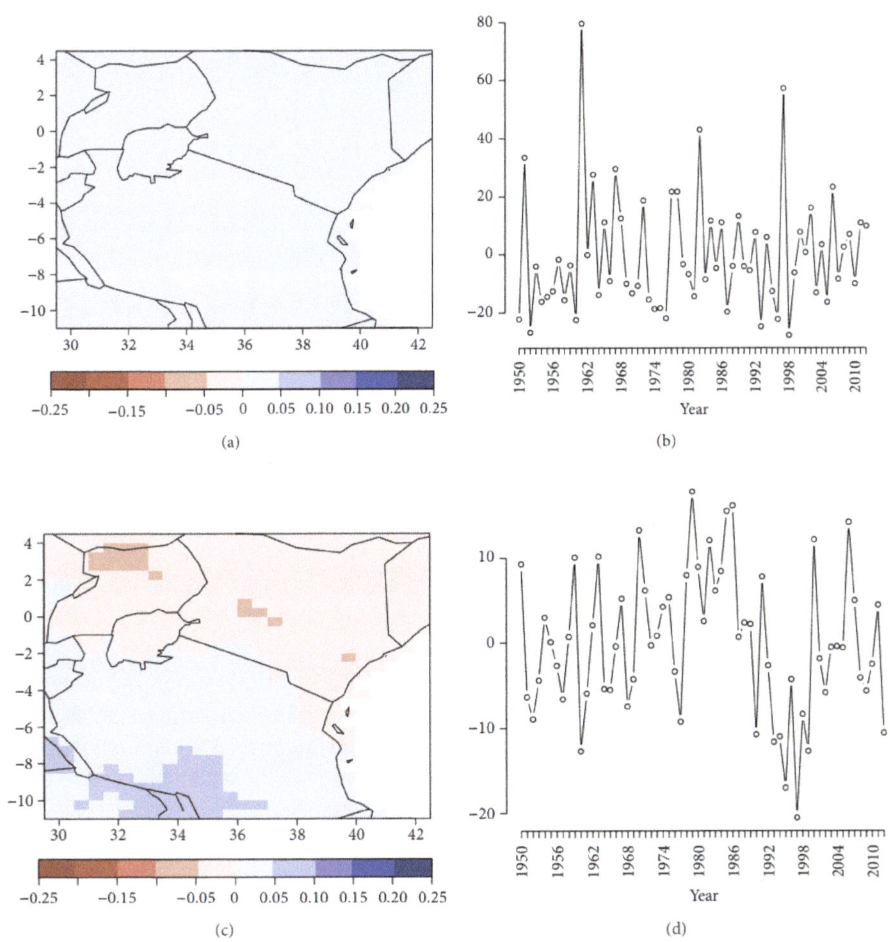

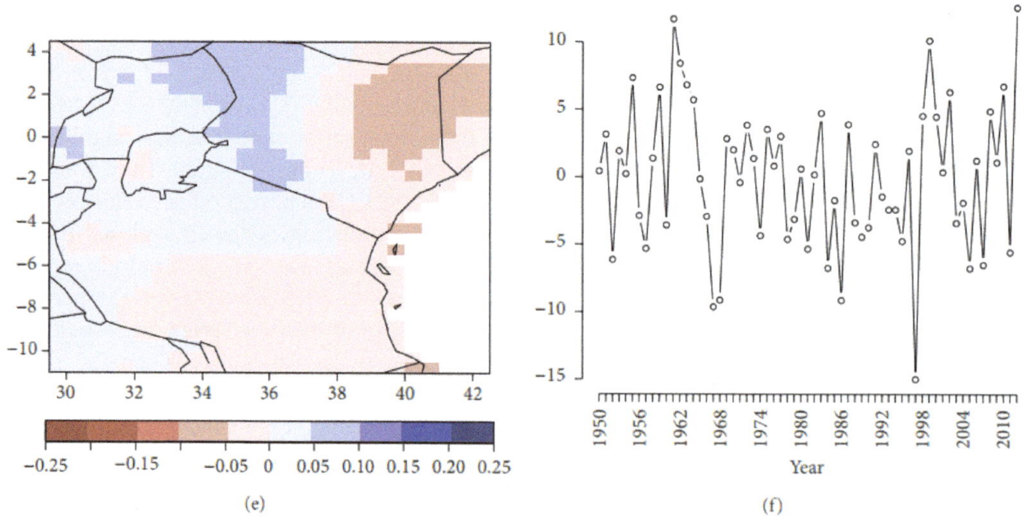

Figure 5: Same as Figure 3, but for short rains (OND) precipitation.

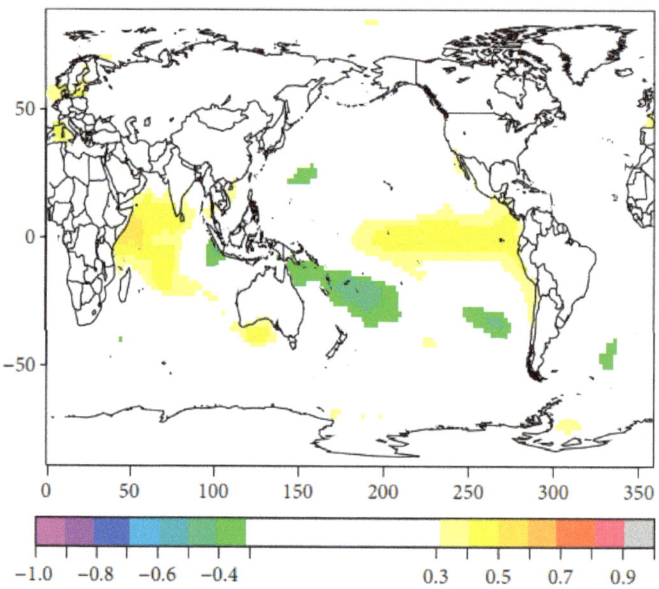

Figure 6: Pearson Correlation of the time series for the first mode of OND season precipitation variability with ERSST sea surface temperatures.

Similar to [3], the second mode of variability for station precipitation is significant. The second mode is a dipole mode with negative loading to the north and positive loading to the south, which split the stations used in the water balance model; the stations to the north of the lake are in an area of negative loading, while those on the southern coast of the lake are in a region with a positive loading (Figure 5(c)). The corresponding time series exhibits both strong interannual variability and low-frequency background variability (Figure 5(d)). Overall, there is an increasing trend in the time series of the second mode from 1949 through 1986, followed by a decreasing trend until 1997, which is then followed by another increasing trend. Interpretation of the distribution of loadings and the corresponding time series suggests that the southern sector of Eastern Africa became wetter during the period from 1961 to 1990, while the northern section has been drying up during the same time period. This is opposite to the dipole described by [3] who claim that the northern sector has been getting wetter while the southern sector has been drying up. While our time series is similar to the time series in [3] over 1961–90, our loadings are the inverse of theirs. We find that their Gauge-EOF2 loading must have been inverted. There is no clear signal when this mode of variability is correlated with global SSTs. This mode has a −0.40 correlation with the second annual mode, which suggests that it might be related to the second annual mode. It has a higher, 0.44 correlation with the third annual mode. The third mode of OND variability is also shown to be statistically significant, accounting for 4.1 percent of variability over the region during the OND season. The loading for this mode shows an East-West dipole, with a positive loading to the west and a negative loading to the east, which is near neutral over Lake Victoria but has sharp contrasts over Kenya (Figure 5(e)); the time series shows high peaks in 1961, 1962, 1999, and 2012, while there are sharp dips in 1967, 1968, 1986, and 1997 (Figure 5(f)). This mode is correlated with SSTs in the Pacific Ocean, off the coast of Mexico (Figure 7). It has a 0.32 correlation with the second mode of annual rainfall variability.

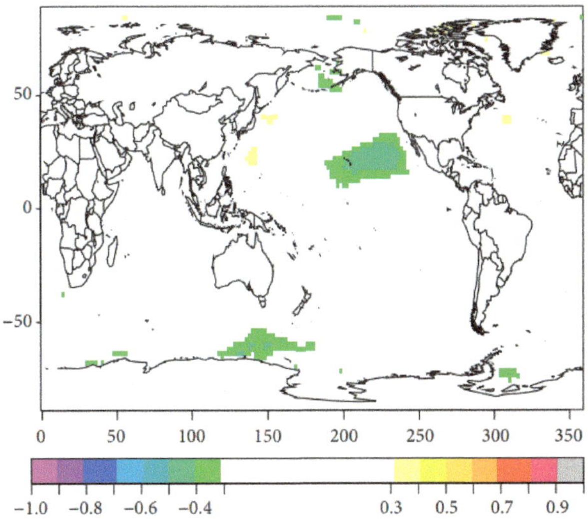

Figure 7: Pearson Correlation of the time series for the third mode of OND season precipitation variability with ERSST sea surface temperatures.

Long Rains (MAM Season)

The first four modes of variability for the long rains, MAM season, are significant and account for just over 51% of the variability for this season. We will examine these modes in detail.

Approximately 42 percent of the total East African rainfall is observed during the MAM season [4]. In the MAM season, there are four significant modes of variability for CRU TS 3.21 precipitation. The first significant mode accounts for 27.5% of the variability for station precipitation. The loading (Figure 8(a)) is mostly negative over the lake basin and has a −0.58 correlation with the first mode of annual variability. The corresponding time series (Figure 8(b)) shows an increasing trend with less variability in recent years. We investigate the source of variability for the mode by correlating it with SST and find a positive correlation with SSTs off the coast of Africa, between South Africa and Madagascar (Figure 9). This mode is examined in more detail since it accounts for a large amount of variability during the MAM season and has a decreasing trend in rainfall similar to that found in [22]. The decrease in trend in our study starts earlier than the abrupt change found in [22]. They relate the decreasing rainfall during this season starting around 1980 to SST anomalies in the Pacific. The difference between positive and negative composites of the time series for this mode with SST anomalies and 850 mb wind anomalies shows an ENSO-like signal similar to [23] (Figure 10); however, the mode is not significantly correlated with ENSO. Song and Zhang [23] attribute the El Niño-like SST warming to cloud feedback, further amplified by water vapor feedback. The corresponding wind anomalies over Eastern and near equatorial Africa (Figure 10) are consistent with Gill-type solution [24]. We believe that this is associated with the trend of reduced midtropospheric diabatic heating associated with the reduced rainfall over the region.

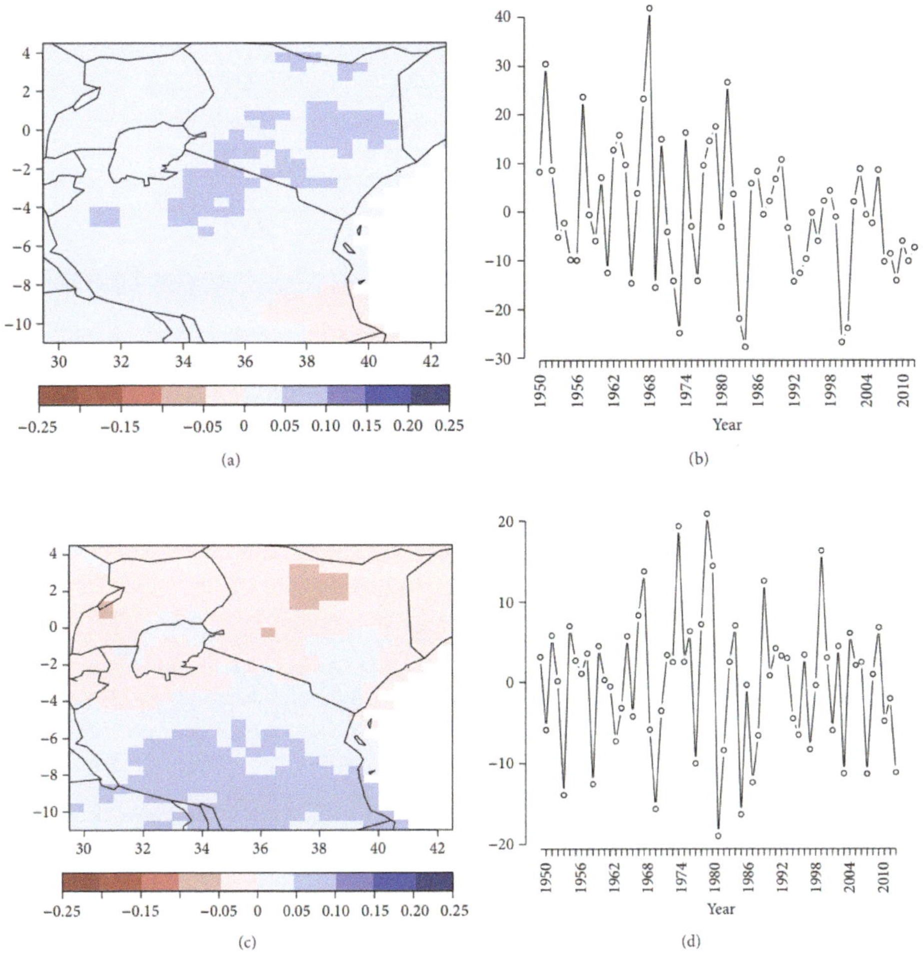

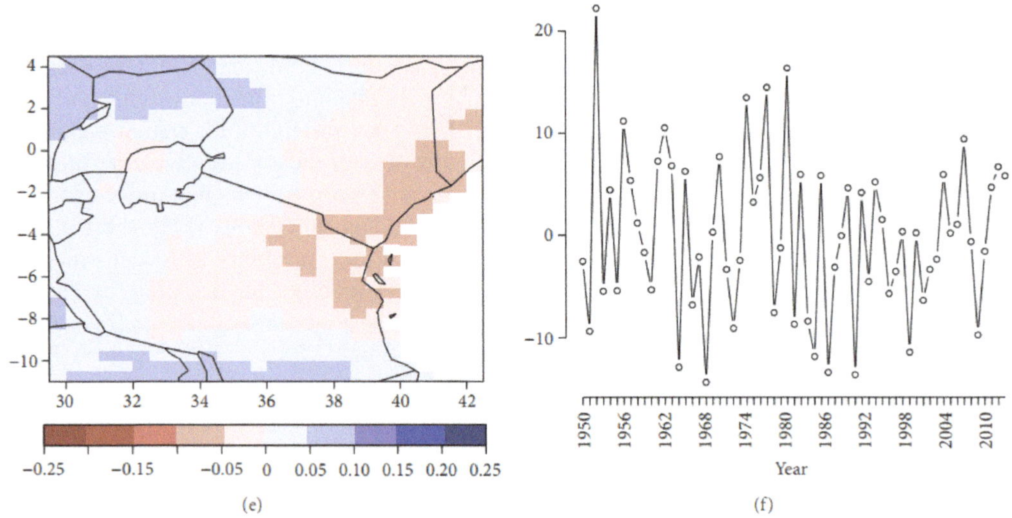

Figure 8: Same as Figure 3, but for long rains (MAM) precipitation.

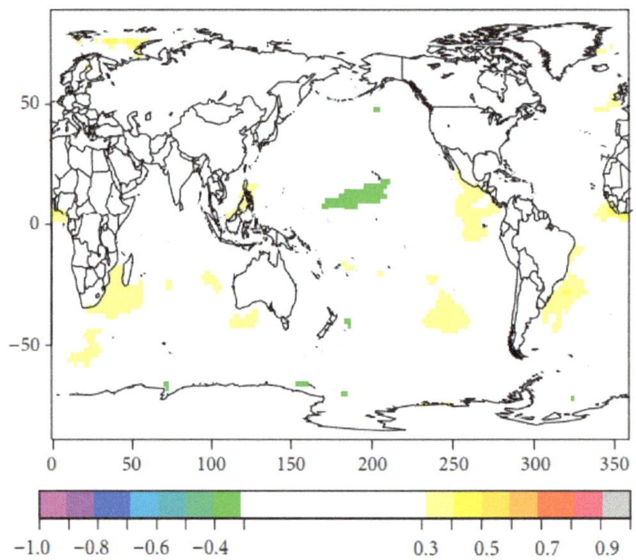

Figure 9: Pearson Correlation of the time series for the first mode of MAM season precipitation variability with ERSST sea surface temperatures.

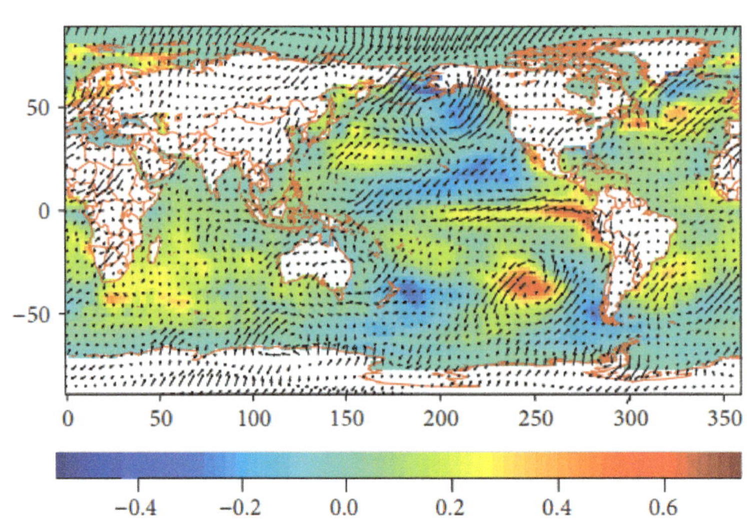

Figure 10: Difference between positive and negative composites of MAM EOF1 with SSTs and 850 mb winds.

The second mode is significant and has a −0.64 correlation with the second mode of annual variability. It accounts for 10.2% of variability in rainfall during the MAM season. The mode is split over Lake Victoria with slightly positive loadings near the northern and western shore and slightly negative loadings over the rest of the lake with stronger positive loadings to the south of the region (Figure 8(c)). There are no clear signals when this mode is correlated with global SSTs. The time series for this mode (Figure 8(d)) is not correlated with the time series of second mode of OND precipitation even though their loadings appear similar. More work needs to be done to identify the physical mechanism for this mode, although it may be related to the north-south movement of the Intertropical Convergence Zone (ITCZ) as it passes over this region during the long rain season.

The third mode, accounting for 8.6% of the variability, has a 0.53 correlation with the third mode of annual variability. This mode has a slightly negative loading in the northwest corner of Lake Victoria but a slightly positive loading over the remainder of the lake. The strongest negative loadings are near the coast, while the strongest positive loadings are in northern Uganda (Figure 8(e)). There are no clear signals when time series for this mode (Figure 8(f)) is correlated with global SSTs. This mode may be related to moisture convergence from the Indian Ocean; however, more work is needed to identify the physical mechanism.

The fourth mode of variability for the MAM season is also significant; however, it is not correlated with any of the significant modes of annual precipitation used in the water balance model for Lake Victoria. Due to this lack of correlation, the fourth mode of MAM precipitation variability is not analyzed in this study.

CONCLUSIONS

We find that the first EOF mode of annual precipitation variability, ENSO, has the highest impact on the annual variability of Lake Victoria level. It has high correlation with the first modes of variability during the short rain (OND) and long rain (MAM) seasons. The first mode of variability during the OND season is found to be a combination of ENSO, while more work needs to be done to determine the physical mechanism behind the first mode of variability during the MAM season, although it is positively correlated with SSTs near Madagascar, suggesting that it is related to the moisture flux in that region.

While the second EOF mode of annual variability accounts for approximately 10 percent of the variability, its effect on the variability in lake levels is negligible. The second EOF mode of annual precipitation variability is significantly correlated with the second and third EOF modes of variability from the OND season and the second mode from the MAM seasons. Based on the EOF analysis of OND season rainfall, the dipoles in the second and third modes are split with very small values over the lake. The dipole in the MAM season also has small loadings over the lake. This suggests that the dipole may have negligible effect on Lake Victoria levels due to the fact that the loadings for the annual dipole are overall small since, in the OND and MAM seasons, there is a combination of positive and negative loadings over the lake that almost completely cancels out once the rainfall is interpolated to the six stations and used as input in the water balance model. The third EOF mode of annual variability accounts for slightly over 7 percent of the variability over East Africa. It is correlated with the second mode of OND rainfall variability and the third mode of MAM variability. Similar to the second annual mode, it has negligible impact on the levels of Lake Victoria, most likely because the dipole modes over the lake are relatively small and may be cancelling each other out when they are combined.

We have sought to estimate the relative contributions of the dominant modes of annual precipitation variability to the modulation of Lake Victoria levels for the present day (1950–2012). The modes found in this study should be used to evaluate climate models over this region. Climate projections should be evaluated to investigate the evolution of these modes in the future. A change in ENSO, or a shift in one of the other modes described in this study, could have a great impact on the levels of Lake Victoria in the future.

APPENDIX

Calculation of Q_{in} and Q_{out} in Water Balance Model

Inflow, Q_{in}, is calculated as a nonlinear function with precipitation, P, as the input. Subcatchment rainfall (P_c) is estimated for each of five tributaries (Nzoia, Yala, Sondu, Awach Kaboun, and Kagera) using regressions of annual total data for the period 1965–1990 (Table 1). The runoff coefficient for each catchment is then estimated from the subcatchment rainfall by linear regression derived by [25] as

$$r_c = 0.0002P_c - 0.1386. \tag{A.1}$$

The runoff from each subcatchment to Lake Victoria is then estimated by multiplying each subcatchment's rainfall by the runoff coefficient and the subcatchment area.The total inflow to Lake Victoria, Q_{in}, is estimated as the runoff of Nzoia, Yala, Sonduuu, and Awach Kaboun scaled up by a factor of 2.7 on an annual basis, plus the flow of the Kagera increased by 10% to account for the flow of the Ngono tributary, which joins it downstream of the gauging station [8].

Table 1: Regressions of rainfall for 1956–1990. P_c is catchment rainfall; P_l is lake rainfall (from Tate)

Catchment	Regression	Correlation, R^2
Nzoia	$P_c = (0.685 * P_l) + 154.5$	0.54
Yala	$P_c = (0.951 * P_l) + 64.7$	0.61
Sondu	$P_c = (0.666 * P_l) + 302.6$	0.39
Awach Kaboun	$P_c = (0.785 * P_l) + 337.4$	0.28
Kagera	$P_c = (0.556 * P_l) + 142.8$	0.60

The outflow, Q_{out}, is calculated from the Agreed Curve based on the current lake level. We use the estimation of the Agreed Curve from [19]

$$Q_{out} = 66.3\left(L_n - 7.96\right)^{2.01}, \tag{A.2}$$

where L_n is the current Lake Victoria level. Since Ln is initially unknown, and as part of the equation to calculate Qout, the water balance model is initially solved setting L_n to the previous year's level to calculate an estimated L_n; the estimated L_n is then used to calculate the outflow in a second iteration of the model for each year.

ACKNOWLEDGMENTS

The authors wish to thank Dr. Emma (Tate) Brown for her assistance with the water balance model. This research was supported by NSF Grant AGS-1043125 and NSF Expeditions in Computing Award no. 1029711.

REFERENCES

1. R. O. Anyah and F. H. M. Semazzi, "Climate variability over the Greater Horn of Africa based on NCAR AGCM ensemble," Theoretical and Applied Climatology, vol. 86, no. 1–4, pp. 39–62, 2006. · ·
2. J. H. Bowden and F. H. M. Semazzi, "Empirical analysis of intraseasonal climate variability over the Greater Horn of Africa," Journal of Climate, vol. 20, no. 23, pp. 5715–5731, 2007. · ·
3. C. J. Schreck III and F. H. M. Semazzi, "Variability of the recent climate of eastern Africa," International Journal of Climatology, vol. 24, no. 6, pp. 681–701, 2004. · ·
4. M. Indeje, F. H. M. Semazzi, and L. J. Ogallo, "ENSO signals in East African rainfall seasons,"International Journal of Climatology, vol. 20, no. 1, pp. 19–46, 2000. · ·
5. G. W. Kite, "Recent changes in level of Lake Victoria," Hydrological Sciences Bulletin, vol. 26, no. 3, pp. 233–243, 1981. · ·
6. B. S. Piper, D. T. Plinston, and J. V. Sutcliffe, "The water balance of Lake Victoria," Hydrological Sciences Journal, vol. 31, no. 1, pp. 25–37, 1986. · ·
7. K. J. Sene and D. T. Plinston, "A review and update of the hydrology of Lake Victoria in East Africa,"Hydrological Sciences Journal, vol. 39, no. 1, pp. 47–63, 1994. · ·
8. E. Tate, J. Sutcliffe, D. Conway, and F. Farquharson, "Water balance of Lake Victoria: update to 2000 and climate change modelling to 2100," Hydrological Sciences Journal, vol. 49, no. 4, pp. 563–574, 2004. · ·
9. X. Yin and S. E. Nicholson, "Interpreting annual rainfall from the levels of Lake Victoria," Journal of Hydrometeorology, vol. 3, no. 4, pp. 406–416, 2002.
10. J. V. Sutcliffe and G. Petersen, "Lake Victoria: derivation of a corrected natural water level series,"Hydrological Sciences Journal, vol. 52, no. 6, pp. 1316–1321, 2007. · ·
11. D. Kull, Connections between Recent Water Level Drops in Lake Victoria, Dam Operations and Drought, 2006, http://www.internationalrivers.org/files/attached-files/full_report_pdf.pdf.
12. D. Kull, "Addendum to: connections between recent water level drops in Lake Victoria, dam operations and drought," 2006, http://www.internationalrivers.org/resources/addendum-lake-victoria-report-4118.
13. I. Harris, P. D. Jones, T. J. Osborn, and D. H. Lister, "Updated high-resolution grids of monthly climatic observations—the CRU TS3.10 Dataset," International Journal of Climatology, vol. 34, no. 3, pp. 623–642, 2013. · ·
14. C. S. Breetherton, C. Smith, and J. M. Wallace, "An intercomparison of methods of finding coupled patterns in climate data," Journal of Climatology, vol. 5, pp. 541–560, 1992.
15. M. Kendall, Multivariate Analysis, Hodder, Arnold, New York, NY, USA, 2nd edition, 1980. View at MathSciNet
16. K. M. Weickmann and R. M. Chervin, "The observed and simulated atmospheric seasonal cycle. Part I: global wind field modes," Journal of Climate, vol. 1, no. 3, pp. 265–289, 1988.
17. F. H. M. Semazzi, B. Burns, N.-H. Lin, and J.-K. Schemm, "A GCM study of the teleconnections between the continental climate of Africa and global sea surface temperature anomalies," Journal of Climate, vol. 9, no. 10, pp. 2480–2497, 1996. · ·
18. G. Capdeville, "A Hermite upwind WENO scheme for solving hyperbolic conservation laws," Journal of Computational Physics, vol. 227, no. 4, pp. 2430–2454, 2008. · · View at MathSciNet ·
19. K. J. Sene, "Theoretical estimates for the influence of Lake Victoria on flows in the upper White Nile,"Hydrological Sciences Journal, vol. 45, no. 1, pp. 125–145, 2000. · ·
20. N. N. Davis, Dynamic and stochastic modeling of various components of the hydrological cycle for East Africa [M.S. thesis], North Carolina State University, 2007.
21. USDA, "Radar altimeter data from NASA/CNES Topex /Poseidon and Jason-1 satellite missions. Time series of altimetric lake level variations from the USDA Reservoir Database," 2012, http://www.pecad.fas.usda.gov/cropexplorer/global_reservoir.
22. B. Lyon and D. G. Dewitt, "A recent and abrupt decline in the East African long rains," Geophysical Research Letters, vol. 39, no. 2, Article ID L02702, 2012. · ·
23. X. Song and G. Zhang, "Role of climate feedback in El Niño-like SST response to global warming,"Journal of Climate, 2014. ·
24. A. E. Gill, "Some simple solutions for heat-induced tropical circulation," Quarterly Journal, vol. 106, no. 449, pp. 447–462, 1980. · ·
25. J. V. Sutcliffe and Y. P. Parks, "The hydrology of the nile," IAHS Special Publication 5, IAHS Press, Institute of Hydrology, Wallingford, UK, 1999.

CHAPTER 4

GRACE DATA APPLICATIONS IN TERRESTRIAL HYDROLOGY MONITORING

Dong Jiang, Jianhua Wang, Yaohuan Huang, Kang Zhou, Xiangyi Ding, and Jingying Fu

INTRODUCTION

Intensified impact of human activities and global change inevitably lead to the changes in the water cycle, including the spatial and temporal distribution and the total amount of water resources [1, 2]. Terrestrial hydrology, which is an important indicator in global change, affects the global climate system in energy, water, and biogeochemical [3]. Traditional monitoring of terrestrial hydrology mainly depends on site measurements or model simulations, which is costly and time-consuming. Furthermore, for the scale effects of the conversion from observation points to large region, additional error will be brought into spatially continuous data obtained by statistical interpolation methods. The accuracy of interpolation results may significantly decrease in areas far away from observation sites. Hydrological models and land surface models solve the problem of simulation of terrestrial hydrology on the plane scale to some degree. However, the lack of systematic global fine-scale hydrological data increases the model uncertainty and lowers the simulation accuracy. This restricts the practical application of hydrological models and land surface models in the monitoring of terrestrial hydrology. With the rapid development of remote sensing technology, remote sensing holds significant potential to the regional hydrological researches. By offering a useful and cost-effective approach to rapidly monitor terrestrial hydrological parameters, remote sensing technology has been widely used in hydrology. Recently, combination models of remote sensing technology, observation network, and hydrological models have been common in terrestrial hydrology studies. However, imaging satellite techniques and satellite altimetry can only be used in the monitoring of surface water that represents just one component of total water storage. They can be hardly applied in the monitoring of other components of terrestrial water such as soil water and groundwater, which impeded further applications of these technologies in terrestrial hydrology.

Launched in March 2002, the Gravity Recovery and Climate Experiment (GRACE) can provide an operational product in the form of global gravity fields every ~30 days [4]. After several gravity effects have been released (e.g., atmospheric mass variations and ocean tides), precise observations of the time-series global gravity field changes make the spatial monitoring of water storage changes at large scale such as basins possible. Through 10 years of development, GRACE satellite data processing and corresponding TWS retrieval algorithms are improved continuously. They are able to detect the changes of TWS within the accuracy of 1.5 cm on a wide range of spatial scales and seasonal time scales [5]. Compared to imagery remote sensing methods, they provide alternative technical methods for the estimation of glaciers, surface snow, soil water, surface water, groundwater, evaporation, and other components in the terrestrial hydrological system. It provides a new technique for the large-scale monitoring of terrestrial water cycle. The objective of this paper is to present an overview of GRACE data applications in terrestrial hydrology monitoring, including GRACE data processing methods and TWS change retrieval and their application in terrestrial hydrology and related fields.

Citation: Dong Jiang, Jianhua Wang, Yaohuan Huang, Kang Zhou, Xiangyi Ding, and Jingying Fu, "GRACE Data Applications in Terrestrial Hydrology Monitoring," Advances in Meteorology, vol. 2014, Article ID 725131, 9 pages, 2014. doi:10.1155/2014/725131.

GRACE DATA

GRACE gravity satellite program was jointly developed by the National Aeronautics and Space Administration (NASA) of the United States and the German Aerospace Center (DLR) with the objective of providing spatiotemporal variations of the Earth's gravity field. The GRACE satellite program can be also used to detect the atmosphere and ionosphere environment. The US Jet Propulsion Laboratory (JPL) is responsible for the project management of the GRACE gravity satellite program. Monthly gravity field solutions are computed at the University of Texas at Austin Center for Space Research (CSR), the German Research Centre for Geosciences Potsdam (GFZ), JPL, Groupe de Recherche de Geodesie Spatiale (GRGS), and the Delft Institute of Earth Observation and Space Systems (DEOS) as well as Delft University of Technology, among others [4, 6–8]. GRACE utilizes a state-of-the-art technique to trace the spatiotemporal gravity field with an increased sensitivity by tracking the micrometer-precise intersatellite range and range-rate observations between two coplanar, low altitude (300 km~500 km) satellites and the distance between which is about 220 km. To precisely measure the distance changes between two satellites at micron meter level accuracy, a K-band ranging (KBR) system based on carrier phase measurements in the K (26 GHz) and Ka (32 GHz) frequencies is provided. Besides, four key instrumentations of a GPS receiver (space-proofed multichannel, two-frequency), capacitive accelerometer, laser retroreflector (LRR), and star camera are equipped on board of each GRACE satellite [4, 7].

By analyzing the relationship between orbits and forces of GRACE satellites, the Earth's gravity field variety is estimated based on the dynamic equations of satellites motion. It monitors the time-varying characteristics of long-wavelength gravity field on a 15~30-day or longer time scale. Large-scale mass redistribution (mass distribution changes over time) in the Earth system reflects the interaction between substances of various forms in the Earth's internal system (atmosphere, oceans and solid crust, viscous mantle, liquid outer core, and solid inner core), which are the important subjects of Earth sciences. Compared to the average Earth's gravity field, the time-varying quantity of the gravity field is very small, but it contains important geophysical information. It reveals the movement, distribution, and changes of all substances in the Earth system and reflects the interaction between the atmosphere, terrestrial water, oceans, and the solid Earth [6].

The time variation of global gravity field caused by the influence of solid Earth (including the inner core and the outer core) is mainly manifested on a 10-year or a longer time scale. However, temporal variation of gravity is mainly caused by the redistribution influence of atmosphere, oceans, and water storages in the surface fluid envelopes of the earth on a seasonal or interannual time scale [9, 10]. The gravity effects of tides (solid Earth, oceans, and atmosphere) and nontidal (atmosphere and oceans) are reduced in the data processing of GRACE gravity field model. Therefore, by excluding the errors in gravity field model and the atmosphere and oceans models, GRACE time-varying gravity field reflects the nonatmospheric and nonocean mass variations due to water mass variations on the continental area. On a seasonal or shorter time scale, it provides information on changes in terrestrial water storage in large river basin [11].

HYDROLOGICAL APPLICATIONS

TWS Changes Monitoring

The TWS is the most direct hydrological parameter obtained by GRACE monitoring. For the main goal of GRACE satellite is to monitor the Earth's gravity field variations, early researches are focused on the feasibility and accuracy of TWS retrieval from GRACE data. Wahr et al. [11] pointed out that the GRACE-based TWS retrieval required the consideration of the impacts from short-wave noises and leakage errors, and the TWS retrieval result is valid for large-scale river basins. Further research of Rodell and Famiglietti showed that TWS accuracy could be improved by increasing the monitoring temporal interval and spatial resolution of the monitored area [12, 13]. For regions larger than 200000 km^2, the TWS changes with intervals on month and longer time scale can be monitored, and the accuracy can reach 1.5 cm and above.

The application of GRACE-based TWS changes data is mainly based on the combination of hydrological models and land surface models. GRACE-based TWS retrieval results offer spatial and temporal distribution of vertically integrated water storage (surface water, soil, groundwater, and snowpack) in large river basins. So the errors due to the use of indirect indicators such as the flow rate and precipitation in some hydrological models can be reduced. In initial researches, the GRACE-based TWS is only considered as a reference to the simulation results from hydrological models (such as GDLAS and CPC), which does not involve the uncertainty of hydrological models in the simulation process. Hu et al. [10] proposed that water storage changes in Yangtze River could reach a magnitude of 3.4 cm equivalent water height, with the maximum occurring in the spring and early autumn. Using 5-year GRACE-based TWS change data in China, Zhong et al. [14] pointed out that the water storage in the north-central region of China diminished at an annual rate of 2.4 cm water equivalent height. By combining GRACE and GPS network observation data, Wang et al. [15] achieved precise retrieval of TWS change in Nordic region and North America. Their studies showed that, over the past decade, a sharp increase appeared in the water storage in North America and Scandinavia and a water recovery process is underway.

With further study, GRACE data is now considered as an important parameter to assess and improve hydrological model simulation. Swenson and Milly [16] and Syed et al. [17] used GRACE data to verify and improve TWS simulation of five climate models and the Global Land Data Assimilation System (GLDAS), respectively. Niu and Yang [18] and Ngo-Duc et al. [19] used GRACE data for the result verification of common land surface model (CLM) of NCAR and ORCHIDEE and corrected the CLM by applying GRACE-based TWS changes data.

Various quantitative analyses between hydrological models and GRACE-based TWS showed that the two simulation results of large areas are in good consistency on a seasonal or longer time scale [16, 20, 21]. Time-series analyses of South and North America, Southeast Asia, the Ganges River Basin, Africa, Eurasia, and other large-scale regions all showed that GRACE-based TWS reflected a significant seasonal variation. However, given the limitations in the spatial resolution of GRACE data, the consistency of hydrological models and

GRACE data in smaller areas is not satisfactory [22]. Excluding the errors of GRACE-based TWS change retrieval, the inconsistency is mainly due to the simulation errors of various TWS components from hydrological models. For example, the generalized deviation in estimating the important elements such as groundwater, surface water, and soil water storage by using hydrological models resulted in smaller magnitudes in TWS changes. The errors in the simulation of rainfall, convergence, and other hydrological processes generated phase deviation in TWS changes time series [17, 23, 24]. In addition, by combination of the measured data of rainfall, recharge, soil water, and groundwater, GRACE-based TWS was used for water storage variation of basin on interannual, annual, seasonal, and monthly time scales [25, 26].

Hydrological Components Evaluation

With the improvement of GRACE-based TWS change accuracy, the GRACE data is further used in hydrology and water resources research. For specific areas, TWS changes can be expressed as [10]

$$\frac{dW}{dt} = P - E - R, \tag{1}$$

where dW/dt represents TWS changes, and the corresponding equivalent water height can be obtained from GRACE. P represents precipitation, E represents evapotranspiration, and Rrepresents surface runoff. By combining GRACE-based TWS changes, hydrological models, and the measured data, the water storage components including groundwater, soil water, ET, and $P-E$ can be estimated, respectively.

Groundwater estimation is difficult in remote sensing applications in hydrology. Optical remote sensing methods can only combine measured data with spectral data to construct an empirical model for the estimation of groundwater changes, whose groundwater retrieval accuracy is poor. Considering the contribution of groundwater changes to TWS variation, groundwater remote sensing using GRACE was feasible by combination with ancillary measurements of surface waters and soil moisture. Rodell and Famiglietti [27] firstly used GRACE-based TWS change data, soil water, and other auxiliary data to analyze the possibility of monitoring groundwater changes. Swenson et al. [25] studied the relation between measured value of groundwater plus soil water and GRACE-based TWS changes and presented the possibility and sources of error in groundwater estimation based on GRACE data on multiple time scales. Strassberg et al. [28] compared the spatiotemporal correlation between GRACEbased TWS changes and hydrological model simulations and field monitoring data and proposed the uncertainty of GRACE-based groundwater retrieval. The above studies suggest that, for aquifers larger than 450000 km^2, the accuracy of GRACE-based groundwater can reach 8.7 mm. Rodell et al. applied GRACE data to the groundwater subsidence monitoring in India and found that excessive irrigation and human activities caused the groundwater in the northwest provinces of India to decline at an annual rate of 3.0 cm~ 4.0 cm [29].

Evapotranspiration (ET) is an important process parameter to studies ofhydrology, which is difficult to measure at a regional or continental scale. Although ET can be indirectly estimated using remote sensing data based on empirical models, energy balance models, or physical models (e.g., PenmanMonteith equation), recent remote-sensed ET estimation is far from satisfactory [30]. For ET is a complex process that related to many variables, its estimation uncertainty based on remote sensing data brings errors to ET retrieval results. Based on terrestrial water balance at scale of river basin, changes of regional ET can be estimated by combining TWS change from GRACE with observed precipitation (P) and runoff data (R). Rodell et al. [31] first used the measured precipitation and runoff data to verify the feasibility of GRACE-based ET retrieval. However, the overdependence on measured data restricts the applicability of this method. Therefore, Ramillien et al. [32] improved this algorithm by using the simulated runoff data from hydrological model. The time-series analysis showed that the ET estimation and WGHM simulation were in good consistency. However, uncertainty of GRACE-based ET estimation increases for its overdependence on accuracy of hydrological model simulated variables such as runoff. In addition, Swenson and Wahr [33] combined GRACE-based TWS changes together with the measured runoff data to estimate the difference between precipitation and evapotranspiration ($P-E$). By comparing with $P-E$ results of a land surface model (GLDAS-Noah), they found that the errors of GLDAS-Noah estimation are mainly due to model force parameter of precipitation.

Drought Analysis

Based on time-series analysis of GRACE-based TWS change with high temporal resolution, extreme hydrological disasters can be monitored and alarmed. Droughts for regional and time-series water storage deficit are the most serious natural hazards that can lead to crop losses and economic havoc in many areas. For droughts can be regarded as terrestrial water storage (TWS) changes that are related to integrated bulk variables, analysis relying upon subcomponents (e.g., precipitation) or proxies (e.g., NDVI and CWSI) of TWS is insufficient. Herewith, another hydrological application of GRACE is severe droughts analysis. Combined with the measurements and hydrological model simulations, Leblanc et al. [34] used GRACE data to detect droughts in Southeastern Australia between 2001 and 2008. Based on time-series analysis of TWS changes in summers from 2002 to 2007, 2005 extreme drought event in the Amazon river basin was detected [35], which was regarded as the worst in over a century. GRACE data is more sensitive to droughts than data-assimilating climate and land surface models such as NECEP and GLDAS, which demonstrated the unique potential of GRACE in monitoring largescale severe drought events. Validated by two independent hydrological estimates of GLDAS and ECMWF and direct gravity observations from superconducting gravimeters, the 2003 excess terrestrial water storage depletion was observed from GRACE, which can be related to the record-breaking heat wave that occurred in central Europe in 2003 [36]. It indicated that GRACE data can be used in heat wave disaster monitoring and evaluation, whose essence is heat wave caused droughts analysis. Combining with imagery remote sensing methods, Frappart and Wilson et al. applied GRACE data to flood monitoring in several flood zones such as the Mekong River Basin and the Amazon River Basin [37, 38]. However, compared to droughts, the water storage changes based on

GRACE data of extreme floods are less responsive, which will bring more uncertainty in GRACEbased floods analysis [39].

Glacier Mass Balance Detection

Driven by global change and population pressure, accelerating of glaciers melting and sea level rise has been a serious global ecological problem. GRACE time-variable gravity field enabled direct measurement of mass loss rates of both mountain glacier and polar glacier systems. As noted by Wouters et al. the GRACE solutions can be used to allow regional estimation of trends, though assessing the mass loss to be dominated by summer events rather than by a linear trend [40]. GRACEbased glacier mass melting volume and rate estimation of Antarctic and Greenland indicate that accelerating polar glacier melting contributes greatly to the current eustatic sea level rise [41–43]. Gardner et al. have also found a rapidly increasing mass loss in the Canadian Arctic Archipelago (CAA) for the period 2004–2009 [44]. Ice loss rate estimation on glaciers of Asia, Alaska, and other global mountains based on GRACE data also agrees with the global tendency of accelerating glacial loss. Accelerated melting of mountain glaciers worldwide might be contributing to the global sea level rise by 0.73 ± 0.10 mm/yr [45, 46]. Moreover, GRACEbased TWS change can be used for researches of different climate variability impact [47, 48] whereas, for the postglacial rebound effects and unique set of gravity data GRACE mission provided, ice sheet mass change estimation includes large uncertainties [49]. Mass change rates estimated by different GRACE solutions may vary by a factor of two or more. So the glacier mass balance detection needs considerable processing to yield usable mass change data [49].

DISCUSSION AND CONCLUSIONS

Through ten years of development, GRACE data has been widely applied in terrestrial hydrology monitoring. Gravity satellite provides new data sets for hydrology researches based on remote sensing technology. Compared with other kinds of hydrological schemes, GRACE can provide realistic spatiotemporal variations of vertically integrated measurement of water storage (groundwater, soil moisture, surface water, snow water, vegetation water, etc.) at the precision of tens of mm of equivalent water height at large scale. However, the hydrological monitoring based on GRACE data needs further research, for example, retrieval accuracy improvement, more quantitative analyses rather than qualitative analyses, and so forth. The focus of future development includes the following aspects.

(1) *Improving the Accuracy of Gravity Satellite Measurements*. Because of the limitations of GRACE satellite sensors in orbital altitude, vertical gravity gradient measurements, high frequency signal aliasing, and accurate measurement of timevarying gravity signals, the accuracy and spatial resolution of time-varying Earth gravity field signals at medium and long wavelength of GRACE is low, which reduced the TWS change retrieval accuracy. Therefore, the development of key technology in gravity satellite sensors to improve accuracy and spatial resolution of satellite monitored gravity field is the basis for a wider application in terrestrial hydrology monitoring.

(2) *Water Storage Retrieval Models Research*. Recently, the appropriate spatial resolution for GRACE-based TWS change is 400 km, and the data accuracy is generally 1.5 cm. Limited by recent constraints of GRACE satellite, improvement of retrieval models is the only method for spatial resolution and accuracy retrieval of TWS increases. For example, it is feasible to improve TWS accuracy by the atmosphere and ocean models improvement in data preprocessing, reducing the gravity field changes noises caused by factors unrelated to terrestrial water such as tides and circulation of atmosphere and ocean. In addition, by applying different filtering methods such as anisotropic Gaussian filtering and spherical radial basis function filtering, wavelet analysis in different research areas can also improve the spatial resolution and accuracy of TWS retrieval. Water storage retrieval techniques involve various aspects (e.g., GRACE data preprocessing, gravity field retrieval, and TWS changes estimation) of the conversion of GRACE gravity field data to TWS changes, which are all important in TWS retrieval improvement.

(3) Further Combination of GRACE Data with Associated Hydrological Models. For GRACE-based TWS changes retrieval does not involve the hydrological mechanisms, it is necessary to complement GRACE-based TWS by hydrological models for furthering its application in terrestrial hydrology monitoring. Further combination of GRACE and hydrological models needs resolving their consistencies of space, time, and component. Spatial consistency can be solved by adjusting the calculation unit of hydrological models to reflect the spatial variability (such as distributed hydrological models). Temporal consistency requires analysis of upscaling GRACE data and downscaling hydrological models simulation. Component consistency can be achieved by GRACE-based TWS signals subdivision or revising parameters of hydrological models to TWS. Through space, time, and component consistency improvement, the gravity field data and measured data together can become basic dataset to force hydrological models in the future.

APPENDICES

A. Water Storage Change Retrieval

The changes in terrestrial water storage result in mass redistribution in the Earth's system, thereby causing changes in the gravity field. For a fixed continental region, the changes in water storage (including soil water and surface snow) come from rainfall, evapotranspiration, river transportation, and deep underground infiltration. Except the rainfall which can cause increased water storage, the remaining three processes all reduce it [10]. Using the FG5 absolute gravimeter, Zhang et al. [50] measured the gravity change of nearly $10^{-7}\ m \cdot s^{-2}$ at the Wuhan University site before and after a rainstorm, which clearly shows the influence of terrestrial water variation on gravity. The Earth's gravity field can be expressed as geoid:

$$N(\theta,\lambda) = a\sum_{n=0}^{\infty}\sum_{m=0}^{n}\left[\overline{C}_{nm}\cos(m\lambda)+\overline{S}_{nm}\sin(m\lambda)\right]\overline{P}_{nm}(\cos\theta), \tag{A.1}$$

where n and m are harmonic degree and order of the gravity field, respectively; a is the Earth's equatorial radius (about 6,371 km); θ and λ are colatitudes (the difference between 90° and latitude) and longitude; $\overline{C}_{nm}$ and $\overline{S}_{nm}$ are spherical harmonic coefficients (dimensionless); $\overline{P}_{nm}$ is the normalized associated Legendre functions. The maximum value N of the order n of ideal gravity field should be infinite ($N \sim \infty$), while the actual order of spherical harmonic coefficients obtained by the gravity satellite has a finite value ($N<\infty$) and the spatial resolution of gravity field data is estimated approximately to be $\pi a/N$ [51]. Geoid height changes ΔN caused by the movement of substances on the Earth's surface can be expressed as

$$\Delta N(\theta,\lambda) = a\sum_{n=0}^{\infty}\sum_{m=0}^{n}\left[\Delta\overline{C}_{nm}\cos(m\lambda)+\Delta\overline{S}_{nm}\sin(m\lambda)\right]\overline{P}_{nm}(\cos\theta), \tag{A.2}$$

where $\Delta\overline{C}_{nm}$ and $\Delta\overline{S}_{nm}$ are the changes of n-degree m-order spherical harmonic coefficients of geoid and can be expressed as [5, 52]

$$\begin{Bmatrix}\Delta\overline{C}_{nm}\\ \Delta\overline{S}_{nm}\end{Bmatrix} = \frac{3}{4\pi a\rho_a(2n+1)} \times\int\Delta\rho(r,\theta,\lambda)\overline{P}_{nm}(\cos\theta)\left(\frac{r}{a}\right)^{n+2}\begin{Bmatrix}\cos(m\lambda)\\ \sin(m\lambda)\end{Bmatrix}\sin\theta d\theta\, d\lambda\, dr, \tag{A.3}$$

where ρ_a is the average density of the Earth (5517 kg/m^3) and $\Delta\rho(r, \theta, \lambda)$ is the change of bulk density at a particular location. In gravity field inversion, because the height change of substances on the Earth's surface H is relatively small compared to the Earth's average radius a in gravity field retrieval ($(r/a) \approx 1$), the changes of the gravity field directly caused by the surface mass can be expressed as

$$\begin{Bmatrix}\Delta\overline{C}_{nm}\\ \Delta\overline{S}_{nm}\end{Bmatrix}_{\text{surf}} = \frac{3}{4\pi a\rho_a(2n+1)} \times\int\Delta\sigma(\theta,\lambda)\overline{P}_{nm}(\cos\theta)\begin{Bmatrix}\cos(m\lambda)\\ \sin(m\lambda)\end{Bmatrix}\sin\theta d\theta\, d\lambda, \tag{A.4}$$

where $\Delta\sigma$ is surface density change $\Delta\sigma = \int\Delta\rho(r, \theta, \lambda)dr$. Furthermore, the surface mass variation in loads will cause deformation to the solid Earth, which in turn can indirectly cause variations to the gravity field. This can be expressed as [5, 52]

$$\begin{Bmatrix}\Delta\overline{C}_{nm}\\ \Delta\overline{S}_{nm}\end{Bmatrix}_{\text{soild}} = \frac{3k'_n}{4\pi a\rho_a(2n+1)} \times\int\Delta\sigma(\theta,\lambda)\overline{P}_{nm}(\cos\theta)\begin{Bmatrix}\cos(m\lambda)\\ \sin(m\lambda)\end{Bmatrix}\sin\theta d\theta\, d\lambda, \tag{A5}$$

where k'_n is load LOVE number of coefficients, and the specific value of k'_n can be found in relevant literature [11]. Thus, the change of the Earth's gravity field caused by mass variations on the Earth's surface is given by

$$\begin{Bmatrix}\Delta\overline{C}_{nm}\\ \Delta\overline{S}_{nm}\end{Bmatrix} = \begin{Bmatrix}\Delta\overline{C}_{nm}\\ \Delta\overline{S}_{nm}\end{Bmatrix}_{\text{surf}} + \begin{Bmatrix}\Delta\overline{C}_{nm}\\ \Delta\overline{S}_{nm}\end{Bmatrix}_{\text{soild}} = \left(1+k'_n\right)\begin{Bmatrix}\Delta\overline{C}_{nm}\\ \Delta\overline{S}_{nm}\end{Bmatrix}_{\text{surf}}. \tag{A.6}$$

If the spherical harmonic expansion is performed on the surface density change $\Delta\sigma$, then

$$\Delta\sigma(\theta,\lambda) = a\rho_w\sum_{n=0}^{\infty}\sum_{m=0}^{n}\left[\widehat{C}_{nm}\cos(m\lambda)+\widehat{S}_{nm}\sin(m\lambda)\right]\overline{P}_{nm}(\cos\theta), \tag{A.7}$$

where ρw is the density of water and $\Delta\sigma/\rho_w$ can be considered as mass variation on the Earth's surface expressed in water equivalent height. Wahr et al. [5] proposed that

$$\begin{Bmatrix}\Delta\widehat{C}_{nm}\\ \Delta\widehat{S}_{nm}\end{Bmatrix} = \frac{\rho_a}{3\rho_w}\frac{2n+1}{1+k'_n}\begin{Bmatrix}\Delta\overline{C}_{nm}\\ \Delta\overline{S}_{nm}\end{Bmatrix}. \tag{A.8}$$

The surface density change $\Delta\sigma$ can be calculated using the following equation:

$$\Delta\sigma(\theta,\lambda) = \frac{a\rho_a}{3}\sum_{n=0}^{\infty}\sum_{m=0}^{n}\frac{2n+1}{1+k'_n}\left[\overline{C}_{nm}\cos(m\lambda)+\overline{S}_{nm}\sin(m\lambda)\right] \times\overline{P}_{nm}(\cos\theta). \tag{A.9}$$

For ocean and atmosphere mass variations are removed based on Parallel Ocean Program (POP) model, the above equation is the basic equation for the retrieval of surface mass variations based on spatiotemporal changes gravity field. The Earth's surface density changes can be derived from the changes of gravity field coefficients obtained from GRACE satellites.

B. Evaluation Errors

Currently, the latest available GRACE dataset product is

the RL05. The accuracy of data provided by CSR and GFZ is better than that provided by JPL. The evaluation errors of GRACE are brought from satellite instruments measurement, retrieval models, and other factors as follows.

(1) Gravity field data measured by GRACE satellite may be contaminated by satellite measurement errors for the influence of satellite orbit, satellite K-band ranging, and accelerometer measurement [20]. The satellite measurement errors also include poor accuracy of C20 due to the insensitivity of track geometry to the gravity field's low degree gravitational variations [5, 20]. It is generally released with removing variation of C20 [10] or replacing it by satellite laser ranging (SLR) substitution [20]. In addition, the missing of first-degree spherical harmonic coefficients will also bring error to GRACE. It can commonly be resolved by substituting value calculated from the term of the seasonal changes of the Earth's mass center [53, 54] or ignoring its impact [55].

(2) In theory, the retrieval of gravity field variation needs to use spherical harmonic coefficients of all degrees from 0 to infinity. However, gravity satellites can only provide definite order data. So the surface density change $\Delta\sigma$ in retrieval models is treated by spherical harmonic expansion to definite orders. For the impact of high-order terms on the Earth's surface density change $\Delta\sigma$ cannot be ignored, it results in truncation errors in gravity field retrieval. RL05 water storage data is estimated by CSR using a retrieval model truncated to degree 60 [8]. Zhu et al. (2008) compared the global water storage retrieval results using models truncated to 15 degrees, 20 degrees, 35 degrees, and 60 degrees [56].They found that although some information of TWS change may be missing the retrieval result of TWS change became more marked with lower truncation degree. Generally, the water storage retrieval truncated to order 60 is widely adopted [17].

(3) For terrestrial water monitoring focusing on the mass changes of a particular area (e.g., river basin), it requires the integral process on density change $\Delta\bar{\sigma} = ((\int \Delta\sigma(\theta,\lambda)u(\theta,\lambda)d\Omega)/\Omega)$. The function of regional characteristics (θ, λ) is equal to 1 inside the particular area and to zero outside. The error will be brought for the discontinuity of (θ, λ) in the domain of integration. In addition, for the influence of rapid increase of the errors of GRACE gravity field model coefficients with increase of spherical harmonic coefficients degree, signal leakage errors, and the striping pollution [23], filtering methods are proposed to smooth GRACE data for noise reducing, which will result in filtering errors of retrieval results. Proposed filtering methods include spatial averaging, symmetric Gaussian filtering, optimized decorrelation filtering, time-series method, global hydrological model correction method, kernel-independent component analysis, and optimal smoothing kernel method [53, 57–62]. Using filtering for the integral treatment of surface density changes can effectively remove the striping to a certain extent. However, the obtained average surface density is critical for it reduces the useful energy of geophysical signals and results in filtering errors. In addition, the existing filtering methods require the support of a priori knowledge (such as filtering radius and truncation degree). Therefore, in the actual retrieval process, the filter selection and parameter calibration require an understanding of the specific regional characteristics [60, 63].

(4) TWS changes retrieval based on spatiotemporal variations of gravity field remains considerable uncertainty, which includes errors produced in the removal of tidal movement and the mass migration because of the atmosphere and ocean circulation [11], as well as the errors in hydrological models used for the estimation of other terrestrial water parameters [6].

ACKNOWLEDGMENTS

This project was supported by National Natural Science Foundation of China (Grant no. 51309210) and the Chinese Academy of Sciences (Grant no. KZZD-EW-08).

REFERENCES

1. Y. Huang, D. Jiang, D. Zhuang, Y. Zhu, and J. Fu, "An improved approach for modeling spatial distribution of water use profit—a case study in Tuhai Majia Basin, China," Ecological Indicators, vol. 36, pp. 94–99, 2014. · ·
2. Y.-H. Huang, D. Jiang, D.-F. Zhuang, J.-H. Wang, H.-J. Yang, and H.-Y. Ren, "Evaluation of relative water use efficiency (RWUE) at a regional scale: a case study of Tuhai-Majia Basin, China," Water Science and Technology, vol. 66, no. 5, pp. 927–933, 2012. · ·
3. J. S. Famiglietti, "Remote sensing of terrestrial water storage, soil moisture and surface waters," in The State of the Planet: Frontiers and Challenges in Geophysics, R. S. J. Sparks and C. J. Hawkesworth, Eds., pp. 197–207, 2004.
4. B. D. Tapley, S. Bettadpur, M. Watkins, and C. Reigber, "The gravity recovery and climate experiment: mission overview and early results," Geophysical Research Letters, vol. 31, no. 9, Article ID L09607, 2004. · ·
5. J. Wahr, S. Swenson, V. Zlotnicki, and I. Velicogna, "Time-variable gravity from GRACE: first results,"Geophysical Research Letters, vol. 31, no. 11, Article ID L11501, 2004. · ·
6. A. Güntner, "Improvement of global hydrological models using GRACE data," Surveys in Geophysics, vol. 29, no. 4-5, pp. 375–397, 2008. · ·
7. G. Ramillien, J. S. Famiglietti, and J. Wahr, "Detection of continental hydrology and glaciology Signals from GRACE: a review," Surveys in Geophysics, vol. 29, no. 4-5, pp. 361–374, 2008. · ·
8. D. P. Chambers, "Converting 11 Release-04 Gravity Coefficients into Maps of Equivalent Water Thickness," 2007, http://gracetellus.jpl.nasa.gov/files/GRACE-dpc200711_RL04.pdf.

9. M. Becker, W. LLovel, A. Cazenave, A. Güntner, and J.-F. Crétaux, "Recent hydrological behavior of the East African great lakes region inferred from GRACE, satellite altimetry and rainfall observations,"Comptes Rendus Geoscience, vol. 342, no. 3, pp. 223–233, 2010. · ·
10. X. Hu, J. Chen, Y. Zhou, C. Huang, and X. Liao, "Seasonal water storage change of the Yangtze River basin detected by Grace," Science in China D, vol. 49, no. 5, pp. 483–491, 2006. · ·
11. J. Wahr, M. Molenaar, and F. Bryan, "Time variability of the Earth's gravity field: hydrological and oceanic effects and their possible detection using GRACE," Journal of Geophysical Research B, vol. 103, no. 12, pp. 30205–30229, 1998. ·
12. M. Rodell and J. S. Famiglietti, "Detectability of variations in continental water storage from satellite observations of the time dependent gravity field," Water Resources Research, vol. 35, no. 9, pp. 2705–2723, 1999. ·
13. M. Rodell and J. S. Famiglietti, "An analysis of terrestrial water storage variations in Illinois with implications for the Gravity Recovery and Climate Experiment (GRACE)," Water Resources Research, vol. 37, no. 5, pp. 1327–1339, 2001. · ·
14. M. Zhong, J. Duan, H. Xu, P. Peng, H. Yan, and Y. Zhu, "Trend of China land water storage redistribution at medi- and large-spatial scales in recent five years by satellite gravity observations,"Chinese Science Bulletin, vol. 54, no. 5, pp. 816–821, 2009. · ·
15. H. Wang, L. Jia, H. Steffen et al., "Increased water storage in North America and Scandinavia from GRACE gravity data," Nature Geoscience, vol. 6, no. 1, pp. 38–42, 2013. · ·
16. S. C. Swenson and P. C. D. Milly, "Climate model biases in seasonally of continental water storage revealed by satellite gravimetry," Water Resources Research, vol. 42, no. 3, Article ID W03201, 2006. · ·
17. T. H. Syed, J. S. Famiglietti, M. Rodell, J. Chen, and C. R. Wilson, "Analysis of terrestrial water storage changes from GRACE and GLDAS," Water Resources Research, vol. 44, no. 2, Article ID W02433, 2008. · ·
18. G.-Y. Niu and Z.-L. Yang, "Assessing a land surface model's improvements with GRACE estimates,"Geophysical Research Letters, vol. 33, no. 7, Article ID L07401, 2006. · ·
19. T. Ngo-Duc, K. Laval, G. Ramillien, J. Polcher, and A. Cazenave, "Validation of the land water storage simulated by Organising Carbon and Hydrology in Dynamic Ecosystems (ORCHIDEE) with Gravity Recovery and Climate Experiment (GRACE) data," Water Resources Research, vol. 43, no. 4, Article ID W04427, 2007. · ·
20. J. L. Chen, C. R. Wilson, B. D. Tapley, and J. C. Ries, "Low degree gravitational changes from GRACE: validation and interpretation," Geophysical Research Letters, vol. 31, no. 22, Article ID L22607, pp. 1–5, 2004. · ·
21. R. Klees, E. A. Revtova, B. C. Gunter et al., "The design of an optimal filter for monthly GRACE gravity models," Geophysical Journal International, vol. 175, no. 2, pp. 417–432, 2008. · ·
22. K.-W. Seo, C. R. Wilson, J. S. Famiglietti, J. L. Chen, and M. Rodell, "Terrestrial water mass load changes from Gravity Recovery and Climate Experiment (GRACE)," Water Resources Research, vol. 42, no. 5, Article ID W05417, 2006. · ·
23. B. D. Tapley, S. Bettadpur, J. C. Ries, P. F. Thompson, and M. M. Watkins, "GRACE measurements of mass variability in the Earth system," Science, vol. 305, no. 5683, pp. 503–505, 2004. · ·
24. D. P. Lettenmaier and J. S. Famiglietti, "Hydrology: water from on high," Nature, vol. 444, no. 7119, pp. 562–563, 2006. · ·
25. S. Swenson, P. J.-F. Yeh, J. Wahr, and J. Famiglietti, "A comparison of terrestrial water storage variations from GRACE with in situ measurements from Illinois," Geophysical Research Letters, vol. 33, no. 16, Article ID L16401, 2006. · ·
26. L. Xavier, M. Becker, A. Cazenave, L. Longuevergne, W. Llovel, and O. C. R. Filho, "Interannual variability in water storage over 2003–2008 in the Amazon Basin from GRACE space gravimetry, in situ river level and precipitation data," Remote Sensing of Environment, vol. 114, no. 8, pp. 1629–1637, 2010. · ·
27. M. Rodell and J. S. Famiglietti, "The potential for satellite-based monitoring of groundwater storage changes using GRACE: the High Plains aquifer, Central US," Journal of Hydrology, vol. 263, no. 1–4, pp. 245–256, 2002. · ·
28. G. Strassberg, B. R. Scanlon, and M. Rodell, "Comparison of seasonal terrestrial water storage variations from GRACE with groundwater-level measurements from the High Plains Aquifer (USA),"Geophysical Research Letters, vol. 34, no. 14, Article ID L14402, 2007. · ·
29. M. Rodell, I. Velicogna, and J. S. Famiglietti, "Satellite-based estimates of groundwater depletion in India," Nature, vol. 460, no. 7258, pp. 999–1002, 2009. · ·
30. X. Lu and Q. Zhuang, "Evaluating evapotranspiration and water-use efficiency of terrestrial ecosystems in the conterminous United States using MODIS and AmeriFlux data," Remote Sensing of Environment, vol. 114, no. 9, pp. 1924–1939, 2010. · ·
31. M. Rodell, J. S. Famiglietti, J. Chen et al., "Basin scale estimates of evapotranspiration using GRACE and other observations," Geophysical Research Letters, vol. 31, no. 20, Article ID L20504, 2004. · ·
32. G. Ramillien, F. Frappart, A. Güntner, T. Ngo-Duc, A. Cazenave, and K. Laval, "Time variations of the regional evapotranspiration rate from Gravity Recovery and Climate Experiment (GRACE) satellite gravimetry," Water Resources Research, vol. 42, no. 10, Article ID W10403, 2006. · ·
33. S. Swenson and J. Wahr, "Estimating large-scale precipitation minus evapotranspiration from GRACE satellite gravity measurements," Journal of Hydrometeorology, vol. 7, no. 2, pp. 252–270, 2006. · ·

34. M. Leblanc, P. Tregoning, G. Ramillien, S. O. Tweed, and A. Fakes, "Basin-scale, integrated observations of the early 21st century multiyear drought in southeast Australia," Water Resources Research, vol. 45, no. 4, Article ID W04408, 2009. ·

35. J. L. Chen, C. R. Wilson, B. D. Tapley, Z. L. Yang, and G. Y. Niu, "2005 drought event in the Amazon River basin as measured by GRACE and estimated by climate models," Journal of Geophysical Research B, vol. 114, no. 5, Article ID B05404, 2009. · ·

36. O. B. Andersen, S. I. Seneviratne, J. Hinderer, and P. Viterbo, "GRACE-derived terrestrial water storage depletion associated with the 2003 European heat wave," Geophysical Research Letters, vol. 32, no. 18, Article ID L18405, pp. 1–4, 2005. · ·

37. F. Frappart, K. Do Minh, J. L'Hermitte et al., "Water volume change in the lower Mekong from satellite altimetry and imagery data," Geophysical Journal International, vol. 167, no. 2, pp. 570–584, 2006. · ·

38. M. D. Wilson, P. Bates, D. Alsdorf et al., "Modeling large-scale inundation of Amazonian seasonally flooded wetlands," Geophysical Research Letters, vol. 34, no. 15, Article ID L15404, 2007. · ·

39. S.-C. Han, C. K. Shum, C. Jekeli, and D. Alsdorf, "Improved estimation of terrestrial water storage changes from GRACE," Geophysical Research Letters, vol. 32, no. 7, Article ID L07302, pp. 1–5, 2005. · ·

40. B. Wouters, D. Chambers, and E. J. O. Schrama, "GRACE observes small-scale mass loss in Greenland,"Geophysical Research Letters, vol. 35, no. 20, Article ID L20501, 2008. · ·

41. I. Velicogna and J. Wahr, "Measurements of time-variable gravity show mass loss in Antarctica,"Science, vol. 311, no. 5768, pp. 1754–1756, 2006. · ·

42. D. C. Slobbe, P. Ditmar, and R. C. Lindenbergh, "Estimating the rates of mass change, ice volume change and snow volume change in Greenland from ICESat and GRACE data," Geophysical Journal International, vol. 176, no. 1, pp. 95–106, 2009. · ·

43. P. L. Svendsen, O. B. Andersen, and A. A. Nielsen, "Acceleration of the Greenland ice sheet mass loss as observed by GRACE: confidence and sensitivity," Earth and Planetary Science Letters, vol. 364, pp. 24–29, 2013. · ·

44. A. S. Gardner, G. Moholdt, B. Wouters et al., "Sharply increased mass loss from glaciers and ice caps in the Canadian Arctic Archipelago," Nature, vol. 473, no. 7347, pp. 357–360, 2011. · ·

45. J. L. Chen, B. D. Tapley, and C. R. Wilson, "Alaskan mountain glacial melting observed by satellite gravimetry," Earth and Planetary Science Letters, vol. 248, no. 1-2, pp. 353–363, 2006. · ·

46. K. Matsuo and K. Heki, "Time-variable ice loss in Asian high mountains from satellite gravimetry,"Earth and Planetary Science Letters, vol. 290, no. 1-2, pp. 30–36, 2010. · ·

47. D. García-García, C. C. Ummenhofer, and V. Zlotnicki, "Australian water mass variations from GRACE data linked to Indo-Pacific climate variability," Remote Sensing of Environment, vol. 115, no. 9, pp. 2175–2183, 2011. · ·

48. W. W. Immerzeel, L. P. H. van Beek, and M. F. P. Bierkens, "Climate change will affect the asian water towers," Science, vol. 328, no. 5984, pp. 1382–1385, 2010. · ·

49. L. S. Sørensen, S. B. Simonsen, K. Nielsen et al., "Mass balance of the Greenland ice sheet (2003–2008) from ICESat data—the impact of interpolation, sampling and firn density," Cryosphere, vol. 5, no. 1, pp. 173–186, 2011. · ·

50. W. Zhang, Y. Wang, and C. Zhang, "The preliminary analysis of effects of the soil moisture on gravity observations," Cartographica Sinica, vol. 30, no. 2, pp. 108–111, 2001.

51. G. Ramillien, J. S. Famiglietti, and J. Wahr, "Detection of continental hydrology and glaciology signals from GRACE: a review," Surveys in Geophysics, vol. 29, no. 4-5, pp. 361–374, 2008. · ·

52. J. L. Chen, C. R. Wilson, R. J. Eanes, and B. D. Tapley, "Geophysical contributions to satellite nodal residual variation," Journal of Geophysical Research B, vol. 104, no. 10, pp. 23237–23244, 1999. ·

53. J. K. Willis, D. P. Chambers, and R. S. Nerem, "Assessing the globally averaged sea level budget on seasonal to interannual timescales," Journal of Geophysical Research C, vol. 113, no. 6, Article ID C06015, 2008. · ·

54. E. W. Leuliette and L. Miller, "Closing the sea level rise budget with altimetry, Argo, and Grace,"Geophysical Research Letters, vol. 36, no. 4, Article ID L04608, 2009. · ·

55. J. L. Chen, C. R. Wilson, J. S. Famiglietti, and M. Rodell, "Attenuation effect on seasonal basin-scale water storage changes from GRACE time-variable gravity," Journal of Geodesy, vol. 81, no. 4, pp. 237–245, 2007. · ·

56. G. Zhu, J. Li, H. Wen, and J. Wang, "Study on variations of global continental water storage with GRACE gravity field models," Journal of Geodesy and Geodynamics, vol. 28, no. 5, pp. 39–44, 2008.

57. S. Swenson and J. Wahr, "Methods for inferring regional surface-mass anomalies from Gravity Recovery and Climate Experiment (GRACE) measurements of time-variable gravity," Journal of Geophysical Research B, vol. 107, no. 9, pp. 3-1–3-13, 2002. ·

58. S. Swenson and J. Wahr, "Monitoring changes in continental water storage with grace," Space Science Reviews, vol. 108, no. 1-2, pp. 345–354, 2003. · ·

59. K.-W. Seo and C. R. Wilson, "Simulated estimation of hydrological loads from GRACE," Journal of Geodesy, vol. 78, no. 7-8, pp. 442–456, 2005. · ·

60. S. Swenson and J. Wahr, "Post-processing removal of correlated errors in GRACE data," Geophysical Research Letters, vol. 33, no. 8, Article ID L08402, 2006. · ·

61. G. Ramillien, F. Frappart, A. Cazenave, and A. Güntner, "Time variations of land water storage from an inversion of 2 years of GRACE geoids," Earth and Planetary Science Letters, vol. 235, no. 1-2, pp. 283–301, 2005. · ·

62. F. Frédéric, G. Ramillien, M. Leblanc, et al., "An independent component analysis filtering approach for estimating continental hydrology in the GRACE gravity data," Remote Sensing of Environment, vol. 115, no. 1, pp. 187–204, 2011. ·

63. S. Werth, A. Güntner, R. Schmidt, and J. Kusche, "Evaluation of GRACE filter tools from a hydrological perspective," Geophysical Journal International, vol. 179, no. 3, pp. 1499–1515, 2009. · ·

CHAPTER 5

QUANTIFYING THE RELATIVE CONTRIBUTIONS OF FOREST CHANGE AND CLIMATIC VARIABILITY TO HYDROLOGY IN LARGE WATERSHEDS

Xiaohua Wei, Wenfei Liu and Peicong Zhou

Keywords: Forest change, climate variability, large watershed hydrology, relative contribution, research methods

INTRODUCTION

Studies on the relationship between forests and water in forested watersheds have been conducted over the last century. Researchers have provided reviews at different times [1,2,3,4,5,6]. The general conclusions are that forest removal increases annual mean flow, peak flow and base flow when soils are not significantly disturbed. This relationship can be reversed if forest change is associated with soil compaction and decreased infiltration [7]. However, the majority of these conclusions is from the studies conducted in small watersheds, usually less than 100 km^2, and cannot simply be extrapolated to large watersheds (more than 1000 km^2). This suggests that more studies on the relationship between forest change and water in large watersheds must be conducted separately. In large watersheds, research targeting the relationship between forest changes and water is limited. There are many reasons contributing to fewer large watershed studies. The first and perhaps the most important reason is that paired watershed approach is not suitable for large watersheds because of the difficulty of locating reference watersheds with similar climate, vegetation and topography. The second reason is that large watersheds have different land uses and their complex interactions. Additionally, large watersheds very rarely coincide with the long-term records of hydrology, land use, and climate that exist for smaller research watersheds. However, with the expanding scope of land use planning and environmental protection, resource managers and planners require appropriate information at large watersheds or landscapes to support management decisions. Therefore, large watershed studies are urgently needed.

It has been widely accepted that climate variability and land cover or land use changes are two critical drivers for influencing watershed hydrological changes. Because of this, it is important to separate their relative contributions to hydrological change so that their individual effects can be quantified. In small watersheds, a paired watersheds approach is usually used to remove the effect of climate variability so that the effects of forest or land cover changes can be quantitatively assessed. However, this experimental paired watersheds approach is generally not feasible for large watersheds. Thus, alternative methods must be used for studying this subject in large watersheds. In the past 20–30 years, researchers have developed and applied diverse techniques for this research subject. These techniques can be broadly classified into statistical and modeling (process-based watershed modeling) categories. The objectives of this review are: (1) to describe various existing research methods and compare their pros and cons; (2) to briefly summarize the progress on quantifying the relative contributions of forest or land cover change and climatic variability to hydrology in large watersheds using

Citation: Wei, X.; Liu, W.; Zhou, P. Quantifying the Relative Contributions of Forest Change and Climatic Variability to Hydrology in Large Watersheds. Water 2013, 5, 728-746. doi:10.3390/w5020728.

available case studies; and (3) to identify research challenges and propose future research priorities.

RESEARCH METHODS

Hydrological Modeling

Hydrological models are frequently used in the large watershed hydrological research. Hydrological models can be divided into lumped, semi-distributed and fully-distributed models. Lumped hydrological models study the watershed as a whole, and do not consider the detailed, spatial representations of watershed elements and processes. Unlike lumped or semi-distributed models, distributed models can well represent a watershed by assigning input data and physical characteristics to grids or elements. Physically based distributed models are able to provide distributed approximations or predictions of hydrological variables across watersheds, and thus have a better representation of reality. However, a fully-distributed model requires a large dataset with data on various processes, components and their interactions. For a large-scale watershed research, a semi-distributed model is commonly used because of a general absence of detailed data at large scales.

The one-factor-at-a-time approach (OFAT) commonly used in sensitivity analysis [8,9,10] is also used in association with hydrological models to distinguish the impact of climate factors and land cover change on watershed hydrology [11,12]. In a hypothetical example, the impacts of climatic variability and land use change on runoff are assessed with the available data in the period of 1960 to 2000. First, we keep the land cover in 1960 unchanged over the simulation period while climate change is allowed from 1960 to 2000. Then we simulate the runoff change (ΔQ_C), which can be treated as the impact of climatic variability on hydrology. Second, keep the climate of 1960 unchanged while land cover is changed, then calculate the runoff change (ΔQ_L) as the impact of land cover change. Finally, we assume the changes of both climate and land cover, then calculate the runoff change (ΔQ_{L+C}). In this way, the relative contributions of forest and land cover changes and climatic variability to hydrology can be computed.

Various distributed hydrological models have been successfully used to quantitatively study the effects of climate change and forest change/land cover change on hydrology, for examples, SWAT [13,14], DHSVM [15,16,17], MIKE-SHE [18], *etc.* In spite of increased applications, hydrological models are still based on our current theories that are deeply rooted in the physics of small-scale processes. This gives rise to difficulties in representing nonlinear hydrological processes and their interactions at all scales across heterogeneous landscapes. In addition, calibrating and testing a model may not always assure its validity, since there are some inherent drawbacks in the approaches of parameter calibration and validation. We often over-parameterize our models to meet high accuracy levels, ignoring the equifinality problem that different parameter sets for a model might yield the same result during calibration, but distinctly different predictions when conditions are altered [19,20].

Trend Analysis

Trend analysis is a commonly used method to qualitatively assess hydrologic responses [21,22,23,24]. The trend analysis uses historic data to fit a curve, which reflects a trend of the interested hydrological variable and can then be used to predict any future values if the trend is statistically significant.

$$b = Median\left(\frac{xj - xl}{j-1}\right)\forall\langle j \tag{1}$$

$$B = 100 * \frac{Tb}{E(x)}(\%) \tag{2}$$

$$r = 1 - \frac{\sum_{i=1}^{n} di^2}{n^3 - n} \tag{3}$$

where the Equation (1) is used to calculate the slope of a time series data; the Equation (2) is used to estimate change percentages of the sample mean $E(x)$; and the Equation (3) is for correlation coefficient at a significant level. When the trend analysis is applied to detect the effects of forest change on hydrology, information on climate change and land use change is needed. If there is no a significant trend in climate variable (e.g., precipitation), the significant trend in hydrological variable is then judged to be caused by forest change or land cover change [24,25]. Clearly, this method is simple and qualitative, but may not be reliable, as the interactions between forest change and climatic variability are not considered. Wei and Zhang detected the offsetting effects of forest change and climatic variability on annual stream flow, which caused an insignificant trend on annual mean flow even though the forest effects on hydrology were statistically significant [26].

Double Mass Curves

Double Mass Curve (DMC) is a simple and intuitive method, widely used in long-term trend analysis of hydro meteorological elements. DMC draws a curve between two cumulative hydro-meteorological variables to test the consistency of two variables or analyze the trend change and its strength [27,28,29,30]. DMC method can be also used to separate the relative influences of forest change and climatic effects to hydrology [31]. For example, a modified DMC (MDMC) between cumulative annual stream flow and cumulative effective precipitation (the difference between total precipitation and evapotranspiration) is constructed for a large forested watershed (Figure 1). In this way, climatic effect on annual stream flow can be eliminated. In the period of no forest disturbance, the curve should produce a straight line, a baseline that describes the linear relation between annual stream flow and annual effective precipitation, and a break in this curve (e.g., year 1986 in Figure 2) would suggest the change of annual stream flow caused by forest disturbance. In other words, a step change or regime shift occurs in the slope of MDMC and the slope before the break is different from that afterwards. However, this visually detected breakpoint needs confirmation of its statistical significance by a non-parametric test or application of ARIMA model [32,33]. The difference between actual observations and the predict line after the change point can be calculated, and is regarded as the cumulative impact of forest changes.

Quasi-Paired Watershed Method

Paired watershed method compares two adjacent basins with similarity in size, morphology, geology, climate and vegetation including one as control and the other as treatment. The method is widely used in studying forest-water relationship in small (<100 km^2) watersheds. For large watersheds, the concept of "quasi-paired watersheds" was proposed [34]. The quasi-paired watersheds method chooses two or more basin, with relative similarity in size, morphology, geology, climate and vegetation, but different forest disturbance levels. In this method, the watershed experienced the least forest disturbance is treated as "control". Double mass curves of runoff coefficients (annual runoff/precipitation) obtained by plotting accumulated coefficients of the "treated" (Ct) against those of the "control" watershed (Cc), are used to detect the effect of forest disturbance on annual mean flow. If the "treated" watershed undergoes insignificant disturbance, a straight line would be expected. Otherwise, a break in the observed line indicates the emerging effect of forest disturbances on a hydrological variable (e.g., annual mean flow), and the difference between the expected and observed lines can be attributed to the difference in forest disturbances between "control" and "treated" watersheds. Statistical significance of the breakpoint found in the DMC slope can be tested using ARIMA intervention analysis [35,36].

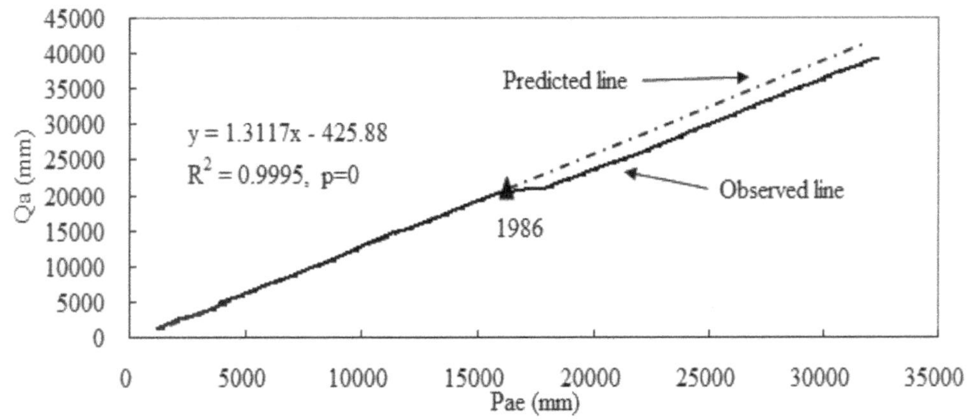

Figure 1: A hypothetical example of application of the MDMC (modified double mass curve) between cumulative annual runoff (Q_a) and cumulative annual effective precipitation (P_{ae}) for assessing the effects of forest change on annual mean flow.

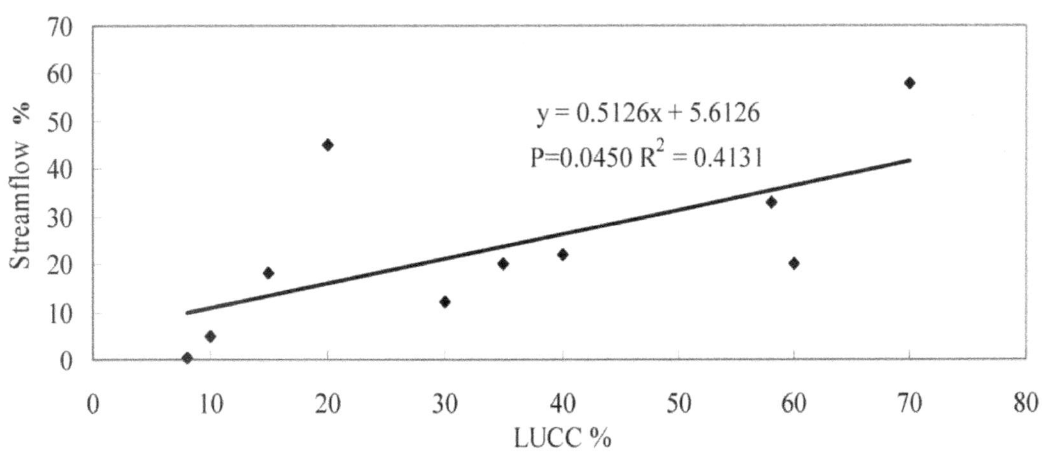

Figure 2: Relationship between forest or land cover changes (LUCC) and their relative contributions to stream flow.

The method is an alternative approach to the paired watersheds method where there are two similar watersheds with contrasted forest disturbance levels or land use changes. If long-term data are available, such a method can provide reliable results as what the paired experimental watersheds do. The challenge of this method is that it requires the "treated" watershed must have two significant long periods with and without severe forest disturbance. In addition, both watersheds should have relative similarity in climate, soil and vegetation. This could impose some difficulties in large watersheds where great spatial heterogeneities and complexities always exist.

Sensitivity-Based Approach

Sensitivity-based approach is similar to elasticity method [37], and is used to calculate the effect of climate variability on stream flow. Perturbations in both precipitation and PET can lead to changes of water balance. It can be assumed that a change in mean annual runoff can be determined using the following expression [31,38]:

$$\Delta Q_{clim} = \beta \Delta P + \gamma \Delta PET \qquad (4)$$

where ΔQ_{clim}, ΔP, and ΔPET are changes in stream flow, precipitation, and potential evapotranspiration, respectively, and β and γ are the sensitivity coefficients of stream flow to precipitation and potential evapotranspiration, expressed as:

$$\beta = \frac{\left(1 + 2x + 3wx^2\right)^2}{\left(1 + w + wx^2\right)^2} \quad (5)$$

$$\gamma = \frac{\left(1 + 2wx\right)}{\left(1 + x + wx^2\right)^2} \quad (6)$$

where x is the mean annual index of dryness (equal to PET/P); and the values of vegetation factor w for forest, grassland and shrub land are 2, 0.5 and 1, respectively [39].

This method is suitable for the analysis of a single basin, and for quantitative calculation of the impact of climate variables on runoff. Once the effects of climatic variability on flow are estimated, the effects of forest disturbance or land use changes can be deducted from total stream flow variations. The method has been successfully used for several case studies [37,38,39]. There may be two challenges in this method. First, it is not easy to determine w values for specific forest vegetation types. Where there are always different types of forests in a large watershed, how to select a specific w value remains challenging. Second, the effect of forest disturbance or land use changes on hydrology is indirectly estimated from total hydrological variations and the effects of climatic variability. Thus, its reliability is dependent on the accuracy of the other two terms.

Simple Water Balance

The water balance methods provide a framework to determine changes in the water balance components [40]. Simple water balance model can be used to determine the influence of climate and vegetation on stream flow at a watershed scale.

$$P = ET + Q + \Delta S \quad (7)$$

where P is precipitation, ET is actual evapotranspiration, Q is stream flow and S is change in catchment water storage. When averaged over a long period, deep percolation (recharge) and change in soil moisture storage is often only 5% to 10% of the annual water balance, and therefore the change in catchment water storage (ΔS) can be neglected [41,42].

Precipitation and actual evapotranspiration constitute the most important variables to influence runoff change at the watershed scale. Precipitation, which varies both in temporal trend and spatial distribution, is regarded as independent of vegetation types [39], which mainly reflects changes of climate. However, actual evapotranspiration is a complex process. There are various ways to estimate watershed-scale evapotranspiration. For example, following the Budyko's hypothesis, the simple two-parameter model for estimating the actual evapotranspiration was developed [43,44]. The model is consistent with the previous theoretical work and shows good agreement with more than 250 catchment-scale measurements from around the world [39,45,46].

$$\frac{ET}{P} = \frac{1 + w\left(\frac{ET}{P}\right)}{1 + w\left(\frac{ET}{P}\right) + \left(\frac{ET_0}{P}\right)^{-1}} \quad (8)$$

where ET is the actual evapotranspiration, ET_0 is reference evapotranspiration as substitute of potential evapotranspiration calculated by the P-M method. The w is the plant-available water coefficient estimated in the same way as in the sensitivity-based approach [39].

The following steps were used to describe how a simple water balance method is implemented to estimate the effects of forest change on hydrology [40]. They are described as follows. First, according to Equation (8), calculate the actual evapotranspiration using the original data including precipitation, air temperature, relative humidity, wind speeds and sunshine hours after calculating ET_0 by the P-M method. Second, estimate annual runoff using the Equation (10). In this step, the change of annual runoff is influenced by both climate variability and vegetation changes, and thus, the calculated annual runoff can be defined as Q_{v+C}; Third, removing the decreasing or increasing trend of precipitation, air temperature, relative humidity, wind speeds and sunshine hours in data series to make them as stationary time series [47], recalculate theET_0 using the detrended climate variables and estimate annual runoff according to the Equations (10) and (11). In this step, the change of annual runoff mainly reflects the influences of vegetation changes, and thus, the recalculated annual runoff can be defined as Q_v. Finally, calculate the difference between Q_{v+C} and Q_v so the change of annual runoff caused by the climate variability can then be estimated [40].

A simple water balance method provides a new way to distinguish the impact of climatic variables and vegetation factors on hydrological change. However, the choice of w values and the difficulty associated with removal of the decreasing or increasing trends in climate data may introduce some errors in this method.

Time Trend Method

In the time trend method, a relationship is established between discharge and climatic variables before the basin's vegetation perturbation occurs and is then used to predict the discharge response of post-perturbation assuming undisturbed basin conditions. The typical time trend approach is to divide the whole study period into the calibration section and prediction section. The model accuracy depends on the length of the calibration or pre-perturbation periods [48].

During the calibration section:

$$Q_1 = aP_1 + b \quad (9)$$

During the forecast section:

$$Q_2 = aP_2 + b \quad (10)$$

$$\Delta Q_{veg} = \overline{Q_2} - \overline{Q'}_2 \quad (11)$$

where P is precipitation, Q is runoff, Q' is the predicted stream flow for the treated catchment (from using Equation

(9) developed during the calibration period), ΔQ_{veg} represents the change in mean annual stream flow because of vegetation change, subscripts 1 and 2 represent the calibration period and the prediction period, and b is the fitted regression coefficient.

This method uses a simple regression to express the relationship between precipitation and runoff both before and after forest cover changes. The method only requires data of precipitation, runoff and other meteorological variables, and the requirement on the detailed forest cover change can be ignored to some extent. The method may accept discontinuous data. Depending on different hydrometeorological characteristics in a study basin, time trend method performance at yearly hydrological variables is better than on the variables at monthly or daily intervals. Guardiola-Claramonte *et al.* proposed to consider the impact of temperature on the relationship between precipitation and runoff [49].

Tomer-Schilling Framework

Tomer and Schilling developed an elegant, coupled water–energy balance framework that requires long-term time series of precipitation (*P*), stream flow (*Q*), and *PET* to assess if unused available energy and water were related to climate and/or to land management in agricultural catchments [25,50]. The conceptual framework that qualitatively discriminates whether the dominant drivers of observed changes are related to land cover change and/or climate. The framework relating changes in land cover change and/or climate to the observed changes in the excess amounts of water (P_{ex}) and excess amounts of energy (E_{ex}) as fractions is illustrated in Peña-Arancibia *et al.* [25].

The Tomer–Schilling framework assumes that land cover change will affect *ET* but not *P* or *PET*, acknowledging that effects of land cover change on *P* and *PET* can be considered indirect at this scale and of second order when compared to changes in *ET* in the woodland environment. Thus, land cover change will cause ecohydrological shifts towards increasedP_{ex} and E_{ex}, or towards decreased P_{ex} and E_{ex}. Changes in climate are required to cause increased P_{ex} and decreased E_{ex}, due to the temporal increase in the *P*/*PET* ratio and vice versa [25].

$$\overline{P_{ex}} = \frac{\left(\overline{P} - \overline{ET}\right)}{\overline{P}} \tag{12}$$

$$E_{ex} = \frac{\left(\overline{PET} - \overline{ET}\right)}{\overline{PET}} \tag{13}$$

Tomer-Schilling framework is an effective and qualitative analysis tool of hydrological processes. The method can not only analyze long-term impacts of climate and forest cover on hydrology, but also explain the main factors of hydrological responses in different time periods. However, as all the qualitative methods, the impact of each variable of climate or vegetation cannot be quantitatively evaluated, and this brings difficulties to study the common law of hydrological responses under different topographies and different vegetation types.

RESEARCH PROGRESS

Table 1 summarizes the progress on quantifying the relative contributions of climate variability and forest or land cover changes to hydrology in large watersheds. As shown in Table 1, more researchers have used statistical methods, while the others have applied the modeling approach. There are two reasons contributing to this difference. First, statistical methods normally analyze historic data for assessing the effects of forest change and climate variability, while models predict the possible effects of future scenarios on hydrology. With increasing availability of long-term data on hydrology, climate and land use change, it is not surprising that statistical methods gain more popularity due to their efficiency and provision of robust inferences. Hydrological models can be used to simulate the relative contributions of forest change and climate variability to hydrology, but they are limited due to the lack of various empirical relationships at large watersheds. Second, data on forest change or land over change are difficult to collect at large spatial scales across extended time periods. Without these data, models are difficult to calibrate. In comparison, some statistical methods such as trend analysis method, sensitivity-based approach and the elasticity method [51,52,53,54,55] can be used to quantify the effects of climatic variability first and then estimate the influence of forest or land cover change from the total variations. Those estimations can be done without long-term data on forest or land cover changes. However, in the watersheds where long-term and high quality data are not available for application of statistical approaches, models can be alternative approaches given that there are sufficient data for model calibration and validations.

Among the applied models (Table 1), SWAT, a semi-distributed model was widely used to evaluate the effects of climate and land use changes on stream flow [56,57]. It simulated stream flow responses to change in climate, land use, or both to distinguish the impact of climate variability from that of land use change on hydrology [35,58,59]. Other distributed hydrological models such as DTVGM model (Distributed Time-Variant Gain Model); VIC model (Variable Infiltration Capacity) and DHSVM model (Hydrology-Soil-Vegetation Model) were also applied to this research topic (Table 1). For example, lumped hydrological model (CHARM) and distributed model (TOP MODEL) were used to simulate the effect of climate variability and land cover change on hydrology in the upper reaches of the Yangtze River [58]. Three applied models show similar results, suggesting that their simulations were reliable.

There is a clear tendency that a combined approach of two or more complementary techniques (particularly the combination of statistical and modeling methods) is superior to application of a single technique alone. For example, both climate elasticity and hydrological modeling methods were applied to distinguish the effects of climate variability and human activities in Baiyangdian Basin [52]. Using the climate elasticity method, the results indicated that climate variations accounted for 40% of decrease in stream flow, while human activities accounted for 60%. Using the hydrological modeling method, the results showed that climate variations accounted for 38% of decrease in stream flow, while human activities accounted for 62%. Because two independent methods achieved relatively consistent results, the result was believed more reliable. The general perception is that climate is dominant in influencing stream flow change [26,60,61]. However, Table 1 shows that the relative hydrological effects of climate variability and forest change or land use change are largely dependent on their own change magnitudes and watershed characteristics. In some severely disturbed watersheds, impacts of forest change or land cover change are as important as those from climate change. For example, large-scale land use changes and practices were found to be the most important factor responsible for the stream flow change in the northwest China's Loess Plateau (Table 1) [21,40]. Based on comparable studies listed in the Table 1, we detected that a significant linear relationship between forest or land cover changes and their contributions to stream flow changes (Figure 2). This significant linear relationship clearly demonstrated that the more severe change of forest or land cover, the greater in their relative contributions to hydrological changes. Another example showed that the forest disturbance (25%–30%) accounted for 45% of decrease in stream flow and is similar to that from climate change (55%) in Willow watershed located in the central interior of British Columbia, Canada [26].

Table 1: Summary of the studies on relative contributions of forest or land cover changes and climate variability to hydrology.

Method	Watershed (km^2)	Forest change or land use change (%) at a watershed	Relative contribution (%)		References
			Climate variability	Forest change/land use change	
Hydrological Modeling	The source regions of Yellow River (1.22×10^5 km^2)	0.13%~34.03% area change in woodland and shrub/grassland	65%–95%	6%–16%	[56]
	Heihe River watershed (1506 km^2)	4.5% of the catchment area was changed mainly from shrubland and sparse	95.8%	9.6%	[57]
	Suomo Watershed (2536 km^2)	Forest area decreased by 17%, open woodland increased by 8%, grassland area increased by 10%	60%–80%	20%	[62]
Hydrological Modeling	The upper Yangtze River watershed (13,721~1,005,501 km^2)	Human activities	55%–73.3%	28.5%–46.7%	[63]
	Chaobai River watershed (57,001 km^2)	Human activities	34%	64%	[64]
	Araguaia River watershed, Brazil (385,000 km^2)	55% Deforestation	33%	67%	[65]
Trend Analysis	Shiyang river watershed (389~1614 km^2)	Area change in Forest, crop and grassland	64.5%–87.9%	12.1%–35.5%	[66]
	Hun–Tai River watershed (1112~11,203 km^2)	Human activities	43%	57%	[67]
	Pyrenees Watershed, Spain (3.25×10^4 km^2)	The increase of agricultural land	70%	30%	[68]
	Wuding River watershed (30,261 km^2)	43% (The soil conservation measures)	13%	87%	[45]
	Qingshui River watershed (436 km^2)	Cropland decreased by 72.1%, woodland increased by 963.60% and residential area increased by 576.07%	46.79%	53.21%	[69]
	Taoer River watershed (41,600 km^2)	Paddy field increased by 2%, upland increased by 15.10%, forest decreased by 10.60%	45%	55%	[61]
	ColumbiaRiver watershed, Canada (385,000 km^2)	Human activities	48%–55%	45%–52%	[70]

Method	Watershed (km²)	Forest change or land use change (%) at a watershed	Relative contribution (%) Climate variability	Forest change/land use change	References
Double Mass Curves	The Willow River watershed, Canada (2,860 km²)	19.7% Deforestation	55%	45%	[26]
	The Upper Zagunao River watershed (2528km²)	15.5% Deforestation	42.5%	57.5%	[71]
Quasi-paired Watershed Method	Yulin, 8.28 km² Xinlin, 28 km²	Forest cover increased from 20% to 80%	/	8.61%	[30]
Sensitivity-Based Approach	Upper catchment of the Yellow River Watershed (222,551 km²)	Human activities	50%	40%	[55]
	Baiyangdian Lake (3,465 km²)	Construction of water conservation facilities	38%–40%	60%–62%	[52]
	The headwaters of the Yellow River Watershed (1.32 × 10⁵ km²)	Water and soil conservation engineering	30%	70%	[72]
Sensitivity-Based Approach	The Loess Plateau (1,279–9,289 km²)	Water and soil conservation engineering	21%–57%	43%–79%	[21]
	Crawford River watershed, Darlot Creek watershed, Tinana Creek watershed (698–1,174 km²)	Afforestation 13.4%–23.5%	21%–49%	61%–64%	[73]
Simple Water Balance	Yiluo River watershed (1.89 × 10⁵ km²)	Water and soil conservation engineering	21.75%	78.26%	[40]
Time Trend Method	Seven paired catchments from Australia, New Zealand, and South Africa (0.18–3.44 km²)	Clearing vegetation 32%–100% Afforestation 67%–83% Forest conversion 100%	10%–72%	28%–98%	[55]
	The upper reach of the Weihe River (1.35 × 10⁵ km²)	Human activities	40%	60%	[64]
	Plata watershed, Paragury (3.2 × 10⁴ km²)	Forest land converted to cropland	41%–53%	51%–59%	[74]
Tomer-Schilling Framework	Four candidate Midwest watersheds	Changes in agricultural land cover	Increase/decreased Pex and Eex	Increased Pex and decreased Eex	[50]

FUTURE RESEARCH CHALLENGES AND RESEARCH PRIORITIES

As mentioned previously, all methods summarized in this paper can be classified into two categories: statistics and simulation. The statistical methods normally provide robust inference about the relationship between hydrological responses and forest change or land cover changes. However, those methods treat a study basin as a black box, and rarely involve the mechanisms or processes. On the other hand, the simulation methods, especially fully-distributed models consider the hydrological processes and their interactions, but due to the lack of data and various empirical relationships in large watersheds, the model calibration and validation are often time consuming and inaccurate. Therefore, it is logical to believe that the combination of statistical methods and modeling can complement their strengths and overcome their shortcomings, and thus could be an improved research strategy. Forest disturbance and climate change are considered to be the two main drivers interactively affecting the hydrological changes in large watersheds (Figure 3). Climate change can affect hydrology directly due to its influence on precipitation and temperature. Climate change can also affect hydrology in an indirect way. For example, more forest fires and pests can be induced by drought, and these forest disturbances in turn affect the hydrological processes. In an opposite way, vegetation can change evapotranspiration and thus stream flow. They can also affect hydrology indirectly. For example, vegetation types and distributions may affect the characteristics of air humidity, temperature, precipitation, and consequently hydrological cycle. As far as we know, the literature is extremely limited on that topic, with some pioneering examples exploring the interactions among forest change, climate and hydrological responses [75,76].

In addition, current studies mainly focused on the changes in annual runoff, with less emphasis on peak and low flows. Assessing the relative contributions of forest or land cover changes and climatic variability to the change in peak and low flows is likely more challenging and may require separate methods such as geochemistry and landscape ecology in large watersheds. The isotope method in geochemistry could help analyze the sources (surface water, groundwater, glacier melt water, *etc.*) and flow pathways in the watersheds, while landscape ecology or geoscience allows us to examine the role of watershed configuration and topographic characteristics in the effects of the forest or land cover changes on hydrology [7,51,77,78,79,80,81]. However, such research is needed to support resource management decisions for protection of watershed functions and public safety in large watersheds or regions.

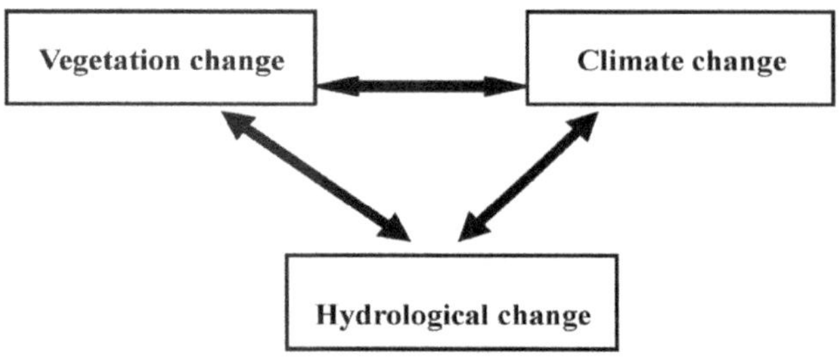

Figure 3: A conceptual framework showing interactions among climate change, forest change and hydrological responses.

Here, we propose the following future research priorities in this subject. First, more case studies are needed. Assessing the relative contributions of forest or land cover changes and climatic variability to hydrology is rather limited in large watersheds. Zhang and Wei compared two adjacent large watersheds located in the interior of British Columbia, and found the contrasted conclusions under the similar forest disturbance levels [35]. They further concluded that the effects of forest change on hydrology in large watersheds are likely watershed specific. This clearly demonstrates that more case studies are needed before generally conclusions can be derived. Second, a research priority should be given on improvement of research methods. Although quite a few research methods are currently available for studying the impacts of forest change and climate change on hydrology, there is no a commonly-accepted method. The lack of commonly-accepted methods may limit our ability to compare the results of different studies. Third, in large watershed studies, analytical results are largely based on data quality and spatial coverage. Large spatial variations in precipitation, temperature, solar radiation, humidity, wind speed, surface albedo, canopy characteristics, *etc.* in these large watersheds constraint our ability to derive robust conclusions. Future studies should be designed to specifically address uncertainties. Finally, more research should be on mechanisms and processes. It is difficult to study the mechanisms and processes in large watersheds, mainly due to lack of the suitable methodology and data for assessing the complicated interactions and cumulative behaviors across various spatial scales. However, such research is critical for explaining and verifying the findings obtained through statistical and modeling approaches.

ACKNOWLEDGMENTS

This work was financially supported by the National Science Foundation of China (No. 31170665) and theYoung Scientific Funding of Nanchang Institute of Technology (2012KJ004).

REFERENCES

1. Andreassian, V. Waters and forests: From historical controversy to scientific debate. *J. Hydrol.* 2004, *291*, 1–27.
2. Bosch, J.M.; Hewlett, J.L. A review of catchment experiments to determine the effects of vegetation changes on water yield and evapotranspiration. *J. Hydrol.* 1982, *55*, 3–23.
3. Hibbert, A.R. Forest Treatment Effects on Water Yield. In Proceedings of International Symposium on Forest Hydrology, Fort Collins, CO, USA, 6–8 September 1967; Sopper, W.E., Lull, H.W., Eds.; p. 813.
4. Jackson, R.B.; Jobbágy, E.G.; Avissar, R.; Roy, S.B.; Barrett, D.J.; Cook, C.W.; Farley, K.A.; le Maitre, D.C.; McCarl, B.A.; Murray, B.C. Trading water for carbon with biological carbon sequestration. *Science* 2005, *310*, 944–1947.
5. MacDonald, L.H.; Stednick, J.D. *Forests and Water: A State-of-the-Art Review for Colorado*; CWRRI Completion Report No. 196; Colorado Water Resources Research Institute, Colorado State University: Fort Collins, CO, USA, 2003.
6. Price, K. Effects of watershed topography, soils, land use, and climate on baseflow hydrology in humid regions: A review. *Progr. Phys. Geogr.* 2011, *35*, 465–492.
7. Price, K.; Jackson, C.R.; Parker, A.J.; Reitan, T.; Dowd, J.; Cyterski, M. Effects of watershed land use and geomorphology on stream low flows during severe drought conditions in the southern Blue Ridge Mountains, Georgia and North Carolina, USA. *Water Resour. Res.* 2011, *47*, W02516.
8. Gao, X.; Sorooshian, S.; Gupta, H.V. Sensitivity analysis of the biosphere-atmosphere transfer scheme. *J. Geophys. Res.*1996, *101*, 7279–7289.
9. Pitman, A.J. Assessing the sensitivity of a land-surface scheme to the parameter values using a single column method.*J. Clim.* 1994, *7*, 1856–1869.
10. Wilson, M.F.; Henderson-Sellers, A.; Dickinson, R.E.; Kennedy, P.J. Sensitivity of the Biosphere-Atmosphere Transfere Scheme (BATS) to the inclusion of variable soil characteristics. *J. Appl. Meteorol. Climatol.* 1987, *26*, 341–362.
11. Wilson, M.F.; Henderson-Sellers, A.; Dickinson, R.E.; Kennedy, P.J. Investigation of the sensitivity of the land-surface parameterization of the NCAR Community Climate Model in regions of tundra vegetation. *Climatology* 1987, *7*, 319–343.
12. Karvonen, T.; Koivusalo, H.; Jauhiainen, M.; Palko, J.; Weppling, K. A hydrological model for predicting runoff from different land use areas. *J. Hydrol.* 1999, *217*, 253–256.
13. Zhang, A.J.; Zhang, C.; Fu, G.B.; Wang, B.D.; Bao, Z.X.; Zheng, H.X. Assessments of impacts of climate change and human activities on runoff with swat for the Huifa River Basin, northeast China. *Water Resour Manage* 2012, *26*, 2199–2217.
14. Chen, J.F.; Li, X.B.; Zhang, M. Simulating the impacts of climate variation and land-cover changes on basin

hydrology: A case study of the Suomo Basin. *Sci. China Ser. D* 2005, *48*, 1501–1509.

15. Cuo, L.; Giambelluca, T.W.; Ziegler, A.D. Lumped parameter sensitivity analysis of a distributed hydrological model within tropical and temperate catchments. *Hydrol. Process.* 2011, *25*, 2405–2421.
16. Sun, W.Y.; Bosilovich, M.G. Planetary boundary layer and surface layer sensitivity to land surface parameters. *Boundary Lay. Meteorol.* 1996, *77*, 353–378.
17. Stonesifer, C.S. Modeling the cumulative effects of forest fire on watershed hydrology: A post-fire application of the Distributed Hydrology-Soil-Vegetation Model (DHSVM). *J. Hydrol.* 2007, *10*, 282–290.
18. Christiaens, K.; Feyen, J. Analysis of uncertainties associated with different methods to determine soil hydraulic properties and their propagation in the distributed hydrological MIKE SHE model. *J. Hydrol.* 2001, *246*, 63–81.
19. Beven, K. The future of distributed models: Model calibration and uncertainty prediction. *Hydrol. Process.* 1992, *6*, 279–298.
20. Kirchner, J.W. Getting the right answers for the right reasons: Linking measurements, analyses, and models to advance the science of hydrology. *Water Resour. Res.* 2006, *42*, W03S04.
21. Zhang, X.P.; Zhang, L.; Zhao, J. Responses of stream flow to changes in climate and land use/cover in the Loess Plateau, China. *Water Resour. Res.* 2008, *44*, W00A07.
22. Sheng, Y.; Michio, H. Statistical interpretation of the impact of forest growth on stream flow of the Sameura Basin, Japan. *Environ. Monit. Assess.* 2005, *104*, 369–384.
23. Wilcox, B.P.; Huang, Y. Woody plant encroachment paradox: Rivers rebound as degraded grasslands convert to woodlands. *Geophys. Res. Lett.* 2010, *37*, L07402.
24. Zhou, G.Y.; Wei, X.H.; Luo, Y.; Zhang, M.F.; Li, Y.; Qiao, Y.; Liu, H.; Wang, C. Forest recovery and river discharge at the regional scale of Guangdong Province, China. *Water Resour. Res.* 2010, *46*, W09503.
25. Peña-Arancibia, J.L.; van Dijk, A.I.J.M.; Guerschman, J.P.; Mulligan, M.; Bruijnzeel, L.A.; McVicar, T.R. Detecting changes in stream flow after partial woodland clearing in two large catchments in the seasonal tropics. *J. Hydrol.* 2012,*416–417*, 60–71.
26. Wei, X.H.; Zhang, M.F. Quantifying stream flow change caused by forest disturbance at a large spatial scale: A single watershed study. *Water Resour. Res.* 2010, *46*, W12525.
27. Buttle, J.M.; Metcalfe, R.A. Boreal forest disturbance and stream flow response, northeastern Ontario. *Can. J. Fish. Aquat. Sci.* 2000, *57*, 5–18.
28. Siriwardena, L.; Finlayson, B.L.; McMahon, T.A. The impact of land use change on catchment hydrology in large catchments: The Comet River, Central Queensland, Australia. *J. Hydrol.* 2006, *326*, 199–214.
29. Wigbout, M. Limitations in the use of double mass curves. *J. Hydrol.* 1973, *12*, 132–138.
30. Yao, Y.F.; Cai, T.J.; Wei, X.H.; Zhang, M.F.; Ju, C.Y. Effect of forest recovery on summer streamflow in small forested watersheds, Northeastern China. *Hydrol. Process.* 2011, *26*, 1208–1214.
31. Koster, R.D.; Suarez, M.J. A simple framework for examining the interannual variability of land surface moisture fluxes. *J. Clim.* 1999, *12*, 1911–1917.
32. Box, G.; Cox, D. An analysis of transformations, J.R. *Stat. Soc. B* 1964, *26*, 211–252.
33. Box, G.; Pierce, D. Distribution of residual autocorrelations in autoregressive integrated moving average time series models. *J. Am. Stat. Assoc.* 1970, *65*, 1509–1526.
34. Buttle, J.M. Identifying Hydrological Responses to Basin Restoration: An Example from Southern Ontario. In*Watershed Restoration Management: Physical, Chemical, and Biological Considerations*; McDonnell, J.J., Stribling, J.B., Neville, L.R., Leopold, D.J., Eds.; American Water Resources Association: Herndon, VA, USA, 1996; pp. 5–13.
35. Zhang, M.F.; Wei, X.H. The effects of forest disturbance on hydrology in two contrasted watersheds. *Hydrol. Processes.*2013. submitted for publication.
36. Zhang, M.F.; Wei, X. Alteration of flow regimes caused by large-scale forest disturbance: A case study from a large watershed in the interior of British Columbia, Canada. *Ecohydrology* 2013. in press.
37. Dooge, J.C.I.; Bruen, M.; Parmentier, B. A simple model for estimating the sensitivity of runoff to long–term changes in precipitation without a change in vegetation. *Adv. Water Resour.* 1999, *23*, 153–163.
38. Jones, R.N.; Chiew, F.H.S.; Boughton, W.C.; Zang, L. Estimating the sensitivity of mean annual runoff to climate change using selected hydrological models. *Adv. Water Resour.* 2006, *29*, 1419–1429.
39. Zhang, L.; Dawes, W.R.; Walker, G.R. The response of mean annual evapotranspiration to vegetation changes at catchment scale. *Water Resour. Res.* 2001, *37*, 701–708.
40. Liu, Q.; Yang, Z.F.; Cui, B.S.; Sun, T. Temporal trends of hydro-climatic variables and runoff response to climatic variability and vegetation changes in the Yiluo River Basin, China. *Hydrol. Process.* 2009, *23*, 3030–3039.
41. Zhang, L.; Dawes, W.R.; Walker, G.R. *Predicting the Effect of Vegetation Changes on Catchment Average Water Balance*; Technical Report 99/12; Cooperative Research Centre for Catchment Hydrology. CSIRO Land and Water: Clayton South, Australia, 1999.
42. Ponce, V.M.; Shetty, A.V. A conceptual-model of catchment water-balance: 1. Formulation and calibration. *J. Hydrol.*1995, *173*, 27–40.
43. Budyko, M.I. *The Heat Balance of the Earth's Surface*; U.S. Department of Commerce: Dordrecht, The Netherland, 1958; p. 259.
44. Budyko, M.I. *Climate and Life*; Academic Press: San Diego, CA, USA, 1974; p. 508.
45. Li, L.J.; Zhang, L.; Wang, H.; Wang, J.; Yang, J.W.; Jiang, D.J.; Li, J.Y.; Qin, D.Y. Assessing the impact of climate

variability and human activities on streamflow from the Wuding River Basin in China. *Hydrol. Proces.* 2007, *21*, 3485–3491.

46. Zhang, L.; Hickel, K.; Dawes, W.R.; Chiew, F.H.S.; Western, A.W.; Briggs, P.R. A rational function approach for estimating mean annual evapotranspiration. *Water Resour. Res.* 2004, *40*, W02502.
47. Xu, C.Y.; Gong, L.B.; Jiang, T.; Chen, D.L.; Singh, V.P. Analysis of spatial distribution and temporal trend of reference evapotranspiration and pan evaporation in Changjiang (Yangtze River) catchment. *J. Hydrol.* 2006, *327*, 81–93.
48. Zhao, F.F.; Zhang, L.; Xu, Z.X.; Scott, D.F. Evaluation of methods for estimating the effects of vegetation change and climate variability on streamflow. *Water Resour. Res.* 2010, *46*, W03505.
49. Guardiola-Claramonte, M.; Troch, P.A.; Breshears, D.D.; Huxman, T.E.; Switanek, M.B.; Durcik, M.; Cobb, N.S. Decreased streamflow in semi-arid basins following drought-induced tree die-off: A counter-intuitive and indirect climate impact on hydrology. *J. Hydrol.* 2011, *406*, 225–233.
50. Tomer, M.D.; Schilling, K.E. A simple approach to distinguish land-use and climate-change effects on watershed hydrology. *J. Hydrol.* 2009, *376*, 24–33.
51. Tootle, G.A.; Singh, A.K.; Piechota, T.C.; Farnham, I. Long lead-time forecasting of US streamflow using partial least squares regression. *J. Hydrol. Eng.* 2007, *12*, 442–451.
52. Hu, S.S.; Zheng, H.X.; Liu, C.M.; Yu, J.J.; Wang, Z.G. Assessing the impacts of climate variability and human activities on streamflow in the water source area of Baiyangdian Lake. *Acta Geograph. Sinica* 2012, *67*, 62–70.
53. Zhang, L.; Zhao, F.F.; Chen, Y.; Dixon, R.N.M. Estimating effects of plantation expansion and climate variability on streamflow for catchments in Australia. *Water Resour. Res.* 2011, *47*, W12539.
54. Zhang, S.R.; Lu, X.X. Hydrological responses to precipitation variation and diverse human activities in a mountainous tributary of the lower Xijiang, China. *Catena* 2009, *77*, 130–142.
55. Zhao, F.F.; Xu, Z.X.; Zhang, L.; Zuo, D.P. Streamflow response to climate variability and human activities in the upper catchment of the Yellow River Basin. *Sci. China Ser. E* 2009, *52*, 1–8.
56. Chen, L.Q.; Liu, C.M. Influence of climate and land-cover change on runoff of the source regions of Yellow River.*China Environ. Sci.* 2007, *27*, 559–565.
57. Li, Z.; Liu, W.Z.; Zhang, X.C.; Zheng, F.L. Impacts of landuse change and climate variability on hydrology in an agriculture catchment on the Loess Plateau of China. *J. Hydrol.* 2009, *377*, 35–42.
58. Guo, H.; Hu, Q.; Jiang, T. Annual and seasonal streamflow responses to climate and land-cover changes in the Poyang Lake Basin, China. *J. Hydrol.* 2008, *355*, 106–122.
59. Hao, X.M.; Chen, Y.N.; Xu, C.C.; Li, W.H. Impacts of climate change and human activities on the surfaces runoff in the Tarim River Basin over the last fifty years. *Water Resour. Manag.* 2008, *22*, 1159–1171.
60. Chen, J.F.; Li, X.B.; Zhang, M. Impacts of climate variability and land cover change on hydrology using model simulation in Suomuo Basin. *Sci. China Ser. D* 2004, *34*, 668–674.
61. Li, L.J.; Li, B.; Liang, L.Q.; Li, J.Y.; Liu, Y.M. Effect of climate change and land use on stream flow in the upper and middle reaches of the Taoer River, northeastern China. *For. Stud. China* 2010, *12*, 107–115.
62. Montenegro, S.; Ragab, R. Impact of possible climate and land use changes in the semi arid regions: A case study from North Eastern Brazil. *J. Hydrol.* 2012, *434–435*, 55–68.
63. Xia, J.; Wang, M.L. Runoff changes and distributed hydrologic simulation in the upper reaches of Yangtze River.*Resour. Sci.* 2008, *30*, 962–967.
64. Wang, G.S.; Xia, J.; Wan, D.H.; Ye, A.Z. A Distributed monthly water balance model for identifying hydrological response to climate changes and human activities. *J. Nat. Resour.* 2006, *21*, 86–91.
65. Coe, M.; Latrubesse, E.; Ferreira, M.; Amsler, M. The effects of deforestation and climate variability on the streamflow of the Araguaia River, Brazil. *Biogeochemistry* 2011, *105*, 119–131.
66. Ma, Z.M.; Kang, S.Z.; Zhang, L.; Tong, L. Analysis of impacts of climate variability and human activity on stream flow for a river basin in arid region of northwest China. *J. Hydrol.* 2008, *352*, 239–249.
67. Zhang, Y.F.; Guan, D.X.; Jin, C.J.; Wang, A.Z.; Wu, J.B.; Yuan, F.H. Analysis of impacts of climate variability and human activity on streamflow for a river basin in northeast China. *J. Hydrol.* 2011, *410*, 239–247.
68. Beguería, S.; López-Moreno, J.I.; Lorente, A.; Seeger, M.; García-Ruiz, J.M. Assessing the effect of climate oscillations and land-use changes on streamflow in the central Spanish Pyrenees. *Ambio* 2003, *32*, 283–286.
69. Tang, L.X.; Zhang, Z.Q.; Wang, X.J.; Wang, S.P.; Zha, T.G. Streamflow response to climate and landuse changes in Qingshui River watershed in the loess hilly-gully region of Western Shanxi Province, China. *Chin. J. Plant Ecol.* 2010,*34*, 800–810.
70. Naik, P.K.; Jay, D.A. Human and climate impacts on Columbia River hydrology and salmonids. *River Res. Appl.* 2011,*27*, 1270–1276.
71. Zhang, M.F.; Wei, X.H.; Sun, P.S.; Liu, S.R. The effect of forest harvesting and climatic variability on runoff in a large watershed: The case study in the Upper Minjiang River of Yangtze River Basin. *J. Hydrol.* 2012, *464–465*, 1–11.
72. Zheng, H.X.; Zhang, L.; Zhu, C.; Liu, C.M.; Sato, Y.; Fukushima, Y. Responses of streamflow to climate and land surface change in the headwaters of the Yellow River Basin. *Water Resour. Res.* 2009, *45*, W00A19.
73. Li, H.Y.; Zhang, Y.Q.; Vaze, J.; Wang, B.D. Separating effects of vegetation change and climate variability using hydrological modelling and sensitivity-based

approaches. *J. Hydrol.* 2012, *420–421*, 403–418.

74. Doylea, M.E.; Vicente, B.R. Attribution of the river flow growth in the Plata Basin. *Int. J. Climatol.* 2011, *31*, 2234–2248.

75. Tague, C.; Grant, G.E.; Farrell, M.; Choate, J.; Jefferson, A. Deep groundwater mediates streamflow response to climate warming in the Oregon Cascades. *Clim. Change* 2008, *86*, 189–210.

76. Van Wateren-de Hoog, B. A regional model to assess the hydrological sensitivity of medium size catchments to climate variability. *Hydrol. Process.* 1998, *12*, 43–56.

77. Vivoni, E.R.; Entekhabi, D.; Bras, R.L.; Ivanov, V.Y. Controls on runoff generation and scale-dependence in a distributed hydrologic model. *Hydrol. Earth Syst. Sci.* 2007, *11*, 1683–1701.

78. Tetzlaff, D.; Soulsby, C.; Waldron, S.; Malcolm, I.A.; Bacon, P.J.; Dunn, S.M. Conceptualization of runoff processes using a geographical information system and tracers in a nested mesoscale catchment. *Hydrol. Process.* 2007, *21*, 1289–1307.

79. Christophersen, N.; Neal, C.; Hooper, R.P.; Vogt, R.D.; Andersen, S. Modeling streamwater chemistry as a mixture of soilwater end-members—A step towards 2nd-generation acidification models. *J. Hydrol.* 1990, *116*, 307–320.

80. McGuire, K.J.; McDonnell, J.J. A review and evaluation of catchment transit time modeling. *J. Hydrol.* 2006, *330*, 543–563.

81. Soulsby, C.; Tetzlaff, D.; Hrachowitz, M. Tracers and transit times: Windows for viewing catchment scale storage?*Hydrol. Process.* 2009, *23*, 3503–3507.

CHAPTER 6 CLIMATE CHANGE IMPACT ON THE HYDROLOGY OF A TYPICAL WATERSHED

Gonghuan Fang, Jing Yang, Yaning Chen, Shuhua Zhang, Haijun Deng, Haimeng Liu, and Philippe De Maeyer

INTRODUCTION

The Tianshan Mountains, regarded as the "water tower of Central Asia" [1], are located in the innermost center of the Eurasian continent. The long distance to the surrounding oceans causes a dry climate, especially for the surrounding basins. Rivers starting in the mountainous regions provide agricultural and domestic water for the surrounding basins and oases. With their distinctive topographic and landscape features, the Tianshan Mountains show a unique energy balance and hydrological cycle and are expected to be sensitive to climate change [2, 3].

Many reports show a widespread climatic and hydrologic change in the Tianshan Mountains during the past few decades [4]. For example, temperature demonstrated a significant rising trend (significant level is smaller than 0.001) at a rate of 0.33~0.34°C/decade during 1960~2010, which is higher than China (0.25°C/decade) and the entire globe (0.13°C/decade) [5, 6]; precipitation increased substantially in most regions especially for the middle and high latitudes; glacier area decreased by 11.5% and the thickness of snowpack has also decreased [5, 7]. Pan evaporation and wind speed have also changed [3]. The annual runoff increased as well, for example, for the Urumqi River, the Kaidu River, and the Aksu River [8, 9].

Future changes in the streamflow and watershed hydrology have become increasingly important to water resource management in the Tianshan Mountains. However, only a limited number of studies currently focus on impact of future climate change on hydrology; for example, Sorg et al. [1] indicated that the total runoff is likely to remain stable or even increase slightly in the near future but it will decrease at the end of the 21st century for Central Asia. There are also researches demonstrating that the annual runoff will decrease to some extent in the first half of the 21st century [2, 10]. Previous studies seldom address implications of climate change on the hydrological cycle and hydrological components (e.g., ET, surface flow, and groundwater). To complement these studies, this paper aims at understanding the future hydrological system and assessing the responses of the hydrologic system to climate change.

In the present study, two sets of climatic data, that is, RCM outputs and SCC data, are used to force SWAT [11] and are applied to the Kaidu River Basin, a typical watershed on the south slope of the Tianshan Mountains, to assess future changes of the hydrologic cycle and the hydrological effects of changes in climate variables. Questions that are addressed include the followings. (1)How will the future climate and hydrology change in this region? (2) What is the effect of climate change on the hydrologic cycling? (3) What is the difference between

Citation: Gonghuan Fang, Jing Yang, Yaning Chen, et al., "Climate Change Impact on the Hydrology of a Typical Watershed," Advances in Meteorology, vol. 2015, Article ID 960471, 10 pages, 2015. doi:10.1155/2015/960471.

simulations with RCM outputs and SCC? Understanding these issues will enable assessing the future hydrological change and its unique hydrometeorological processes better. The remaining is constructed as follows: Section 2 introduces the study area and data; Section 3 describes the hydrological model, bias correction methods, and analysis procedures; Section 4 provides the results and discussion, followed by conclusions in Section 5.

STUDY AREA AND DATA

The Kaidu River Basin, with a drainage area of 18,634 km² above the Dashankou hydrological station, is considered to be a very typical watershed in the Tianshan Mountains (Figure 1). Its altitude ranges from 1,342 m to 4,796 m above sea level (asl) with an average elevation of 2,995 m. The climate here is temperate continental with alpine climate characteristics and obvious seasonal variation. This river provides water resources for agricultural activity and the ecological environment of the oasis, with an area of over 70,000 km² and a population of over 1.15 million, which is mainly stressed by water scarcity [12]. Therefore, projecting the impact of future climate change on water resources is urgent for the sustainable development of this region and it also provides information on the implications of climate change on the water tower in Central Asia.

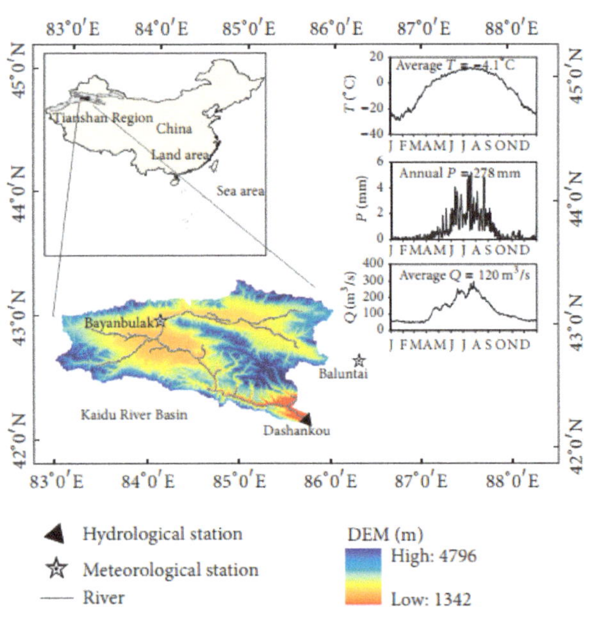

Figure 1: Location (top left), topography, and hydrometeorologic stations (bottom) of the Kaidu River Basin and the daily average temperature (T) and precipitation (P) at the Bayanbulak station and streamflow (Q) at the Dashankou station (top right).

The daily observed meteorological data (including precipitation, maximum/minimum temperature, wind speed, and relative humidity of two meteorological stations Bayanbulak and Baluntai) from 1970 to 2005 are from the China Meteorological Data Sharing Service System (http://www.cma.gov.cn/2011qxfw/2011qsjgx/). The annual mean temperature at the Bayanbulak meteorological station amounts to −4.1°C and the mean annual precipitation is 278 mm (Figure 1). Generally, precipitation falls as rain from May to September each year and as snow from October to April. The observed streamflow data at the Dashankou hydrologic station are from the Xinjiang Tarim River Basin Management Bureau. The average flow at the Dashankou hydrological station amounts to approximately 120 m³/s (equivalent to 202 mm runoff/year), ranging from 15 m³/s to 973 m³/s.

METHODOLOGY

Regional Climate Model and RCP Scenarios

The outputs of a regional climate model (RegCM4.0) [13] forced by a global climate model (Beijing Climate Center Climate System Model, BCC_CSM1.1) [14, 15] at a horizontal resolution of 50 km are used as future climate data. Firstly, the RCM model was validated with the observational data set over China for the period from 1976 to 2005 and then it was used to predict the future climate change under the new emission scenarios of the Representative Concentration Pathways, RCP4.5 (lower emission scenario) and RCP8.5 (higher emission scenario). RCP4.5 is a stabilization scenario with the total radiative forcing rising until 2070 and then remaining at a stable centration of 4.5 W/m². In contrast, RCP8.5 is a continuously rising radiative forcing pathway (at a target of 8.5 W/m² in 2100) with a further enhanced residual circulation and significant CH_4 increases [16, 17]. The RCM validation shows reasonable simulations of temperature and precipitation were obtained over China and, compared to the BCC_CSM1.1 model, marked improvement of the RCM was achieved in reproducing present day precipitation and temperature (for more details refer to [18]).

Bias Correction Methods

Five precipitation and three temperature correction methods were selected to bias-correct the raw RCM outputs. Precipitation correction methods include linear scaling, local intensity scaling, power transformation, distribution mapping, and quantile mapping. Temperature correction methods include linear scaling, variance scaling, and distribution mapping. They are combined into 15 schemes to evaluate their performances in simulating streamflow. It turns out that the precipitation correction methods have more significant influence than the temperature correction methods on streamflow simulation, and the power transformation and quantile mapping perform best in terms of frequency based statistics. Thereafter, the quantile mapping method (for precipitation) and the distribution mapping (for temperature) are selected to correct the raw RCM outputs for the future period (for more details see [19]).

Hydrologic Model and Uncertainty Analysis Method

SWAT has been extensively used for the comprehensive modeling of the impact of management practices and climate change. SWAT simulates the hydrologic and sedimentary processes, plant growth, river routing, and in-stream water quality process, among which the surface runoff is calculated from daily rainfall and snowmelt with a modified Soil Conservation Service (SCS) curve number method [20]; water routing is simulated using variable storage or the Muskingum river routing method [11].

The SWAT model input includes the digital elevation model (DEM), soil textural and physicochemical properties, and land use data. The meteorological variables including daily precipitation, max./min. temperature, relative humidity, solar radiation, and wind speed were used to force the SWAT model. SWAT uses elevation bands to represent the topographic effects on precipitation and temperature. Within each elevation band, the precipitation and temperature are estimated based on their lapse rates. For more details, refer to the SWAT manual (http://www.brc.tamus.edu/).

The SWAT model (forced by the observed meteorological data) was calibrated against the observed streamflow. The calibration period is from 1986 to 1989 and the validation period from 1990 to 2005 [21]. The calibrated optimal parameters are then kept fixed in the following simulations. The evaluation indices for the hydrological model include NS, PBIAS, and the determination coefficient R^2. Consider

$$\mathrm{NS} = 1 - \frac{\sum_{i=1}^{n}\left(Y_i^{\mathrm{obs}} - Y_i^{\mathrm{sim}}\right)^2}{\sum_{i=1}^{n}\left(Y_i^{\mathrm{obs}} - Y^{\mathrm{mean}}\right)^2},$$

$$\mathrm{PBIAS} = \frac{\sum_{i=1}^{n}\left(Y_i^{\mathrm{obs}} - Y_i^{\mathrm{sim}}\right)}{\sum_{i=1}^{n}\left(Y_i^{\mathrm{obs}}\right)}, \quad (1)$$

where Y_i^{obs} and Y_i^{sim} are the ith observed and simulated flows, Ymean is the mean of the observed data, and n is the number of observations. Normally NS > 0.50, |PBIAS| < 25%, and $R^2 > 0.6$ are taken as the criteria for satisfactory modeling of the river discharge and the model performance can be evaluated as excellent if NS > 0.75 and |PBIAS| < 10% [22]. GLUE (generalized likelihood uncertainty estimation) [23] is an uncertainty analysis technique, in which the parameter uncertainty accounts for all sources of uncertainty, such as input uncertainty, structure uncertainty, parameter uncertainty, and response uncertainty [24]. In GLUE, the parameter uncertainty is described as a set of discrete "behavioral" parameter sets with corresponding "likelihood weights.

The procedure of a GLUE analysis consists of three steps. Firstly, after the definition of the "generalized likelihood measure," (θ), a large number of parameter sets are randomly sampled from the prior distribution and each parameter set is assessed as either "behavioral" or "nonbehavioral" by comparing its value of (θ) to the threshold value. Secondly, each behavioral parameter set is given a "likelihood weight" and we gave them equal weights in this study. Finally, prediction uncertainty is represented by 5% and 95% quantiles of the cumulative distribution of the behavioral parameter sets. Two indices are used to quantify the quality of the uncertainty performance. Those indices are the percentage of measurements bracketed by the 95% prediction uncertainty band (P-factor) and width of band (R-factor, calculated by the average width of the band divided by the standard deviation of the corresponding measured variable).

SCC Data Description and Analysis Procedures

In the following section, temperature and precipitation are denoted as T and P and the absolute and relative changes are represented by Δ and δ. For example, ΔT refers to an absolute temperature change and δP a relative precipitation change. The hydrological processes analyzed in this study include streamflow, surface runoff, subsurface runoff, and evapotranspiration, which are denoted as Q, R_s, R_g, and ET, and their relative changes are described as δQ, δR_s, δR_g, and δET, respectively.

The SCC was constructed to represent a wide range of changes in climatic variables and how these changes might translate in streamflow and other hydrological components and also to analyze the differences between simulations with RCM outputs and SCC. For SCC, perturbations of the corrected RCM simulated P and T from 1986~2005 (control period) are set; that is, for T, an additive change (Δ) is used: $\Delta T = -1, 0, 1, 2, 3, 4, 5$, and 6° C. For P, a relative change (δ) is used: δP = −20%, −10%, 0, 10%, 20%, 30%, 40%, 50%, and 60%. They are put into 81 SCC scenarios, with $\Delta T = 0$ and $\delta P = 0\%$ being the climate for control period.

By investigating the transient evolution of climate change in the corrected RCM outputs on decadal scales, five periods (each spanning 20 years) are defined: 1986~2005 (control period) and 2020~2039, 2040~2059, 2060~2079, and 2080~ 2099. Due to the intra-annual characteristics of the hydrometeorology in the Kaidu River Basin (Figure 1), the wet season (from April to September) and dry season (from October to March next year) are defined based on the intra-annual distribution of P and Q; for example, P and Q in the wet season account for 88% and 73% of their annual amounts. The climatic and hydrological changes are classified into three categories, that is, a significant change, small change, and insignificant change, to clearly demonstrate the changing magnitude according to the values of relative change for precipitation and hydrological components and absolute change of temperature. These categories are presented in Table 1.

Table 1: Classification of magnitude for climatic and hydrological changes. δ and Δ represent relative change and absolute change

	Precipitation & hydrological components (%)	Temperature (°C)
Significant change	$\lvert\delta\rvert \geq 20$	$\lvert\Delta\rvert \geq 2$
Small change	$10 \leq \lvert\delta\rvert < 20$	$1 \leq \lvert\Delta\rvert < 2$
Insignificant change	$\lvert\delta\rvert < 10$	$\lvert\Delta\rvert < 1$

RESULTS AND DISCUSSION

Validation of the Hydrological Model and the Bias Correction Methods

Performance of the hydrological model forced by observed meteorological data and the 95% prediction uncertainty bands are shown in Figure 2. The simulated streamflow agrees quite well with the observation for both calibration period (1986~1989) and validation period (1990~ 2002). For the uncertainty analysis, NS is used as (θ) and 0.70 as threshold value with 10,000 initial parameter sets; 288 sets were selected as behavioral points. The results show that most of the observations are bracketed by the 95% prediction uncertainty band (P-factor being 87% and 80% for calibration and validation periods and R-factor being 1.18 and 1.19, resp.). The lower P-factor for the validation period can be partly attributed to operation of hydropower station since 1991 (Figure 2) which leads to great fluctuation in winter streamflow. Statistics of model efficiency (Table 2) indicate excellent performances for both calibration and validation periods, with "NS"s and "R^2 "s over 0.80, which is highly acceptable according to Moriasi et al. [22]. Concerning the monthly streamflow, the "NS" is 0.89 during 1986~2005 and it indicates that the SWAT model captured the natural monthly streamflow variability adequately. The performances of bias-corrected RCM outputs (compared to observed meteorological data) are listed in Table 2. The "NS"s are −0.57 (0.57) and 0.77 (0.95) for daily (monthly) precipitation and temperature for 1990~2005, respectively. And the statistics of the streamflow simulated with the biascorrected RCM outputs shows acceptable results with "NS"s equal to 0.46 and 0.62 and PBIAS within 10% for daily and monthly streamflows.

RCM Projected Hydrometeorologic Changes

Changes in Temperature and Precipitation

Temperature is highly likely to increase in the future, with a basin warming of 1.0~2.2°C and 1.6~4.6°C under RCP4.5 and RCP8.5 in the 21st century (Table 3). Temperature increases continuously under both scenarios but the magnitude is larger under RCP8.5 (Figure 3). Precipitation shows an overall increasing trend in the 21st century with an annual increase of 2%~16% and 7%~ 24% under RCP4.5 and RCP8.5, which confirms the previous arguments of Sorg et al. [1]. However, precipitation change varies substantially among seasons (Figure 3). Normally a small increase in the wet season (−2%~16%) and a significant increase during dry season (18%~78%) are projected.

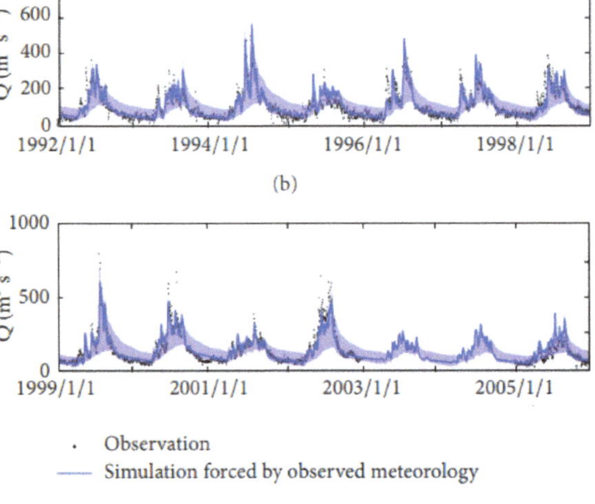

Figure 2: Time series of daily observed streamflows (dots) and simulated streamflows forced by observed meteorological data (blue line) for calibration period (1986–1989) and validation period (1990–2005) with 95% prediction uncertainty bands (blue shaded area).

Note that the relative increase (not the absolute increment) of precipitation for the dry season is much bigger than for the wet season, which is in line with the climate changes in other regions, for example, the semiarid Colorado River Basin [25] and the wet Ganges-Brahmaputra-Meghna basin [26].

Changes in the Hydrological Cycle

The changes in precipitation and temperature cause changes in potential streamflow. The average annual streamflow rises by −1%~ 18% and 4~20% under the RCP4.5 and RCP8.5 in the 21st century, based on the average annual streamflow of 194 mm for the control period (1986~2005) (Table 3). Note that the streamflow stopped increasing in 2080~2099 (end of 21st century) under RCP8.5 despite the rise in precipitation, which confirms the finding of Sorg et al. [1] and may aggravate water scarcity in this region. Figure 3 also shows the projected changes in surface runoff (R_s), subsurface runoff (R_g), and evapotranspiration (ET) under RCP4.5 and RCP8.5. Overall, changes of hydrologic components are bigger for RCP8.5 than for RCP4.5. The annual change of Rs is insignificant (<5%) but with obvious seasonal variability; for example, changes of Rs range from −22% to 2% for the wet season and in 4%~78% for the dry season under RCP8.5.

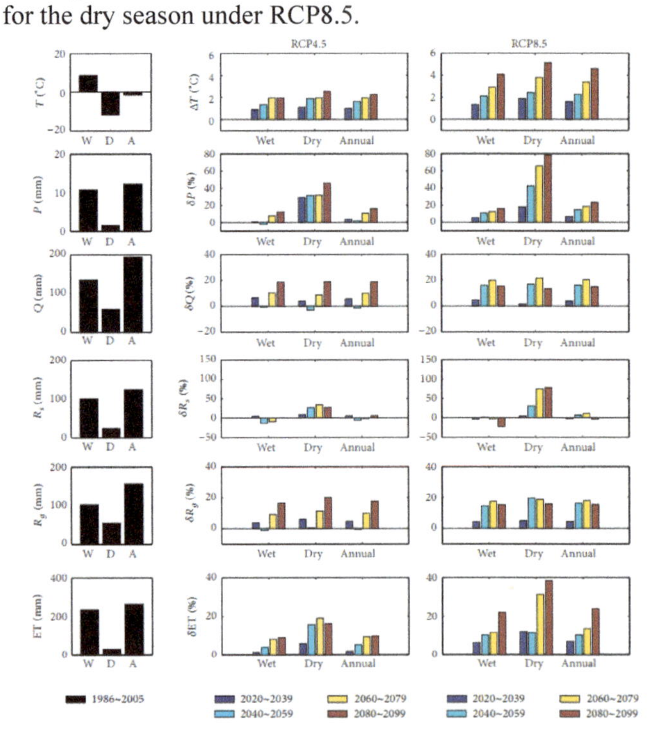

Figure 3: Summary of future climate inputs (P and T) and simulated hydrologic components (Q, R_s, R_g, and ET) under RCP4.5 and RCP8.5, compared to their values in the control period (1986~2005). All these hydrometeorologic factors are presented in terms of wet season, dry season, and annual values.

The annual Rg changes by −0.7%~17% and 4%~18% for RCP4.5 and RCP8.5, which is consistent with the changes of Q. ET increases continuously in the 21st century with average increases of 2%~10% and 7%~24% under RCP4.5 and RCP8.5.

Response of Hydrological Cycle to Climate Change

The response of the hydrological cycle to climate change is estimated by running the hydrological model forced by SCC. The responses of Q, R_s, R_g, and ET to climate change are demonstrated with response surfaces in Figure 4. Q is positively related to P and negatively related to T.The relationship of δQ and δP is almost linear with the streamflow elasticity ($\delta Q/\delta P$) being about 1.0 when $\Delta T < 2$°C; that is, a 1% change in the mean annual precipitation results in a 1% change in the mean annual streamflow. $\delta Q/\delta P$ is lower than that for other arid regions, for example, 2.0~3.5 for Australia [27].The possible reasons are mentioned as follows:(1)the Kaidu River Basin, located in the south slope of the Tianshan Mountains with a high average altitude (2,995 m), is characterized by a cold climate (average annual temperature is −4.1°C for the Bayanbulak station) and accordingly there is a low amount of energy available for ET, which results in a relatively high runoff coefficient (Q/P = 0.51) and consequently a low streamflow elasticity; (2) the streamflow is also influenced by temperature dominated snowmelt (snowfall accounts for about 17% of watershed precipitation), which reduces the dependence of streamflow on precipitation and therefore results in a low streamflow elasticity [27].

Table 2: Statistics of bias-corrected RCM outputs and SWAT simulated streamflows forced by the observed climate variables and biascorrected RCM outputs.

Statistics	NS	PBIAS	R^2
"RCM simulated precipitation with bias correction"[a]			
Validation period 1990~2005 (daily)[b]	−0.57	−6.80%	0.00
Validation period 1990~2005 (monthly)	0.57	−6.80%	0.60
"RCM simulated maximum temperature with bias correction"[a]			
Validation period 1990~2005 (daily)	0.77	3.80%	0.80
Validation period 1990~2005 (monthly)	0.95	4.00%	0.90
"Streamflow simulated with observed meteorological data"			
Calibration period 1986~1989 (daily)	0.80	0.01%	0.80
First validation period 1990~2002 (daily)	0.81	2.94%	0.81
Second validation period 1986~2005 (monthly)	0.89	2.86%	0.90
"Streamflow simulated with bias-corrected RCM outputs"			
Validation period 1990~2002 (daily)	0.46	−6.98%	0.47
Validation period 1986~2005 (monthly)	0.62	−7.85%	0.63

[a]Bias correction methods used are quantile mapping for precipitation and distribution mapping for temperature [19].
[b]"Daily" or "monthly" in the brackets means the time step used to calculate the statistics.

Table 3: RCM projected precipitation change (δP), temperature change (ΔT), and streamflow change (δQ) for the 21st century under RCP4.5 and RCP8.5 compared to the control period (1986~2005).

		2020~2039	2040~2059	2060~2079	2080~2099
RCP4.5	δP (%)	4.0	2.0	11.0	16.0
	ΔT (°C)	1.0	1.6	2.0	2.2
	δQ (%)	6.0	−1.0	10.0	18.0
RCP8.5	δP (%)	7.0	15.0	19.0	24.0
	ΔT (°C)	1.6	2.3	3.3	4.6
	δQ (%)	4.0	16.0	20.0	15.0

The response of Q toΔTdepends on the magnitude of ΔT. Q decreases slightly when $0 < \Delta T \leq 2.0$°C while it decreases dramatically when $\Delta T > 2.0$°C for both wet and dry seasons. For example, when $\Delta T = 2.0$°C, a 40% precipitation increase results in an average value of Q being 240 mm (23% increase compared to the average streamflow of 194 mm) but when $\Delta T = 4.0$°C, the same precipitation increase only generates an average Q of 180 mm (about 7% decrease) (Figure 4). The responses of R_s, R_g, and ET to climate change are also demonstrated in Figure 4. For R_s, the responses of R_s to ΔT are quite different for the wet and dry seasons: the higher ΔT, the lower Rs for the wet season but the higher R_s for the dry season. Since R_s in the dry season only accounts for 13% of the annual R_s, the response of annual R_s is consistent with that of the wet season. For R_g, the responses of R_g to ΔT and δP are similar to the responses of Q due to the dominant role of groundwater recharge in water yield in the Kaidu River Basin. For ET, it is mainly influenced by ΔT with temperature sensitivity (δET/ΔT) being 7.3%/∘ C. To verify this result, we firstly investigated basin-scale energy and water budget using the Budyko method [28, 29]. It is shown that ET is mainly energy limited rather than water limited (average ET/P = 0.67 and PET/P = 0.88). Secondly, the high determinate coefficient $R^2 = 0.75$ (significant level is smaller than 0.01) between the mean annual T and ET also indicates that ET has a strong correlation with T. This is consistent with previous studies, which have shown that a significant variation in ET is expected to follow changes in air temperature [30, 31]. In addition, simulations with RCM outputs are shown in Figure 4 to analyze the differences between simulations of these two data sets. Two typical periods of RCM simulations are selected, that is, 2020~2039 under RCP4.5 and 2080~2099 under RCP8.5, to represent mild and intense climate change scenarios (shown as blue and red stars in Figure 4). It is indicated that the simulations of hydrological components with RCM outputs for 2020~2039 under RCP4.5 (mild climate change) are close to these of the nearby contour lines (simulations with SCC), which suggests that similar results of Q, R_s, R_g, and ET are obtained for RCM outputs and for SCC under mild climate change scenarios. However, for 2080~2099 under RCP8.5 with δP = 24% and ΔT = 4.6°C, the simulated values of Q, R_s, R_g, and ET deviate from the simulations of SCC. There are two possible reasons: (1) changes of other meteorological inputs, that is, solar radiation, wind speed, and humidity, are slightly smaller for 2020~2039 under RCP4.5 than those for 2080~2099 under RCP8.5 (−0.8%, 2.6%, and 0.9% compared to −2.2%, 4.1%, and 1.4%); (2) for 2080~2099 under RCP8.5, precipitation increases by 24% with great seasonal variation, which may alter the hydrological regime; for example, precipitation increases by 139% for March, April, and May while it decreases by −0.1% for June, July, and August. Since changes of solar radiation, wind speed, and humidity are within ±5%, the second reason, that is, the shift of the precipitation temporal distribution, contributes a lot to the deviation of simulations with RCM outputs from simulations with SCC.

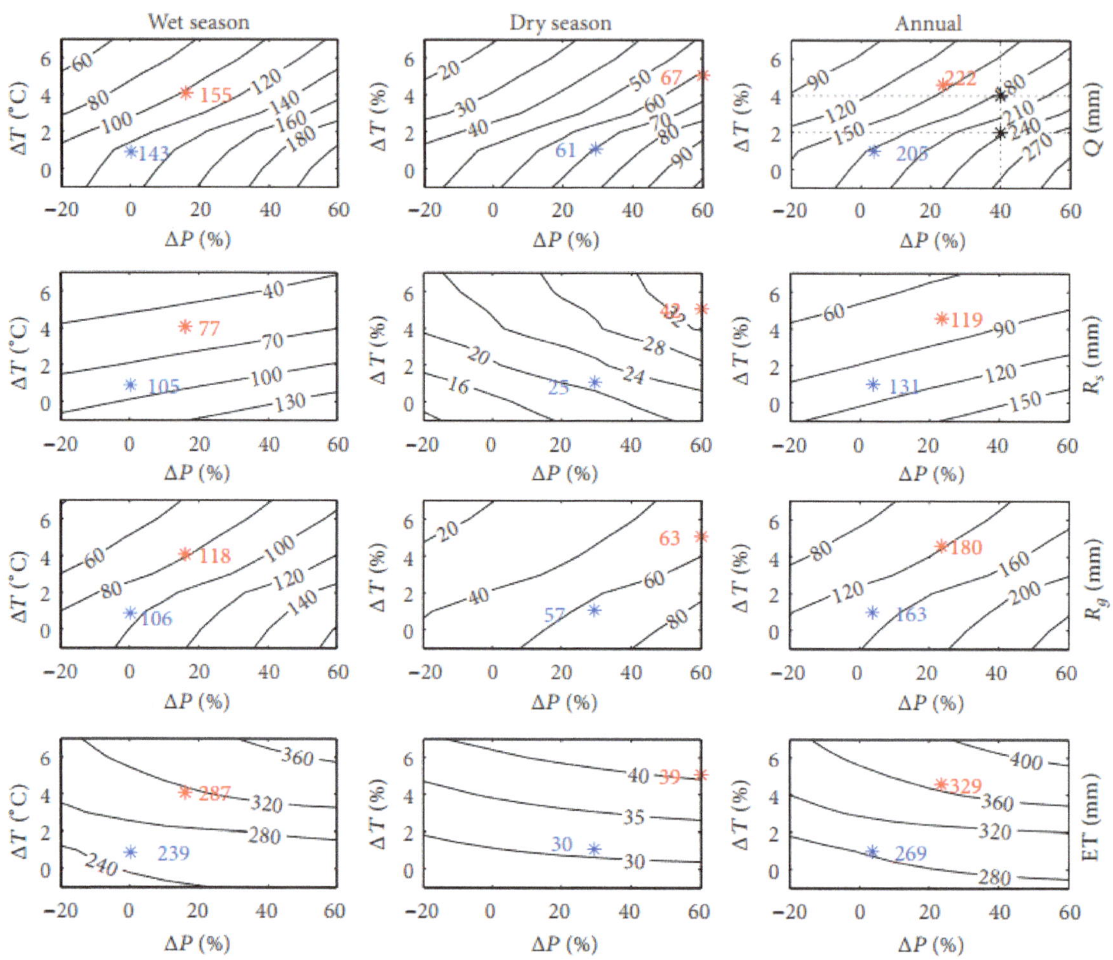

Figure 4: Response surfaces of streamflow (Q), surface runoff (R_s), subsurface runoff (R_g), and evapotranspiration (ET) to climate change. The simulations with RCM outputs for 2020~2039 under RCP4.5 and for 2080~2099 under RCP8.5 (their corresponding meteorological changes are listed in Table 3) are indicated using blue and red stars with labels.

Furthermore, the exceedance probability curves of the annual runoff in response to climate change are demonstrated in Figure 5. The exceedance probability curves are almost parallel when ΔT ranges in 0~6°C. However, the responses of Q to δP are not the same for each exceedance probability: high sensitivity of Q with probabilities less than 0.1 and low sensitivity of Q with probabilities larger than 0.9. A comparison of the simulations with RCM outputs (four future periods under RCP4.5 and RCP8.5) and with SCC (the same changes in T or P with the corresponding RCM outputs) indicates that differences between simulation with RCM outputs and SCC are becoming greater as climate change gets more intense; for example, the simulation with RCM for 2080~2099 under RCP8.5 overestimates the corresponding simulations with SCC (Figure 5(h)), which collaborates the conclusion that under intense climate change scenarios the simulated hydrology with RCM deviates from that simulated with SCC.

The contributions of hydrologic components to water yield are displayed by the De Finetti diagram in Figure 6. For the control period, the averages of R_s, R_g, and ET are 0.22, 0.28, and 0.50. For SCC, as ΔT increases from 0 to 6°C, the contribution of ET increases rapidly from 0.49 to 0.73 and the contributions of R_s and R_g decrease from 0.22 to 0.11 and from 0.29 to 0.16. ΔT has a more significant influence on the proportion change than δP. As δP changes from −20% to 60%, ET decreases from 0.71 to 0.58 and R_s and R_g increase from 0.13 to 0.15 and from 0.16 to 0.27. For simulations with RCM outputs, proportions of hydrological components do not change significantly under RCP4.5, while the proportion of ET shows a significant increase under RCP8.5.

Sources of Uncertainty and Other Considerations

There are uncertainties in estimating climate change impact on hydrology. As indicated by previous studies [32], the sources of uncertainty may rise from climate models, emission scenarios, downscaling, and the hydrological model.

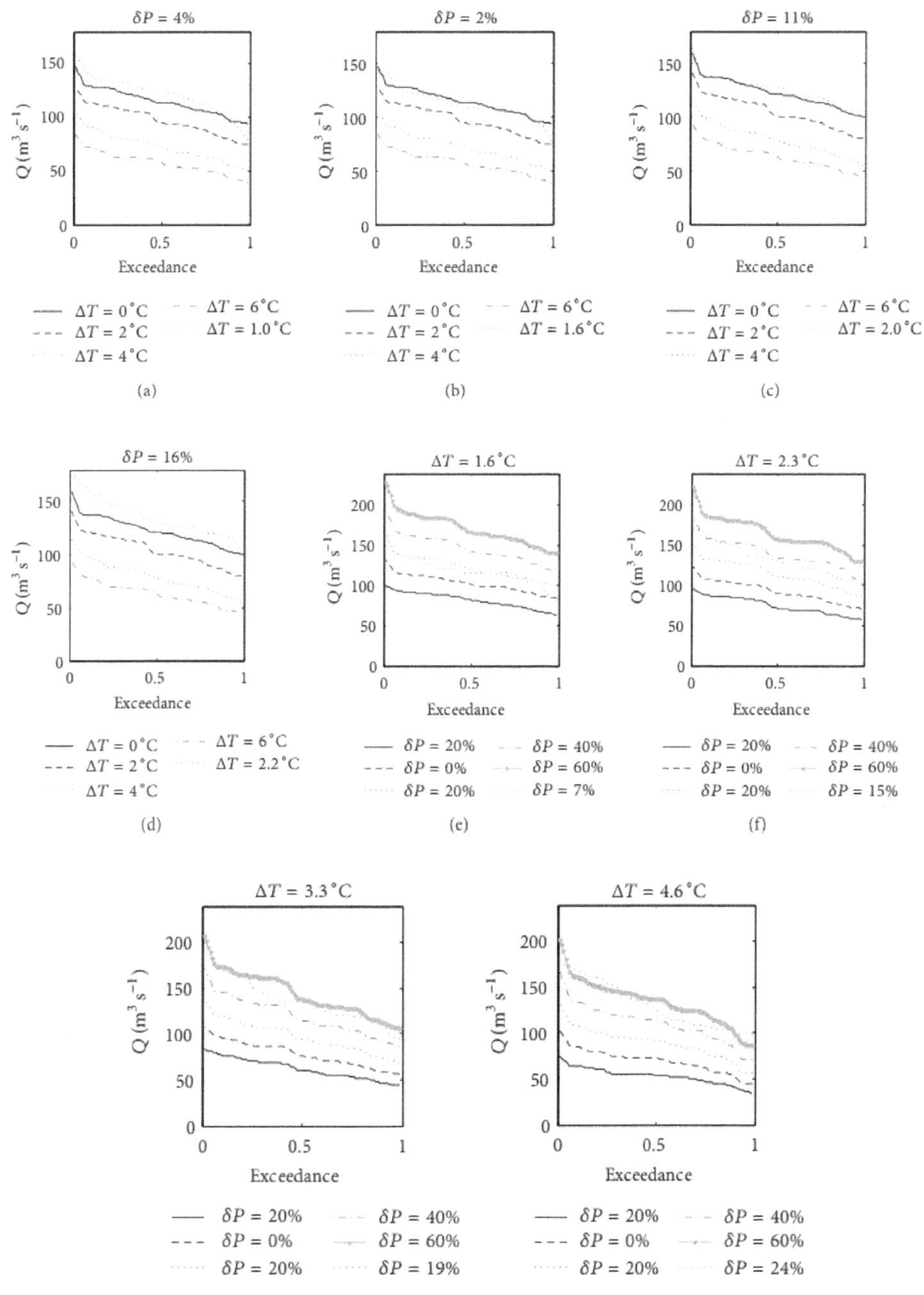

Figure 5: Exceedance probability curves of average annual streamflow (Q) in response to temperature change and precipitation change based on SCC with each plot either with fixed precipitation change (a~d) or with fixed temperature change (e~h). Dotted blue line in each plot denotes exceedance probability curves of average annual streamflow for the simulation with RCM outputs given fixed δP and ΔT as summarized in Table 3.

For hydrological modeling itself, the effect of future climate in any specific catchment is difficult to project due to the possibility that the hydrological system may not be stationary with complex feedbacks [33]. For example, the same land cover and soil data were used for both control period and future climate change period, which may not well represent the land surface under the future climate changes. Effects of land cover change on streamflows and other components of the hydrological cycle are not considered.

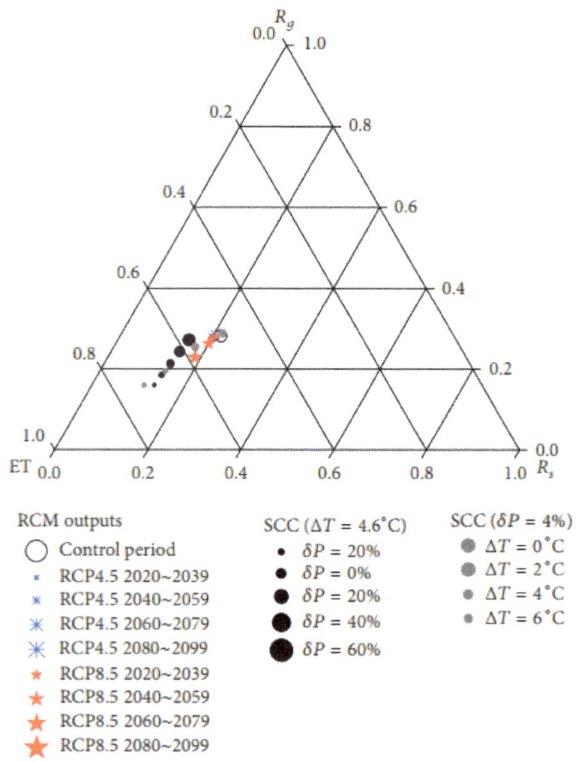

Figure 6: De Finetti diagram (ternary plot) of evapotranspiration (ET), surface runoff (R_s), and subsurface runoff (R_g) for SCC (shown as dots) and RCM outputs (shown as stars; for details of the projected changes in RCM outputs refer to Table 3).

Though uncertainty in hydrologic modeling was quantified with the GLUE method, it only accounts for part of the total uncertainty in climate change impact studies [34]. Uncertainties associated with the climate model and downscaling were not considered here although two emission scenarios were included. Any uncertainty associated with them may cause the results to deviate from reality. However, we are dedicated to pursuing a thorough investigation of the response of the hydrological cycle to future climate change for this region and we believe this study is an important first step in achieving this goal.

CONCLUSIONS

This study assessed the implications of climate change on hydrology in a typical watershed in the Tianshan Mountains with two sets of climatic data, that is, RCM outputs and SCC, loosely coupled to a hydrological model (SWAT). Major conclusions can be summarized as follows:

1. The hydrological model shows excellent performance with "NS"s over 0.8 for the daily streamflow for both calibration and validation periods. And the selected bias correction methods were effective in downscaling RCM outputs, with "NS"s being 0.57 and 0.95 regarding monthly precipitation and temperature.

2. T increases by 1.0∘ C~2.2°C and by 1.6°C~4.6°C under RCP4.5 and RCP8.5 in the 21st century. For P, it shows an overall increasing trend (2%~24%) with significant increase for the dry season (18%~78%) and relatively small change for the wet season (−2%~16%). The projected Q shows an overall increasing trend (−1%~ 18% and 4~20% for RCP4.5 and RCP8.5) in the 21st century.

3. Q increases almost linearly with P while the response of Q to T depends on the magnitude of ΔT and Q decreases significantly when ΔT is greater than 2∘ C.

4. Similar responses of Q, R_s, R_g, and ET to P and T are obtained for the RCM outputs and for SCC under mild climate change scenarios. However, for intense climate change scenarios, simulations of Q, Rs, Rg, and ET with RCM outputs (e.g., for 2080~2099 under RCP8.5) deviate from simulations with SCC.

5. ΔT has more significant influence on the proportion change of each hydrologic component than δP does. As ΔT increases from 0 to 6°C, the contribution of ET increases rapidly from 0.49 to 0.73 and R_s and R_g decrease by 0.11 and 0.13. As δP changes from −20% to 60%, ET, Rs, and Rg change by −0.13, 0.02, and 0.11 as a result.

It is valuable to quantify the future responses of hydrology to climate change in the Tianshan Mountains. This study will provide useful information for water resource management and will serve as a basis for further climate change impact studies.

ACKNOWLEDGMENTS

Research was supported by the "Thousand Youth Talents Plan" (Xinjiang Project), the National Natural Science Foundation of China (41471030), and the Foundation of State Key Laboratory of Desert and Oasis Ecology (Y371163). The authors wish to thank Professor Xuejie Gao at the National Climate Center (China) for providing the outputs of the regional climate model used in this paper.

REFERENCES

1. A. Sorg, T. Bolch, M. Stoffel, O. Solomina, and M. Beniston, "Climate change impacts on glaciers and runoff in Tien Shan (Central Asia)," Nature Climate Change, vol. 2, no. 10, pp. 725–731, 2012. · ·
2. Z. Liu, Z. Xu, J. Huang, S. P. Charles, and G. Fu, "Impacts of climate change on hydrological processes in the headwater catchment of the Tarim River basin, China," Hydrological Processes, vol. 24, no. 2, pp. 196–208, 2010. · ·
3. Z. Li, Y. Chen, Y. Shen, Y. Liu, and S. Zhang, "Analysis of changing pan evaporation in the arid region of Northwest China," Water Resources Research, vol. 49, no. 4, pp. 2205–2212, 2013. · ·
4. Y. Shi, Y. Shen, E. Kang et al., "Recent and future climate change in Northwest China," Climatic Change, vol. 80, no. 3-4, pp. 379–393, 2007. · ·
5. S. Wang, M. Zhang, Z. Li et al., "Glacier area variation and climate change in the Chinese Tianshan Mountains since 1960," Journal of Geographical Sciences, vol. 21, no. 2, pp. 263–273, 2011. · ·
6. B. Li, Y. Chen, and X. Shi, "Why does the temperature

rise faster in the arid region of northwest China?"Journal of Geophysical Research D: Atmospheres, vol. 117, no. 16, Article ID D16115, 2012. · ·

7. V. B. Aizen, E. M. Aizen, J. M. Melack, and J. Dozier, "Climatic and hydrologic changes in the Tien Shan, central Asia," Journal of Climate, vol. 10, no. 6, pp. 1393–1404, 1997. · ·
8. B. Ye, D. Yang, K. Jiao et al., "The Urumqi River source Glacier No. 1, Tianshan, China: changes over the past 45 years," Geophysical Research Letters, vol. 32, no. 21, Article ID L21504, 2005. · ·
9. H. Wang, Y. Chen, W. Li, and H. Deng, "Runoff responses to climate change in arid region of northwestern China during 1960–2010," Chinese Geographical Science, vol. 23, no. 3, pp. 286–300, 2013. · ·
10. T. Liu, P. Willems, X. L. Pan et al., "Climate change impact on water resource extremes in a headwater region of the Tarim basin in China," Hydrology and Earth System Sciences, vol. 15, no. 11, pp. 3511–3527, 2011. · ·
11. J. G. Arnold, R. Srinivasan, R. S. Muttiah, and J. R. Williams, "Large area hydrologic modeling and assessment part I: model development," Journal of the American Water Resources Association, vol. 34, no. 1, pp. 73–89, 1998. · ·
12. Y. Chen and Q. Du, Sustainable Water Use in the Bosten Lake Basin, Science Press, Beijing, China, 2013.
13. F. Giorgi and L. O. Mearns, "Introduction to special section: regional climate modeling revisited," Journal of Geophysical Research, vol. 104, no. D6, pp. 6335–6352, 1999. ·
14. T. Wu, W. Li, J. Ji et al., "Global carbon budgets simulated by the Beijing climate center climate system model for the last century," Journal of Geophysical Research D: Atmospheres, vol. 118, no. 10, pp. 4326–4347, 2013. · ·
15. X. Xin, T. Wu, J. Li, et al., "How well does BCC_CSM1. 1 reproduce the 20th Century climate change over China?" Atmospheric and Oceanic Science Letters, vol. 6, pp. 21–26, 2013.
16. D. P. van Vuuren, J. Edmonds, M. Kainuma et al., "The representative concentration pathways: an overview," Climatic Change, vol. 109, no. 1, pp. 5–31, 2011. · ·
17. H. Kawase, T. Nagashima, K. Sudo, and T. Nozawa, "Future changes in tropospheric ozone under Representative Concentration Pathways (RCPs)," Geophysical Research Letters, vol. 38, no. 5, 2011. · ·
18. X. Gao, M. Wang, and F. Giorgi, "Climate change over China in the 21st century as simu-lated by BCC_CSM1.1-RegCM4.0," Atmospheric and Oceanic Science Letters, vol. 6, pp. 381–386, 2013.
19. G. H. Fang, J. Yang, Y. N. Chen, and C. Zammit, "Comparing bias correction methods in downscaling meteorological variables for hydrologic impact study in an arid area in China," Hydrology and Earth System Sciences Discussions, vol. 11, no. 11, pp. 12659–12696, 2014. ·
20. Soil Conservation Service, "Urban hydrology for small watershed," Technical Release 55, U.S. Department of Agriculture, 1968.
21. G. Fang, J. Yang, Y. Chen, C. Xu, and P. de Maeyer, "Contribution of meteorological input in calibrating a distributed hydrologic model in a watershed in the Tianshan Mountains, China," Environmental Earth Sciences, 2015. ·
22. D. N. Moriasi, J. G. Arnold, M. W. van Liew, R. L. Bingner, R. D. Harmel, and T. L. Veith, "Model evaluation guidelines for systematic quantification of accuracy in watershed simulations," Transactions of the ASABE, vol. 50, no. 3, pp. 885–900, 2007. · ·
23. K. Beven and A. Binley, "The future of distributed models: model calibration and uncertainty prediction," Hydrological Processes, vol. 6, no. 3, pp. 279–298, 1992. · ·
24. J. Yang, P. Reichert, K. C. Abbaspour, J. Xia, and H. Yang, "Comparing uncertainty analysis techniques for a SWAT application to the Chaohe Basin in China," Journal of Hydrology, vol. 358, no. 1-2, pp. 1–23, 2008. · ·
25. N. S. Christensen and D. P. Lettenmaier, "A multimodel ensemble approach to assessment of climate change impacts on the hydrology and water resources of the Colorado River Basin," Hydrology and Earth System Sciences, vol. 11, no. 4, pp. 1417–1434, 2007. · ·
26. M. Masood, P. J. Yeh, N. Hanasaki, and K. Takeuchi, "Model study of the impacts of future climate change on the hydrology of Ganges–Brahmaputra–Meghna basin," Hydrology and Earth System Sciences, vol. 19, no. 2, pp. 747–770, 2015. ·
27. F. H. S. Chiew, "Estimation of rainfall elasticity of streamflow in Australia," Hydrological Sciences Journal, vol. 51, no. 4, pp. 613–625, 2006. · ·
28. J. A. Jones, I. F. Creed, K. L. Hatcher et al., "Ecosystem processes and human influences regulate streamflow response to climate change at long-term ecological research sites," BioScience, vol. 62, no. 4, pp. 390–404, 2012. · ·
29. M. I. Budyko, Climate and Life, Academic Press, New York, NY, USA, 1974.
30. K. C. Abbaspour, M. Faramarzi, S. S. Ghasemi, and H. Yang, "Assessing the impact of climate change on water resources in Iran," Water Resources Research, vol. 45, no. 10, 2009. · ·
31. S. G. Setegn, D. Rayner, A. M. Melesse, B. Dargahi, and R. Srinivasan, "Impact of climate change on the hydroclimatology of Lake Tana Basin, Ethiopia," Water Resources Research, vol. 47, no. 4, Article ID W04511, 2011. · ·
32. S. Hagemann, C. Chen, D. B. Clark et al., "Climate change impact on available water resources obtained using multiple global climate and hydrology models," Earth System Dynamics, vol. 4, no. 1, pp. 129–144, 2013. · ·
33. R. P. Silberstein, S. K. Aryal, J. Durrant et al., "Climate change and runoff in south-western Australia,"Journal of Hydrology, vol. 475, pp. 441–455, 2012. · ·

34. D. G. Kingston and R. G. Taylor, "Sources of uncertainty in climate change impacts on river discharge and groundwater in a headwater catchment of the Upper Nile Basin, Uganda," Hydrology and Earth System Sciences, vol. 14, no. 7, pp. 1297–1308, 2010. · ·

CHAPTER 7

IMPACTS OF FOREST FIRES AND CLIMATE VARIABILITY ON THE HYDROLOGY OF AN ALPINE MEDIUM SIZED CATCHMENT

Johanna Springer Ralf Ludwig, and Stefan W. Kienzle

Keywords: Alpine watershed, hydrological modeling, WaSiM, climate change, runoff trend, wild-fires, remote sensing of burned areas

INTRODUCTION

The southern Rocky Mountains in Alberta, Canada, are an important region for the generation of freshwater runoff especially due to their contribution of melt water in spring months [1,2]. Climate change affects the watersheds in the southern Rocky Mountains with an increase in air temperatures, especially during winter months, and thus a decrease in snow cover. As a consequence, spring melt runoff has been declining [3]. Precipitation changes are more ambiguous in pattern and amplitude, but also contribute to changes in runoff variability. The impact of climate change on southern Albertan watersheds has been investigated by several authors [1,2,3,4,5]. The trend in runoff due to climate variability in the Castle River watershed was analyzed by Rood [5] and Byrne [1].

Land cover changes can also impact the watershed's runoff regime and discharge amounts. Wildfires are an important natural occurrence on the eastern slopes of Alberta [6] and have always played an integral role in the functioning of ecosystems in Canadian forests [7]. However, over the past three decades there has been a distinct trend in increase of areas burned by wildfire [8]. Flannigan *et al.* [9] examined the effect of climate change on wildfires and emphasize that fire projection models combined with climate change predictions indicate not only an increase in the area burned but also in fire season length, the intensity of fires and thus in burn severity [7]. Pierson, *et al.* [10] investigated the impact of wildfires on watershed hydrology and detected a reduced infiltration and an increased erosion in a steep rangeland study side. According to Silins [11], after the 2003 Lost Creek fire in southern Alberta, important changes in the hydrology of the affected rivers could be measured; mean annual flows increased due to reduced actual evapotranspiration, as did peak discharges and sediment concentration.

To analyze changes in discharge patterns due to environmental change, hydrological modeling is broadly applied. Currently, special focus in environmental modeling is set on the investigation of the impacts of climate variability, land cover changes or human induced land degradation [12]. To provide input data for hydrological modeling or to detect changes in land cover, satellite remote sensing is a recommended and widely used method. Landsat data are available at high spatial and temporal resolution and deliver information in a broad range of wavelengths especially suited to distinguish different land cover [13]. Thus, different types of patterns in land cover can be detected as well as variations when applying a multi-temporal change detection method. Particularly changes of vegetation characteristics can be distinguished by using infrared bands [13]. In this study, the physically based Water Balance Simulation Model (WaSiM)

Citation: Springer, J.; Ludwig, R.; Kienzle, S.W. Impacts of Forest Fires and Climate Variability on the Hydrology of an Alpine Medium Sized Catchment. Hydrology 2015, 2, 23-47.

is applied to simulate the hydrological processes in the Castle watershed. The effect of the fire on the entire Castle River watershed has neither been investigated nor modeled. Thus, this study provides an analysis whether the Castle River shows a reaction in streamflow after the fire and if the model is able to simulate runoff behavior with a changed land cover after wildfires.

The study intends answer to the following questions: How has the hydrology of Castle River been changing during the last decades and what are the reasons for that? Is the hydrological model WaSiM able to represent the hydrological conditions? How severe was the wildfire in 2003 and can the effects on hydrology be observed and simulated with an adjusted parameterization?

The objectives of this study are to (i) analyze characteristics and trends of measured climate and streamflow data; (ii) calibrate and validate the hydrological model WaSiM for the watershed; (iii) map burn severity from satellite imagery for a severe wildfire in 2003; and (iv) simulate the impact of climate variability and forest fires on runoff behavior in the Castle River watershed.

STUDY AREA

The Castle River watershed is situated at the eastern slopes of the Rocky Mountains in southwestern Alberta (Figure 1). It is located between the Waterton Lakes National Park in the south, the Crowsnest River watershed in the north and the border to British Columbia in the west. In the east, the Castle River drains into the plains and provides water for a wide variety of purposes, in particular for irrigated agriculture [14].

The gauge is situated near the town of Beaver Mines (at 49°29'19' North and 114°8'39' West) and defines the watershed for this study with a size of 820.70 km^2 [15]. The watershed has an elevation difference of more than 1600 m as the gauge is at 1187 m and the Castle Mountain has an altitude of 2766 m [16]. The mean annual streamflow of the Castle River at the Beaver Mines gauge is 15.57 m^3·s^{-1} (runoff data from 1960 to present available), with an explicit maximum in May and June and minimum runoff in the winter months. The peak of the hydrograph in early summer reflects the convergence of the period of peak melt and the most rainy period of the year in June [11].

The Castle basin has frequently been affected by wildfires [11]. The Lost Creek wildfire in summer 2003 was one of the most severe in the upper eastern slopes in many decades and burned 21,000 ha in both the headwater regions of Crowsnest and northern Castle River [17], where it spread in the catchment areas of Lynx Creek and Lost Creek. The 2003 fire perimeter is indicated in Figure 1. The fire burned in all structural forest strata as a so-called crown fire and expanded quite quickly so that the landscape was changed significantly. As the Lost Creek fire took place along two of the tributaries of the Oldman River, *i.e.*, the Castle and the Crowsnest Rivers, local and regional water-based resources were impacted through these forest disturbances [11].

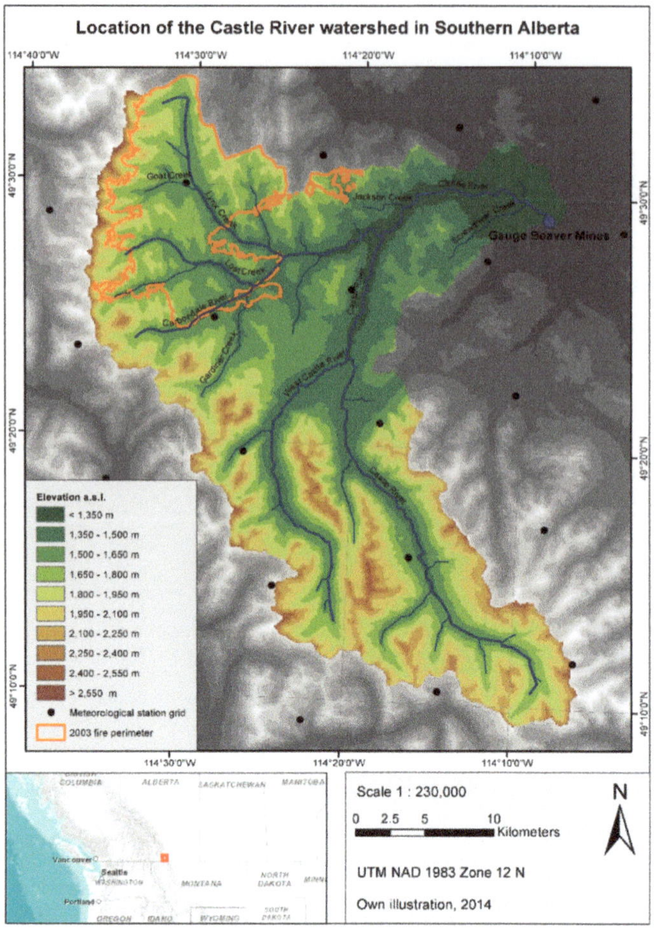

Figure 1: Location of the Castle River watershed in the Southern Rocky Mountains, indicating the burned area of the 2003 Lost Creek fire and the applied meteorological station grid available from AAFC [18].

DATA AND METHODS

Climate and Streamflow Observations and Trends

Historically, meteorological monitoring in the eastern slopes of the Rocky Mountains is extremely sparse [4]. The only available data in the area are interpolated daily minimum and maximum temperatures (in °C) as well as daily precipitation sums (in mm) from AAFC [18] from 1950 to 2010. This data is organized in a regular station grid with a spacing of 10 km, which contains interpolated point estimates. It is derived from an Anusplin interpolation method of Environment Canada climate stations south of 60°N in which a thin-plate smoothing spline surface fitting method was implemented using longitude, latitude and elevation [19]. The situation of the station grid points is represented in Figure 1. Hence, the meteorological data used in this study represent variations from real measured values but they demonstrate general patterns and trends. Climate features are analyzed over a time period of 61 years to investigate the data regarding a climate change trend. To analyze mean temperature characteristics in the region of Castle River watershed, a mean of all 26 stations values is calculated for every day. The same procedure is applied to the precipitation data. Additionally, to analyze spatial variability in the watershed, spatially distributed analyses are accomplished for which the individual station values are considered. Stream discharge data from Castle River are available at the gauge near the town of Beaver Mines, operated by Environment Canada [15]. The location of the gauge is indicated in Figure 1. Daily runoff values from 1960 to 2010 are analyzed regarding runoff regime, quantitative or temporal changes and trends.

Hydrological Modeling with WaSiM ETH

The hydrological modeling in this study is executed by the mainly physically based, distributed and deterministic hydrological model WaSiM (Water Balance Simulation Model) [20]. It runs in continuous time steps (one day in this study) and the spatial resolution is a raster of constant grid cell size, which is 100 m in this study. WaSiM allows short-term flood event runoff simulations as well as long-term water balance applications [21]. According to Beckers, *et al.* [22], it delivers very high model functionality and complexity for forest management and climate change applications and is, therefore, applied in this study. A detailed description of all modules, algorithms and the functioning of the model is specified in the WaSiM Model Description by Schulla [20]. In this application the model version 9.1.0 is applied, which integrates the Richard-equation.

WaSiM Input Data

WaSiM requires meteorological and spatially distributed input data. The main meteorological input data for an optimal configuration are spatially explicit temperature (°C), precipitation (mm/time step), wind speed ($m \cdot s^{-1}$), relative humidity (1/1) or vapor pressure (mbar) and either sunshine duration (1/1) or global radiation ($Wh \cdot m^{-2}$) for each time step. The data are acquired from several meteorological station points distributed in and around the considered watershed so that an interpolation in WaSiM delivers a comprehensive meteorological information.

As AAFC [18] provides minimum and maximum temperature and precipitation data in daily resolution [19] but WaSiM requires daily mean temperature, the mean values are calculated with the equation (Tmax + Tmin)/2, according to Klein Tank, *et al.* [23].

For the study area wind speed, global radiation and relative humidity are only available as monthly means from a spline interpolation of climate normals from the National Climate Data Information Archive [24]. Considering wind speed, in this study the available monthly mean values are used instead of daily ones. For relative humidity and global radiation an estimation of daily data based on air temperature and precipitation data is applied using equations after Thornton, *et al.* [25] and Bristow and Campbell [26]. The meteorological data are thus supplemented by information about the geographical situation and altitude of each station.

Temporal constant geographical data in a grid format are required for a digital elevation model (DEM), land use and soil types. The DEM [27] contributes information about the elevation, the slope, the exposition and the curvature of the surface. Additionally, hydrologic information is derived from the DEM using the program Topographic Analysis (TANALYS) from Schulla [20]. Land cover information in 100 m spatial resolution for the Castle watershed are provided by the GeoBase Land Cover Product [28]. However, an official soil map with soil texture for the Castle River area is not available. WaSiM requires information about the soils in order to calculate infiltration and evaporation processes which depend essentially on the soil texture [20]. Thus, a soil texture map is generated, mainly based on the land cover information and the predominant soil types in different vegetated areas, according to the Canadian Society of Soil Science [29]. In the application of the model, the different soil texture classes obtain specific attributes in several soil layers like the hydraulic conductivity, the saturated and residual water content and other parameters that depend on the grain-size composition of the soil [20].

Calibration and Validation

Calibration of WaSiM is executed gradually. Sensitive model parameters to be calibrated in the Richards-Approach version of WaSiM are the soil model parameters of the unsaturated zone, especially the scaling parameter for interflow and the recession parameter for base discharge for saturated hydraulic conductivity with increasing soil depth (controls peak runoff and base flow) [12].

The calibration period is defined from 1 November 1960 until 31 October 1970, so it equals 10 hydrological years. A comparison of the modeled and observed data sets is conducted and the hydrological goodness of fit criteria NSE and R^2 are calculated in daily and monthly resolution [30,31,32]. Furthermore, the mean annual runoff (MQ), the monthly mean runoff (MMQ), as well as mean high (MHQ) and mean low flow (MLQ) values are compared. Calibration steps are iteratively conducted until a maximum in model efficiency is reached.

An elevation dependent correction for precipitation data is applied separately for rain and snow to derive altitudinal gradients. A threshold temperature between rain and snow have to be chosen for which the value 1 °C showed the best model performance, which is also the threshold temperature for rain in the snow model. The method used is a combination of inverse distance weighting and an elevation dependent regression with internal processing making use of all available meteorological observations. The stomatal resistance parameter was calibrated and needed to be increased to obtain realistic values for actual evapotranspiration. During this calibration process the model efficiency is increased until a stable parameter set is gained.

In order to validate the established model the calibrated parameter set is applied for the following 30 years. Validation runs are executed for each of the decades from 1970 to 1980, from 1980 to 1990, from 1990 to 1999 and from 2000 to 2010.

Remote Sensing of Forest Fires

Spatial wildfire data from ESRD Alberta [33] delivers perimeters of historical big fire events in Alberta. The shapefile polygons are derived from aerial photos and show, beside other fire events, the expansion of the wildfire in 2003, however they lack information on burn severity, which is critical to assess the impact of fires on hydrological variables.

Data

To determine the grade of damage of the burned land cover, infrared Landsat satellite data are applied for vegetation change detection as these wavelengths react sensitive for water content and chlorophyll [34]. Landsat Thematic Mapper 5 (TM5) and Enhanced Thematic Mapper+ (ETM+) data, provided in GeoTIFF format, are used at a spatial resolution of 30 m. Pre- and post-fire images from the 2003 wildfire are chosen according to smallest cloud cover and seasonal similarity to ensure analogous phenology and minimized sun angle effects. Pre-fire imagery is taken from 2002, post-fire imagery stems from shortly after the wildfire was extinguished.

Normalized Burn Ratio and Change Detection Algorithm

The Normalized Burn Ratio (NBR) has been developed for the detection of fire scars [35]; it is explained in detail by Miller and Thode [13] and Miller, *et al.* [36]. The NBR is computed from Landsat bands 4 and 7, the near and short wave infrared bands [36]. Analyzing pixels from pre- and post-fire images, both bands demonstrate considerable changes especially in forested regions. The reflectance in band 4 decreases because of a loss of the photosynthetic activity while the reflectance of band 7 increases due to less water absorption.

Miller and Thode [13] emphasize that vegetated pre-fire areas have different NBR values as the value refers to the density in vegetation cover with its chlorophyll and water content. Hence, to generate comparable fire severity values, a change detection method (differenced Normalized Burn Ratio (dNBR)) is applied in which each post-fire NBR is subtracted from the pre-fire NBR [36]. Low dNBR values indicate unchanged landscape or low fire severity, high dNBR values represent severe damage to vegetation through the fire [13].

The dNBR is meant to normalize the results to pre-fire vegetation cover. However, in several studies the relationship between measured severity values and calculated dNBR has been tenuous [37,38]. Miller and Thode [13] finally present a relative version of the dNBR which includes the previous vegetation density before the fire and minimizes the error caused by pre-fire vegetation. The RdNBR is defined as the relationship between the vegetation killed through the fire and the amount of pre-fire vegetation. Thus, positive RdNBR values represent a decrease in vegetation cover and consequently a high burn severity; negative values indicate an increase of vegetation and thus characterize unburned areas [36]. According to existing thresholds derived from measured data after fire events from Miller and Thode [13] four burn severity classes are defined from the RdNBR calculation.

Impacts of Forest Fires on Runoff

Background

Fire severity is the greatest determinant on the impact of a fire on streamflow generation. The changes in annual water budgets can last for decades in forested regions [39]. Severe fire events can reduce infiltration rates as a consequence of the generation of a hydrophobic soil layer caused by high temperatures or soil sealing forming crusts after a fire [10,39]. The decreased infiltration has an immense effect on post-fire runoff especially for extreme precipitation events during convective storms, when overland flow rises. The soils generally recover within a few years [39]. The long-term effects of forest fires on hydrology are caused by a reduction in evapotranspiration due to the loss of biomass, similar to those effects occurring after harvesting. Interception of rain and snow is also reduced. During the winter the snow melt rates are higher due to reduced shading by canopy [39].

The Southern Rockies Watershed Project [11] analyzed the impact of disturbance by the 2003 Lost Creek wildfire on hydrology, water quality and ecology on several reaches affected by the fire. The results are an increase in snow packs (54% higher in burned forest stands compared to unburned) and snow density, higher runoff values, especially in low flow periods during summer due to less evaporation and interception (by 14% compared to mean values from the 1990s), and peak runoffs are produced more rapidly in burned watersheds. Furthermore, an earlier spring streamflow peak was identified as well as a second peak during the highest precipitation in June. To analyze whether wildfire impacts at smaller reaches propagate in the runoff behavior of the whole Castle River watershed further investigation is carried out.

Data and Methods

To analyze whether the established WaSiM model can reproduce the observed runoff under changed land cover conditions after the wildfire, further analyses are conducted. The pixels identified as *burned* using the RdNBR method are now reclassified to depict the land cover conditions after a wildfire. Previously, land cover in the burned area was mainly coniferous forest which contributes immensely to evapotranspiration. Due to the lack of on-site information about the conditions after the fire, four replacement options are tested in the burned area: (1) no change; (2) barren soil; (3) grass and (4) shrubs, for which individual land use parameter settings are provided in the hydrological model. Specific values for albedo, stomata resistance, leaf area index and root depth must be provided to calculate evapotranspiration [20]. The parameter values for the four classes are presented in Table 1. According to Schulla [20] these are the most important parameters and the values vary between the four classes. It is notable, that barren land has mostly constant values as vegetation is missing.

First a model run from August 1993 to July 2003 is carried out with the original land cover to set the initial state of the hydrological conditions for the time the fire began. With these initial conditions, model runs for the first year after the fire from August 2003 to October 2004 are executed with the four land cover properties applied for the burned area. Subsequently, the different applied land cover classes in the burned areas are assessed regarding the model efficiency of the produced runoff.

Table 1: Selected default parameter values defined in the WaSiM landuse table for the four different land cover classes in the burned area. A seasonal evolution of vegetation can be defined with a high temporal resolution for the parameters in the table.

Parameter in (Landuse_Table)	Explanation	Previous Land Cover (Coniferous Forest)	Barren Land	Grass	Shrub
Albedo	Albedo	0.12	0.25	0.25	0.2
rsc (s/m)	Leaf surface resistance	220–320	400	200–360	200–320
rs_interception (s/m)	Evaporation of intercepted water	5	0.5	5	5
rs_evaporation (s/m)	Evaporation of water from the soil surface	1000	200	600	1000
LAI (1/1)	Leaf Area Index	6–10	1	2–3	3–5
Z0 (m)	Roughness length	3	1	0.03–0.04	0.2
VCF	Vegetation covered fraction	0.9–0.95	0.2	0.8–0.9	0.9–0.95
RootDepth (m)	Root depth	1.2	0.2	0.4	0.5

RESULTS AND DISCUSSION

Observed Climate and Streamflow Trend Results

AAFC data of monthly mean air temperatures (1950–2010) in the study basin range from −8.53 °C in January to 14.24 °C in July. The annual mean temperature is 2.66 °C with a standard deviation of 9.65 °C. Within the watershed there are extreme variations from these annual mean values as the elevation difference exceeds 1600 m, with measured values ranging from 4.51 °C (1213 m) to 0.01 °C (2139 m).

An indication of a warming climate is shown in Figure 2 by a linear increasing trend in yearly temperature by *ca.* 0.03 °C*i.e.*, 1 °C in 30 years. A general temperature increase is evident throughout the whole watershed. Temperatures in the lower section of Castle River watershed rose more considerably (2 °C) as compared to the mountainous parts (1 °C). Minimum and maximum temperatures are rising at a greater rate than the annual means which may indicate a shift towards less cold extremes and increasing hot day events.

Considering the monthly mean temperatures during the last 60 years the winter temperatures, especially in January have increased by about 4 °C in the last three decades. Monthly mean temperature data are shown in Figure 3. The spring temperatures also rose by *ca.* 2 °C in the 1990s and 2000s compared to the middle of the 20th century. In July, the monthly mean temperature in the 2000s was about 2 °C warmer than in the other decades and the August temperatures have also risen during these last decades. There is no distinct temperature trend visible in the fall.

The illustrated temporal and spatial trends of the temperature data from AAFC [18] do generally agree with the conclusions from the IPCC [40]. In the southern Rockies, especially temperatures between December and April have increased by up to 2.5 °C during the last six decades. This trend is also reflected by data from the Castle area in Figure 2.

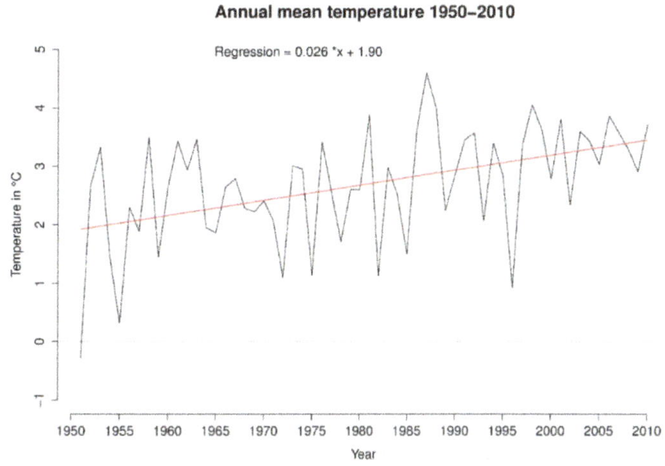

Figure 2: Annual mean temperatures between 1950 and 2010. The linear regression shows an increasing trend in temperatures

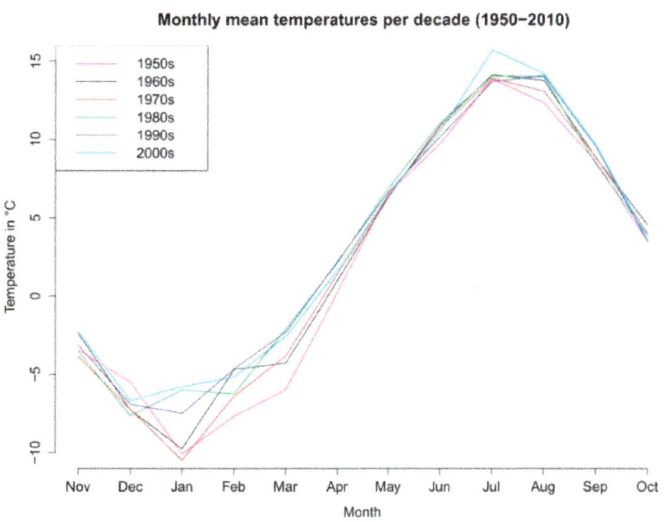

Figure 3: Monthly mean temperatures per decade (1950–2010).

Mean annual precipitation in the Castle River watershed is estimated at 686 mm·a^{-1} [18], with a high variability within the watershed ranging from *ca.* 600 mm·a^{-1} in the lower valleys to over 1000 mm·a^{-1} in the mountains [6]. Highest precipitation is observed in June (average of 86.8 mm) with a minimum in July

(46.4 mm) or during winter [18]. Analyzing precipitation data from 1950 to 2010 reveals the following trends: The annual precipitation sum is generally slightly increasing with a high interannual variability. The seasonal distribution of precipitation is changing, even though the patterns are more ambiguous than the temperature data. Winter precipitation has been decreasing in the last six decades while the summer precipitation is rising with increasing temperatures. The 1980s showed significantly lower precipitation compared to other decades (annual mean: 655 mm·a^{-1} in the 1980s *vs.* 692 mm·a^{-1}). In the 1980s, winters, springs and summers were drier and warmer. The reasons for this relative drought were probably weaker west winds and a series of stable high pressure fields which blocked out storms [41].

Spatially distributed precipitation is high in the mountainous areas of the watershed in the south and west and is declining towards the northeast. The trend from 1950 to 2010 amplifies this pattern, as the mountains became more humid with increases of the annual precipitation sum by about 50 mm and the lower lying areas became drier.

The derived precipitation trend results are similar to those presented in the IPCC [40], which show an annual precipitation increase by 12% on average but in the southern Rockies this trend is weaker [42]. Significant changes can be observed in terms of temporal distribution of precipitation which conform to those of Castle watershed. Heavy precipitation frequencies increased as well in the last 60 years [42,43].

Considering gauge data at Beaver Mines from Environment Canada [15], the mean runoff has not been constant throughout the last 45 years. In the 1960s the mean annual flow was 15.1 m^3·s^{-1} (580 mm·a^{-1}), then it declined until, in the 1980s, it was 13.2 m^3·s^{-1} (507 mm·a^{-1}). In the 1990s it increased to 14.8 m^3·s^{-1} (569 mm·a^{-1}) and then decreased again to 13.3 m^3·s^{-1} (511 mm·a^{-1}) between 2000 and 2010. The monthly low and high flows show a corresponding behavior through the decades.

Overall, when applying a linear regression, the data shows a trend of decreasing annual mean runoff values over the considered decades. The observed annual mean runoff and the derived trend in a 5-year moving average for the whole period is shown in Figure 4. From 1962 to 1999, runoff decreased about 20%, which corresponds to an annual decrease of *ca.*0.53%. This reduction of annual river flow conforms to the investigations from Byrne [1] who analyzed streamflow of the Castle and Oldman Rivers. They derived a general decline in mean streamflow of the Oldman River watershed by about 26% since 1949. The reason for this trend is attributed to the reduction of snow packs in the watershed due to the identified rising temperatures. Although precipitation in winter is increasing, this cannot compensate for the warming and thus more precipitation falls as rain than as snow. Due to less melting spring runoff is reduced [1], even at a stronger rate than in other comparable watersheds in the North American Rocky Mountain region.

Considering the runoff trend for the decade 2000 to 2010 a slight increase is detected. This can mainly be explained by extraordinary high precipitation in 2002 with 800 mm·a^{-1} and over 1050 mm·a^{-1} in the year 2005 [18].

Figure 5 shows the monthly mean runoff values per decade. The declining runoff values in May and June are an indicator of reduced snowpack, as less snow supplies less melt water for runoff generation. It is noticeable that the 1980s reflected a much lower runoff rate than the other decades, especially between May and August due to the relative dryness in this decade. Apart from the 1980s, the peak streamflow in June has constantly declined. In the other seasons, a distinct trend is not visible.

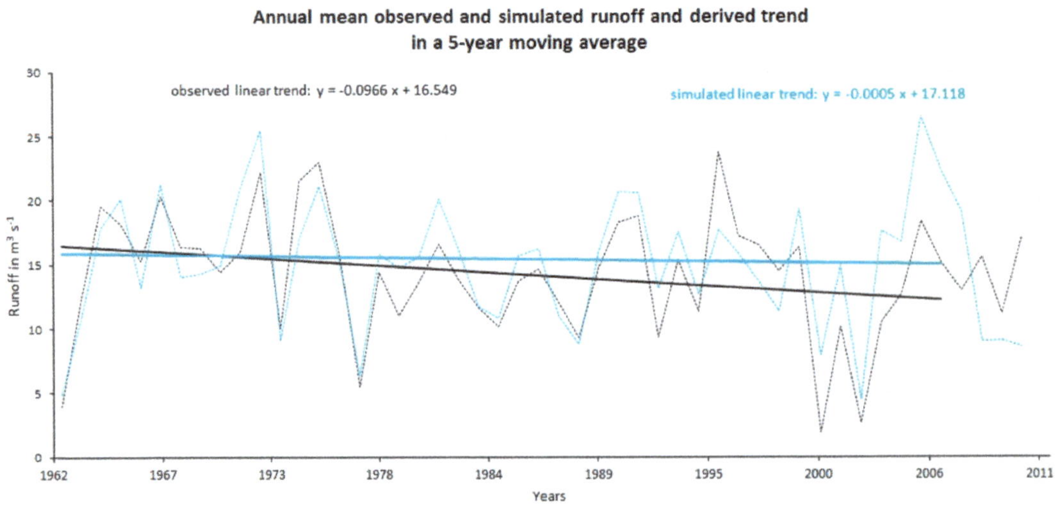

Figure 4: Annual mean values of observed and simulated streamflow at the gauge Beaver Mines at Castle River. The derived linear trends are calculated with a 5-year moving average from 1962 to 2010.

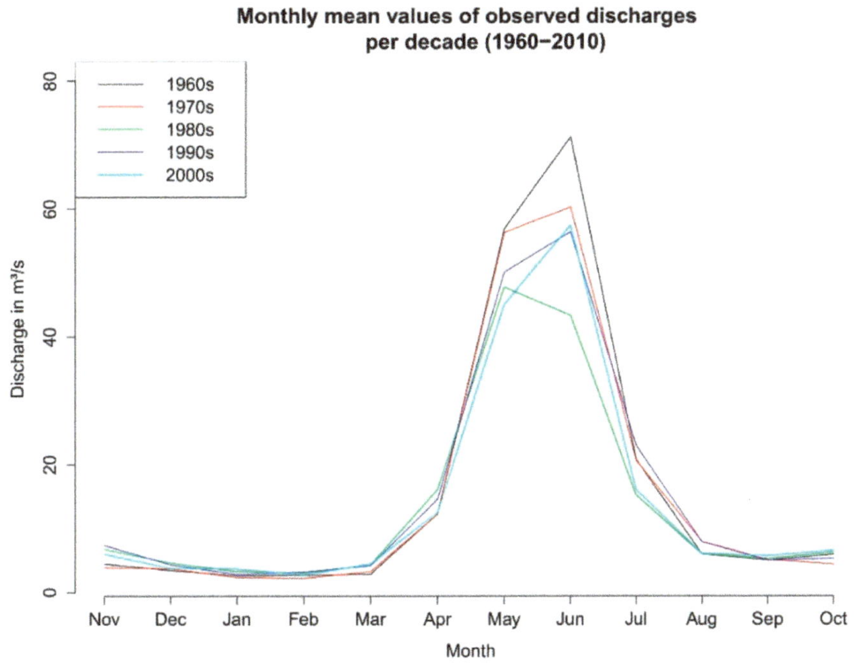

Figure 5: Monthly mean values of observed discharges per decade (1960–2010).

Calibration and Validation Results of WaSiM

First, time series of the simulated runoff in calibration and validation periods are discussed. Second, an analysis of the spatial output grids for water balance calculation and runoff components is executed.

Calibration and validation results are listed in Table 2. The result of the best calibration run from 1960 to 1970 is shown in Figure 6. Daily and monthly mean observed and simulated data are represented as well as the precipitation.

Table 2: Calibration and Validation goodness of fit with MQ (mean runoff), MHQ (mean high flow) and MLQ (mean low flow) in $m^3 \cdot s^{-1}$ (first two years of initial time in model run are not considered in efficiency estimation).

	Period	Data	MQ	MQ Relative Error	MHQ	MLQ	NSE	R^2
Calibration:	1960–1970	observed	15.47		87	1.20		
		simulated	14.75	−5%	106	3.26	0.81	0.82
Validation:	1970–1980	observed	13.38		94	1.61		
		simulated	13.58	+1%	95	3.18	0.81	0.82
Validation:	1980–1990	observed	12.16		93	2.09		
		simulated	12.98	+6%	110	3.29	0.76	0.87
Validation:	1990–1999	observed	14.87		91	1.65		
		simulated	14.13	−5%	123	4.01	0.76	0.76
Validation:	2000–2010	observed	13.27		90	1.36		
		simulated	15.55	+15%	89	4.92	0.45	0.63

Model efficiency is assessed according to Moriasi [30] who suggested a performance rating in four NSE classes: (i) Very good ($0.75 < NSE \leq 1.00$); (ii) Good ($0.65 < NSE \leq 0.75$); (iii) Satisfactory ($0.50 < NSE \leq 0.65$) and (iv) Unsatisfactory ($NSE \leq 0.50$). Generally, the efficiency of the established WaSiM model is very good in daily (NSE: 0.81, R^2: 0.82) and in monthly mean temporal resolution (NSE: 0.88, R^2: 0.89), as is listed in Table 2. The mean annual flow is underestimated while mean high and low flows are overestimated. For the model validation run from 1970 to 1980, the accuracy is the same as in the calibration. R^2 also stays high between 1970 and 1999; from 1980 to 1990 it goes up to 0.87. It is conspicuous that the model efficiency decreases considerably to an NSE of 0.45. This result shows that a further investigation of modeling in this decade with the severe wildfire in 2003 has to be accomplished (3.3 and 4.5). The mean low flows are constantly overestimated in the validation periods. The mean high flows are overestimated but are modeled in the magnitude of the observed values.

Generally the set up WaSiM model is able to reproduce runoff time series of the Castle River watershed and provides consistent simulation quality in a *ca.* 40 year period with very high accuracy.

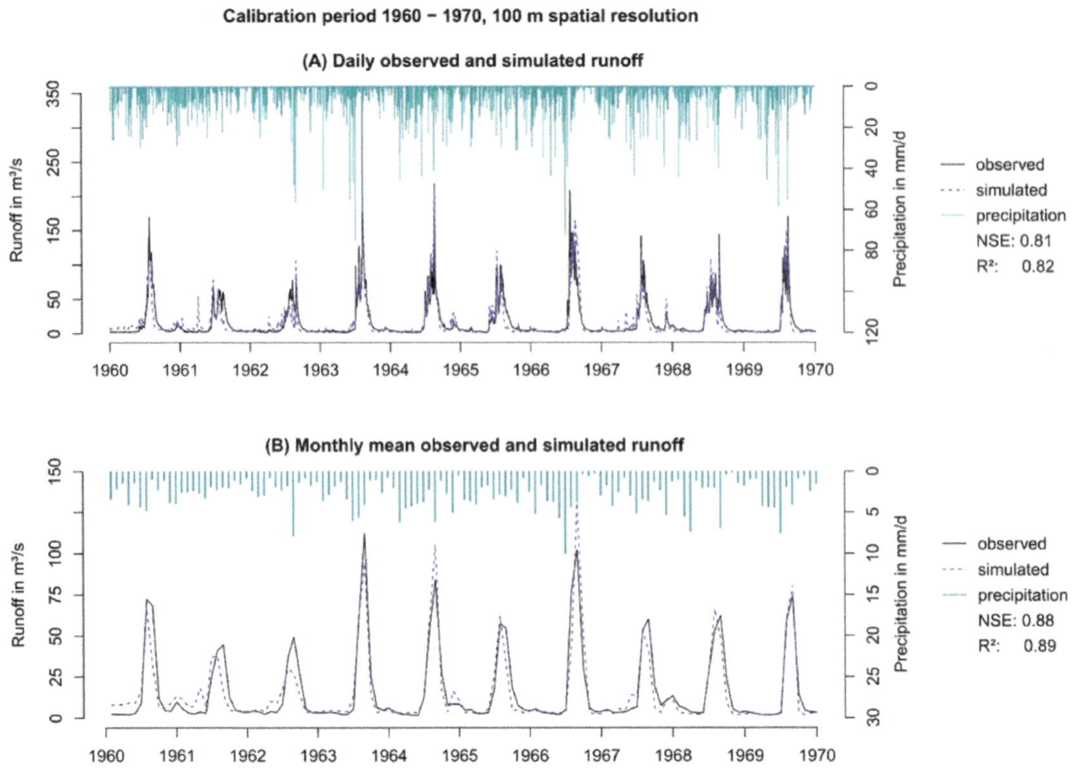

Figure 6: (**A**) Daily and (**B**) monthly observed and simulated runoff in the calibration period 1960–1970.

In addition to the temporal and the quantitative accuracy of the model, spatial output data of the three runoff components is analyzed. Especially the direct flow but also the interflow is mainly generated in the mountainous parts, where precipitation is higher. Direct flow is very high in non-vegetated rocky summit regions. Base flow is mainly built in regions where the soil is more developed and is especially visible in the valleys. These spatial patterns seem to reflect real conditions very well. The interflow contributes the largest amount of water to the streamflow with over 440 $mm \cdot a^{-1}$, whereas the direct flow and the base flow together only account for around 150 $mm \cdot a^{-1}$. Base flow contributes the highest amount of runoff in the low flow periods. Interflow is especially high during snow melt in the spring to early summer whereas the direct flow reacts to high precipitation events.

A water balance [44] is calculated for the calibration period from 1960 to 1970 and presented in Figure 7. The mountainous precipitation is increased after the precipitation correction method in the model. According to Silins [11] who conducted a few meteorological measurements in the northern reaches of the Castle watershed and detected annual precipitation values of over 1200 $mm \cdot a^{-1}$ the derived precipitation in the mountains can be claimed to be appropriate. However, the precipitation in the lower regions of the watershed is thereby overestimated. The mean annual precipitation at the meteorological station Beaver Mines, close to the gauge, amounts to around 650 $mm \cdot a^{-1}$ [24] whereas the model output defines at least 910 $mm \cdot a^{-1}$. Nevertheless, the precipitation correction is applied, as without it the model efficiency is significantly lower (NSE: 0.40, R^2: 0.68) and the mean flow of the Castle River is significantly underestimated (5.5 $m^3 \cdot s^{-1}$ compared to the observed 15.5 $m^3 \cdot s^{-1}$). Most of the runoff is produced in mountainous areas of the watershed, where precipitation correction delivers more appropriate values. Furthermore, most of the snow is produced in the mountains and the evapotranspiration has its maximum in the lower areas. Thus, the derived mistake by an overestimation of precipitation downstream can be neglected, as its contribution to the streamflow generation is low.

The evapotranspiration varies between values of *ca.* 500 $mm \cdot a^{-1}$ to more than 750 $mm \cdot a^{-1}$. Due to low transpiration the calculated evapotranspiration is lower in non-vegetated areas such as the higher altitudes. In the forested valleys and in the agricultural used areas in the northeast, very high evapotranspiration values can be detected. The mean runoff for the period 1960 to 1970 is 596 $mm \cdot a^{-1}$. It is obvious that the highest runoff values (up to over 850 $mm \cdot a^{-1}$) are generated in the mountain tops as precipitation and thus snow cover is higher there. The lowest contribution to the runoff is generated in the agricultural areas because of lower precipitation and higher evaporation.

The results of the established WaSiM model show that it is able to reproduce runoff with high accuracy over a period of 40 years. This also implies that the partially calculated or generated input data is relatively accurate and that calibration has been successful in this study.

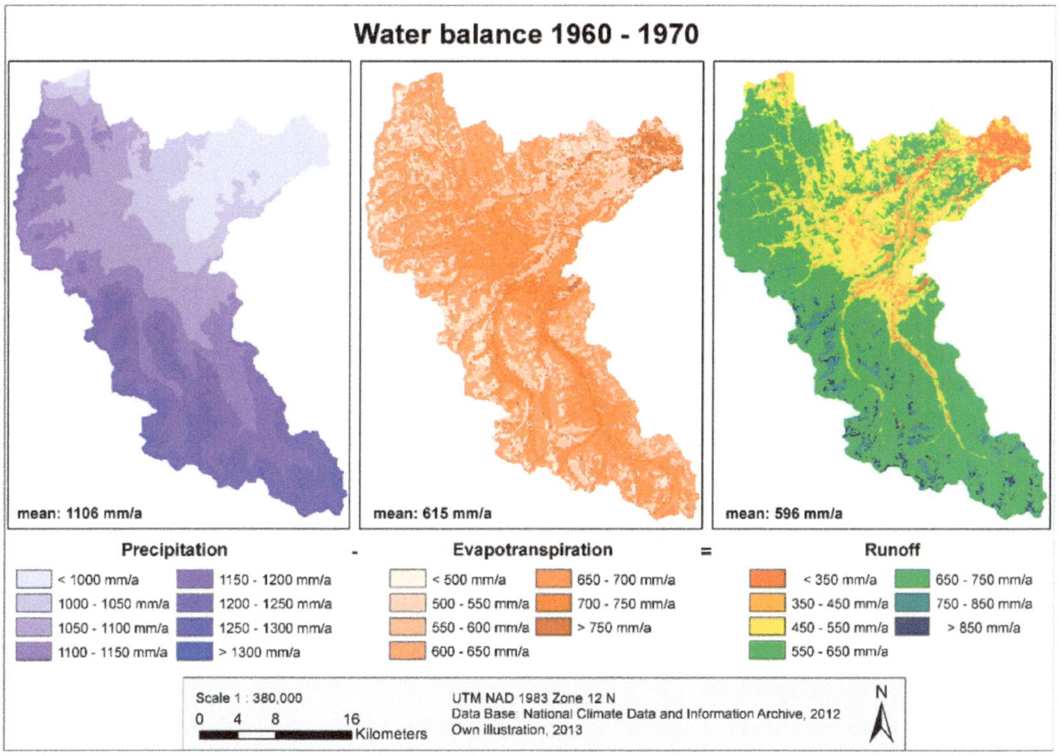

Figure 7: Simulated mean annual water balance in the calibration period from 1960 to 1970. The values are derived from the spatial output data of WaSiM and represent interpolated precipitation, as well as the modelled evaporation and runoff.

Comparison of Observed and Simulated Declining Runoff Trend

The observed and simulated runoff in an annual mean is demonstrated in Figure 4.

Analyzing the trends in total runoff between 1960 and 1999, the observed data show a linear decrease of *ca.* 1 $m^3 \cdot s^{-1}$ (38 mm) in 12 years. Therefore, in the 40 years considered, the observed streamflow declined by about 3.1 $m^3 \cdot s^{-1}$ which equals a decrease of 119 $mm \cdot a^{-1}$, although in the four decades the trend has not been continuous. Regarding monthly mean values the simulated runoff trend has a very similar inclination compared the observed. A decrease of 1 $m^3 \cdot s^{-1}$ in 9 years and a total decline of 4.1 $m^3 \cdot s^{-1}$ (157 $mm \cdot a^{-1}$) between 1960 and 1999 can be derived from the simulation. Low and high flow trends are also decreasing. The decline in snow cover since 1960 is captured by the model. WaSiM generates a total snow storage output which sums the snow in mm per day. The snow cover is varying but the linear trend shows a distinct decline of 53 $mm \cdot a^{-1}$ in an annual mean.

Considering the observed and simulated runoff between 2000 and 2010 a different behavior of the model is evident. Runoff is slightly increasing in observed values while WaSiM produces a decrease of 9.1 $m^3 \cdot s^{-1}$ in the decade when regarding monthly mean values. Figure 4 shows that between 2002 and 2007 runoff is overestimated considerably. The assumption is that the changed condition in the basin after the 2003 wildfire is the reason for the decline in model efficiency (compare Table 2). Thus, further investigation in this decade is conducted in chapter 4.5.

The water balance for the period 1990 to 1999 is calculated and then compared to the water balance from 1960 to 1970. The change maps for simulated precipitation, evapotranspiration and runoff are demonstrated in Figure 8. Considering precipitation, a spatially differenced distribution and an increased amplitude is evident, as the decrease in the lower parts is up to 30 $mm \cdot a^{-1}$, and the increase, especially in the northern and western mountains, is up to 99 $mm \cdot a^{-1}$. This observation conforms with the projections from Field [42] which show an increasing variability in precipitation. Considering evapotranspiration values between the 1960s and 1990s, an overall increase can be seen due to increased temperatures. Nevertheless, in the higher mountainous parts a decreasing evapotranspiration was simulated. A possible explanation for this development could be the decline in snow cover which is represented by the model. Mac Donald, *et al.* [45] emphasized the importance of sublimation of snow cover for the mass balance in a watershed. Therefore, the reduction in evapotranspiration could possibly be explained by the loss of sublimation which contributes to evapotranspiration. As a consequence of the developments in the distribution of precipitation and evapotranspiration, runoff is on average declining between the 1960s and 1990s. In the western, high altitude regions, more precipitation produces more runoff, especially direct streamflow and interflow. In the valleys, increasing evapotranspiration reduces the runoff generation and causes its declining trend.

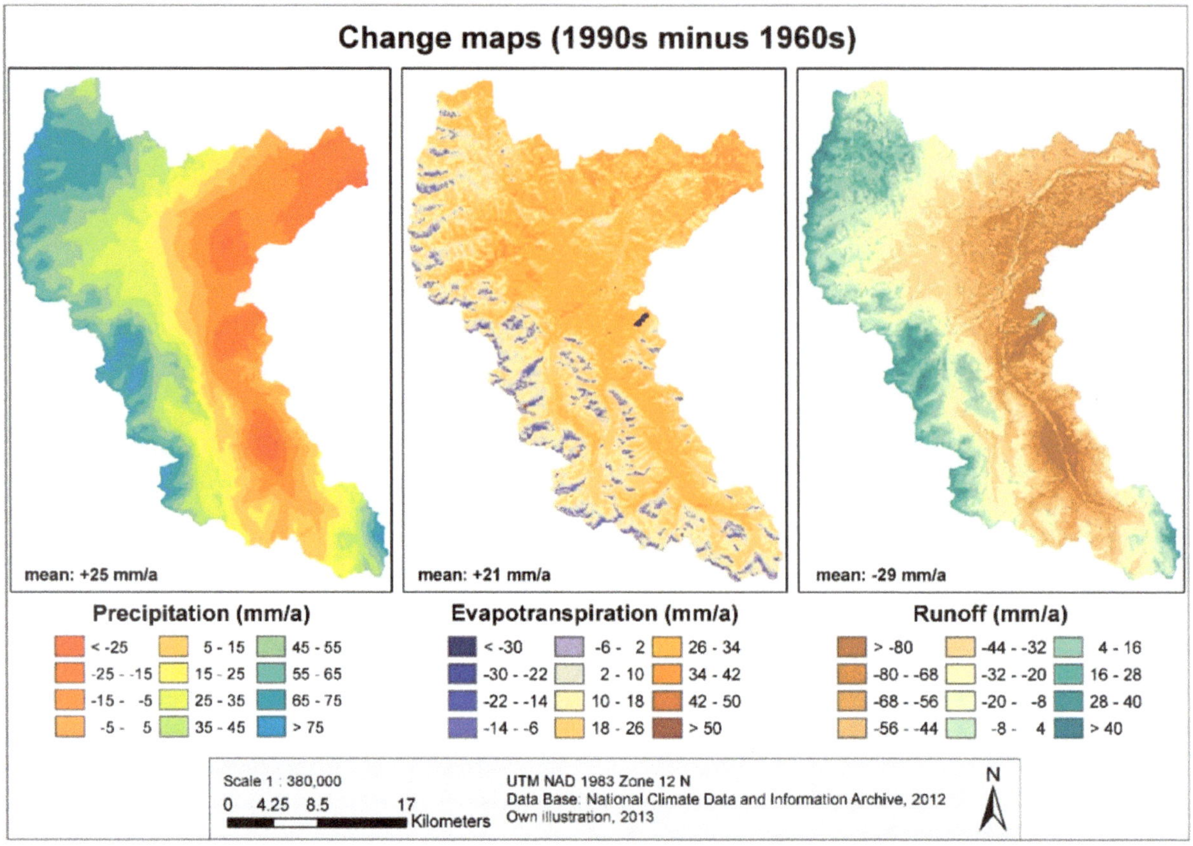

Figure 8: Change maps of precipitation, evapotranspiration and runoff generation between the 1960s and the 1990s.

It can be claimed that WaSiM is able to reproduce quite well runoff values under changing climate conditions. The decline in snow cover and, consequently, in total runoff is reproduced in the model results. Changes in precipitation distribution and quantity, an increase in evapotranspiration and trends in runoff generation in which mainly the interflow is declining are demonstrated with the established model. The intensity of decline is slightly overestimated. Further investigation with measured input data or a tested and approved precipitation correction factor would be necessary to describe the changes to the watershed more accurately.

Results of Burn Ratio and Derived Burned Areas

The result of the Relative differenced NBR (RdNBR) is presented in Figure 9. High RdNBR values signify a decreased vegetation cover and thus a high burn severity. Low or negative RdNBR values indicate an increase in vegetation cover and show unchanged areas [13]. The derived image shows a clear zoning of the Lost Creek fire mainly in the high severity class.

Field measurements and derived burn severity values for the wildfire in the Castle River watershed are not available. Hence, the dNBR and RdNBR thresholds derived from Miller and Thode [13], who analyzed 14 fires in the Sierra Nevada Mountain range in California, USA, are applied. They compared measured CBI (Composite Burn Index) values with dNBR and RdNBR from pre- and post-fire Landsat data and defined thresholds, derived from a regression model.

Applying the RdNBR shows that the derivation of burned areas from Landsat satellite imagery is generally feasible in the Castle River watershed. The area derived by this method agrees quite well with the size collected by ESRD Alberta [33], with a discrepancy of 14%. There are two possible reasons for this difference. Either, the burned area was collected very generally and the RdNBR calculation gives a more precise result for the burned areas, or else there are mistakes within the calculation or the application of the thresholds from [13] are used. Therefore, the definition of burned areas in the severity categories can only be an approximation. For a correct delineation of burn severity categories in Castle River watershed, location-dependent CBI-values would have been necessary. Nevertheless, the presented method has several advantages. The Landsat data are freely available, and long time-series exist so that change detection and post-fire succession analyses are possible. Compared to the simple burned area map from ESRD Alberta [33], Landsat data deliver valuable information about the plant characteristics and changes of vital vegetation occurring with a fire event [46]. Most importantly, the RdNBR method delivers information about different grades of burn severity which are, according to Luce [39], the supreme determinants for the dimension of the impact of a wildfire on the hydrology of a watershed.

For further analysis of the impacts of the extensive forest fire on the hydrology of Castle River watershed, moderate and high severity RdNBR values are included in the land cover map and defined in a "burned" land cover class. These newly derived land cover maps are later applied as input data sets for the hydrological model WaSiM.

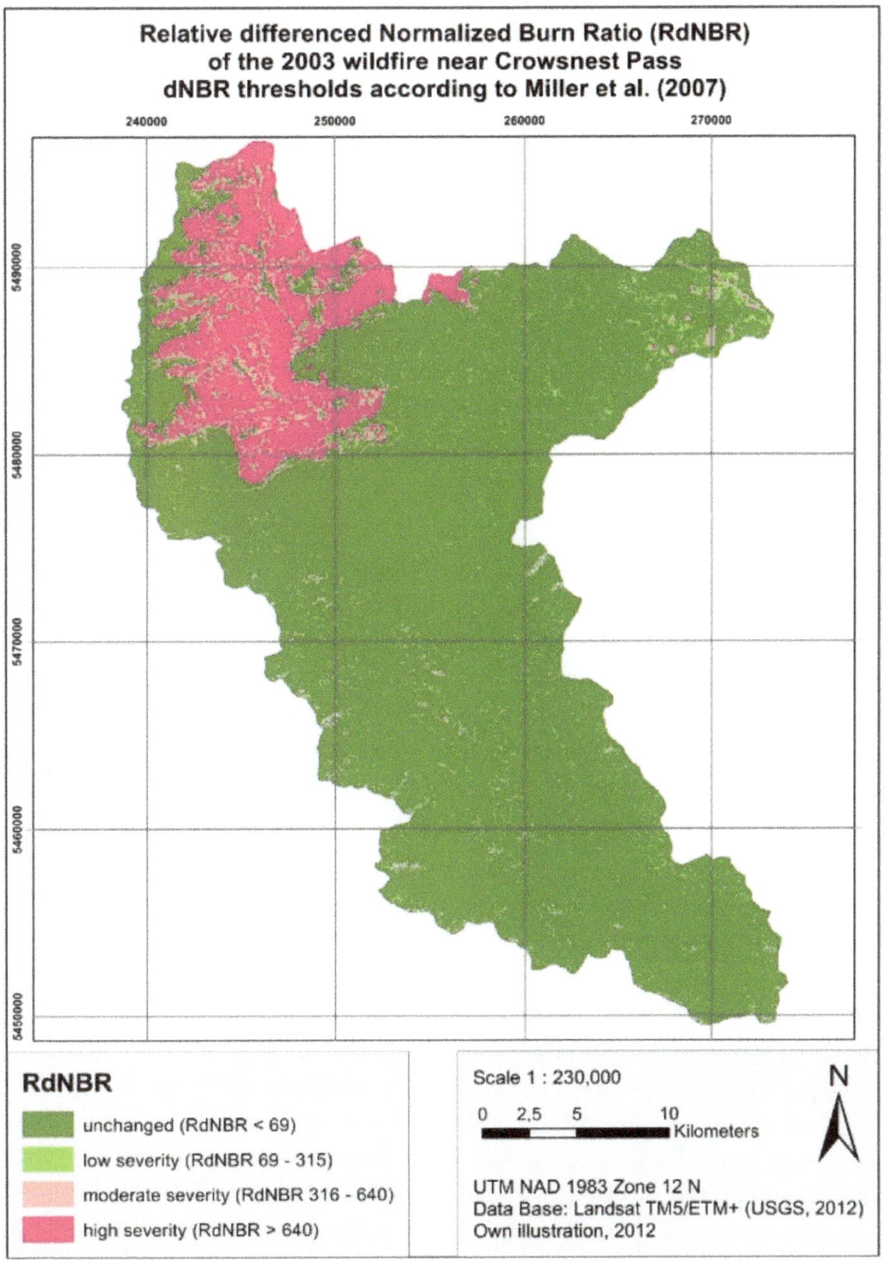

Figure 9: Relative differenced Normalized Burn Ratio (RdNBR) of the Lost Creek fire; thresholds according to [13].

Results of Simulated Runoff Reaction to Fires

The model run from August 1993 to July 2003, with the purpose to produce an initial state for the time the fire began, does not have a considerable lower efficiency than the model runs in the decades before (NSE: 0.76, R^2: 0.82). Simulating runoff after the fire from August 2003 to November 2004 does not lead to accurate results applying the original land cover input for the burned area (NSE: 0.51, R^2: 0.86). The mean annual flow is overestimated considerably by 18%. Thus, it can be assumed that the model allows an accurate runoff simulation only until the fire changed the land cover conditions.

Applying the three other land cover characteristics in the burned area as explained in 3.4.2, NSE efficiency results range between 0.44 with shrub and 0.56 for barren soil characteristics in the burned area. The relative error of the mean annual runoff is around 20%. Barren soil characteristics in the burned area deliver the best results for this time period (NSE: 0.56, R^2: 0.90, MQ relative error: +17%). The simulation results compared with the observed data are shown in Figure 10. Although the mean annual runoff is overestimated under this scenario, satisfactory results can still be achieved, regarding the categorization from Moriasi [30]. Consequently, barren soil characteristics seem to represent a burned vegetation in the first year after the fire more accurately than the other soil characteristics applied. Although dead tree trunks and burned organic litter are present in burned areas, all vegetation based hydrological processes are obviously interrupted. Especially low flow is represented quite well. Melt runoff peaks are characterized quite well but the runoff in July, when highest precipitation values are present, is overestimated by WaSiM.

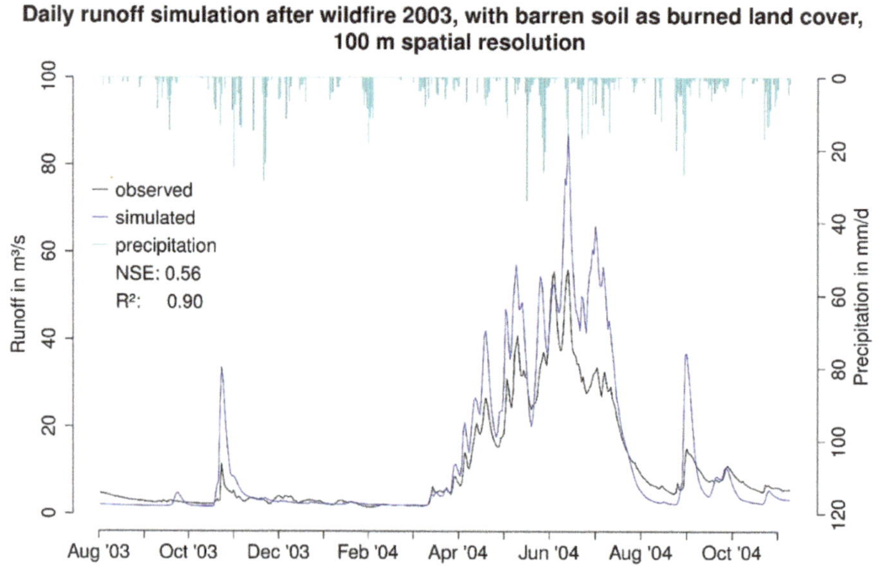

Figure 10: Daily runoff simulation after the wildfire in summer 2003. This graph represents the model run from August 2003 to November 2004 with barren soil as burned land cover classification.

Regarding the simulated total runoff, it is evident that in the burned region runoff generation is slightly increased compared to the unburned area. This is caused by an increase of interflow by *ca.* 8% in the burned area. Due to an absent transpiration without vegetation cover, more water can infiltrate and contribute to the total runoff. This is especially caused by the parameterization of low LAI, low evaporation (rs_interception and rs_evaporation) and high rsc values which are demonstrated in Table 1.

Despite the modifications it is conspicuous that WaSiM still produces less accurate results for the period after the forest fire than in the four decades before. A reason for this can be traced to the change in land cover after the fire. In addition, as the model results showed a reaction to changes in the burned area land cover parameterization it demonstrates that this approach is correct. Hence a new land cover class for burned areas needs to be established.

Another reason why a high model accuracy cannot be achieved for the burned simulation scenario is the climate variability. As shown before, WaSiM reproduces a declining runoff trend between 1960 and 2010 due to changes in climate patterns, but the runoff and the decreasing trend are also overestimated in this period. In the 2000s, precipitation was lower in the winter months but higher in the summer months than in other decades. The applied precipitation correction likely resulted in an overestimation of precipitation in summers in the watershed, which, in turn, alters the annual streamflow and unrealistically increases the peak flows in June and July.

Therefore, generally, a superposed situation is given in Castle River watershed after the 2003 fire. On the one hand a distinct climate trend reduces the freshets, which mainly contribute to runoff generation in southern Albertan alpine catchments. On the other hand, the general behavior after wildfires is an increase in runoff which results in an earlier spring melt flow as well as higher peak flows. Both characteristics can be detected in the observed and modeled runoff values. It is demonstrated by a higher low flow between August and March and an earlier begin of melt possibly due to the fire effects as well as a significantly decreased peak flow as a consequence of reduced snow packs due to climate variability.

CONCLUSIONS AND OUTLOOK

This study investigated the hydrology of Castle River watershed in the southern Albertan Rocky Mountains with particular focus on climate change induced alterations and the impacts of forest fires on streamflow. The initially raised questions and objectives could mainly be achieved. The declining runoff trend, predominantly a consequence of higher evapotranspiration and decreasing snow cover due to rising temperatures, is evident in the Castle River. This trend could also be reproduced by the applied hydrological model WaSiM, which resulted in high accuracy in runoff simulations in the watershed when compared to recorded streamflow records.

The fire severity of the Lost Creek fire in 2003 was detected with pre- and post-fire Landsat satellite images using a differencing burn ratio method (RdNBR). Thus, the area burned by the fire as well as severity classes could be defined very precisely. The impact on the discharge in the watershed is evident in the observed runoff. Nevertheless, a decrease in spring peak flow due to climate induced decreasing snow cover was dominant over the increased peak flow as a consequence of the fire.

Applying the model after the wildfire in 2003 lead to a significantly lower model efficiency. This concludes that WaSiM output is sensitive to land cover changes due to forest fires. An attempt to imitate burned land cover characteristics resulted in increasing accuracy in runoff simulations. Though it is anticipated that a new land cover class which contains parameter values more representative of a burned forest would result in an improved simulation of runoff with WaSiM after a fire. Depending on burn severity a land cover class similar to barren soil would be appropriate, but including features such as e.g. a burned organic layer, changed soil characteristics and dead tree trunks which can still intercept some water. In the following years a dynamic forest succession in the burned areas has to be applied.

This study also shows that it is possible to undertake satisfactory hydrological modeling even if principle input data is missing. The meteorological input data relative humidity and global radiation were estimated based on air temperature and precipitation data, soil information in the watershed was deviated from land use and the temperature and precipitation data were also not measured but interpolated in a broad scale. This confirms the quality of the model WaSiM, which yields plausible results with a coherent parameterization even with poor input data availability.

Further investigation of the outlined research questions is recommended. Future trends are tending towards lower discharge due to decreasing snow cover as a result of climate change [1] as the trend from 1960 to 2010 already represents. Furthermore, an increase in forest fires is expected due to rising temperatures and an extended fire season length [47]. Additionally, changes in land cover either by forest fires or other partly anthropogenic interventions will impact the hydrology in the Castle area in the future. The establishment of gauges of sub-basins and meteorological stations at different altitudes in the watershed would provide significant hydro-meteorological data that is currently estimated [18]. A site-specific elevation dependent regression for correcting measured precipitation could be derived and applied in the hydrological model to obtain spatially distributed meteorological information. A precise soil texture mapping would possibly further improve the conditions to work with the model and produce more accurate results as infiltration processes could be reproduced more precisely.

However, the presented methodological approach is considered robust to implement a scenario simulations framework for projecting the impacts of future climate and land cover change in the vulnerable region of Alberta's Rocky Mountains and comparable mountainous watersheds.

ACKNOWLEDGMENTS

Our thanks go to the Department of Geography, Ludwig-Maximilians-Universitaet Muenchen, and to the Department of Geography, University of Lethbridge, for providing data, methodical support and a working environment with the required facilities and programs. We would also like to mention the enriching cooperation between the two universities and in the project ABBY-Net. The idea for this study was initiated during an ABBY-Net summer school in Alberta.

REFERENCES

1. Byrne, J.; Kienzle, S.; Johnson, D.; Duke, G.; Gannon, V.; Selinger, B.; Thomas, J. Current and future water issues in the Oldman River Basin of Alberta, Canada. *Water Sci. Technol.* 2006, *53*, 327–334. [PubMed]
2. Forbes, K.A.; Kienzle, S.W.; Coburn, C.A.; Byrne, J.M.; Rasmussen, J. Simulating the hydrological response to predicted climate change on a watershed in southern Alberta, Canada. *Clim. Chang.* 2011, *105*, 555–576.
3. Lapp, S.; Byrne, J.; Townshend, I.; Kienzle, S. Climate warming impacts on snowpack accumulation in an alpine watershed. *Int. J. Climatol.* 2005, *25*, 521–536.
4. Kienzle, S.W. Water Yield and Streamflow Trend Analysis for Alberta Watersheds. Available online: http://www.albertawater.com/index.php/projects-research/dynamics-of-alberta-s-water-supply/41-water-research/dynamics-of-alberta-s-water-supply/518-water-yield-and-streamflow-trend-analysis-for-alberta-s-watersheds (accessed on 1 March 2013).
5. Rood, S.B.; Samuelson, G.M.; Weber, J.K.; Wywrot, K.A. Twentieth-century decline in streamflows from the hydrographic apex of North America. *J. Hydrol.* 2005, *306*, 215–233.
6. Sheppard, D.H.; Parkstrom, G.; Taylor, C. Bringing It Back: A Restoration Framework for the Castle Wilderness. Available online: http://www.ccwc.ab.ca/files/BRINGING_IT_BACK.pdf (accessed on 17 December 2012).
7. Wulder, M.A.; White, J.C.; Alvarez, F.; Han, T.; Rogan, J.; Hawkes, B. Characterizing boreal forest wildfire with multi-temporal Landsat and LIDAR data. *Remote Sens. Environ.* 2009, *113*, 1540–1555.
8. Stocks, B.J.; Mason, J.A.; Todd, J.B.; Bosch, E.M.; Wotton, B.M.; Amiro, B.D.; Flannigan, M.D.; Hirsch, K.G.; Logan, K.A.; Martell, D.L.; *et al.* Large forest fires in Canada, 1959–1997. *J. Geophys. Res.* 2003.
9. Flannigan, M.D.; Logan, K.A.; Amiro, B.D.; Skinner, W.R.; Stocks, B.J. Future area burned in Canada. *Clim. Chang.*2005, *72*, 1–16.
10. Pierson, F.B.; Robichaud, P.R.; Spaeth, K.E. Spatial and temporal effects of wildfire on the hydrology of a steep rangeland watershed. *Hydrol. Process.* 2001, *15*, 2905–2916.
11. Silins, U.B.; Kevin, D.; Stone, M.; Emelko, M.B.; Boon, S.; Williams, C.; Wagner, M.J. Howery, Jocelyn Southern Rockies Watershed Project: Impact of Natural Disturbance by Wildfire on Hydrology, Water Quality, and Aquatic Ecology of Rocky Mountain Watersheds Phase I (2004–2008) Final Report. Available online: http://oldmanbasin.org./files/Publications/SouthernRockiesWatershedProject-FinalReport-(Phase-I-2004-2008).pdf (accessed on 17 December 2012).
12. Wainwright, J.; Mulligan, M. *Environmental Modelling: Finding Simplicity in Complexity*; Wiley: Chichester, UK, 2004.
13. Miller, J.D.; Thode, A.E. Quantifying burn severity in a heterogeneous landscape with a relative version of the delta Normalized Burn Ratio (dNBR). *Remote Sens. Environ.* 2007, *109*, 66–80.
14. Silins, U.; Stone, M.; Emelko, M.B.; Bladon, K.D. Sediment production following severe wildfire and post-fire salvage logging in the Rocky Mountain headwaters of the Oldman River Basin, Alberta. *CATENA* 2009, *79*, 189–197.
15. Environment Canada. Water Survey of Canada-Castle River near Beaver Mines: Station ID 05AA022. Available online:http://www.wsc.ec.gc.ca/applications/H2O/report-eng.cfm?yearb=&yeare=&station=05AA022&report=daily&year=2011 (accessed on 17 December 2012).
16. Alberta Wilderness Association. Areas of Concern-Castle. Available online: http://albertawilderness.ca/

issues/wildlands/areas-of-concern/castle (accessed on 10 December 2012).

17. Kulig, J.C.; Edge, D.; Reimer, W.; Townshend, I.; Lightfoot, N. Levels of Risk: Perspectives of the Lost Creek Fire. Available online: https://www.uleth.ca/dspace/bitstream/handle/10133/1266/Levels%20of%20Risk.pdf?sequence=1(accessed on 17 December 2012).

18. Agriculture and Agri-Food Canada. Gridded Climate Data. Available online: http://www4.agr.gc.ca/AAFC-AAC/display-afficher.do?id=1227620138144&lang=eng (accessed on 10 October 2012).

19. University of Waterloo. Climate Dataset (Daily 10km Grids). Available online: http://www.lib.uwaterloo.ca/locations/umd/digital/documents/ClimateDatasetDaily10kmGrids.html (accessed on 25 February 2013).

20. Schulla, J. Model Description WaSiM (Water Balance Simulation Model). Available online: http://www.wasim.ch/downloads/doku/wasim/wasim_2012_ed2_en.pdf (accessed on 17 December 2012).

21. Jasper, K.; Calanca, P.; Gyalistras, D.; Fuhrer, J. Differential impacts of climate change on the hydrology of two alpine river basins. *Clim. Res.* 2004, *26*, 113–129.

22. Beckers, J.; Smerdon, B.; Wilson, M. Review of hydrologic models for forest management and climate change applications in British Columbia and Alberta. In *FORREX Forum for Research and Extension in Natural Resources, Kamloops*; Forrex Forum for Research and Extension in Natural Resources: Kamloops, BC, Canada, 2009; Forrex Series 25.

23. Klein Tank, A.M.G.; Wijngaard, J.B.; Können, G.P.; Böhm, R.; Demarée, G.; Gocheva, A.; Mileta, M.; Pashiardis, S.; Hejkrlik, L.; Kern-Hansen, C.; *et al.* Daily dataset of 20th-century surface air temperature and precipitation series for the European Climate Assessment. *Int. J. Climatol.* 2002, *22*, 1441–1453.

24. National Climate Data Information Archive Climate Data Online. Available online: http://climate.weatheroffice.gc.ca/climateData/canada_e.html (accessed on 1 January 2013).

25. Thornton, P.E.; Hasenauer, H.; White, M.A. Simultaneous estimation of daily solar radiation and humidity from observed temperature and precipitation: An application over complex terrain in Austria. *Agric. For. Meteorol.* 2000,*104*, 255–271.

26. Bristow, K.L.; Campbell, G.S. On the relationship between incoming solar radiation and daily maximum and minimum temperature. *Agric. For. Meteorol.* 1984, *31*, 159–166.

27. GeoBase Canadian Council on Geomatics. Digital Elevation Model Data. Available online: http://www.geobase.ca/geobase/en/find.do?produit=cded (accessed on 2 November 2012).

28. GeoBase National Land Cover Project Team; DB Geoservices Inc. GeoBase Land Cover Product: User Needs Assessment. Available online: http://www.geobase.ca/geobase/en/data/landcover/csc2000v/description.html (accessed on 5 February 2013).

29. CSSS Canadian Society of Soil Sciences; Department of Soil Science, University of Saskatchewan. Soils of Canada: Orders. Available online: http://www.soilsofcanada.ca/orders/index.php (accessed on 2 February 2013),.

30. Moriasi, D.N.; Arnold, J.G.; van Liew, M.W.; Bingner, R.L.; Harmel, R.D.; Veith, T.L. Model evaluation guidelines for systematic quantification of accuracy in watershed simulations. *Trans. ASABE* 2007, *50*, 885–900.

31. Nash, J.E.; Sutcliffe, J.V. River flow forecasting through conceptual models, Part I-A discussion of principles. *J. Hydrol.* 1970, *10*, 282–290.

32. Krause, P.; Boyle, D.P.; Bäse, F. Comparison of different efficiency criteria for hydrological model assessment. *Adv. Geosci.* 2005, *2005*, 89–97.

33. ESRD Alberta Spatial Wildfire Data: Historical Wildfire Perimeter Data: 1931–2011. Available online: http://srd.alberta.ca/Wildfire/WildfireStatus/HistoricalWildfireInformation/SpatialWildfireData.aspx (accessed on 2 February 2013).

34. Soverel, N.O.; Perrakis, D.D.B.; Coops, N.C. Estimating burn severity from Landsat dNBR and RdNBR indices across western Canada. *Remote Sens. Environ.* 2010, *114*, 1896–1909.

35. Key, C.H.; Benson, N.C. Landscape assessment: Remote sensing of severity, the normalized burn ratio. In *FIREMON: Fire Effects Monitoring and Inventory System: General Technical Report*; Lutes, D.C., Ed.; USDA Forest Service, Rocky Mountain Research Station: Ogden, UT, USA, 2006; Volume 2006, pp. 219–273.

36. Miller, J.D.; Knapp, E.E.; Key, C.H.; Skinner, C.N.; Isbell, C.J.; Creasy, R.M.; Sherlock, J.W. Calibration and validation of the relative differenced Normalized Burn Ratio (RdNBR) to three measures of fire severity in the Sierra Nevada and Klamath Mountains, California, USA. *Remote Sens. Environ.* 2009, *113*, 645–656.

37. Epting, J.; Verbyla, D.; Sorbel, B. Evaluation of remotely sensed indices for assessing burn severity in interior Alaska using Landsat TM and ETM+. *Remote Sens. Environ.* 2005, *96*, 328–339.

38. Verbyla, D.L.; Kasischke, E.S.; Hoy, E.E. Seasonal and topographic effects on estimating fire severity from Landsat TM/ETM+ data. *Int. J. Wildland Fire* 2008, *17*, 527–534.

39. Luce, C.H. Land use and land cover effects on runoff processes: Fire. In *Encyclopedia of Hydrological Sciences*; Anderson, M.G., McDonnel, J.J., Eds.; Wiley: Hoboken, NJ, USA, 2005; Volume 3, pp. 1831–1837.

40. IPCC. *Climate Change 2007-Impacts, Adaptation and Vulnerability: Working Group II Contribution to the Fourth Assessment Report of the Intergovernmental Panel on Climate Change*; Parry, M.L., Canziani, O.F., Palutikof, J.P., van der Linden, P.J., Hanson, C.E., Eds.; IPCC Secretariat: Geneva, Switzerland, 2007.

41. Nkemdirim, L.; Weber, L. Comparison between the Droughts of the 1930s and the 1980s in the Southern Prairies of Canada. *J. Clim.* 1999, *12*, 2434–2450.

42. Field, C.B.M.; Brklacich, M.; Forbes, D.L.; Kovacs, J.A.; Running Steven, W.; Scott, M.J. Chapter 14: North America. In*Climate Change 2007-Impacts, Adaptation and Vulnerability: Working Group II Contribution to the Fourth Assessment Report of the Intergovernmental Panel on Climate Change*; Parry, M.L., Canziani, O.F., Palutikof, J.P., van der Linden, P.J., Hanson, C.E., Eds.; IPCC Secretariat: Geneva, Switzerland, 2007.

43. Lemmen, D.S.; Warren, F.J.; Lacroix, J.; Bush, E. *From Impacts to Adaptation: Canada in a Changing Climate 2007*; Government of Canada: Ottawa, ON, Canada, 2007; Available online: http://www.nrcan.gc.ca/sites/www.nrcan.gc.ca.earth-sciences/files/pdf/assess/2007/pdf/full-complet_e.pdf (accessed on 5 February 2013).

44. Brutsaert, W. *Hydrology: An Introduction*; Cambridge University Press: Cambridge, UK/New York, NY, USA, 2005.

45. MacDonald, M.K.; Pomeroy, J.W.; Pietroniro, A. On the importance of sublimation to an alpine snow mass balance in the Canadian Rocky Mountains. *Hydrol. Earth Syst. Sci.* 2010, *14*, 1401–1415.

46. Roeder, A.; Hill, J.; Duguy, B.; Alloza, J.; Vallejo, R. Mapping fire events and post fire succession using long time-series of Landsat-TM and -MSS data. In Proceedings of the 5th International Workshop on Remote Sensing and GIS Applications to Forest Fire Management: Fire Effects Assessment, Zaragoza, Spain, 16–18 June 2005; La Riva Fernández, J.R.D., Pérez Cabello, F., Chuvieco Salinero, E., Eds.; Universidad de Zaragoza, Servicio de Publicaciones: Zaragoza, Spain, 2005; pp. 287–290.

47. Gillett, N.P. Detecting the effect of climate change on Canadian forest fires. *Geophys. Res. Lett.* 2004, *31*, L18211.

CHAPTER 8

SCENARIO-BASED IMPACT ASSESSMENT OF LAND USE/COVER AND CLIMATE CHANGES ON WATERSHED HYDROLOGY

Feng Wu, Jinyan Zhan, Hongbo Su, Haiming Yan, and Enjun Ma

INTRODUCTION

Climate and land use/land cover (LULC) changes are amongst the greatest global environmental pressures resulting from anthropogenic activities, both of which greatly impact the hydrological cycle [1–3]. Their impact on the hydrological cycle at the basin scale has become an important research issue in hydrology community owing to the increasingly serious water scarcity [4, 5]. The hydrological response of watersheds to climate and LULC changes is an important issue of water resource planning and management [6, 7], and the potential impacts of LULC change on the hydrological cycle must be considered by water resource managers [8, 9]. For example, the LULC changes due to urbanization and deforestation can alter the hydrological processes and lead to change of flood frequency and annual mean discharge by impacting the evapotranspiration, soil infiltration capacity, and surface and subsurface flow regimes [8, 10, 11], while climate change can alter the flow routing time and peak flows [10, 12]. It is crucial to the long-term water resource planning and management to better understand the potential impacts of climate and LULC changes on the runoff and streamflow in basins [9, 13]. In particular, effective water resource management under changing conditions requires reliable information about flows and modes that can be used to simulate flow regimes under different scenarios of changing land use and climate [6].

Citation: Feng Wu, Jinyan Zhan, Hongbo Su, Haiming Yan, and Enjun Ma, "Scenario-Based Impact Assessment of Land Use/Cover and Climate Changes on Watershed Hydrology," Advances in Meteorology, vol. 2015, Article ID 410198, 11 pages, 2015. doi:10.1155/2015/410198.

Separation of impacts of climate and LULC changes on the hydrological cycle is of great importance to improving the land use planning and water resource management [10, 14], especially in arid and semiarid regions where the climate change may significantly affect the hydrological cycle [15]. The effects of LULC and climate changes on the streamflow are more evident in the arid and semiarid regions. One typical example is the Heihe River Basin in Northwest China, which is characterized by limited water resources and special hydroclimatic and physiographic conditions [16]. Understanding the hydrological responses to potential climate change is very important for developing sustainable water resource management strategies in this region. However, the impacts of urbanization and deforestation on the hydrological cycle in arid and semiarid regions have been rarely documented [17–19]. Overall, there is still very limited understanding of the separate as well as combined impacts of LULC and climate changes on regional water and energy cycles, and therefore more in-depth research is needed, especially in the arid and semiarid regions [17].

There are many studies about the hydrological impacts of land use change or climate change at the basin scale, and most of them were conducted with a hydrological model based on a series of land use data extracted from satellite images [9, 20, 21]. For example, the impacts of land use change scenarios in the Wutu watershed, North Taiwan, were assessed using the conversion of a land use model (CLUE-s) and a generalized watershed loading functions model [16]. A Soil Water Assessment Tool (SWAT) and multiple General Circulation Models (GCMs) were used to investigate the relationship between climatic and hydrological changes in the Upper Mississippi River [22]. Most of these studies assume no change in LULC [4, 23], but the impacts of climate change on hydrology vary among regions and should be investigated with regional climate change scenarios [4]. Besides, the hydrological impacts of LULC change also vary with the climatic conditions [9]. For example, water balance variables might add or subtract the impacts of climate change under varying land cover conditions. In particular, the regional LULC change can offset or magnify the changes in global average temperature and can significantly alter the impacts associated with global warming [17, 24]. In addition, some studies about the combined effects of climate and LULC changes on streamflow showed that climate change was generally more significant than LULC change in determining the basin hydrological response [25–27]. For example, climate dominates the changing streamflow in the Xinjiang River Basin of Poyang Lake, China [25]. However, the hydrological cycle in a basin is a complex process influenced by climate and the physical properties of the catchment and human activities together [4, 5]. The complexity of these factors complicates the separation between effects of land use and climatic variability on streamflow [26, 28]. Therefore, it is still a challenge to distinguish the effects of LULC change from that of concurrent climate variability [14].

This study aims to separate the impacts of climate and LULC changes on the hydrological cycle in the Heihe River Basin under future scenarios to provide some useful reference information that can be used to improve the water resource management and guarantee the sustainable development. The climatic data predicted by General Circulation Models (GCMs) under RCP 4.5 scenario were used to represent the climate change scenarios for 2006–2030, and the land use data simulated with the Dynamic Land Use System (DLS) model were used to represent the land use change scenarios. The future hydrological cycle was simulated with the SWAT based on the scenario data of climate change and land use change, the hydrological impacts of which were analyzed by comparing the simulation results under different scenarios. The results of this study can provide valuable information for guiding future water resource management in the Heihe River Basin as well as other arid and semiarid regions in China.

METHODS

Study Area

The Heihe River Basin is the second largest inland river basin in China, which lies between 37°43′–42°41′N and 97°23′–102°72′E with a total area of 127,96 thousand km^2. This basin expands across Qinghai Province, Gansu Province, and Inner Mongolia Autonomous Region in Northwest China (Figure 1). With a total length of 821 km, the Heihe River is divided into the upper, middle, and lower reaches, where the natural and socioeconomic characteristics differ significantly. For example, the average annual precipitation is between 200 and 500 mm, less than 200 mm, and less than 50 mm in these reaches, respectively, while the annual evaporation ranges from 700 mm in the upper reach to more than 3000 mm in the lower reach [29]. Besides, the annual average temperature is 9.4°C over the last 30 years, and this basin enjoys a dry continental climate. The altitude ranges from 869 to 5542 m, with an average of 1778 m. The main land cover types are desert (57.15% of total basin area), mountains (33.16%), and oasis (8.19%) [30]. The ecosystems from the upper reach to the lower reach are linked by the hydrological cycle, but the hydrological cycle has significantly changed due to the land use change and climate change in the past decades. For example, about 65% of the irrigation water in the middle reach was extracted from the river runoff, which greatly influences the hydrological cycle of the whole basin. Therefore, a detailed and integrated simulation analysis of the water resources is critical and urgent for better water resource management in the Heihe River Basin.

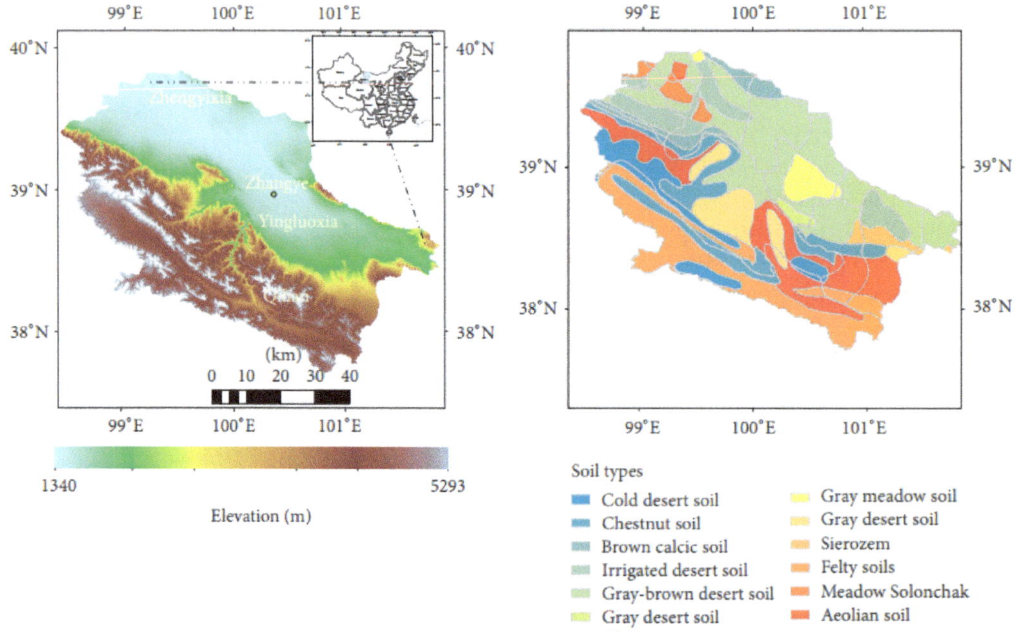

(a) (b)

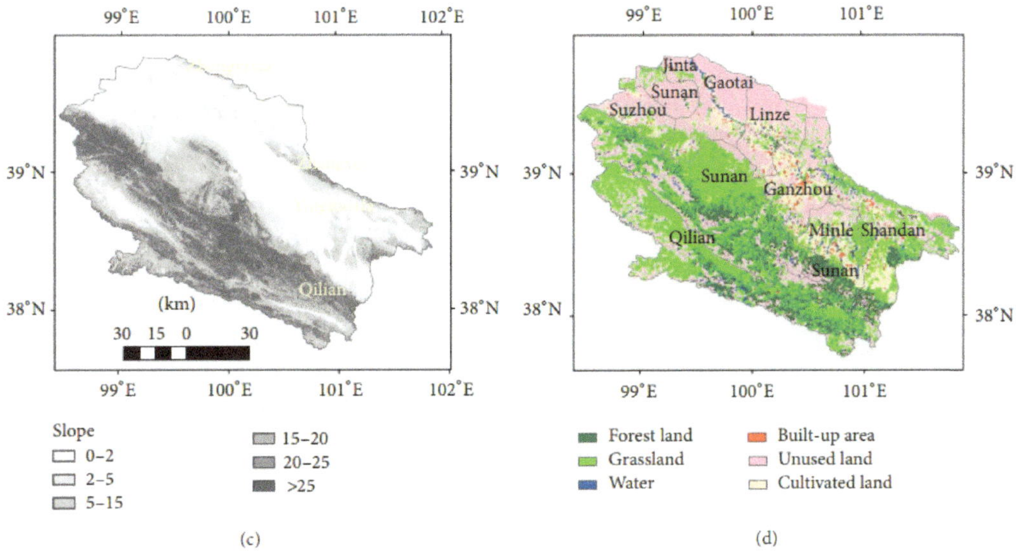

Figure 1: The upper and middle Heihe River Basin: (a) digital elevation model, (b) soil type, (c) slope, and (d) land use in 2010.

Data for Model Simulation

The spatial data (i.e., topography, soil and land use), historical climate data, and hydrological data for the watershed were first prepared for the SWAT model and DLS model. The topography was represented with the 90 m resolution digital elevation model (DEM) of Shuttle Radar Topography Mission (SRTM) (http://srtm.csi.cgiar.org/) [31]. The soil data, including texture, depth, and drainage attributes, were from the Harmonized World Soil Database (HWSD) supplied by the Environmental and Ecological Science Data Center for West China (WestDC) (http://westdc.westgis.ac.cn/). The historical land use data including 25 land use types, which were derived from Landsat TM/ETM images, were provided by Data Center of Chinese Academy of Sciences (CAS) [32]. In particular, the glacier data were obtained from WestDC (http://westdc.westgis.ac.cn/), and land use properties were directly obtained from the SWAT model database (Table 1). The historical hydrological data for SWAT model calibration and validation include the river flow data from four hydrological stations. The hydrological observation data, including annual data in Heihe River Basin during 1980–2010, were obtained from the hydrological yearbook provided by WestDC. The river flow data used for the model calibration and validation were provided by the Data Center of CAS. The historical daily weather data during 1980–2010 were collected from China Meteorological Administration (CMA), including daily precipitation, maximum and minimum temperatures, solar radiation, humidity, wind speed, and wind direction from 13 weather stations in or near the Heihe River Basin.

Table 1: Data used and sources

Data type	Data sources	Scale	Description
DEM	Shuttle Radar Topography Mission (SRTM)	90 m	Elevation
Land use	Data Center of Chinese Academy of Science	1:100000	The classification system contains 25 categories
Soil	WestDC (http://westdc.westgis.ac.cn/)	1:1000000	Some parameters were calculated using a (Soil-Plant-Atmosphere-Water) Field and Pond Hydrology model
Weather	China Meteorological Administration	Daily	5 weather stations
Hydrological observation	Hydrological yearbook	Daily	4 stations
River flow	Data Center of Chinese Academy of Science	1:250000	River network diversion
Glacier	WestDC (http://westdc.westgis.ac.cn/)	1:100000	The attributes include width, length, and depth

Simulation of Land Use Changes with the DLS Model

The DLS model is a set of applications used to simulate the changing process of land use system and is an effective tool for simulation of spatial-temporal land use changes to assist land management [33]. The DLS model consists of three modules: a spatial regression module that identifies the relationships between land uses and the influencing factors; a scenario analysis module of land use changes that determines the land demands at the regional level; a spatial disaggregation module that allocates land use changes from a regional level to the disaggregated grid cells [34, 35].

The DLS model simulates the spatial-temporal land use changes with three processes: scenario analysis of land use change, spatial regression analysis, and spatial allocation of land use changes. The first process was carried out with the scenario analysis module, which provides the data of total land demands at the annual scale during a given period. By including the scenario analyses of land use changes, a set of spatially explicit simulation results of land use change can be exported by the DLS model [36]. The total land use demands can be set by several approaches, such as trend analysis methods (e.g., linear interpolations or more sophisticated econometric models) and economic models. In this study, the total land demands during the simulation period were first determined using trend analysis methods and then were used to establish a scenario of land use change during 2006–2030.

In the spatial regression module, the relationships between land uses and influencing factors were analyzed via stepwise logistic regression analysis of past land use changes and their drivers [37]. For each grid cell, the total probability for each land use type is calculated on basis of the multinomial logistic regression at the pixel scale as follows:

$$P_{i,k} = \begin{cases} \dfrac{1}{1+\sum_{k\neq i} e^{x_k\beta_k}} & i = k \\ \dfrac{e^{x_k\beta_k}}{1+\sum_{k\neq i} e^{x_k\beta_k}} & i \neq k, \end{cases} \quad (1)$$

where $P_{i,k}$ is the probability of conversion from land type i to k in the cells under given driving factors; $X_1, X_2,\ldots,X_n$ represent the driving factors of climate, landform, location, population, economic growth, policy, and other categories; $\beta_1, \beta_2 \cdots \beta_n$ are the regression analysis coefficients of driving factors for further estimation.

In this study, all the data of land use and the influencing factors were prepared at the annual scale. The land use data in 2000 and 2005 were used in the logistic regression. The assumed driving factors were categorized into five groups: climate, geophysics, transportation, location, and socioeconomics. The data of these factors in the corresponding years were also prepared (Table 2). The spatial disaggregation module is used to spatially and explicitly convert the land demands into land use changes at various locations of the study area. The spatial disaggregation is carried out in an iterative procedure based on the probability maps, conversion rules, historical land use maps, and land demands under the scenarios. The probability maps of each land use type were prepared with the logistic regression results. Besides, the rules of land use conversion were set for each land use type, whose value ranges from 0 to 1. A smaller value means one land use type is more likely to be converted to another type, and vice versa. The development-restricted areas in the study area were also specified.

Table 2: Results of logistical regression for the six land use types with 13 driving factors.

	Cultivated land	Forest land	Grassland	Water	Built-up area	Unused land
Slope	−0.000049	0.000121	−7.42*E* − 05	−7.76*E* − 06	−1.1*E* − 06	0.0000105
	(−27.52)***	(−63.65)***	(−18.89)***	(−7.24)***	(−2.07)*	(−2.64)**
Elevation	−0.000136	−3.02*E* − 05	0.000116	−0.0000149	−5.9*E* − 06	0.0000711
	(−53.82)***	(−11.13)***	(−20.8)***	(−9.76)***	(−7.85)***	(−12.56)***
Rain	−0.000045	0.0000228	0.0000419	0.0000136	1.4*E* − 06	−0.0000347
	(−16.09)***	(−7.61)***	(−6.79)***	(−8.05)***	−1.7	(−5.54)***
Radiation	−0.000407	0.000243	−0.000923	−8.64*E* − 06	1.16*E* − 05	0.00108
	(−24.27)***	(−13.55)***	(−24.98)***	(−0.86)	(−2.34)*	(−28.91)***
>0°C accumulated temperature	−0.0000106	0.0000127	0.000026	−3.88*E* − 06	−7.2*E* − 07	−0.0000234
	(−24.03)***	(−26.78)***	(−26.59)***	(−14.55)***	(−5.52)***	(−23.67)***
Soil organic matter	0.0685	0.232	−0.186	0.0152	−0.00655	−0.123
	(−9.38)***	(−29.64)***	(−11.58)***	(−3.46)***	(−3.03)**	(−7.52)***
pH value	0.00769	−0.0131	−0.00541	0.00171	0.000719	0.00834
	(−7.15)***	(−11.32)***	(−2.28)*	(−2.64)**	(−2.26)*	(−3.47)***
Population density	0.000247	−7.82*E* − 05	−3.16*E* − 05	0.0000206	−3.2*E* − 06	−0.000154
	(−32.18)***	(−9.52)***	(−1.87)	(−4.46)***	(−1.40)	(−9.00)***
Per capita GDP	0.000548	−3.08*E* − 05	−0.000185	−0.00002	0.000411	−0.000723
	(−76.02)***	(−3.99)***	(−11.62)***	(−4.61)***	(−192.54)***	(−44.87)***
Distance to railway	0.000193	0.000229	0.000979	0.00000653	1.72*E* − 05	−0.00142
	(−13.55)***	(−14.96)***	(−31.16)***	−0.76	(−4.07)***	(−44.71)***
Distance to road	−0.000221	−0.000335	0.00111	−0.000066	−1.4*E* − 05	−0.000477
	(−18.45)***	(−26.06)***	(−42.08)***	(−9.13)***	(−3.79)***	(−17.79)***
Distance to water source	−0.000338	0.0000239	0.000926	−0.000346	−5.2*E* − 06	−0.000261
	(−14.46)***	−0.95	(−17.96)***	(−24.55)***	(−0.75)	(−4.99)***
Distance to city	−0.000125	−0.000108	−0.000567	0.0000996	−2*E* − 05	0.00072
	(−8.11)***	(−6.58)***	(−16.76)***	(−10.78)***	(−4.31)***	(−20.98)***

Note: *t* statistics in parentheses; *significant at 10%; **significant at 5%; ***significant at 1%.

Downscaling of GCM Climate Data

GCMs are arguably the best available tools for modeling future climate. Yet GCMs provide information at a resolution that is too coarse to be directly used in hydrological modeling [38]. Therefore downscaling is required to transform the low resolution GCM outputs to the high resolution climate features needed for hydrological simulation. The downscaling procedure is as follows. First, the average annual precipitation and temperature of 30 years (1980–2010) were calculated, which were adopted as a baseline for selecting the GCMs. We considered future climate change scenarios for the basin (Figure 2) by using the spatially distributed outputs from 10 GCMs under RCP 4.5 scenario. The climate projections of Max Planck Institute (MPI) were downscaled to the 3 km × 3 km grid in the study area, and bias was corrected and climate change scenarios were developed by MPI for Meteorology. The annual mean values of 10 GCM spatial data from 2006 to 2030 were calculated according to the basin perimeter, and one out of ten GCMs was also selected based on the historical trends and annual averages of temperature and precipitation. The MPI model was finally chosen through comparison. The result of MPI model originates in MPI, and the spatial resolution is 1.865° (LAT) × 1.875° (LON). Then the parameters in MPI GCM were transformed into the forcing data of a regional climate model in weather research and forecasting (WRF) simulation, and thereafter dynamical downscaling simulation was performed at the spatial resolution of 3 km for the period of 2006–2030. We considered the impacts of land use changes on regional climate, and the land cover data before WRF simulation were dynamically replaced with land use change data based on the simulation with the DLS model. Finally, the data simulated by the regional climate model were matched with meteorological sites, and the meteorological site data were prepared for the simulation with SWAT.

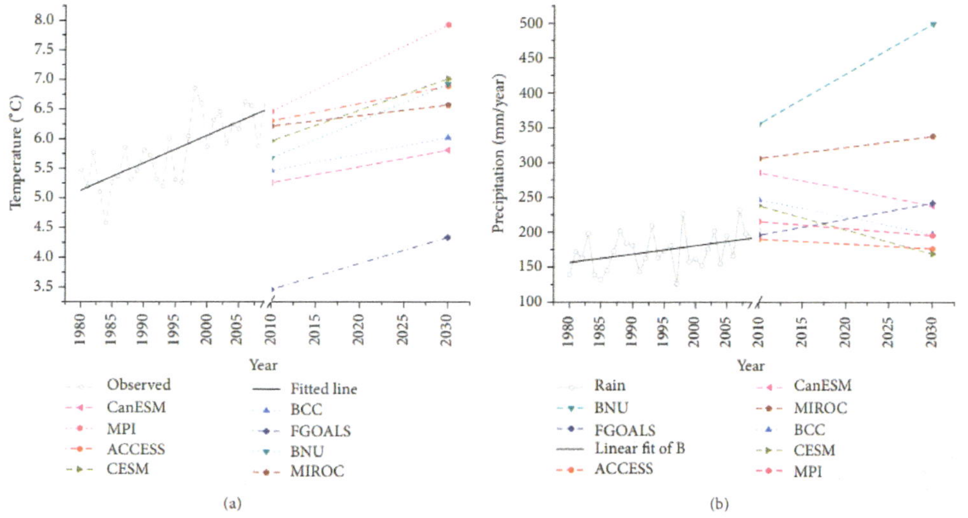

Figure 2: The trends in (a) observed (grey line; running average in black) and projected (10AR5 GCMs; colored dash lines) temperature, (b) observed and projected precipitation.

Simulation of Hydrological Cycle with the SWAT Model

The study area was first divided into subwatersheds, which were subdivided into hydrological response units (HRUs). Besides, for each subwatershed, the climate data used are taken from the GCM grid point that is the closest to its centroid. To improve performance, the SWAT model was calibrated and validated by adjusting several parameters and comparing the simulated streamflow with observed values. The most sensitive parameters were identified with the built-in sensitivity analysis tool in SWAT [39]. The daily streamflow observation data from Yingluoxia Hydrological Station in 2004 were used for calibration, and the observation data in 2005 were used for validation. It should be noted that the first three years were used as a warm-up period to mitigate the effects of unknown initial conditions, which were then excluded from the subsequent analysis. The ability of the SWAT model to replicate the temporal trends in the historical hydrological observations was assessed using the coefficient of determination (R^2), the Nash and Sutcliffe (1970) model efficiency (NSE), and the root mean square error (RMSE).

RESULT AND DISCUSSION

Calibration and Validation

The SWAT model was calibrated for 2004 and validated for 2005 using the daily streamflow observation data from four gauging stations within the study area. Finally, fifteen parameters were selected for the calibration (Table 2), which are associated with snow (SFTMP, SMTMP, SMFMX, SMFMN, and TIMP), runoff (CN2), groundwater (ALPHA_BF and GW_DELAY), soil (SOL_AWC), channel (CH_N and CH_K2), and evaporation (ESCO) processes. After the sensitivity analysis, 9 relatively more sensitive parameters were identified for the calibration. Most of the parameters were adjusted based on multiple trials, and the SWAT model was calibrated using an automatic calibration technique with the program Sequential Uncertainty Fitting Version (SUFI-2). With SUFI-2, sensitive initial and default parameters related to hydrology varied simultaneously until an optimal solution was achieved. The most sensitive parameters with their best ranges and best-fitted values are shown in Table 3. Finally, these best-fitted values were used to adjust the initial model inputs for the simulation during 2006–2030. The model was validated using daily streamflow observation data from the Yingluoxia Hydrological Station in 2005. The validation results show that the NSE is 0.78 and R^2 of the observed and simulated data is 0.81 (Figure 3), demonstrating the high behavioral performance of the SWAT model.

Table 3: List of calibration parameters and the optimized values

Parameter	Description	Range	Optimized value
TLAPS	Temperature lapse rate [°C/km]	0, −10	−3.8
PLAPS	Precipitation lapse rate [mm H_2O/km]	0, 100	5.8
SFTMP	Snowfall temperature [°C]	−2, +2	0.9
SMTMP	Snow melt base temperature [°C]	−5, +5	2.1
SNOEB	Initial snow water content in elevation band [mm]	50, 230	100
TIMP	Snowpack temperature lag factor	0.38–0.62	0.49
SMFMN	Melt factor for snow on December 21 [mm H_2O/°C-day]	3.05–3.51	3.25
SMFMX	Melt factor for snow on June 21 [mm H_2O/°C-day]	5.85–6.27	6.02
SURLAG	Surface runoff lag time [days]	4.18–5.19	4.68

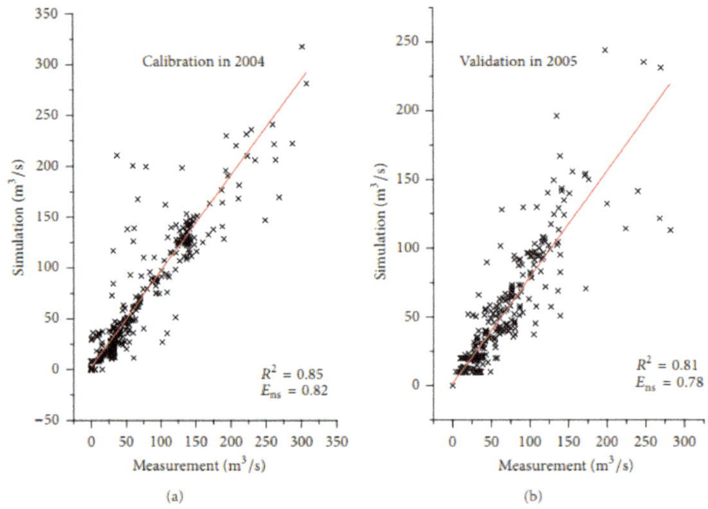

Figure 3: The calibration and validation on streamflow of the SWAT.

Future Climate under the RCP 4.5 Scenario

Based on the downscaled GCM climate data, we calculated the mean temperature and precipitation of the 9 lattice points around the grid that included Qilian meteorological station. The results were compared with the mean monthly temperatures and precipitation of the meteorological station during 1981–2005. The monthly mean temperatures of the 25 years ranged from −3 to 3°C and increased by around 0.8°C. The mean monthly precipitation ranges from −0.3% to 10% and increased by around 7.8% (Figure 4). The increase range of mean monthly precipitation is large, while the range of reduction is smaller.

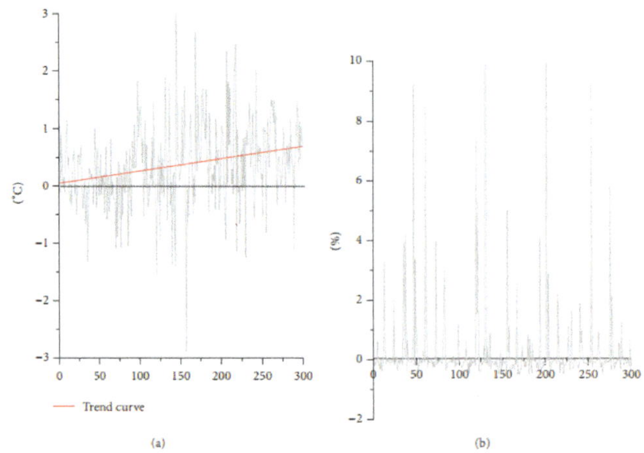

Figure 4: The difference of average monthly temperature and precipitation between 1981–2005 and 2006–2030 in Qilian Weather observed station.

Future Land Use Change Simulated with DLS

The results suggested the change in one land use type was influenced by multiple factors, and the 13 driving factors can reasonably explain the spatial patterns of all land use types. For example, the existence of forest land was significantly influenced by all the 13 driving factors, while the existence of cultivated land and grassland was affected by the altitude, distance, and soil factors. The future land uses for 2006–2030 were simulated with the DLS model by combining the probability maps prepared with logistic regression analysis, the land demands under different scenarios, and the map of development-restricted areas. The simulation results indicated that the most dramatic land use changes during 2006–2030 will mainly occur in the upper reach and some parts of the middle reach of Heihe River Basin. Compared to 2005, the areas of forest land and unused land in 2030 will decrease by 6.2% and 1.6%, respectively, while the areas of built-up land, cultivated land, and grassland will increase by 1.7%, 1.3%, and 4.8%, respectively (Figure 5). The significant increase of grassland area may mainly result from the steady pasture construction, and this uptrend may continue in the future owing to the increasing demand for pasture products.

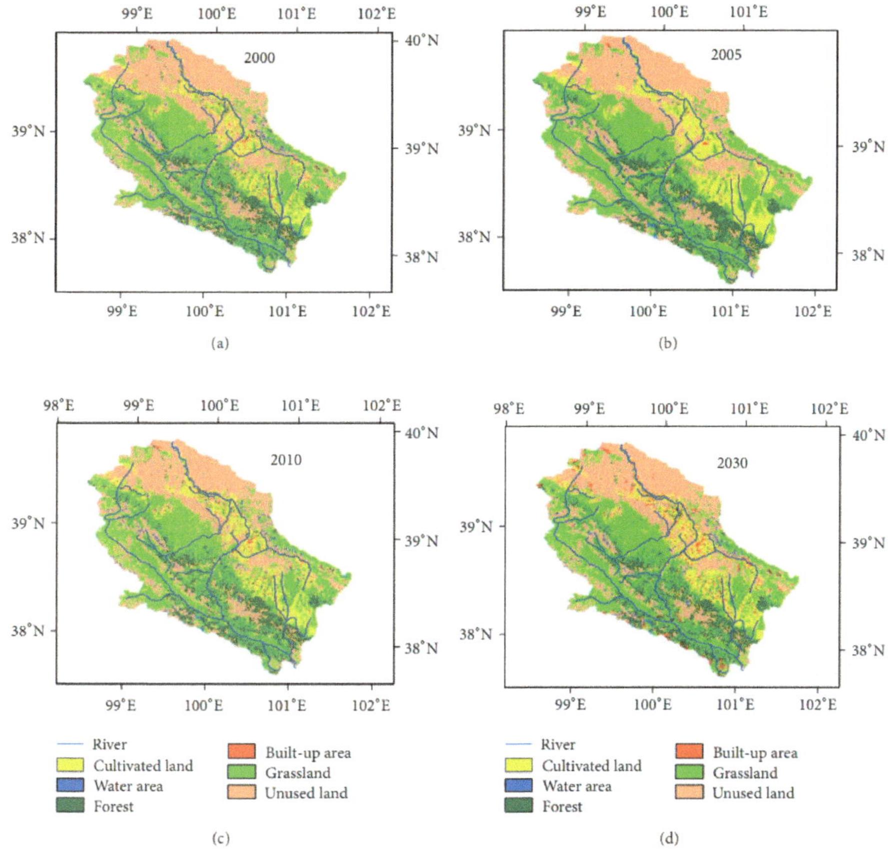

Figure 5: Comparison of the land uses interpreted in 2000, 2005 and simulated in 2010, 2030.

Impacts of Climate and Land Use Changes on Watershed Hydrology

Four simulation experiments were designed based on the land use data and climate data. In the baseline experiment for the period during 1981–2005, the water yield was simulated with the land use data in 2000, 2005 and the weather station observations during 1981–2005 (Figure 6(b)). Then three scenarios for the period during 2006–2030 were designed based on the land use and climate change (Figure 6(a)), the results from which were compared with that in the baseline experiment. In the first scenario during 2006–2030, the water yield was simulation with the land use data in 2010 and 2030, temperature data during 2006–2030, and the precipitation data during 1981–2005. The simulation result shows that the impacts of future land use change on the water yield vary with seasons, and the land use change will have negative overall influence on the water yield, with an influence degree of −1.8% according to the annual mean water yield.

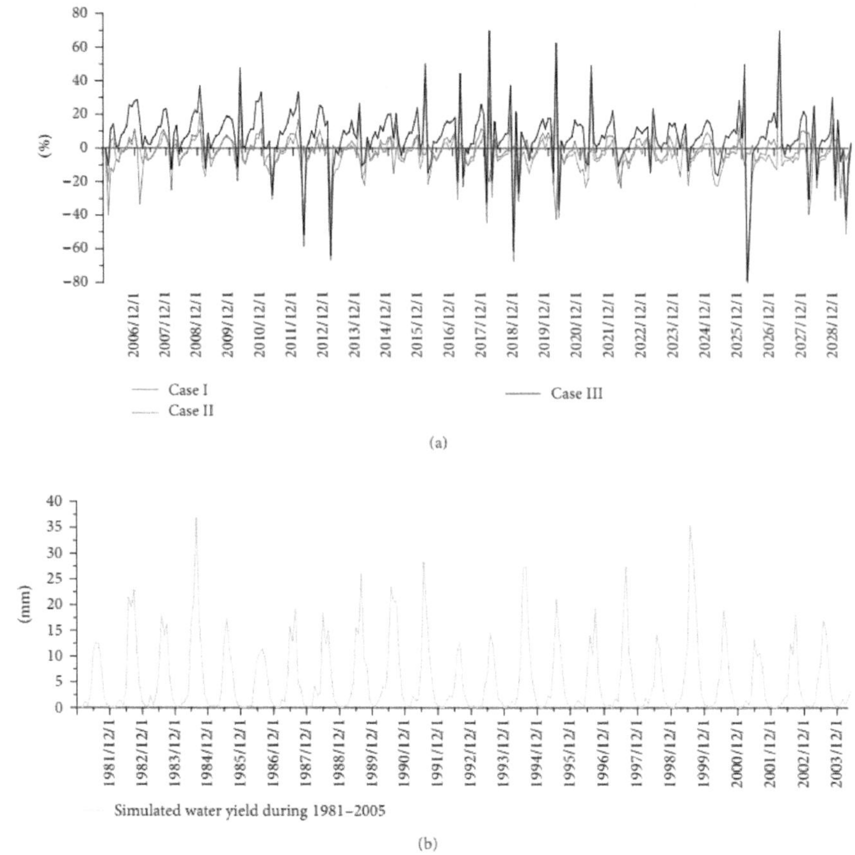

Figure 6: Comparison of water yields under three scenarios.

The second scenario during 2006–2030 was based on scenarios of temperature and land use changes. The second experiment used the land use data in 2010 and 2030, scenario data of temperature during 2006–2030, and precipitation data during 1981–2005. The analysis of climate change scenarios shows that the average temperature will rise by 0.8°C between 1981–2005 and 2006–2030. The simulation result in the second experiment shows that the land use and temperature changes will make the water yield change by 0.6%–1.1%, the change range of which is relatively smaller compared to the simulation results under the scenario with only land use change. The reasons may be that the temperature rise and melting of a small amount of snow slightly offset the adverse effects of land use change. At the same time, the higher temperatures will result in more winter precipitation in the form of rain rather than snow, leading to the hydrologic consequences including increased winter discharge, a shift in the spring snowmelt peak to earlier in the season, and decreased summer discharge.

The third scenario during 2006–2030 involves scenarios of changes in all of land use, temperature, and precipitation. The land use data in 2010 and 2030 and temperature and precipitation data during 2006–2030 were used under the third scenario. The simulation result shows that these three factors jointly have positive impacts on the water yield, making the basin water yield increase by about 9.8%. The increase of the basin water yield is mainly caused by the change in precipitation, which will increase by around 10.8% during 2006–2030 in comparison to that during 1981–2005. Overall, the simulation results suggest that the basin water yield will increase in the future under different scenarios of climate and land use changes.

DISCUSSION AND CONCLUSIONS

In the Heihe River Basin, the upper reaches are featured with the generation and use of blue water, while the lower reach and surrounding areas are characterized by natural ecosystems and a low population density. LULC is defined as syndromes of human activities such as agriculture, forestry, and building construction, and most of previous studies only focused on the hydrological influence of LULC change in the upper reach. The separation between hydrological impacts of land use and climate changes has never been studied in the upper and middle reaches of the Heihe River Basin. However, we argue that studying the hydrological processes in the upper and middle reaches is essential since water supply to the lower reach is impacted by both the climate change and human activities in the upper and middle reaches.

In this study, we analyzed the impacts of potential climate and land use changes on the water yield in the upper and middle reaches of Heihe River Basin based on the simulation with the SWAT model. The results show that the water yield was more affected by climate change than by land use change. This indicates that the predicted increase in precipitation will exert more significant impacts on the watershed hydrology than the predicted land use changes will. However, the analysis of the projected streamflow changes shows that there are higher uncertainties in the dry season compared with the wet season in the simulation with the hydrological model and GCMs climate data. It is difficult to accurately project the hydrological changes since there are various uncertainties associated with the future Green House Gas (GHG) emission scenarios, GCM structure, downscaling method, LULC, and hydrological models. In particular, water resource managers are generally confronted with complex problems in sustainable management and conservation of water resources due to the uncertainties in the future hydrological projection under climate and land use changes. It is therefore crucial to consider both land use and climate changes in water resource planning for the Heihe River Basin so as to mitigate their negative hydrological impacts, and more valuable information may be provided to the water resource managers if these uncertainties in the future hydrological projection can be effectively reduced through advanced modeling and research.

ACKNOWLEDGMENTS

This research was financially supported by the major research plan of the National Natural Science Foundation of China (Grant no. 91325302), National Basic Research Program of China (973 Program) (no. 2010CB950904), and the National Natural Science Funds of China for Distinguished Young Scholar (Grant no. 71225005).

REFERENCES

1. Z. Žalud, M. Trnka, M. Dubrovský, P. Hlavinka, D. Semerádová, and E. Kocmánková, "Climate change impacts on selected aspects of the Czech agricultural production," Plant Protection Science, vol. 45, pp. S11–S19, 2009. ·
2. E. Lorencová, J. Frélichová, E. Nelson, and D. Vačkář, "Past and future impacts of land use and climate change on agricultural ecosystem services in the Czech Republic," Land Use Policy, vol. 33, pp. 183–194, 2013. · ·
3. R. Mahmood, A. I. Quintanar, G. Conner, et al., "Impacts of land use/land cover change on climate and future research priorities," Bulletin of the American Meteorological Society, vol. 91, no. 1, pp. 37–46, 2010. · ·
4. Z. Ma, S. Kang, L. Zhang, L. Tong, and X. Su, "Analysis of impacts of climate variability and human activity on streamflow for a river basin in arid region of northwest China," Journal of Hydrology, vol. 352, no. 3-4, pp. 239–249, 2008. · ·
5. B. R. Scanlon, I. Jolly, M. Sophocleous, and L. Zhang, "Global impacts of conversions from natural to agricultural ecosystems on water resources: quantity versus quality," Water Resources Research, vol. 43, no. 3, Article ID W03437, 2007. · ·
6. L. M. Mango, A. M. Melesse, M. E. McClain, D. Gann, and S. G. Setegn, "Land use and climate change impacts on the hydrology of the upper Mara River Basin, Kenya: results of a modeling study to support better resource management," Hydrology and Earth System Sciences, vol. 15, no. 7, pp. 2245–2258, 2011. · ·
7. C. J. Vörösmarty, P. Green, J. Salisbury, and R. B. Lammers, "Global water resources: vulnerability from climate change and population growth," Science, vol. 289, no. 5477, pp. 284–288, 2000. · ·
8. S. Qi, G. Sun, Y. Wang, S. G. McNulty, and J. A. M. Myers, "Streamflow response to climate and landuse changes in a coastal watershed in North Carolina," Transactions of the ASABE, vol. 52, no. 3, pp. 739–749, 2009. · ·
9. J. Kim, J. Choi, C. Choi, and S. Park, "Impacts of changes in climate and land use/land cover under IPCC RCP scenarios on streamflow in the Hoeya River Basin, Korea," Science of the Total Environment, vol. 452-453, pp. 181–195, 2013. · ·
10. Z. Li, W.-Z. Liu, X.-C. Zhang, and F.-L. Zheng, "Impacts of land use change and climate variability on hydrology in an agricultural catchment on the Loess Plateau of China," Journal of Hydrology, vol. 377, no. 1-2, pp. 35–42, 2009. · ·
11. R. W. Skaggs, D. M. Amatya, G. Chescheir, C. Blanton, and J. Gilliam, "Effect of drainage and management practices on hydrology of pine plantation," in Proceedings of the International Conference on Hydrology and Management of Forested Wetlands, 2006.
12. T. D. Prowse, S. Beltaos, J. T. Gardner et al., "Climate change, flow regulation and land-use effects on the hydrology of the Peace-Athabasca-Slave system; Findings from the Northern Rivers Ecosystem Initiative," Environmental Monitoring and Assessment, vol. 113, no. 1–3, pp. 167–197, 2006. · ·
13. B. Dixon and J. Earls, "Effects of urbanization on streamflow using SWAT with real and simulated meteorological data," Applied Geography, vol. 35, no. 1-2, pp. 174–190, 2012. · ·
14. E. Lioubimtseva, R. Cole, J. M. Adams, and G. Kapustin, "Impacts of climate and land-cover changes in arid lands of Central Asia," Journal of Arid Environments, vol. 62, no. 2, pp. 285–308, 2005. · ·
15. Z. Wang, D. L. Ficklin, Y. Zhang, and M. Zhang, "Impact of climate change on streamflow in the arid Shiyang River Basin of northwest China," Hydrological Processes, vol. 26, no. 18, pp. 2733–2744, 2012. · ·
16. D. R. D›Agostino, L. G. Trisorio, N. Lamaddalena, and R. Ragab, "Assessing the results of scenarios of climate and land use changes on the hydrology of an Italian catchment: modelling study," Hydrological Processes, vol. 24, no. 19, pp. 2693–2704, 2010. · ·
17. V. Mishra, K. A. Cherkauer, D. Niyogi, et al., "A regional scale assessment of land use/land cover and climatic changes on water and energy cycle in the upper Midwest United States," International Journal of Climatology, vol. 30, no. 13, pp. 2025–2044, 2010. · ·
18. J. A. Vano, J. A. Foley, C. J. Kucharik, and M. T. Coe, "Evaluating the seasonal and interannual variations in water balance in Northern Wisconsin using a land surface model," Journal of Geophysical Research G: Biogeosciences, vol. 111, no. 2, 2006. · ·
19. D. Mao and K. A. Cherkauer, "Impacts of land-use change on hydrologic responses in the Great Lakes region," Journal of Hydrology, vol. 374, no. 1-2, pp. 71–82, 2009. · ·
20. J.-Y. Park, M.-J. Park, M.-J. H.-K. Joh et al., "Assessment of MIROC3.2 hires climate and clue-s land use change impacts on watershed hydrology using Swat," Transactions of the ASABE, vol. 54, no. 5, pp. 1713–1724, 2011. · ·

21. S. J. Kim, H. J. Kwon, G. A. Park, and M. S. Lee, "Assessment of land-use impact on streamflow via a grid-based modelling approach including paddy fields," Hydrological Processes, vol. 19, no. 19, pp. 3801–3817, 2005. · ·
22. E. Lu, E. S. Takle, and J. Manoj, "The relationships between climatic and hydrological changes in the upper Mississippi river basin: a SWAT and multi-GCM study," Journal of Hydrometeorology, vol. 11, no. 2, pp. 437–451, 2010. · ·
23. L. Liu, Z. Liu, X. Ren, T. Fischer, and Y. Xu, "Hydrological impacts of climate change in the Yellow River Basin for the 21st century using hydrological model and statistical downscaling model," Quaternary International, vol. 244, no. 2, pp. 211–220, 2011. · ·
24. S. Fall, D. Niyogi, A. Gluhovsky, R. A. Pielke, E. Kalnay, and G. Rochon, "Impacts of land use land cover on temperature trends over the continental United States: assessment using the North American Regional Reanalysis," International Journal of Climatology, vol. 30, no. 13, pp. 1980–1993, 2010. · ·
25. H. Guo, Q. Hu, and T. Jiang, "Annual and seasonal streamflow responses to climate and land-cover changes in the Poyang Lake basin, China," Journal of Hydrology, vol. 355, no. 1–4, pp. 106–122, 2008. · ·
26. L. Tang, D. Yang, H. Hu, and B. Gao, "Detecting the effect of land-use change on streamflow, sediment and nutrient losses by distributed hydrological simulation," Journal of Hydrology, vol. 409, no. 1-2, pp. 172–182, 2011. · ·
27. M. D. Tomer and K. E. Schilling, "A simple approach to distinguish land-use and climate-change effects on watershed hydrology," Journal of Hydrology, vol. 376, no. 1-2, pp. 24–33, 2009. · ·
28. P. F. Juckem, R. J. Hunt, M. P. Anderson, and D. M. Robertson, "Effects of climate and land management change on streamflow in the driftless area of Wisconsin," Journal of Hydrology, vol. 355, no. 1–4, pp. 123–130, 2008. · ·
29. Q. Feng, G. D. Cheng, and K. N. Endo, "Towards sustainable development of the environmentally degraded River Heihe basin, China," Hydrological Sciences Journal, vol. 46, no. 5, pp. 647–658, 2001. · ·
30. G.-D. Cheng, H.-L. Xiao, Z.-M. Xu, J.-X. Li, and M.-F. Lu, "Water issue and its countermeasure in the inland river basins of Northwest China—a case study in Heihe River Basin," Journal of Glaciology and Geocryology, vol. 28, no. 3, pp. 406–413, 2006.
31. A. Jarvis, H. I. Reuter, A. Nelson, and E. Guevara, Hole-Filled Seamless SRTM Data V3, International Centre for Tropical Agriculture (CIAT), 2006.
32. X. Deng, Q. Jiang, H. Su, and F. Wu, "Trace forest conversions in Northeast China with a 1-km area percentage data model," Journal of Applied Remote Sensing, vol. 4, no. 1, Article ID 041893, pp. 1–13, 2010. · ·
33. X. Deng, H. Su, and J. Zhan, "Integration of multiple data sources to simulate the dynamics of land systems," Sensors, vol. 8, no. 2, pp. 620–634, 2008. · ·
34. X. Deng, J. Liu, Y. Lin, and C. Shi, "A framework for the land use change dynamics model compatible with rcms," Advances in Meteorology, vol. 2013, Article ID 658941, 7 pages, 2013. · ·
35. X. Deng, C. Zhao, and H. Yan, "Systematic modeling of impacts of land use and land cover changes on regional climate: a review," Advances in Meteorology, vol. 2013, Article ID 317678, 11 pages, 2013. · ·
36. X. Deng, F. Yin, Y. Lin, Q. Jin, and R. Qu, "Equilibrium analyses on structural changes of land uses in Jiangxi Province," Journal of Food, Agriculture and Environment, vol. 10, no. 1, pp. 846–852, 2012. ·
37. X. Deng, Q. Jiang, J. Zhan, S. He, and Y. Lin, "Simulation on the dynamics of forest area changes in Northeast China," Journal of Geographical Sciences, vol. 20, no. 4, pp. 495–509, 2010. · ·
38. X. Deng, C. Zhao, Y. Lin et al., "Downscaling the impacts of large-scale LUCC on surface temperature along with IPCC RCPs: a global perspective," Energies, vol. 7, no. 4, pp. 2720–2739, 2014. · ·
39. S. Neitsch, J. Arnold, J. E. A. Kiniry, R. Srinivasan, and J. Williams, "Soil and water assessment tool user›s manual version 2000," GSWRL Report 02-06, 2002.

CHAPTER 9 HYDROLOGIC RESPONSES TO LAND USE CHANGE IN THE LOESS PLATEAU

Zhixiang Lu, Songbing Zou, Zuodong Qin, Yonggang Yang, Honglang Xiao, Yongping Wei, Kai Zhang,5 and Jiali Xie

INTRODUCTION

Changing land use represents the greatest human impact on the surface of the earth, and between one-third and one-half has been altered over the last 300 years [1–3]. Land use directly impacts the hydrologic components of a watershed, such as evapotranspiration, surface runoff, groundwater recharge, and streamflow [4–7]. Many previous studies from around the world have demonstrated that land use significantly impacts hydrological processes [8–15]. For example, Ghaffari et al. (2010) found that a 34.5% decrease of grassland with increases of shrubland (13.9%) and rain-fed agriculture (12.1%) led to a 33% increase in the amount of surface runoff and a 22% decrease in the groundwater recharge in northwest Iran [8]. Nie et al. (2011) revealed the effects of urbanization and agricultural development on the runoff in North America [9], and Wang et al. (2014), Li et al. (2015), and Yao et al. (2015) revealed the effects of urbanization, land use patterns, and changes on the runoff, temperature, evapotranspiration, and surface climate in Chinese landscapes [10–12]. López-Vicente et al. (2013) and Ohana-Levi et al. (2015) modeled the effects of land use change on the runoff and sediment yield in Europe's Pyrenees and rainfall-runoff relationships in an Eastern Mediterranean watershed, respectively [13, 14]. Wang et al. (2014) assessed the impacts of land use changes on evapotranspiration, quickflow, infiltration, and baseflow in the upstream regions of the Three Gorges reservoir in China [15]. Assessing impacts of land use and land cover (LULC) changes on hydrology is essential for watershed management and ecological restoration [7, 15, 16].

In general, regional impacts of land use change on hydrology vary from place to place and need to be considered for specific circumstances and environments [15, 16]. The Loess Plateau of western China is infamous for serious soil erosion and is the main source of sediment in the middle reaches of the Yellow River and has become the focus of world attention. As early as the 1950s, China started soil and water conservation works on the Loess Plateau in order to improve the local environment and mitigate sediment accessions to the middle reaches of the Yellow River, and works were expanded after the 1970s [17, 18]. As a consequence, land use in the Loess Plateau has undergone great changes. Likewise, in the past century, a range of management measures were carried out in the Fenhe River Basin on the eastern margin of the Loess Plateau. Some research has been done on the impacts of land use change on basin hydrology [19–22], but research on the contribution of individual LULC to the total runoff and the impacts of LULC changes on watershed hydrology is lacking. In particular, there is a lack of information for evaluating the benefits of soil and water conservation in the Fenhe River basin and the whole of the Loess Plateau, where it is difficult to distinguish the impacts of LULC changes on hydrology. Greater understanding of the contribution of individual LULC change to runoff and the impacts of LULC changes on the hydrology at different scales is needed to guide comprehensive natural resources management in this region.

Methods for assessing hydrological impacts of land use changes in watersheds include comparisons of paired catchments, statistical analysis, and hydrological modeling

Citation: Zhixiang Lu, Songbing Zou, Zuodong Qin, et al., "Hydrologic Responses to Land Use Change in the Loess Plateau," Advances in Meteorology, vol. 2015, Article ID 676030, 10 pages, 2015. doi:10.1155/2015/676030.

[5, 23, 24]. Among these approaches, hydrological modeling is the most suitable one for use in scenario studies at different scales. Widely used hydrological models in studies on the impacts on watershed hydrology include the Hydrologic Simulation Program, FORTRAN, the Soil and Water Assessment Tool (SWAT), and WaTEM/SEDEM [4, 5, 9, 15]. The SWAT model is widely used to assess hydrology and water quality in agricultural catchments around the world (https://www.card.iastate.edu/swat_articles/). It is readily available and user-friendly for data input [25].

The overall objectives of this study are to investigate the contributions of individual LULC change to runoff and to determine the impacts of LULC changes on the hydrology of the Upper Fenhe River watershed by an integrated approach that combines hydrological modeling and remotely sensed digital maps of LULC changes. To avoid the impacts of large-scale water resource development projects on hydrological processes, particularly the water diversion project from the Yellow River to Fenhe River at Wanjiazhai in the study area after 2002, the period from 1990 to 2000 was selected as the study period, during which sufficient data are available. The specific objectives are (1) to investigate the LULC changes of the watershed from 1995 to 2010 with LULC maps determined by remote sensing images in 1995 and 2010; (2) to calibrate and validate the SWAT model in terms of monthly streamflow based on the DEM, soil map, land use in 1995, and climate data from 1990 to 2000; (3) to assess the hydrological effects of individual land uses, and (4) to simulate responses of hydrologic components to land use changes at basin and subbasin scales under the same climatic conditions. The results should assist decision-makers in target water resources planning and vegetation restoration on the Loess Plateau.

STUDY AREA

The Fenhe River is located in the eastern part of the Loess Plateau, North China, and runs north to south through almost all of Shanxi Province [26]. The study area is the Upper Fenhe River watershed above the Hecha hydrological control station, which measures the amount of water and sediment entering Fenhe reservoir (Figure 1). The Upper Fenhe River watershed has an area of about 3242 km^2 and lies between latitudes 38°00′ and 39°00′ and longitudes 111°40′ and 112°40′. Elevations in the basin range from 1,124 to 2,762 m and it has an arid to semiarid climate with a mean annual precipitation of 430 mm. The precipitation is concentrated between the months of June and September and accounts for approximately 70% of total annual precipitation, coming in the form of high-intensity storms that often cause extreme soil erosion. The average annual temperature is 9.0°C, with a minimum average temperature of −5°C in winter and a maximum average temperature of 20°C in summer. The land use classes of the watershed include farmland, forest, grassland, water or wetland, built-up land, and unused land; their percentages in 2010 were 30%, 28.3%, 39.57%, 1.39%, 0.73%, and 0.01%, respectively. The region is covered with highly erodible loess-derived soils. Major soil series include loess soil, cinnamon soil, brown soil, and calcium skeletal soil.

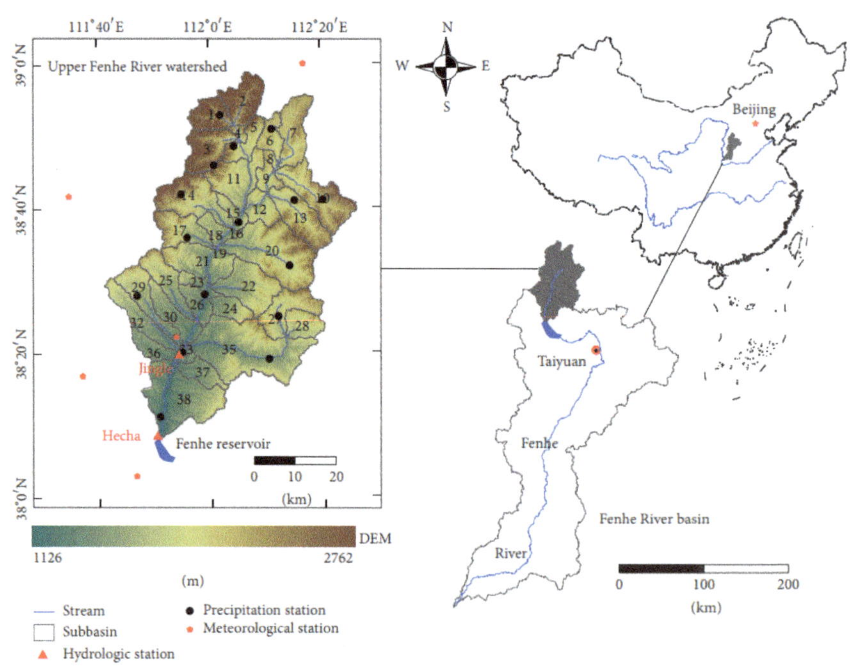

Figure 1: Locations of study area, hydrologic stations, and meteorological stations.

METHODOLOGY

A conceptual framework consisting of three stages was developed (Figure 2): (1) transition matrix analysis to describe the transformations of land use from 1995 to 2010; (2) hydrological modeling to simulate hydrological components for two LULC maps under the same climatic conditions; and (3) comparison of the contribution of each land use class to the watershed's water yield and their hydrological effects on subbasins.

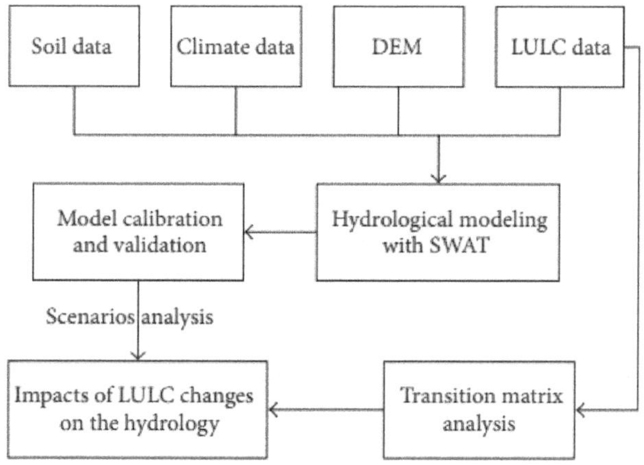

Figure 2: The conceptual framework for analyzing the impacts of LULC changes on the hydrology.

Transition Matrix

The transition matrix of land use change was calculated as

$$B_{ij} = \begin{bmatrix} B_{11} & B_{12} & \cdots & B_{1n} \\ B_{21} & B_{22} & \cdots & B_{2n} \\ \vdots & \vdots & \cdots & \vdots \\ B_{n1} & B_{n2} & \cdots & B_{nn} \end{bmatrix}, \quad (1)$$

where B is the area of a certain kind of land use, n is the number of land use types in the study area, i and j ($i, j = 1, 2, \ldots, n$) are the types of the land use before and after transition, respectively, and B_{ij} is area of the land use i turned into the land use j. The study period was defined as 1995 to 2010 because soil and water conservation practices were first implemented from 1988 to 1997, and a second ten-year conservation project was carried out from 1998 to 2007 [19, 20, 27]. The LULC maps in 1995 and 2010 represent the surface cover conditions in the latter part of the first conservation project and after the projects, respectively. The LULC map for 1995 was used as input data to calibrate the model.

Hydrological Modeling

SWAT Model Description

The SWAT 2005 model [28] was applied to data of the Upper Fenhe River watershed to assess the impacts of LULC changes on hydrological components. The SWAT model is a continuous, long-term, physically based distributed model developed to assess impacts of climate and land management on hydrological components, sediment loading, and pollution transport in watersheds [25]. In the SWAT model, a watershed is divided into subbasins. Subbasins are further divided into a series of uniform hydrological response units (HRUs) based on slope, soil, and LULC changes. Hydrological components, sediment yield, and nutrient cycles are simulated for each HRU and then aggregated for the subbasins.

Hydrological components simulated by the SWAT model include evapotranspiration (ET), surface runoff, percolation, lateral flow, groundwater flow (return flow), and transmission losses [25]. Evaporation and transpiration are simulated using exponential functions of soil depth and water content and a linear function of potential evapotranspiration (PET) and leaf area index. Hargreaves method was used to estimate PET, which was originally derived from eight years of cool-season Alta fescue grass lysimeter data from Davis, California [29], and its reference crop is similar to grassland which is the dominant land use type in this watershed. The surface runoff is estimated using a modification of the SCS curve number method with daily rainfall amounts. Percolation is calculated using the combination of a storage routing technique and a crack flow model. The lateral flow is estimated simultaneously with percolation using a kinematic storage model. The groundwater flow (base flow) into the channel is calculated from hydraulic conductivity of the shallow aquifer, distance from subbasin to main channel, and water table height.

SWAT Model Inputs and Outputs

The input data used in the SWAT model included a digital elevation model (DEM), soil data, digital LULC maps, and climate data. The DEM was derived from the USGS National Elevation Dataset with a resolution of about 90 m, and the soil data were obtained from the Resources and Environment Science Data Center, Chinese Academy of Sciences, with the scale of 1 : 1,000,000. The LULC data for two times (1995 and 2010) were used to assess the impacts of LULC change on hydrology and were derived from Landsat TM images with a spatial resolution of 30 m using an interpretation method. The climate data, including daily values of precipitation and minimum-maximum temperature from January 1st, 1990, to December 31, 2000, were derived from 3 meteorological stations and 16 precipitation stations located in and around the Upper Fenhe River watershed (Table 1).

Table 1: Hydrometeorological stations in the study area

Types of stations	Name	Factors	Data series
Hydrological stations	Jingle, Hecha	Monthly streamflow	1990–2000
Precipitation stations	Baergou, Haizibei, Dongzhai, Qianmalong, Gedongzi, Huaidao, Dongmafang, Ninghuabao, Xinbao, Tanger, Duanjiazhai, Ximafang, Suopo, Jingle, Kangjiahui, and Hecha	Daily precipitation	1990–2000
Meteorological stations	Ningwu, Kelan, and Jingle	Daily precipitation and maximum and minimum air temperature	1990–2000

The outputs of the SWAT model used in this study were hydrological components, including precipitation (P), PET, actual evapotranspiration (AET), water yield (WY), and soil water (SW).

SWAT Model Calibration and Validation

Simulations using the 1995 LULC map were used to calibrate monthly streamflows from January 1992 to December 1995 at two hydrological stations (Jingle and Hecha, Figure 1). After model calibration, simulations with the same model parameters and LULC map data were used to validate monthly streamflow from January 1996 to December 2000 (Table 2).

Table 2: The warm-up, calibration, and validation period for the model

Periods	Warm-up period	Calibration period	Validation period
Time series	1990-1991	1992–1995	1996–2000

Performance Evaluation Criteria

The model was calibrated manually and three criteria were used to evaluate performance of model calibration and validation. The Nash-Sutcliffe efficiency (NSE) is a normalized statistic that determines the relative magnitude of the residual variance compared to the measured data variance. NSE indicates how well the plot of observed versus simulated data fits the 1 : 1 line. Coefficient of determination (R^2) describes the degree of colinearity between simulated and measured data and the proportion of the variance in the measured data explained by the model. Percent bias (PBIAS) measures the average tendency of the simulated data to be larger or smaller than their observed counterparts [29]. The calibration and validation performance for the SWAT model were considered acceptable when R^2 and NSE were greater than 0.5. The SWAT model performance is satisfactory when NSE is larger than 0.5, adequate when NSE ranges from 0.54 to 0.65, and very good when NSE is larger than 0.65. The SWAT model is rated as satisfactory, when the absolute value of PBIAS ranges from 15 to 25, rated good when from 10 to 15, and very good when smaller than 10 [30].

Analyzing Hydrological Effects of Land Use Classes

The calibrated and validated hydrological model was run for each of the LULC maps (1995 and 2010) with constant DEM and soil data, from January 1990 to December 2000 (11 years), to quantify contributions of individual LULC classes to the total discharge and to evaluate impacts of LULC changes on hydrology at the basin and subbasin scales.

RESULTS AND DISCUSSION

LULC Changes

The LULC maps of the Upper Fenhe River watershed in 1995 and 2010 are shown in Figure 3. Grassland, farmland, and forest were the main land use classes in the Upper Fenhe River watershed.

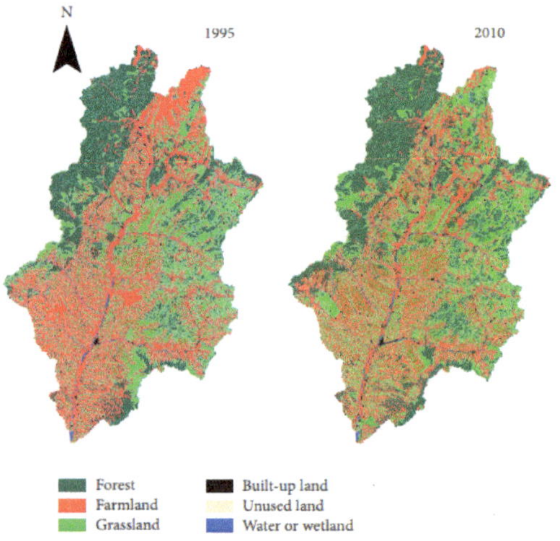

Figure 3: Maps for land uses and land cover in 1995 and 2010 in the Upper Fenhe River watershed.

The results of the transition matrix of LULC changes between 1995 and 2010 are shown in Table 3. The transformations among farmland, forest, and grassland were the main forms of land use changes in the Upper Fenhe watershed; the area of forest and grassland increased and the area of farmland decreased. Farmland showed the largest change. It was transformed into grassland in the northern part of the watershed due to the implementation of the "grain for green" government conservation program that compensated farmers for converting arable land to grassland. Meanwhile, due to the social development, the area of built-up land increased markedly at the expense of farmland. Finally, a tiny proportion of it was changed into water or wetland and unused land. In the southeast, a part of the watershed grassland was turned into farmland. Overall, the vegetation cover increased from 1995 to 2010, during periods of increased soil and water conservation measures.

Table 3: Transition matrix of land uses from 1995 to 2010 (km^2).

		2010						
	Types	Farmland	Forest	Grassland	Water or Wetland	Built-up land	Unused land	Total
	Farmland	908.26	17.74	79.66	1.13	6.99	0.29	1014.07
	Forest	9.68	840.49	45.88	0.48	0.82	0.02	897.37
	Grassland	54.04	59.11	1157.32	0.09	2.07	0.08	1272.71
1995	Water or wetland	0.95	0.06	0.11	43.32	0.24	0.00	44.68
	Built-up land	0.03	0.04	0.01	0.00	13.44	0.00	13.52
	Unused land	0.00	0.00	0.00	0.00	0.00	0.00	0.00
	Total	**972.96**	**917.44**	**1282.98**	**45.02**	**23.56**	**0.39**	**3242.35**

Model Calibration and Validation

Parameter sensitivity analysis was carried out using the model parameters identified from the instructions for the calibration of the SWAT model, as given in the user's manual [29]. The model was calibrated with a comprehensive model calibration method [30]. The results of sensitivity analysis for the Upper Fenhe River watershed and the value of the parameters are shown in Table 4.

Table 4: The results of parameter sensitivity analysis and calibration

Parameter	Definition	Hydrologic process or variable affected	Rank	Value
Alpha_Bf	Baseflow recession constant	Baseflow	1	0.084
Cn2	Moisture condition curve number	Surface runoff	2	30~92
Ch_K2	Effective hydraulic conductivity in main channel alluvium (mm/h)	Concentration of channel	3	10
Esco	Soil evaporation compensation factor	Soil water and soil evaporation	4	0.6
Sol_K	Saturated hydraulic conductivity of first layer (mm/h)	Infiltration and soil water	5	0~2000
Sol_Z	Depth from soil surface to bottom of layer (mm)	Soil water	6	0~3500
Canmx	Maximum canopy storage (mm)	Interception	7	0~10
Sol_Awc	Available soil water capacity (mm/mm)	Soil water	8	0.0125~0.5

Comparisons between simulated and observed monthly streamflow values in the periods of calibration and validation are shown in Figure 4. Good matches were shown between simulated and observed values, except in 1997 and 2000. In these two years, the simulated runoff was overestimated for both Jingle and Hecha. For 1997, this may be because the water stored in the watershed in two consecutive wet years (1995 and 1996) was released in the subsequent year. For 2000, the overestimation may have happened because 1999 was an extremely dry year, and water bodies such as small swamps and lakes were dry, and the soil was also very dry. This was followed by heavy rain in 2000 that firstly wetted the soil, filled the dry water bodies, and did not form surface runoff. This study did not take into account impounding of water in small water bodies which could have resulted in the overestimation of the actual runoff in 2000.

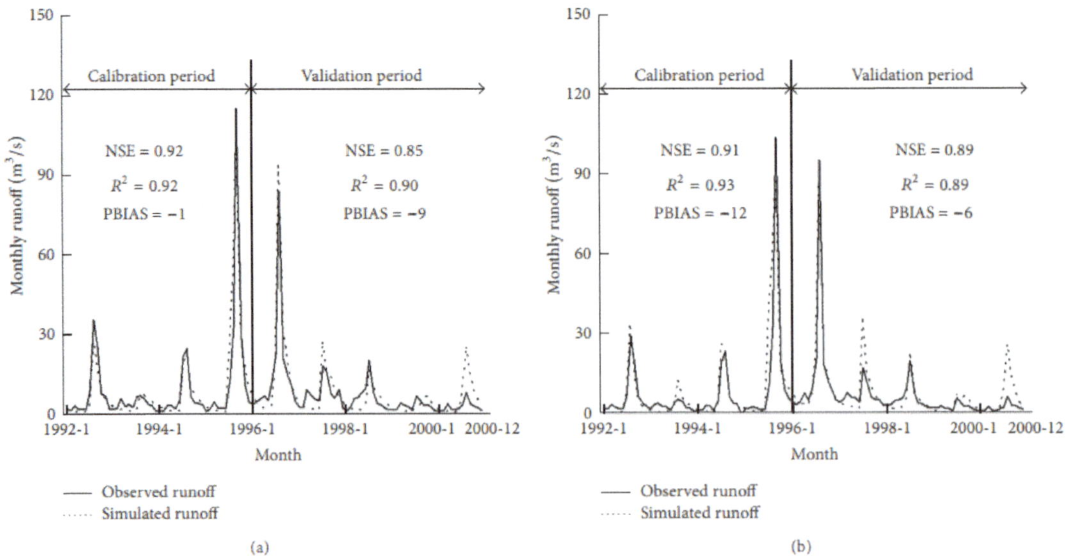

Figure 4: Monthly simulated and observed runoff in the Upper Fenhe River watershed. ((a) Hecha station; (b) Jingle station.)

All NSE and R^2 values for the monthly calibration and validation were above 0.80, and PBIAS values were in the range of −15% to 0 (most PBIAS values were in the range of −10% to 0), suggesting very good model performance [30]. The overall performance of the model was very good as shown in Figure 4.

Contribution of Individual LULC Changes to Watershed Water Yield

The proportional contributions of five LULC changes (i.e., farmland, forest, grassland, water or wetland, and built-up land) and simulated water yield (WY) are shown in Figure 5. Grassland, farmland, and forest were the main LULC changes in the Upper Fenhe River watershed, and the sum of their areas accounted for 98% of the total area. Their contributions to water yield were up to about 98% but there was no obvious positive correlation between the area of individual LULC changes and their proportional contributions to watershed WY. The grassland class includes mainly moderately dense and sparse grassland and very little dense grassland, so its water-yielding capacity was large. Although the area of forested land was much less than that of the grassland, its contribution to watershed WY was slightly more than the grassland's, because forests in this watershed are mainly shrubs and sparse trees. In addition, the rain mostly came in the form of high-intensity storms, which could reduce the amount of canopy interception compared with light rain. This is similar to the results obtained by Hao et al. (2004) in the upper Luohe River watershed and Song and Ma (2008) in the Nanhe River basin in the Longxi Loess Plateau [6, 7]. The area of farmland was between that for forests and grasslands, but its water-yielding capacity was much less. The farmland in the Upper Fenhe River watershed is terraced and used to grow close and straight row-seeded wheat and maize which greatly reduce the water yield.

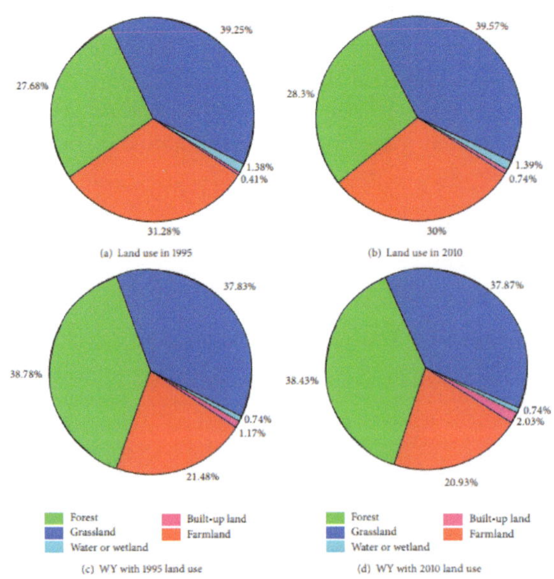

Figure 5: Percentage of each LULC change and its contribution to the total average annual water yields (WY) for the Upper Fenhe River watershed from 1992 to 2000.

By comparing the percentages of the area of each land use type and their contribution to water yield, it was found that an increase of built-up land was the main cause of increases of runoff, although the area of built-up land was very small. This is because its water-yielding capacity was the highest due to little initial loss and infiltration and rapid rainfall-runoff processes [9, 31].

As LULCs changed from 1995 to 2010, the contribution of individual LULC change to watershed WY changed too, but the change process was complex, because the WY is an integrated result of LULCs, soil, topography, and climate.

Impacts of LULC Changes on Hydrology at the Basin Scale

The average basin values of P, PET, AET, WY, and SW simulated from each LULC map in 1995 and 2010 are shown in Table 5. Compared to the LULC change baseline year of 1995, the average annual WY over the watershed was 0.13 mm higher in 2010. Similar to WY, average annual PET with LULC in 1995 was 1026 mm; it increased to 1028 mm by 2010. The AET increased from 391 mm for LULC in 1995 to 393 mm for LULC in 2010. On the contrary, the average annual SW for LULC in 2010 was 0.71 mm lower than that in 1995.

Table 5: Annual basin values of hydrologic features for the Upper Fenhe River watershed on the different land use from 1992 to 2000 (mm).

Year	P	Scenario with land use in 1995				Scenario with land use in 2010			
		PET	ET	WY	SW	PET	ET	WY	SW
1992	540.96	987.47	423.23	66.42	41.10	989.30	426.06	66.40	40.37
1993	395.4	1017.43	359.38	34.74	36.10	1019.25	361.25	34.90	35.66
1994	493.39	1029.46	409.19	67.96	38.68	1031.40	412.00	68.04	37.63
1995	709.2	1000.44	418.90	215.99	54.46	1002.42	421.12	215.94	53.76
1996	618.65	966.47	427.45	166.85	53.57	968.45	430.58	166.51	52.48
1997	437.26	1048.62	359.69	76.13	49.58	1050.60	361.02	76.82	49.03
1998	448.72	1081.87	403.59	50.94	43.10	1083.77	405.50	51.27	42.78
1999	327.71	1086.06	294.28	33.34	45.65	1088.06	297.30	33.41	45.17
2000	501.18	1020.10	424.42	58.84	54.54	1021.87	426.83	59.09	53.52
Mean	496.94	1026.44	391.12	85.69	46.31	1028.35	393.52	85.82	45.60

Note: P: precipitation; PET: potential evapotranspiration; ET: actual evapotranspiration; WY: water yield; SW: soil water.

Furthermore, compared to the LULC baseline year 1995, the annual PET, AET, and SW for each year from 1992 to 2000 for LULC change in 2010 changed in a similar manner as the average annual PET, AET, and SW. But in contrast to the size of the average annual WY for LULC in 1995 and 2010, the annual WY for each year in this period did not increase consistently. For example, in years of abundant precipitation, such as 1995 and 1996, the annual WY for the LULC change in 1995 was higher than that in 2010. This indicated that the precipitation could affect the impacts of the LULC changes on the hydrology in this region, which was also found by Hao et al. (2004) in the upper Luohe River watershed [6].

The comparison of variations of PET and AET and changes in LULCs suggested that the increase of annual PET could be mainly attributed to returning cropland to forest and grassland and to urban expansion from 1995 to 2010. Further comparison between changes in WY and changes of LULCs indicated that an increase of WY was mainly due to increases of urbanization, which can increase the area of impervious surfaces, increase runoff, and decrease infiltration [9, 31]. An association between the decreases of SW and forest, grassland, and built-up land expansion from 1995 to 2010 could be indicated from the comparison between variations of average annual SW and changes in LULC from 1995 to 2010. Expansion of forest, grassland, and built-up land by replacing them with farmland could promote water infiltration and drainage because of well-developed root systems and prevention of infiltration due to increases in the areas of impervious surfaces.

Impacts of LULC Changes on Hydrology at the Subbasin Scale

Three subbasins (numbers 7, 17, and 35 in Figure 1) with obvious LULC changes were selected for analysis of the impacts of LULC changes on hydrology at the subbasin scale, and the results are shown in Table 6. In subbasin number 7, the area of farmland decreased sharply, and the areas of forest, grassland, water, and built-up land increased from 1995 to 2010. The PET, AET, and WY increased, and, in contrast, the SW decreased. In subbasin number 17 only the forested area decreased and grassland area increased a little, so the impacts on hydrology were that AET increased clearly and WY and SW decreased. In subbasin number 35, areas of farmland, forest, and built-up land increased and grassland decreased, and WY increased and SW decreased. By comparing the LULC changes and the hydrological responses to them in these three sun-basins, obvious LULC changes may not show obvious impacts on hydrology, which means different combinations of LULCs may produce similar hydrological effects. Furthermore, the impacts of the same LULC changes on hydrology may be different under the conditions of different precipitation intensity and distribution.

Table 6: LULC changes and average annual hydrologic features from 1992 to 2000 of three subbasins

Subbasin	Number 7		Number 17		Number 35	
LULC scenario	1995	2010	1995	2010	1995	2010
Total area (km^2)	109.9	109.9	97.5	97.5	340	340
Farmland (%)	53.54	26.16	19.56	19.59	29.87	32.45
Forest (%)	9.31	17.42	49.6	46.51	15.57	16.78
Grassland (%)	36.98	55.35	28.1	31.18	52.75	48.23
Water or wetland (%)	0.01	0.72	2.29	2.29	1.25	1.24
Built-up land (%)	0.16	0.35	0.45	0.43	0.56	1.3
P (mm)	492.01	492.01	599.03	599.03	473.68	473.68
PET (mm)	918.93	926.55	1077.47	1077.5	1072.05	1072.18
ET (mm)	396.89	403.63	457.99	469.51	380.7	380.08
WY (mm)	64.74	66.57	124.43	116.62	71.17	72.17
SW (mm)	74.84	72.22	71.47	58.01	24.93	23.12

CONCLUSIONS

In this study we aimed to investigate the impacts of LULC changes on hydrology in China's Loess Plateau. Major research findings and their implication on practices and future research are as follows.

From 1995 to 2010, the main land use changes were the transformations of farmland to forests and from farmland to grassland, and the area of built-up land increased at the expense of farmland. These changes were due to the implementation of watershed management measures and social and economic development. For the contribution of each LULC to the total WY of the watershed, the forest, grassland, and farmland areas were the largest contributors, with up to about 39%, 38%, and 21%, respectively. The land use that produced the greatest water yield was shown to be built-up land, which was higher than any other land use, followed by forest, grassland, and farmland.

Accompanying the LULC changes in the Upper Fenhe River watershed, increases in PET, AET, and WY indicated soil and water conservation practices increased runoff, while the expansion of the forest and grassland increased the PET and AET. In addition, the consumption of SW was increased by vegetation combined with decreased infiltration because increases of the built-up land resulted in decreases in SW. Furthermore, the precipitation could affect the impacts of LULC changes on the hydrologic components in this arid and semiarid area.

The approach used in this study simply determined contributions of individual LULC classes to the total discharge, providing quantitative information for decision-makers to make better choices for land and water resource planning and management. This approach also provides a solid example of the potential of hydrologic modeling using remotely sensed digital LULCs in understanding the impacts of landscape change on water provisioning, a vital ecosystem service in the Loess Plateau of China. It can be widely applied to a variety of watersheds, where time-sequenced digital land cover data are available, and to predict hydrological consequences to LULC changes.

ACKNOWLEDGMENTS

This work was supported by the International Science & Technology Cooperation Program of China (2012DFA20770 and 2013DFG70990), the National Natural Science Foundation of China (41201043 and 91125007), the National Key Technology R&D Program (2011BAC07B05), and Shanxi Province Science Foundation for Youths (2012021026-3).

REFERENCES

1. J. A. Foley, R. DeFries, G. P. Asner et al., "Global consequences of land use," Science, vol. 309, no. 5734, pp. 570–574, 2005. · ·
2. E. F. Lambin, B. L. Turner, H. J. Geist et al., "The causes of land-use and land-cover change: moving beyond the myths," Global Environmental Change, vol. 11, no. 4, pp. 261–269, 2001. · ·
3. N. Ramankutty and J. A. Foley, "Estimating historical changes in global land cover: croplands from 1700 to 1992," Global Biogeochemical Cycles, vol. 13, no. 4, pp. 997–1027, 1999. · ·
4. L. M. Mango, A. M. Melesse, M. E. McClain, D. Gann, and S. G. Setegn, "Land use and climate change impacts on the hydrology of the upper Mara River Basin, Kenya: results of a modeling study to support better resource management," Hydrology and Earth System Sciences, vol. 15, no. 7, pp. 2245–2258, 2011. · ·
5. D. N. Khoi and T. Suetsugi, "The responses of hydrological processes and sediment yield to land-use and climate change in the Be River Catchment, Vietnam," Hydrological Processes, vol. 28, no. 3, pp. 640–652, 2014. · ·
6. F. Hao, L. Chen, C. Liu, and D. Dai, "Impact of land use change on runoff and sediment yield," Journal of Soil and Water Conservation, vol. 18, no. 3, pp. 5–8, 2004.
7. Y.-H. Song and J.-H. Ma, "SWAT-Aided research on hydrological responses to ecological restoration; a case study of the Nanhe River Basin in Huajialing of Longxi Loess Plateau," Acta Ecologica Sinica, vol. 28, no. 2, pp. 636–644, 2008. ·
8. G. Ghaffari, S. Keesstra, J. Ghodousi, and H. Ahmadi, "SWAT-simulated hydrological impact of land-use change in the Zanjanrood basin, Northwest Iran," Hydrological Processes, vol. 24, no. 7, pp. 892–903, 2010. · ·
9. W. Nie, Y. Yuan, W. Kepner, M. S. Nash, M. Jackson, and C. Erickson, "Assessing impacts of Landuse and

Landcover changes on hydrology for the upper San Pedro watershed," Journal of Hydrology, vol. 407, no. 1–4, pp. 105–114, 2011. · ·

10. Y. Wang, W. Ji, X. Yu et al., "The impact of urbanization on the annual average temperature of the past 60 years in Beijing," Advances in Meteorology, vol. 2014, Article ID 374987, 9 pages, 2014. ·
11. Y. Li, J. Fan, Z. Hu, Q. Shao, L. Zhang, and H. Yu, "Influence of land use patterns on evapotranspiration and its components in a temperate grassland ecosystem," Advances in Meteorology. In press.
12. X. Yao, Z. Wang, and H. Wang, "Impact of urbanization and land-use change on surface climate in middle and lower reaches of the Yangtze River, 1988–2008," Advances in Meteorology. In press.
13. M. López-Vicente, J. Poesen, A. Navas, and L. Gaspar, "Predicting runoff and sediment connectivity and soil erosion by water for different land use scenarios in the Spanish Pre-Pyrenees," Catena, vol. 102, pp. 62–73, 2013. · ·
14. N. Ohana-Levi, A. Karnieli, R. Egozi, A. Givati, and A. Peeters, "Modeling the effects of land-cover change on rainfall-runoff relationships in a semi-arid, Eastern Mediterranean watershed," Advances in Meteorology. In press.
15. G. Wang, H. Yang, L. Wang, Z. Xu, and B. Xue, "Using the SWAT model to assess impacts of land use changes on runoff generation in headwaters," Hydrological Processes, vol. 28, no. 3, pp. 1032–1042, 2014. · ·
16. X. Li, "A review of the international researches on land use/land cover change," Acta Geographica Sinica, vol. 51, no. 6, pp. 553–558, 1996. ·
17. Q. F. Zhang, F. Q. Wu, L. Wang, L. Yuan, and L.-S. Zhao, "Application of PCA integrated with CA and GIS in eco-economic regionalization of Chinese Loess Plateau," Ecological Economics, vol. 70, no. 6, pp. 1051–1056, 2011. · ·
18. C. Shi, Y. Zhou, X. Fan, and W. Shao, "A study on the annual runoff change and its relationship with water and soil conservation practices and climate change in the middle Yellow River basin," Catena, vol. 100, pp. 31–41, 2013. · ·
19. J. Zhang, "Comprehensive treatment for large basin and harmonious development of upstream and downstream-survey on comprehensive harness engineering of soil and water conservation for upper Fenhe River," Shanxi Water Resources, vol. 21, no. 4, pp. 5–9, 2005.
20. X. Wang and X. Nie, "Analysis on the benefit of soil and water conservation in the upper-reach of Fenhe reservoir-a case study of comprehensive management to Fenhe during the second phase in Jingle county,"Journal of Shanxi Agricultural University (Social Science Edition), vol. 6, no. 3, pp. 237–240, 2007.
21. B. Yang, "Viewing the function of integrated treatment of soil and water conservation conducted in the upper Fenhe River from sediment discharge variations at Jingle hydrometric station," Soil and Water Conservation in China, no. 10, pp. 32–34, 2007.
22. W. Meng, S. Wang, and J. Niu, "Upstream Fenhe River the Basin humanity moves under the forest cover change," Journal of Taiyuan Normal University (Natural Science Edition), vol. 6, no. 1, pp. 75–79, 2007.
23. Z. Li, W.-Z. Liu, X.-C. Zhang, and F.-L. Zheng, "Impacts of land use change and climate variability on hydrology in an agricultural catchment on the Loess Plateau of China," Journal of Hydrology, vol. 377, no. 1-2, pp. 35–42, 2009. · ·
24. H. Li, Y. Zhang, J. Vaze, and B. Wang, "Separating effects of vegetation change and climate variability using hydrological modelling and sensitivity-based approaches," Journal of Hydrology, vol. 420-421, pp. 403–418, 2012. · ·
25. J. G. Arnold, R. Srinivasan, R. S. Muttiah, and J. R. Williams, "Large area hydrologic modeling and assessment part I: model development," Journal of the American Water Resources Association, vol. 34, no. 1, pp. 73–89, 1998. · ·
26. J. Sun, Y. Liu, Y. Wang, G. Bao, and B. Sun, "Tree-ring based runoff reconstruction of the upper Fenhe River basin, North China, since 1799 AD," Quaternary International, vol. 283, pp. 117–124, 2013. · ·
27. X. Wang, "Studies on ecology and environment improvement of Fenhe reservoir and its upstream water sources," China Water Resources, vol. 16, pp. 44–47, 2008.
28. S. Neitsch, J. Arnold, J. Kiniry, J. Williams, and K. King, Soil and Water Assessment Tool: Theoretical Documentation, Version 2005, Texas A&M University, College Station, Tex, USA, 2005.
29. Z. Lu, S. Zou, Z. Yin, A. Long, and B. Xu, "A new suitable method for SWAT model calibration and its application in data-scarce basins," Journal of Lanzhou University (Natural Sciences), vol. 48, no. 1, pp. 1–7, 2012.
30. D. N. Moriasi, J. G. Arnold, M. W. Van Liew, R. L. Bingner, R. D. Harmel, and T. L. Veith, "Model evaluation guidelines for systematic quantification of accuracy in watershed simulations," Transactions of the ASABE, vol. 50, no. 3, pp. 885–900, 2007. · ·
31. J. Franczyk and H. Chang, "The effects of climate change and urbanization on the runoff of the Rock Creek basin in the Portland metropolitan area, Oregon, USA," Hydrological Processes, vol. 23, no. 6, pp. 805–815, 2009. · ·

CHAPTER 10

AN ANALYSIS OF LAND USE CHANGE DYNAMICS AND ITS IMPACTS ON HYDROLOGICAL PROCESSES

Tao Zhang, Xingnan Zhang, Dazhong Xia and Yangyang Liu

Keywords: Land use, land use change model, hydrological response, VIC model, real change, final change, average effect, compensation effect

INTRODUCTION

With the intensification of several global issues, including increasing population, food shortages, environmental pollution and climate change, global change has become one of the most popular areas of research, within which land use change is particularly important [1]. To survive and develop, humans have explored new forms of land use, resulting in land use changes.

Many studies have shown that land use changes are closely linked to hydrological processes. The main effects of land use changes on the water cycle are changes in evapotranspiration (ET), changes in the soil's ability to hold water and changes in the abilities of vegetation to intercept precipitation [2]. In addition, land use change extensively modifies the pathway and the surface roughness and then affects the timing of runoff, which leads to changes in the river flow (for example, flood peak appearance time and volume).

Dunn and Mackay [3], Matheussen *et al.* [4], Olchev *et al.* [5], Jin [6] and Mao and Cherkauer [7] have detected changes in ET due to land use changes and have shown that decreases in cover or damage to woodlands result in clear decreases in ET. Andraski [8], Shi and Li [9] and Krause [10] studied soil water variation in the field. Normally, after vegetation is removed, the distribution of roots becomes shallower, the soil porosity (especially the non-capillary porosity) decreases and the soil moisture-holding capacity decreases. Legesse *et al.* [11], Coe *et al.* [12], Savary *et al.* [13] and Schilling *et al.* [14] have performed related studies regarding the changes in runoff under different land use conditions and have obtained similar results. These authors observed that ET is stronger from woodlands than from other types of land. In addition, forested catchments have greater infiltration rates, which may decrease catchment runoff. Many other researchers have studied the effects of land use changes on river flow [15,16,17,18,19], and most of them have indicated that intensified afforestation will reduce both runoff peak and total runoff volume.

The methods used in these studies primarily included comparative analyses and hydrological modeling. With the development of physically-based and distributed/semi-distributed hydrology models, such as the Systeme Hydrologique Europeen (SHE), Soil and Water Assessment Tool (SWAT) and variable infiltration capacity (VIC) models, more researchers [20,21,22,23] are increasingly utilizing hydrological models to interpret and predict hydrological responses to land use changes. Although these studies provide a feasible method for drawing conclusions or addressing hydrological and water resource problems that result from land use change, they do not focus on land use dynamics.

The Jialing River is an important upstream watershed on the Yangtze River drainage basin, and its surrounding area is undergoing dramatic population growth with rapid economic development. In addition, various land use policies are being developed in this region, including the "Yangtze River Management" project and the return of croplands to forest and grasslands (RCFG), which were implemented in the 1980s and at the end of the 1990s, respectively. The land use dynamics along the Jialing River have changed significantly. Although

Citation: Zhang, T.; Zhang, X.; Xia, D.; Liu, Y. An Analysis of Land Use Change Dynamics and Its Impacts on Hydrological Processes. Water 2014, 6, 3758-3782.

many researchers [24,25,26,27,28,29] have examined the dynamics of land use change in this area, few studies have assessed the effects of these changes from a hydrological perspective. Thus, it is important to analyze land use changes and their impacts on the hydrological processes in this area.

This study uses land use change models to analyze and evaluate temporal and spatial land use changes. In addition, the VIC model was applied in this study to simulate hydrological elements under different land use types. Finally, this study assesses the impacts of land use changes on the hydrological regimes.

The main objectives of this study are as follows:

- To clarify how land use has changed in the Jialing River Basin and to identify hot spots of land use change;
- To present a new viewpoint for examining hydrological regime changes due to land use changes.

MATERIALS AND METHODS

Study Area and Data

Study Area

The Jialing River is the largest branch of the Yangtze River and is approximately 1120 km long. The Jialing Basin covers an area of approximately 160,000 km^2 within a geographical range of 29°17'30" N–34°28'11" N and 102°35'36" E–109°01'08" E (Figure 1). This region is located in the subtropical zone and has a humid monsoon climate. It has an annual average daily maximum temperature of 19.4 °C, an annual average daily minimum temperature of 4.3 °C, an average wind speed of 1.1 m/s and an average annual precipitation of 931 mm.

Materials

Three periods (1985, 1995, 2000) of land use were obtained from the Data Center for Resources and Environmental Sciences, Chinese Academy of Sciences (RESDC) [30]. This dataset was created by the Chinese Academy of Sciences (CAS) at 1-km resolution with the National Land Resource Classification System of China based on land use maps for the 1980s, 1995 and 2000 provided by the National 1:100,000 Land Use Database. The land use type interpretation resolution of the National 1:100,000 Land Use Database is as high as 92.92% according to assessments based on field surveys. These data were first used to analyze the dynamics of land use changes in the study area and were then used to obtain land use change scenarios, using the VIC model to assess the effects of land use changes on subwatershed hydrological regimes. Because the CAS classification system differs from the University of Maryland land cover classification (UMD), which is compatible with the VIC model, the conversion method used by Zhang *et al.* [31] was adopted. Land use types were reclassified into cropland, woodland, shrubland, grassland, water, settlement and bare ground.

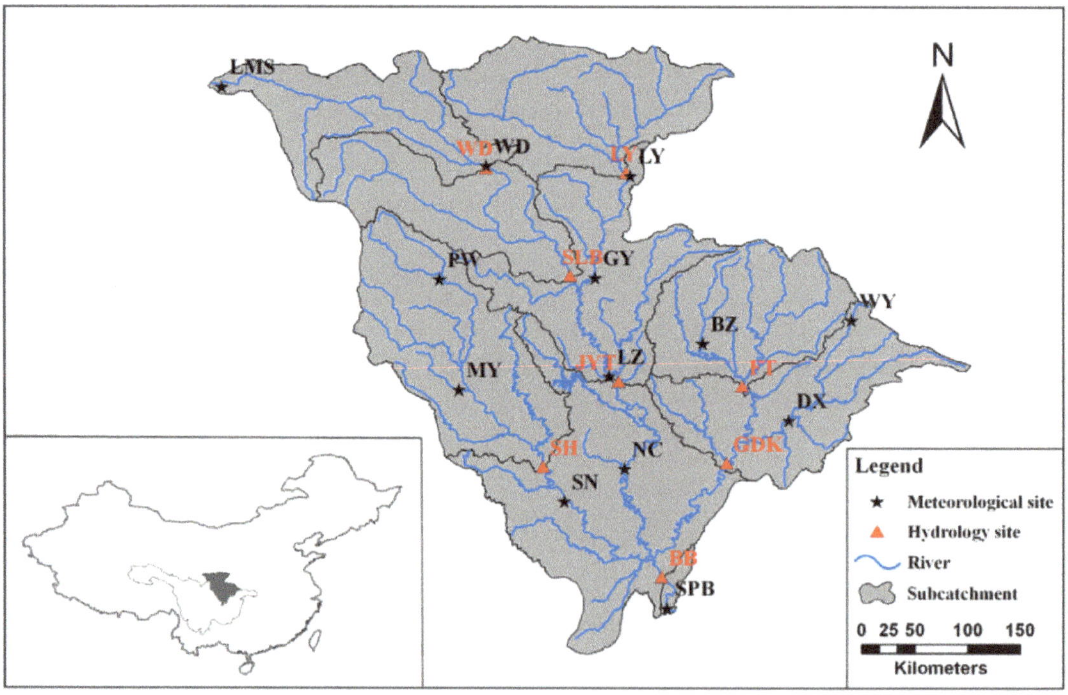

Figure 1: Location of the Jialing River Basin.

The following data are required for the VIC model:

- The digital elevation model (DEM) with a spatial resolution of 3 arc-seconds was obtained from the Shuttle Radar Topography Mission (SRTM) website [32].
- Spatial soil data with a resolution of 5 min were obtained from the United States Department of Agriculture (USDA) Agricultural Research Service and were published by the NOAA National Geophysical Data Center in Boulder, CO, USA [33].

The standard forcing data for the VIC model include precipitation (mm), maximum and minimum air temperature (Tmax and Tmin, °C) and mean wind speed (m/s). A series of daily climate data was collected from 14 of the National Meteorological Observatory (NMO) sites located in the Jialing River Basin between 1980 and 1987. In addition, a series of daily precipitation and streamflow data from 373 precipitation stations and 8 hydrological stations in the Jialing River Basin between 1980 and 1987 was obtained from the Annual Hydrological Report of the Jialing River Basin. The forcing data series was interpolated onto 4112 divided grid cells using the inverse distance interpolation method.

Land Use Change Model

Land use change models can be divided into stochastic-empirical models, dynamic models and integrated models [34,35]. Due to their simple structure and easy operation, stochastic-empirical models are widely used to evaluate rates of change, regulation and the other characteristics of the different land use types. In this article, the land use conversion matrix and the land use dynamic degree model were used to analyze land use change processes.

Land Use Conversion Matrix

The transitions among land use types can be described by a conversion matrix that depicts the structural characteristics and the direction of land use changes. This method is widely used by researchers [36] and is expressed in the following form:

$$S_{ij} = \begin{vmatrix} S_{11} \; S_{12} \; S_{13} \cdots S_{1n} \\ S_{21} \; S_{22} \; S_{23} \cdots S_{2n} \\ \cdots \; \cdots \; \cdots \; \cdots \\ S_{n1} \; S_{n2} \; S_{n3} \cdots S_{nn} \end{vmatrix} \tag{1}$$

where n is the number of different land use types and S_{ij} is the area converted from the i-th land use type to the j-th land use type.

Land Use Dynamic Degree Model

Single Land Use Dynamic Degree Model

The single land use dynamic degree model [35] describes the change in the quantity of one land use type during a specified period.

$$K = \frac{U_b - U_a}{U_a} \times \frac{1}{T} \times 100\% \tag{2}$$

here, K is the dynamic degree of one land use type; U_a and U_b are the areas of this land use type at the beginning and end of the study period, respectively; and T is the study period.

For a given land use type, the model only considers the differences in the areas between the beginning and end of the study period and neglects the transition. Consequently, we adopted the following revised version of this equation [37]:

$$K = \frac{U_i + U_o}{U_a} \times \frac{1}{T} \times 100\% \tag{3}$$

where K is the revised dynamic degree of a single land use; U_i and U_o are the areas of the land use type that are converted from one land use type to another land use type; U_a is the area of the land use type at the beginning of the study period; and T is the study period.

Dynamic Degree of Regional Comprehensive Land Use

When considering the process and not the results of land use change, the comprehensive dynamic degree of land use can represent the land use change intensity, which can be used to identify land use change hot spots [35,36]. The dynamic degree of the land use can be calculated as follows:

$$LC = \left[\frac{\sum_{i=1}^{n} \Delta LU_{i-j}}{2\sum_{i=1}^{n} LU_i} \right] \times \frac{1}{T} \times 100\% \tag{4}$$

where LU_i is the area of the i-th land use type; ΔLU_{i-j} is the area of the i-th land use type that is converted to another land use type; T is the study period; and n is the number of different land use types.

VIC Model

The large-scale VIC model [38,39,40] was developed by Washington University, the University of California at Berkeley and Princeton University. The key characteristics of the grid-based VIC model include its representation of vegetation heterogeneity, multiple soil layers with variable infiltration and non-linear base flow.

The VIC model is based on grid cells. In each grid cell, the land surface is described by $N + 1$ land cover types, where $n = 1, 2,\ldots, N$ represents N different tiles of vegetation and $n = N + 1$ represents bare soil. The corresponding ET (canopy evaporation, vegetation transpiration and bare soil evaporation) is calculated based on the Penman–Monteith equation. The VIC model was first designed as a 2 soil layer model, which lacks a moisture diffusion process between soil layers and cannot capture the dynamic behavior of the soil moisture content. Thus, Liang *et al.*, added a top thin layer to the VIC-2L [39]. The top two soil layers are designed to represent the dynamic response of soil to the infiltrated rainfall, with diffusion allowed from the middle layer to the upper layer when the middle layer is wetter. The bottom soil layer receives moisture from the middle layer through gravity drainage, which is regulated by a Brooks–Corey relationship for the unsaturated hydraulic conductivity [41].

The soil surface is characterized by a variable infiltration curve, which is expressed as:

$$i = i_m[1 - (1 - A)^{1/b_i}] \tag{5}$$

where i and i_m are the infiltration capacity and maximum infiltration capacity, respectively; A is the fraction of an area for which the infiltration capacity is less than i; and b_i is the infiltration shape parameter, a measure of the spatial variability of the infiltration capacity [39]. The VIC model is widely used for modeling and assessing the effects of land use changes on hydrological processes [7,42,43,44,45].

As in most physically-based hydrologic models, many parameters must be specified in the VIC model. However, most of the parameters can be derived from *in situ* measurements and remote sensing. Generally, 6 parameters need to be calibrated: bi, Ds, Dmax, Ws, D2 and D3. Detailed information about these 6 parameters is listed in Table 1. This table cites the study of Gao *et al.* [41].

Table 1: Parameters that need to be calibrated in the VIC model

Parameters	Description	Range
bi	Describes the variable infiltration curve. A higher value gives lower infiltration and yields higher surface runoff.	0–0.4
Ds	Represents the fraction of the Dsmax parameter at which non-linear baseflow occurs. With a higher value of Ds, the baseflow will be higher at lower water content in the lowest soil layer.	0–1
Dmax	The maximum velocity of baseflow for each grid cell.	0–30
Ws	The fraction of maximum soil moisture where non-linear baseflow occurs. A higher value will raise the water content required for rapidly increasing, non-linear baseflow, which will tend to delay runoff peaks	0–1
D2/D3	The thickness of the second/third soil layer. D2 represents the dynamic response of soil to the infiltrated rainfall. D3 characterizes seasonal soil moisture behavior. In general, for runoff considerations, greater soil depths slow down (baseflow dominated) seasonal peak flows and increase the loss due to evapotranspiration. The deeper the soil, the less runoff is generated.	0.1–1.5

In this paper, calibration was conducted manually using trial and error. The water balance was controlled by the relative error (RE), and the goodness of fit was evaluated with the Nash–Sutcliffe model efficiency coefficient (NSE) [46]. The RE and NSE are defined as follows:

$$RE = (\overline{Q}_{sim} - \overline{Q}_{obs}) / \overline{Q}_{obs} \tag{6}$$

$$NSE = 1 - \frac{\sum_{i=1}^{n} (Q_{obs,i} - Q_{sim,i})^2}{\sum_{i=1}^{n} (Q_{obs,i} - \overline{Q}_{obs})^2} \tag{7}$$

where $Q_{obs,i}$ and $Q_{sim,i}$ are the observed and simulated streamflows, respectively; $\overline{Q}_{obs}$ and $\overline{Q}_{sim}$ represent the mean values of the observed and simulated streamflows, respectively; and n is the length of the time series.

Hydrological Simulation Scheme

In order to detect the impacts of land use changes on the hydrological regime, the simulation results for 1980 to 1987 were used as background values. By holding the hydrometeorological conditions constant and by substituting the land use parameters derived from the land use maps of 1995 and 2000, the VIC model was used again to model the hydrological regimes under different land use scenarios. The hydrological changes that were induced by land use changes in scenarios S1-S0 and S2-S1 are discussed in the following sections. The simulation scheme is shown in Table 2.

Table 2: Hydrological simulation scheme

Land Use	Hydrometeorological Conditions	Scenario
1985		S0
1995	1980–1987	S1
2000		S2

Notes: S1-S0 is the hydrological difference between scenario S1 and S0; S2-S1 is the hydrological difference between scenario S2 and S1.

RESULTS

In this paper, two important concepts are defined: real change and final change. Real change is defined as the total variation in a parameter during a fixed period, including its increase and decrease (*i.e.*, the sum of the absolute values of the decrease and increase). Final change is defined as the difference in one factor between the beginning and end of a defined period. These two concepts were first used to analyze land use change and were then used to analyze the changes in hydrological regimes under different land use scenarios.

Land Use Change

Dynamic land use change was analyzed based on three land use maps for 1985, 1990 and 2000.

Quantitative Analysis of Land Use Change

Final Land Use Change

The final land use changes for the Jialing River Basin are listed in Table 3.

Table 3: Final land use change in the Jialing River Basin

Land Use	Area in 1985		Area in 1995		Area in 2000		Change 1985–1995		Change 1995–2000	
	km^2	%	km^2	%	km^2	%	km^2	%	km^2	%
Cropland	74,420	46.88	73,549	46.33	74,411	46.88	−871	−1.17	862	1.17
Woodland	31,670	19.95	32,193	20.28	31,731	19.99	523	1.65	−462	−1.44
Shrubland	17,193	10.83	14,430	9.09	16,219	10.22	−2,763	−16.07	1,789	12.40
Grassland	34,090	21.48	37,127	23.39	34,807	21.93	3,037	8.91	−2,320	−6.25
Water	652	0.41	639	0.40	681	0.43	−13	−1.99	42	6.57
Settlement	313	0.20	378	0.24	489	0.31	65	20.77	111	29.37
Bare ground	400	0.25	422	0.27	400	0.25	22	5.50	−22	−5.21
Total	158,738	100	158,738	100	158,738	100	-	-	-	-

The land uses in the Jialing River Basin primarily included croplands, woodlands, shrublands and grasslands, which represented more than 99% of the total area during all three periods (1985, 1995 and 2000). Croplands represented the greatest area, followed by the grasslands, woodlands and shrublands.

From 1985 to 1995, the most notable total area changes were observed in settlement, shrublands and grasslands, with change rates of 20.77%, −16.07% and 8.91%, respectively. Although the changes in the croplands and woodlands were larger, their change rates were low due to their large areas. From 1995 to 2000, settlement and shrubland area changed significantly (29.37% and 12.40%), and the change rates of the grasslands, water and bare ground were similar (−6.25%, 6.57% and −5.21%, respectively). From 1985 to 2000, the areas of the croplands, shrublands and water decreased and then increased. In contrast, the areas of the woodlands, grasslands and bare ground increased and then decreased. Furthermore, settlement increased throughout the entire period.

Due to the transitional relationships between the land use types, the total change in the area of one land use type was not obvious. Therefore, real land use change analysis was necessary.

Real Land Use Change

The land use conversion matrix was calculated by using Equation (1), and the results are shown in Table 4 and Table 5.

Table 4: Conversion matrix among the land use types from 1985 to 1995 in the Jialing River Basin (km^2)

Land Use	Crop	Wood	Shrub	Grass	Water	Settlement	Bare Ground	Year 1985	Out
Cropland	72,325	784	206	1,015	16	74	0	74,420	2,095
Woodland	386	28,299	885	2,089	2	1	8	31,670	3,371
Shrubland	386	2,116	13,228	1,461	0	2	0	17,193	3,965
Grassland	407	990	111	32,558	2	8	14	34,090	1,532
Water	28	2	0	3	618	1	0	652	34
Settlement	17	2	0	1	1	292	0	313	21
Bare ground	0	0	0	0	0	0	400	400	0
Year 1995	73,549	32,193	14,430	37,127	639	378	422	158,738	11,018
In	1,224	3,894	1,202	4,569	21	86	22	11,018	-

Notes: "In" indicates the increased area of one land use type that is converted from another land use type during a specific time period, here from 1985 to 1995; "Out" indicates the decreased area of one land use type that is converted into another land use type during a specific time period, here from 1985 to 1995; and real change is the sum of "In" and "Out". This note applies to the following tables, as well.

Table 5: Conversion matrix among the land use types from 1995 to 2000 in the Jialing River Basin (km²)

Land Use	Crop	Wood	Shrub	Grass	Water	Settlement	Bare Ground	Year 1995	Out
Cropland	72,404	408	315	254	50	118	0	73,549	1,145
Woodland	736	28,362	1,809	1,278	1	7	0	32,193	3,831
Shrubland	197	939	12,960	334	0	0	0	14,430	1,470
Grassland	1,057	2,011	1,134	32,917	6	2	0	37,127	4,210
Water	7	2	0	2	624	4	0	639	15
Settlement	10	1	1	8	0	358	0	378	20
Bare ground	0	8	0	14	0	0	400	422	22
Year 2000	74,411	31,731	16,219	34,807	681	489	400	158,738	10,713
In	2,007	3,369	3,259	1,890	57	131	0	10,713	-

Notes: "In" indicates the increased area of one land use type that is converted from another land use type during a specific time period, here from 1985 to 1995; "Out" indicates the decreased area of one land use type that is converted into another land use type during a specific time period, here from 1985 to 1995; and real change is the sum of "In" and "Out". This note applies to the following tables, as well.

Regarding real change, the three largest change rates were 34.2% (settlement), 30.1% (shrubland) and 22.9% (woodland) from 1985 to 1995. Compared to the final change, the change rate of the woodland was highlighted by the real change analysis (final change rate: 1.65%; real change rate: 22.9%). Woodlands had the largest real change area (7265 km²), followed by grasslands (6101 km²) and shrublands (5167 km²). The change in the woodland area was also highlighted by real change analysis (final change area: 523 km²; real change area: 7265 km²). From 1995 to 2000, the settlement change rate was the largest (39.9%), followed by shrublands and woodlands, with change rates of 32.8% and 22.4%, respectively. The ranks of the area changes during this period were the same as those in the previous period (1985–1995), namely, woodland (7200 km²), grassland (6100 km²) and shrubland (4729 km²) (Table 5). The most significant land use transitions occurred among the croplands, woodlands, shrublands and grasslands throughout this period.

Spatial Analysis of Land Use Changes

The spatial changes in land use were analyzed using GIS. Due to the resolution limitations of the pictures, the transition areas that were less than 50 km² were neglected. The spatial transition results are shown in Figure 2.

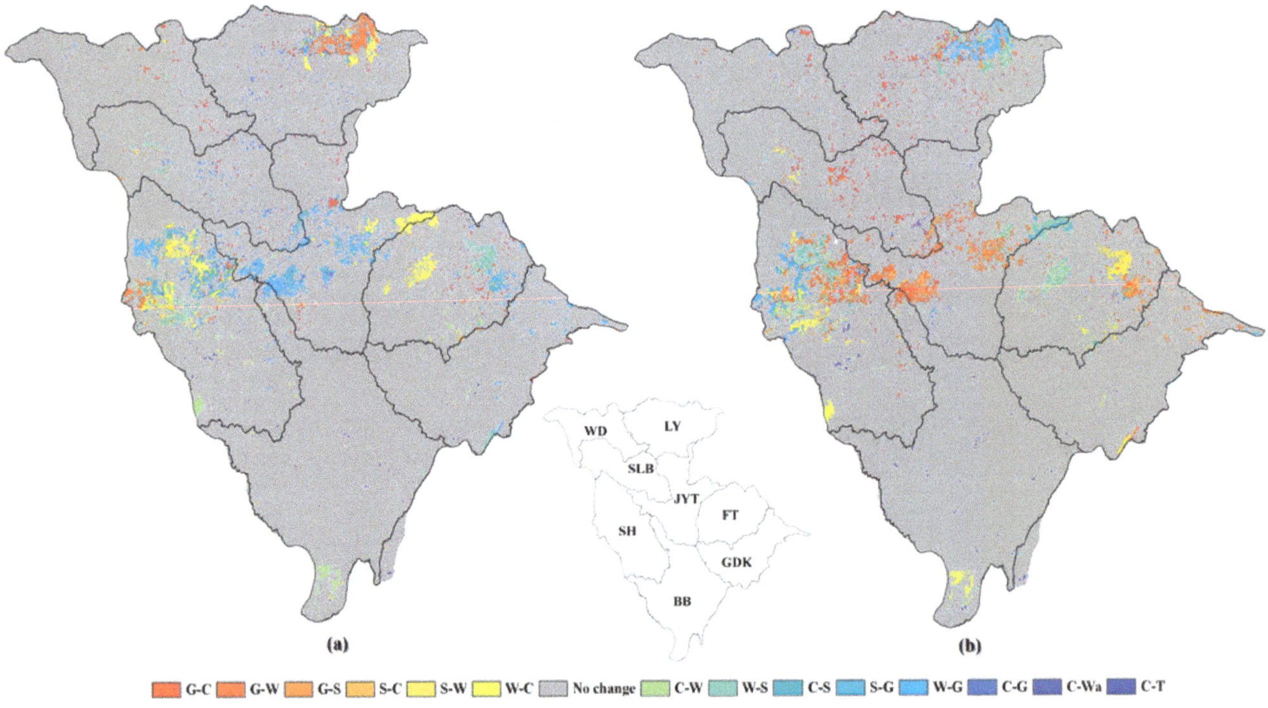

Figure 2: Spatial land use changes in the Jialing River Basin: (**a**) land use changes from 1985 to 1995 and (**b**) land use changes from 1995 to 2000.

From Table 4 and Table 5 and Figure 2, we obtained a clear picture of the quantity and spatial positions of land use changes. Most of the changes occurred in the Lueyang (LY), Shehong (SH), Fengtan (FT) and Jinyintai (JYT) subcatchments. In the LY subcatchment, the changes mainly occurred in the northeast and corresponded with the conversion of G-W (G, grass; W, wood) and S-W (S, shrub) between 1985 and 1995 and the conversion of W-G and W-S between 1995 and 2000. In SH, the changes occurred in the upper to middle area with the conversion of W-G, S-W and G-W between 1985 and 1995 and with the conversion of G-W, S-W, W-G and G-S between 1995 and 2000. Most changes in the JYT subcatchment occurred in the middle, with the conversion of W-G, S-G, S-W and G-C (C, crop) from 1985 to 1995 and with the conversion of G-W, G-S and W-S from 1995 to 2000. In the FT subcatchment, the changes mainly occurred in the middle area, with the conversion of S-W, W-S, W-G and S-G from 1985 to 1995 and with the conversion of G-W, G-S, S-W and W-G from 1995 to 2000.

Dynamic Analysis of Land Use Change

The dynamic degrees of the single land use were calculated using Equation (3) and are listed in Table 6.

Table 6: Dynamic degrees of the single land use of the Jialing River Basin

Period	Crop	Wood	Shrub	Grass	Water	Settlement	Bare Ground
1985–1995	0.45	2.29	3.01	1.79	0.84	3.42	0.55
1995–2000	0.86	4.47	6.55	3.29	2.25	7.99	1.04

During each period, the dynamic land use changes of the woodlands, shrublands and settlement were the most active, with 2.29%, 3.01% and 3.42%, respectively, for 1985 to 1995, and with 4.47%, 6.55% and 7.99%, respectively, for 1995 to 2000.

The degrees of the land use dynamics for 1995 to 2000 were larger than those for 1985 to 1995. In addition, the woodlands, shrublands and settlement had the largest increases of 2.18%, 3.54% and 4.57%, respectively.

To reflect the spatial differences in the degrees of land use changes, the dynamic degrees of the single land use were calculated at the subcatchment scale (Table 7). Overall, water and settlement changed most in the Wudu (WD) and LY subcatchments. The most dramatic changes that occurred in the SLB (Sanleiba) were for the croplands, shrublands and settlement for 1985–1995. Between 1995 and 2000, the croplands, water and settlement primarily changed. The woodland, shrubland and grassland areas changed most in JYT. Furthermore, the shrubland and bare ground areas changed significantly in the SH subcatchment, with greater land use changes for the other types of land use relative to the other subcatchments. In the FT subcatchment, woodland, shrubland and settlement were the first three dynamic land use types. The grassland and settlement areas changed more dramatically in the Goudukou (GDK) and Beibei (BB) subcatchments. Throughout the entire catchment, the settlement changes were relatively more dramatic.

The degrees of the regional comprehensive land use dynamics (LC) were calculated using Equation (4) and are listed inTable 8.

From 1985 to 1995, the LC of the Jialing River Basin was 0.35%. The land use dynamic change was greater from 1995 to 2000, with LC = 0.68%. At the subcatchment level, comprehensive land use dynamic changes in LY, JYT, SH and FT were greater than the average (0.35%), and the LC in the SH subcatchment was 0.79%. During the period from 1995 to 2000, the LC values of all of the subcatchments were larger than during the previous period (1985–1995). In the JYT and SH subcatchments, the LC reached 1.04% and 1.54%, respectively, which indicated that the land use changed significantly in these subcatchments. Based on the analysis above, the SH, FT and LY subcatchments were identified as land use change hot spots and were used to explore the impacts of the dynamics of land use changes on the hydrological regimes. Land use changes in the JYT subcatchment, which was not included in hot spots, were obvious. However, because JYT is an internal catchment that is affected by LY, WD and SLB, it was difficult to distinguish the hydrological changes that were only caused by the land use change within its borders.

Table 7: Degrees of the single land use dynamics by subcatchment

Subcatchment	Period	Crop	Wood	Shrub	Grass	Water	Settlement	Bare Ground
WD	1985–1995	0.83	0.09	0.19	0.22	15.00	7.5	0
	1995–2000	1.81	0.15	0.34	0.53	40.00	13.33	0
LY	1985–1995	0.63	3.06	3.68	1.57	13.33	17.14	NONE
	1995–2000	1.32	5.04	11.38	3.27	20.00	23.64	NONE
SLB	1985–1995	1.40	0.46	1.32	0.60	0.83	8.33	0
	1995–2000	3.94	0.92	2.54	1.19	60.00	16.00	0
JYT	1985–1995	0.70	2.99	4.04	3.11	1.05	1.67	0.26
	1995–2000	0.97	6.50	11.01	4.56	2.35	2.22	0.50
SH	1985–1995	0.98	4.91	5.94	5.49	2.37	4.31	31.67
	1995–2000	1.94	9.59	15.31	8.34	3.06	8.99	15.20
FT	1985–1995	0.61	3.88	3.37	1.77	0	3.33	NONE
	1995–2000	1.12	7.69	7.01	2.23	0	4.17	NONE
GDK	1985–1995	0.11	0.72	0.74	2.37	0.96	2.50	0
	1995–2000	0.18	1.42	1.56	3.95	2.63	3.90	0
BB	1985–1995	0.13	1.95	1.34	9.73	0.34	4.12	0
	1995–2000	0.25	3.47	2.83	10.54	0.21	13.49	0

Note: NONE, this land use type does not occur in this area.

Table 8: Comprehensive dynamic degrees of land use (%)

Period	WD	LY	SLB	JYT	SH	FT	GDK	BB	JLJ
1985–1995	0.12	0.44	0.18	0.58	0.79	0.50	0.11	0.08	0.35
1995–2000	0.13	0.89	0.40	1.04	1.54	0.97	0.21	0.15	0.68

Note: JLJ: Jialing River Basin.

Hydrological Response to Land Use Change

Hydrological Modeling Based on the VIC Model

Based on the hydrometeorological data for 1980–1987 as the background climate and by adopting the land use map from 1985, the VIC model was applied to the study area. The period of 1980 to 1984 was used as a calibration period, and the period of 1985 to 1987 was used as a verification period. The simulation results are shown in Figure 3. The VIC modeled the hydrological processes for both the calibration and validation periods with an RE of less than 10% and an NSE of more than 0.9.

Hydrological Changes under Different Land Use Scenarios

The calibrated VIC model was used to simulate the hydrological process of different land use scenarios (1985, 1995, 2000) with the same background climate (1980–1987).

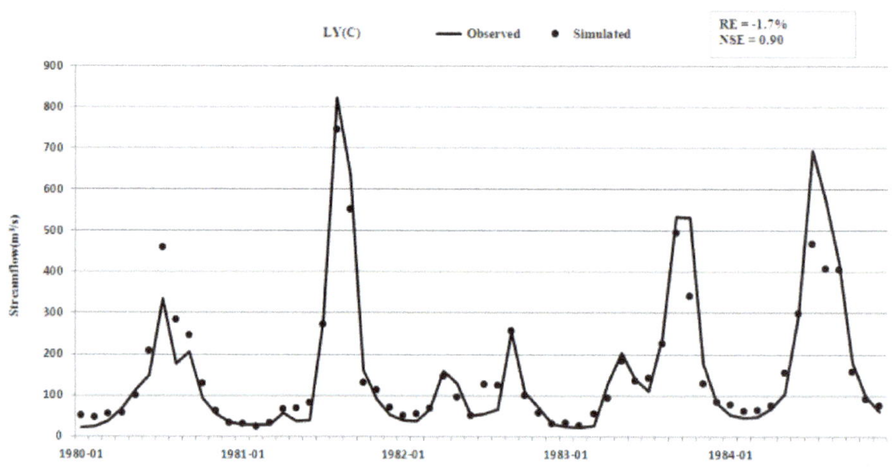

(a)

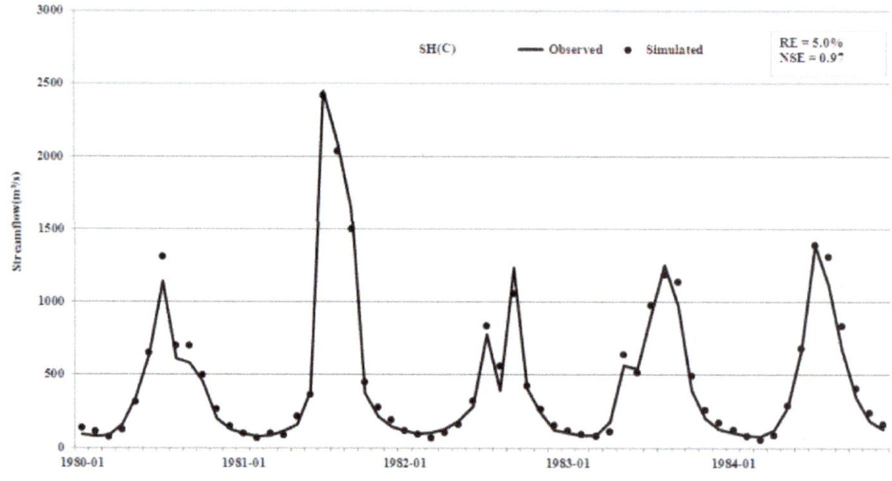

(b)

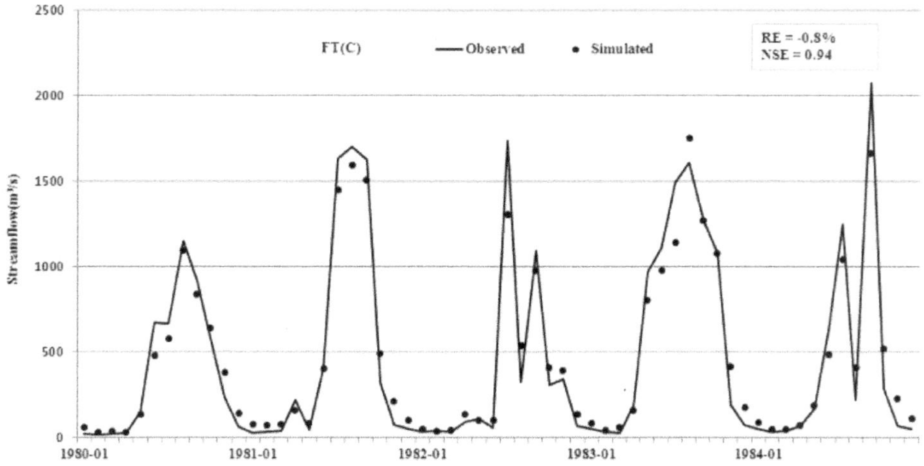
FT(C)
Observed
Simulated
RE = -0.8%
NSE = 0.94
Streamflow(m³/s)
2500
2000
1500
1000
500
0
1980-01
1981-01
1982-01
1983-01
1984-01

(c)

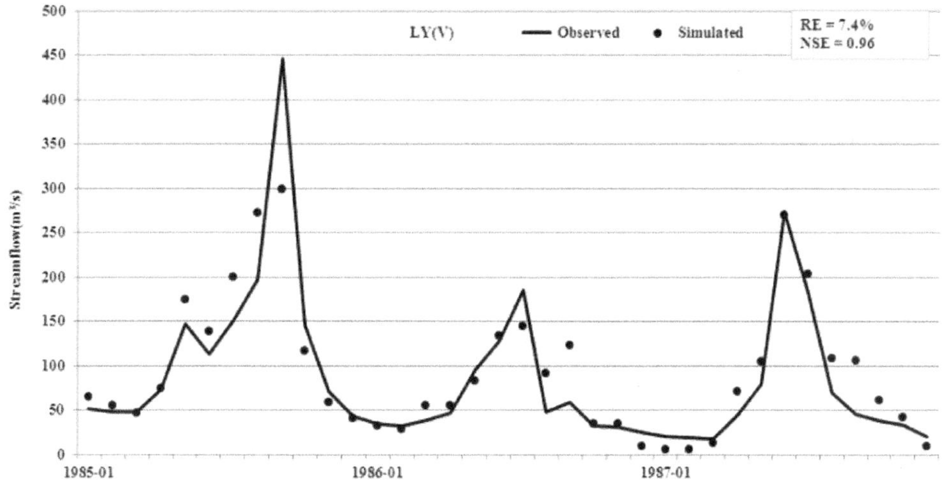
LY(V)
Observed
Simulated
RE = 7.4%
NSE = 0.96
Streamflow(m³/s)
500
450
400
350
300
250
200
150
100
50
0
1985-01
1986-01
1987-01

(d)

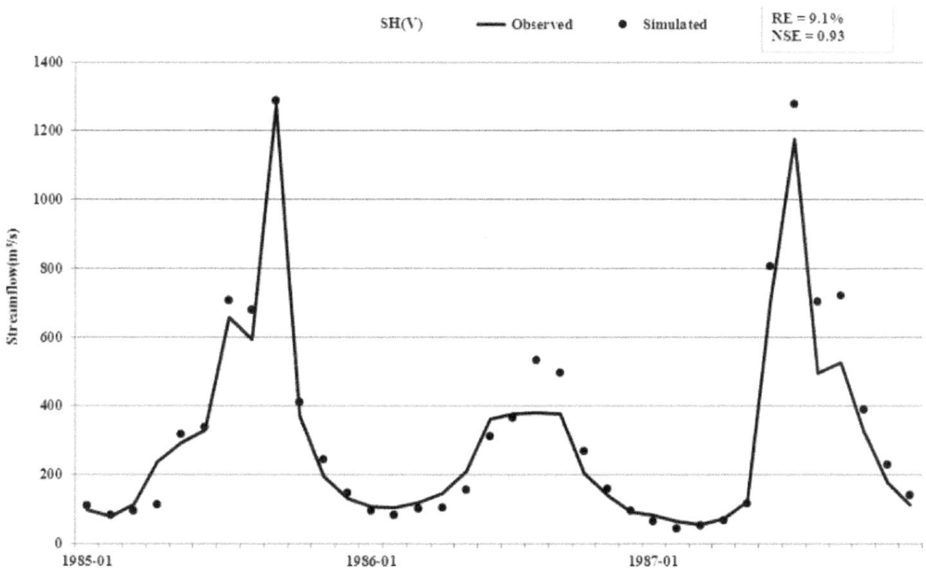
SH(V)
Observed
Simulated
RE = 9.1%
NSE = 0.93
Streamflow(m³/s)
1400
1200
1000
800
600
400
200
0
1985-01
1986-01
1987-01

(e)

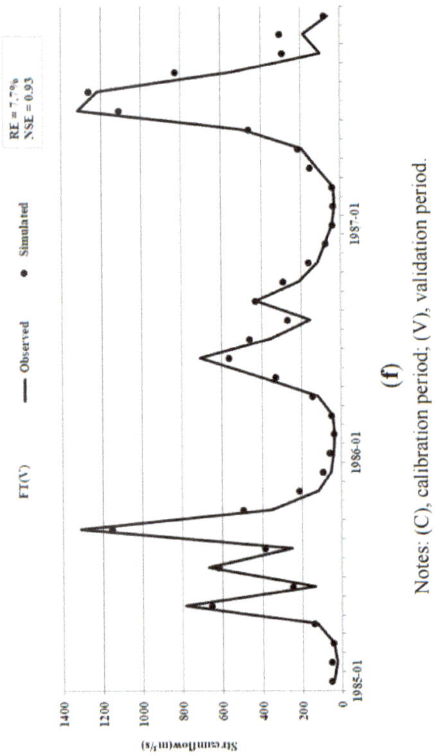

Figure 3: Monthly hydrographs for the calibration and validation periods. RE: relative error; NSE: Nash–Sutcliffe model efficiency coefficient. (**a**) Monthly hydrographs of LY for the calibration period; (**b**) Monthly hydrographs of SH for the calibration period; (**c**) Monthly hydrographs of FT for the calibration period; (**d**) Monthly hydrographs of LY for the validation period; (**e**) Monthly hydrographs of SH for the validation period; (**f**) Monthly hydrographs of FT for the validation period.

Because, in a subcatchment, one hydrological element may increase in one place and decrease elsewhere, which are caused by land use in-and-out conversion, this adds difficulty to the detection of the variation of this element in the whole subcatchment scale if the magnitudes of increase and decrease are similar. We call this phenomenon the "compensation effect". To eliminate this effect, the real change concept was introduced here again.

Final Change

The ET, surface water generation, baseflow, soil water content and canopy interception in the LY, SH and FT subcatchments are listed in Table 9.

In the S1-S0 scenario, increases in the baseflow, surface water generation and soil water contents occurred in LY with a variation of approximately 1 mm. In addition, the ET decreased by 2 mm. In Scenario S2-S1, the baseflow, surface water generation and soil water content decreased by approximately 1 mm, and the ET increased by 2.1 mm. All of the hydrological elements in the SH and FT changed little in Scenarios S1-S0 and S2-S1. However, the land use changed extensively during these periods. (The reasons for this change will be discussed in the next section.)

Table 9: Hydrological regimes for the 1985, 1995 and 2000 land use scenarios

Subcatchment	Hydrological Elements	S0	S1	S2	S1-S0	S2-S1
		mm	mm	mm	mm	mm
LY	ET	444.1	442.1	444.2	−2.0	2.1
	RS	158.6	159.5	158.6	0.9	−0.9
	Bf	50.8	51.7	50.7	0.9	−1.0
	Sw	168.7	169.6	168.7	0.9	−0.9
	W	35.4	35.7	35.4	0.3	−0.3
SH	ET	336.0	335.2	335.4	−0.8	0.2
	RS	355.6	356.1	356.0	0.5	−0.1
	Bf	199.9	200.3	200.1	0.4	−0.2
	Sw	122.2	122.3	122.3	0.1	0
	W	41.2	41.0	41.0	−0.2	0
FT	ET	568.1	568.4	568.1	0.3	−0.3
	RS	552.2	552.2	552.3	0	0.1
	Bf	239.2	239.0	239.1	−0.2	0.1
	Sw	209.6	209.5	209.6	−0.1	0.1
	W	44.2	44.0	44.1	−0.2	0.1

Note: Bf: baseflow; ET: evapotranspiration; Rs: surface water generation; Sw: soil water; W: canopy interception. The same abbreviations are used in the following tables.

Real Change

We got real change by three steps. Firstly, we processed the VIC modeling outputs (ET, Rs, baseflow, SW and W) into grid format (Figure 4 and Figure 5). Then, the "Raster Calculator" Tool of ArcGIS 9.2 [47] was used to get the absolute value of each grid of the outputs. Finally, we got the average value of each subcatchment (LY, SH and FT) by using the "Zonal Statistic as Table" Tool of ArcGIS 9.2.

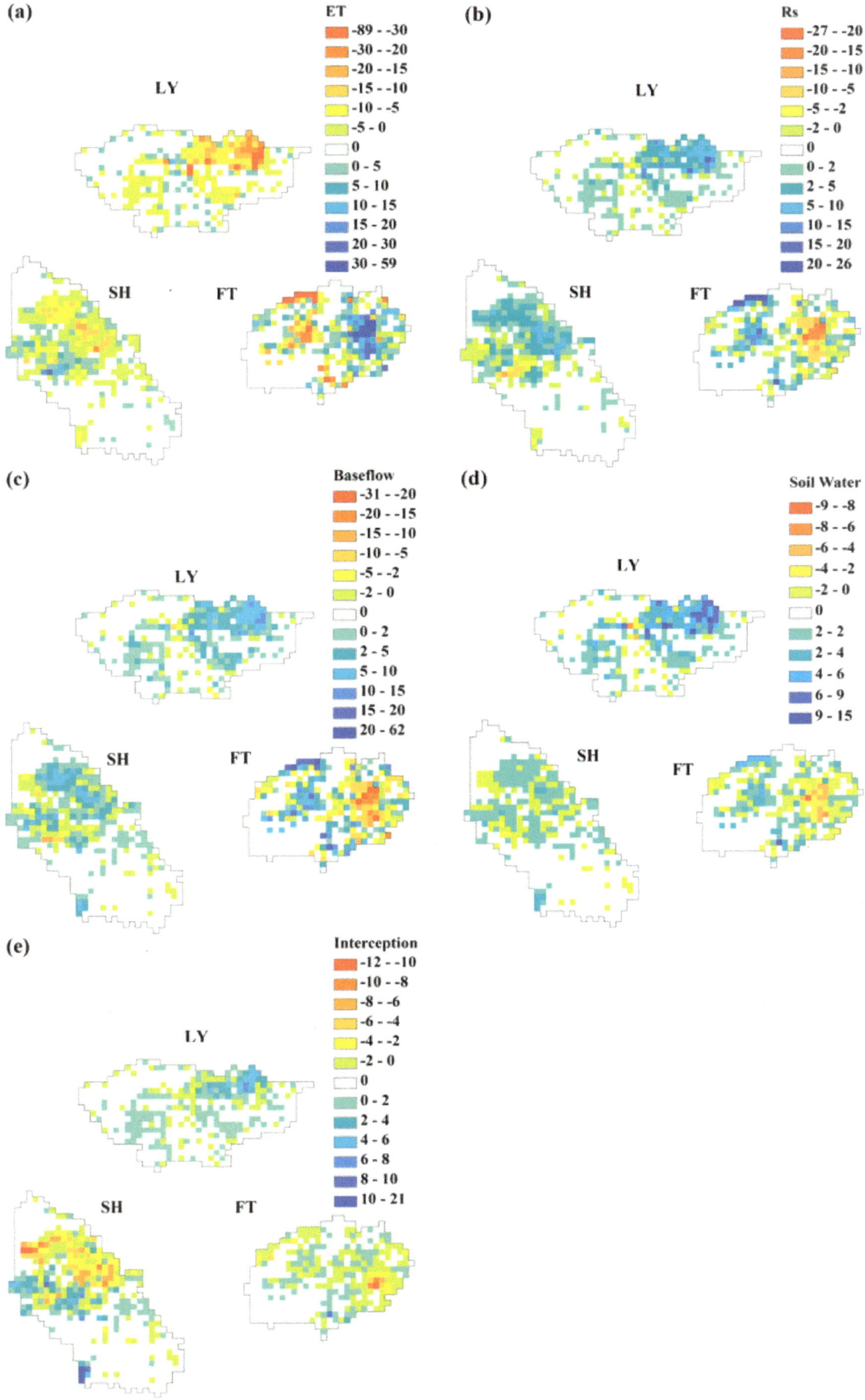

Figure 4: Differences in the hydrological elements in Scenario S1-S0: (**a**) change in evapotranspiration; (**b**) change in surface water generation; (**c**) change in baseflow; (**d**) change in soil water; and (**e**) change in canopy interception.

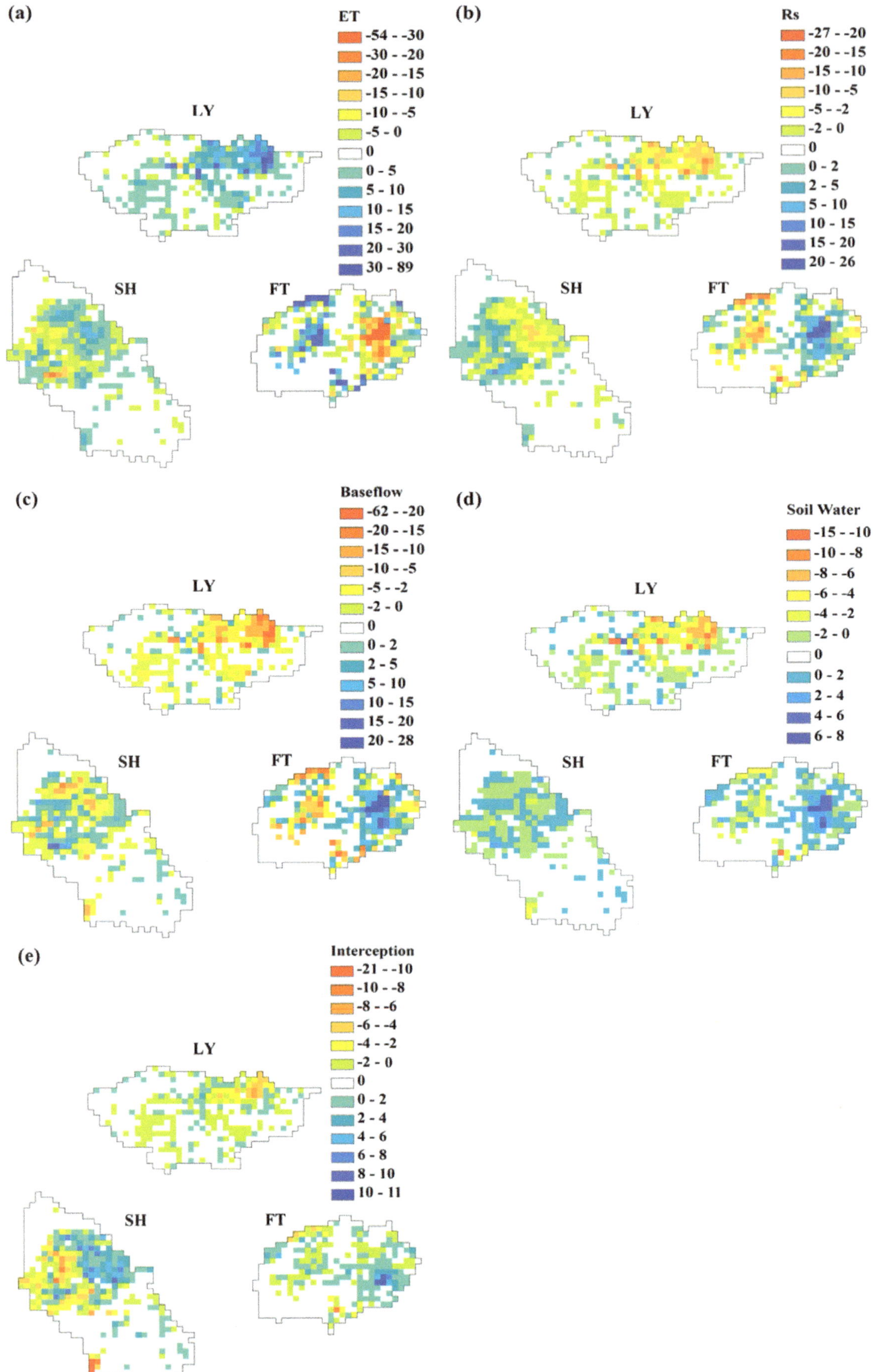

Figure 5: Differences in the hydrological elements in Scenario S2-S1: (**a**) change in evapotranspiration; (**b**) change in surface water generation; (**c**) change in baseflow; (**d**) change in soil water; and (**e**) change in canopy interception.

The real change of the hydrological elements is shown in Table 10.

Table 10: Real change of the hydrological elements under different land use scenarios

Subcatchment	Scenario	Et	Rs	Baseflow	SW	W
		mm	mm	mm	mm	mm
LY	S1-S0	2.4	1.1	1.1	1.1	0.4
	S2-S1	2.4	1.1	1.2	1.1	0.4
SH	S1-S0	1.6	0.8	0.9	0.3	1.2
	S2-S1	1.4	0.7	0.8	0.3	1.1
FT	S1-S0	5.8	2.4	3.4	0.9	0.5
	S2-S1	5.5	2.2	3.2	0.9	0.5

As shown in Table 10, the real changes in the hydrological elements in the hot spots were more significant than the final changes (Table 9). In LY, the real change was slightly larger than the final change. The real change in SH was clearly greater than the final change. The greatest difference between the real change and the final change occurred in FT (S1-S0: ET, Rs and baseflow changed by 5.8 mm, 2.4 mm and 3.3 mm, respectively, and by 5.5 mm, 2.2 mm and 3.2 mm for S2-S1, respectively). Real change is the sum of increase and decrease. It represents all of the variations related to one hydrological component. Thus, it is a figure that reflects the change in the whole area, eliminating the compensation effect, but not a figure that represents the quantity of change of the whole area. Accordingly, real change can be used as an index to detect actual changes in the magnitude of the hydrological elements in a catchment.

DISCUSSION

Land Use Change

To reflect the dynamic character of spatial land use changes during a specific period, an in-and-out phenomenon is described in Section 3.1. If only the final change is used in the analysis, the real change will be hidden in the process of land use change. For example, this study found that the woodland area in the Jialing River Basin was 31,670 km^2 in the 1980s (land use 1985) and 32,193 km^2 in the 1990s (land use 1995), which resulted in a final change of 523 km^2 (Table 2). From 1985 to 1995, 3894 km^2 of woodlands were converted into other land use types, and 3371 km^2 of other types of land use were converted to woodland (Table 3). Therefore, the actual change related to woodlands (real change) was up to 7265 km^2. In this paper, both the change rate and the change in the area of woodlands were highlighted. This in-and-out phenomenon directly affected the detectability of changes in the hydrological regimes that were caused by land use changes, because the changes in the hydrological elements were normally expressed as an overall value for a catchment or a statistical unit in which the in-and-out phenomenon existed.

From the analysis in Section 3.1, it is clear that the land use along the Jialing River significantly changed. The forces driving the frequent transfer and conversion primarily included economic, population and macro-policy forces. In pursuit of economic development, humans have engaged in extensive land use changes, including building and grazing. These changes have resulted in reduced croplands and the degradation of grasslands, which have resulted in environmental deterioration, but also in changes in land use type. In addition, farmers have adjusted the structure of agricultural production to increase their income, and more young people in rural areas have sought better jobs in large cities. All of these factors have indirectly affected land use patterns. Since 1980, the rapid economic development of the Jialing River has accelerated the urbanization process, which has resulted in the conversion of cropland to other types of land use and has resulted in a greater amount of land under construction. Population is an important factor that could lead to land use change. A population increase results in greater residential land. From the 1980s to 2000, the area of settlement in the Jialing River Basin increased steadily from 313 km^2 in the 1980s to 378 km^2 in 1995 and to 489 km^2 in 2000. However, additional croplands are required to feed the growing population. Since 1985, due to the continuing population growth in the Jialing River Basin (43.9 million in 1995, 44.8 million in 2000, calculated from 1 km $\times$ 1 km grid datasets of population, China), industrial development and the expansion of farming areas inevitably resulted in decreasing woodland and grassland areas. Population pressure and human activities are the main reasons for the dramatic changes in the woodland area. In addition, macro-policy is a factor that cannot be neglected. In the 1980s, the government of the P.R. of China implemented a policy to construct a huge shelter-forest system upstream of the Yangtze River. This project was called the "Yangtze River Management" project. The affected area was approximately 200,000 km^2, covering Sichuang, Yunnan, Guizhou and Hubei provinces, and included the Tuo River, Min River, Jialing River, Wu River and the mainstream of the Three Gorges [48]. Since the 1990s, a national policy of returning cropland to woodlands and grasslands (RCFG) has been implemented. Furthermore, maintaining a dynamic balance of arable land is currently a basic land use policy in China. All of these factors contributed to dynamic changes in land use within the Jialing River Watershed.

Hydrological Response

In LY, woodlands increased by 944 km^2 (croplands, shrublands and grasslands decreased by 193 km^2, 452 km^2 and 303 km^2, respectively) from 1985 to 1995. In addition, from

1995 to 2000, 936 km² of woodlands were converted into croplands, shrublands and grasslands, which increased by 216 km², 453 km² and 267 km², respectively. As the woodland area increased, the ET decreased (S1-S0) and *vice versa* (Table 9). In terms of the hydrological effects of the woodlands, many researchers [3,4,7] have concluded that harvesting forests or converting them into grasslands or croplands decreases the ET because the canopy interception decreases. However, we observed an increase in the ET in this study. Actually, the effects of land use change on hydrological processes are extremely complex. In addition to the biophysical characteristics of vegetation, such as LAI, albedo, stoma resistance and root distribution, differences in the background climate are also important. Actual ET is similar to potential ET for the humid climate characteristic of the upstream region of the Yangtze River [49]. In addition, the lower temperature, higher moisture and slower air exchange in woodlands will decrease ET [48]. Therefore, a decrease in forestland and shrubland may not result in a decrease in ET. The research institute of the Yangtze River Basin's planning office has observed the same results in their study of paired subcatchments [49]. Furthermore, a study by Zheng *et al.* [50] of the Pingtong River, one of the subcatchments in the Jialing River, has supported this conclusion.

In Scenarios S1-S0 and S2-S1, the hydrological regimes did not change noticeably, as we previously mentioned. However, the land use changed extensively. Lahmer *et al.* [51] applied a GIS-based modeling approach to two river basins and concluded that moderate land use change results in only small changes in various water balance components. Lu Zhixiang *et al.* [52] also did not detect obvious changes in hydrological elements, even in two of their three studied subcatchments, where land use changed significantly. Take Subcatchment 7 in their study, for example: from 1995 to 2000, the area of cropland decreased 26.94%, while woodland and grassland area increased 8.69% and 18.55%. Land use changes during this period only led to a 0.47- and 1.64-mm increase in ET and water yield and a 2.37-mm decrease in soil water. Thus, the current paper attempts to explain why this change occurred by combining the concepts of real change and final change with the spatial distribution maps (Figure 4 and Figure 5) and statistic distribution tables of the hydrological elements (Table 11,Table 12, Table 13, Table 14 and Table 15).

Table 11: Statistic distribution of ET variation in Scenario S1-S0 (area percentage, %)

ET	−89–−30	−30–−20	−20–−15	−15–−10	−10–−5	−5–0	0	0–5	5–10	10–15	15–20	20–30	30–59
LY	0.2	1.6	1.8	3.7	6.5	22.6	54.3	8.9	0.2	0.2	0	0	0
SH	0	0.2	0.3	2.4	6.9	22.7	53.0	12.7	0.9	0.7	0.2	0	0
FT	2.2	2.5	2.2	4.2	4.7	10.6	44.2	12.3	6.1	4.9	2.0	2.2	2.0

Table 12: Statistic distribution of Rs variation in Scenario S1-S0 (area percentage, %)

Rs	−27–−20	−20–−15	−15–10	−10–−5	−5–−2	−2–0	0	0–2	2–5	5–10	10–15	15–20	20–26
LY	0	0	0	0.2	0.4	9.1	54.5	21.3	8.1	4.5	1.4	0.4	0
SH	0	0	0.2	0.9	1.7	11.4	56.1	16.9	10.3	2.4	0.2	0	0
FT	0.5	1.5	1.0	4.4	7.6	12.0	45.5	11.1	7.6	5.2	1.5	0.5	1.7

Table 13: Statistic distribution of baseflow variation in Scenario S1-S0 (area percentage, %)

Baseflow	−31–−20	−20–−15	−15–−10	−10–−5	−5–−2	−2–0	0	0–2	2–5	5–10	10–15	15–20	20–62
LY	0	0	0	0.2	1.0	8.1	54.5	18.7	10.4	5.9	1.2	0	0
SH	0	0	0.2	0.7	2.1	13.9	52.8	18.2	7.9	4.1	0	0	0
FT	1.2	1.5	2.0	7.9	10.8	6.4	44.5	5.2	8.1	7.1	2.7	2.0	0.7

Table 14: Statistic distribution of soil water variation in Scenario S1-S0 (area percentage, %)

Soil water	−9–−8	−8–−6	−6–−4	−4–−2	−2–0	0	0–2	2–4	4–6	6–9	9–15
LY	0	0.2	0.2	0.8	8.1	54.5	19.5	6.5	5.7	3.9	0.6
SH	0	0	0	0.3	15.7	57.3	25.1	1.4	0.2	0	0
FT	0.2	0.2	2.2	5.7	21.1	44.5	17.4	6.4	2.0	0.0	0.2

Table 15: Statistic distribution of interception variation in Scenario S1-S0 (area percentage, %)

Interception	−12–−10	−10–−8	−8–−6	−6–−4	−4–−2	−2–0	0	0–2	2–4	4–6	6–8	8–10	10–21
LY	0	0	0	0	0.2	14.2	55.9	24.0	3.0	2.2	0.4	0	0
SH	0.3	0.7	2.9	2.9	7.1	15.0	52.8	11.0	4.3	1.7	0.3	0.3	0.5
FT	0.2	0.2	0.5	0.7	2.5	28.7	47.2	18.9	0.7	0	0	0.2	0

Change in the hydrological elements in S1-S0 can be used as an example, because in this scenario, the hydrological element changed the most. In Figure 4, we can see that the local hydrological elements changed significantly. ET changed from −89 to 59 mm. The generation of surface water changed from −27 to 26 mm. The baseflow changed from −31 to 62 mm, and the soil water and interception changed from −9 to 15 mm and from −12 to 21 mm, respectively. However, the magnitudes of change in the hydrological components in each subcatchment (LY, SH and FT) were not obvious (Table 9). This finding may be explained by the compensation and average effects.

In LY, the difference between the final change and the real change in the hydrological elements was not great (Table 9 andTable 10), which indicated that compensation effect was not obvious. This outcome resulted from the unique directions of change in LY (Figure 4 and Table 11, Table 12, Table 13, Table 14 and Table 15). There were 36.4% of the grids with decreased ET values. Only in 9.3% of area did ET increase. However, other elements have the opposite variation characteristics. Rs, baseflow, soil water content and interception increased in 35.8%, 36.2%, 36.2% and 29.7% and decreased in 9.8%, 9.3%, 9.3% and 14.4% of the area, separately. Thus, the compensation effect of each element was small. Meanwhile, because many grids exhibited no change (around 54.5% of grids) or small change (ET: 22.6% of grids with a range of [−5–0); Rs: 29.4% of grids with a range of (0–5]; baseflow: 29.1% of grids with a range of (0–5]; soil water content: 31.7% of grids with a range of (0–6]; and interception: 29.3% of grids with a range of (0–6]), the average effect was great. Essentially, little overall change occurred in the hydrological regimes in LY.

SH and FT had more significant real changes than LY. In the SH and FT, positive and negative changes occurred in each element (Figure 4 and Table 11, Table 12, Table 13, Table 14 and Table 15). The compensation effect played an important role that made it difficult to detect the change. The FT showed the largest difference between real change and final change (Table 9 and Table 10). In FT, there was 26.3% of the area with decreased ET and 29.5% with increased ET. The proportions of area for other hydrological components (Rs, baseflow, soil water content and interception) with decreased values were 27.0%, 29.7%, 29.5% and 32.9%, separately. Correspondingly, the area percentages with increased values were 29.5%, 27.5%, 25.8%, 26.0% and 19.9%. The increased area and decreased area were similar. This made it difficult to detect variations that result from land use changes in the hydrological elements.

To detect obvious changes in hydrological components caused by land use change, it is necessary to pay attention to the scale problem. In a large watershed, the compensation and average effects may decrease the detectability.

CONCLUSIONS

Based on the analysis above, we reached the following three conclusions:

- The land use quantities and spatial configuration in the Jialing River Basin changed dramatically due to economic development, population growth and national macro-policy. Settlement and shrubland areas changed significantly, with final change rates of 20.77% and −16.07%, respectively, from 1985 to 1995; real change rates of 34.2% and 30.1%, respectively, from 1985 to 1995; final change rates of 29.37% and 12.40%, respectively, from 1995 to 2000; and real change rates of 39.9% and 32.8%, respectively, from 1995 to 2000. Due to an in-and-out conversion relationship, the actual land use change was hidden in the final change analysis. In the real change analysis, the change rate and change in woodland area were highlighted. These changes were more dynamic in the LY, SH and FT subcatchments;
- The decrease in woodland area in the LY subcatchment resulted in greater ET; however, the other hydrological elements decreased;
- Although the hydrological elements changed little overall due to the average and compensation effects, they changed significantly at the local scale. Real change analysis can reduce the compensation effects and enhance the detectability of change.

This study considered the concepts of real change and final change and the average and compensation effects, which are important for deeply analyzing land use changes and their impacts on hydrological regimes.

ACKNOWLEDGMENTS

The study is supported by the National Natural Science Foundation of China (41030636), the National High Technology Research and Development Program of China (2012AA12A309), National Natural Science Foundation of China (51420105014) and the Research and Innovation Project for College Graduates of Jiangsu Province (CXZZ12_0239).

REFERENCES

1. Zhang, L.T.; Ding, W.R.; Zhou, Y.; Zhang, L. Study on land use and land cover change. *Environ. Sci. Surv.* 2007,*26*, 7–10.
2. Chen, J.; Li, X. Simulation of hydrological response to land-cover changes. *Chin. J. Appl. Ecol.* 2004, *15*, 833–836.
3. Dunn, S.M.; Mackay, R. Spatial variation in evapotranspiration and the influence of land use on catchment hydrology. *J. Hydrol.* 1995, *1995*, 49–73.
4. Matheussen, B.; Krichbaum, R.L.; Goodman, I.A.;

O'Donnell, G.M.; Lettenmaier, D.P. Effects of land cover change on streamflow in the interior Columbia River Basin. *Hydrol. Process.* 2000, *14*, 867–885.

5. Olchev, A.; Ibrom, A.; Priess, J.; Erasmi, S.; Leemhuis, C.; Twele, A.; Radler, K.; Kreilein, H.; Panferov, O.; Gravenhorst, G. Effects of land-use changes on evapotranspiration of tropical rain forest margin area in Central Sulawesi (Indonesia): Modelling study with a regional SVAT model. *Ecol. Model.* 2008, *212*, 131–137.
6. Jin, C.; Zhang, B.; Song, K.; Wang, Z.; Yang, G. RS-based analysis on the effects of land use/cover change on regional evapotranspiration: A case study in Qian'an County, Jilin Province. *Arid Zone Res.* 2009, *26*, 734–743.
7. Mao, D.Z.; Cherkauer, K.A. Impacts of land-use change on hydrologic responses in the Great Lakes region. *J. Hydrol.* 2009, *374*, 71–82.
8. Andraski, B.J. Soil-water movement under natural-site and waste-site conditions: A multiple-year field study in the Mojave Desert, Nevada. *Water Resour. Res.* 1997, *33*, 1901–1916.
9. Shi, P.L.; Li, W.H. Influence of forest cover change on hydrological process and watered runoff. *J. Nat. Resour.*2001, *16*, 481–487.
10. Krause, P. Quantifying the impact of land use changes on the water balance of large catchments using the J2000 model. *Phys. Chem. Earth* 2002, *27*, 663–673.
11. Legesse, D.; Vallet-Coulomb, C.; Gasse, F. Hydrological response of a catchment to climate and land use changes in Tropical Africa: case study South Central Ethiopia. *J. Hydrol.* 2003, *275*, 67–85.
12. Coe, M.T.; Costa, M.H.; Soares-Filhoc, B.S. The influence of historical and potential future deforestation on the stream flow of the Amazon River—Land surface processes and atmospheric feedbacks. *J. Hydrol.* 2009, *369*, 165–174.
13. Savary, S.; Rousseau, A.N.; Quilbe, R. Assessing the effects of historical land cover changes on runoff and low flows using remote sensing and hydrological modeling. *J. Hydrol. Eng.* 2009, *14*, 575–587.
14. Schilling, K.E.; Chan, K.S.; Liu, H.; Zhang, Y.K. Quantifying the effect of land use land cover change on increasing discharge in the Upper Mississippi River. *J. Hydrol.* 2010, *387*, 343–345.
15. Storck, P.; Bowling, L.; Wetherbee, P.; Lettenmaier, D. Application of a GIS-based distributed hydrology model for prediction of forest harvest effects on peak stream flow in the Pacific Northwes. *Hydrol. Process.* 1998, *12*, 889–904.
16. Fohrer, N.; Haverkamp, S.; Eckhardt, K.; Frede, H.G. Hydrologic response to land use changes on the catchment scale. *Phys. Chem. Earth Part B Hydrol. Oceans Atmos.* 2001, *26*, 577–582.
17. Sun, H.N. Progress of research on the role of forest during past 20 years. *J. Nat. Resour.* 2001, *16*, 407–412.
18. Hundecha, Y.; Bardossy, A. Modeling of the effect of land use changes on the runoff generation of a river basin through parameter regionalization of a watershed model. *J. Hydrol.* 2004, *292*, 281–295.
19. Archer, D.R.; Climent-Soler, D.; Holman, I.P. Changes in discharge rise and fall rates applied to impact assessment of catchment land use. *Hydrol. Res.* 2010, *41*, 13–26.
20. Cuo, L.; Lettenmaier, D.P.; Alberti, M.; Richey, J.E. Effects of a century of land cover and climate change on the hydrology of the Puget Sound basin. *Hydrol. Process* 2009, *23*, 907–933.
21. Liu, Y.; Zhang, X.; Xia, D.; You, J.; Rong, Y.; Bakir, M. Impacts of land-use and climate changes on hydrologic processes in the Qingyi River watershed, China. *J. Hydrol. Eng.* 2013, *18*, 1495–1512.
22. Ghaffari, G.; Keesstra, S.; Ghodousi, J.; Ahmadi, H. SWAT-simulated hydrological impact of land-use change in the Zanjanrood basin, Northwest Iran. *Hydrol. Process* 2010, *24*, 892–903.
23. Ashagrie, A.G.; de Laat, P.; de Wit, M.; Tu, M.; Uhlenbrook, S. Detecting the influence of land use changes on discharges and floods in the Meuse River Basin—The predictive power of a ninety-year rainfall-runoff relation? *Hydrol. Earth Syst. Sci.* 2006, *10*, 691–701.
24. Xian, W.; Shao, H.Y.; Zhou, W.C. The dynamic monitoring and prediction of land use pattern in the area of middle and lower reaches of the Jialingjiang River. *Res. Soil Water Conserv.* 2005, *12*, 61–64.
25. Tang, Q.; Shao, H.; Xian, W.; Zhou, W. Dynamic change in land use/cover in middle and lower reaches of the Jialing River. *Soils* 2006, *38*, 36–41.
26. Shao, H.; Xian, W.; Yang, W. Contrastive research on the process of land use change in main valleys of the upper reaches of the Yangtze River. *Ecol. Environ.* 2008, *17*, 792–797.
27. Wu, X.; Shen, Z.; Liu, R. Study on the reginal differntation of land use/land cover change in the upper reaches of Yangtze River. *J. Basic Sci. Eng.* 2008, *16*, 819–829.
28. Leng, S.; Zhang, X.; Xia, D.; Liu, Y. Dynamic analysis of landscape structure of land use in Jialingjiang Watershed. *J. China Three Gorge Univ. Nat. Sci.* 2012, *34*, 38–42.
29. Qi, S.; Zhang, X.; Xia, D.; Liu, Y.; Liu, B. Characteristic and driving forces of land use/cover change in Jialingjiang River Basin. *J. Yangtze River Sci. Res. Inst.* 2013, *30*, 1–7.
30. Data Center for Resources and Environmental Sciences, Chinese Academy of Sciences (RESDC). Available online: http://www.resdc.cn (accessed on 19 June 2013). (In Chinese).
31. Zhang, X.N.; Liu, Y.Y.; Fang, Y.H.; Liu, B.J.; Xia, D.Z. Modeling and assessing hydrologic processes for historical and potential land-cover change in the Duoyingping watershed, southwest China. *Phys. Chem. Earth*2012, *53–54*, 19–29.
32. Shuttle Radar Topography Mission (SRTM) Website. Available online: http://srtm.csi.cgiar.org/ (accessed on 1 September 2012).
33. Reynolds, C.A.; Jackson, T.J.; Rawls, W.J. Estimating

available water content by linking the FAO soil map of the world with global soil profile databases and pedotransfer functions. In Proceedings of the AGU 1999 Spring Conference, Boston, MA, USA, 1–4 June 1999.

34. Lin, Q.; Luo, G.P.; Chen, X. Review of land use model. *Prog. Geogr.* 2005, *24*, 81–89.
35. Wang, X.L.; Bao, Y.H. Study on the methods of land use dynamic change research. *Prog. Geogr.* 1999, *18*, 83–89.
36. Zhu, H.Y.; Li, X.B. Discussion on the index method of redional land use change. *J. Geogr. Sci.* 2003, *58*, 643–650.
37. Liu, Y. Characteristics of Land Cover Change and Hydrologic Responses at the Watershed Scale. Ph.D. Thesis, Hohai University, Nanjing, China, 31 May 2012.
38. Liang, X.; Lettenmaier, D.P. A simple hydrologically based model of land surface water and energy fluxes for general circulation models. *J. Geophys. Res.* 1994, *99*, 14415–14428.
39. Liang, X.; Wood, E.F.; Lettenmaier, D.P. Surface soil moisture parameterization of the VIC-2L model: Evaluation and modification. *Glob. Planet. Change* 1996, *13*, 195–206.
40. Nijssen, B.; Lettenmaier, D.P.; Liang, X.; Wetzel, S.W.; Wood, E.F. Streamflow simulation for continental-scale river basins. *Water Resour. Res.* 1997, *33*, 711–724.
41. Gao, H.; Tang, Q.; Shi, X.; Zhu, C.; Bohn, T.; Su, F.; Sheffield, J.; Pan, M.; Lettenmaier, D.; Wood, E.F. Water Budget Record from Variable Infiltration Capacity (VIC) Model. http://www.hydro.washington.edu/Lettenmaier/Models/VIC/Documentation/References.shtml (accessed on 13 March 2013).
42. VanShaar, J.; Haddeland, I.; Lettenmaier, D.P. Effects of land-cover changes on the hydrological response of interior Columbia River basin forested catchments. *Hydrol. Process* 2002, *12*, 2499–2520.
43. Hurkmans, R.T.W.L.; Terink, W.; Uijlenhoet, R.; Moors, E.J.; Troch, P.A.; Verburg, P.H. Effects of land use changes on streamflow generation in the Rhine basin. *Water Resour. Res.* 2009, *45*, 1–15.
44. Mishra, V.; Cherkauer, K.A.; Niyogi, D.; Lei, M.; Pijanowski, B.C.; Ray, D.K.; Bowling, L.C.; Yang, G. A regional scale assessment of land use/land cover and climatic changes on water and energy cycle in the upper Midwest United States. *Int. J. Climatol.* 2010, *30*, 2025–2044.
45. Cuo, L.; Zhang, Y.; Gao, Y.; Hao, Z.; Cairang, L. The impacts of climate change and land cover/use transition on the hydrology in the upper Yellow River Basin, China. *J. Hydrol.* 2013, *502*, 37–52.
46. Nash, J.E.; Sutcliffe, J.V. River flow forecasting through conceptual models part I—A discussion of principles. *J. Hydrol.* 1970, *10*, 282–290.
47. ArcGIS for Desktop. Available online: http://www.esri.com/software/arcgis/arcgis-for-desktop (accessed on 12 October 2009).
48. Cheng, G.W.; Sun, Y. Forest change: Hydrological effects in the upper Yangtze River Valley. *Ambio* 1999, *28*, 457–459.
49. Liu, C.M.; Chung, C.H. The influence of forest cover upon annal runoff in the loess plateau of China. *J. Geogr. Sci.* 1978, *33*, 112–127.
50. Zheng, S.; Mu, C.; Chen, Z.; Gong, G.; Li, Y. Simulations and analysis on the effects of forest on the hydrological process in the upper reaches of Yangtze River. *Acta Ecol. Sinica* 2010, *30*, 3046–3056.
51. Lahmer, W.; Pfützner, B.; Becker, A. Assessment of land use and climate change impacts on the mesoscale.*Phys. Chem. Earth Part. B Hydrol. Oceans Atmos.* 2001, *26*, 565–575.
52. Lu, Z.X. A study of the land use change and its hydrologic response in the upper reaches of the Fen River. *J. Glaciol. Geocryol.* 2014, *36*, 192–199.

CHAPTER 11

GEOGRAPHIC CONCERNS ON FLOOD CLIMATE AND FLOOD HYDROLOGY IN MONSOON

Sandipan Ghosh and Biswaranjan Mistri

INTRODUCTION

The flowing water finds the lowest parts of the earth's surface, forming small rivulets that merge into larger ones and eventually find their way into a river. When a river carries much more water than usual, it is said to be "in spate," a term of uncertain origin but which may be derived from the Dutch verb spuiten, meaning "to flood" [1]. A flood is an unusual high stage of a river due to runoff from rainfall and/or melting of snow in quantities too great to be confined in the normal water surface elevation of the river or stream, as the result of unusual meteorological combination [2]. "Flood" means that the flow in the river is in such an excess as to raise the level of the river at places so that it overflows the banks and rises to a level higher than the adjacent countrysides, thus inundating the areas adjacent to the channel [3, 4]. The river fails at that time to transmit the excess flow keeping it within the confines of its bank heights. A flood thus comprises two main phases, namely, "the land phase" where rain rejected by soil becomes runoff and generates flood and "the channel phase" which follows the land phase closely and starts when surface runoff enters the stream channels. Ordinarily, when speaking of floods, we mean that the channel phase is the most spectacular and on the whole the most destructive [4]. The term "flood climate" is probably introduced by Hayden (1988) who has developed a global classification of flood-producing climates based on the mean seasonal state of the atmosphere [5]. Hirschboeck (1988) introduces the discipline of "flood hydroclimatology" which gives focus of hydrometeorological-scale atmosphere activity, while, at the same time, seeking to place this activity within a broader spatial and temporal climatic perspective of flood [6]. Tropical climates including monsoon climate have high potentiality of floods because, here, the atmosphere (i.e., primary reservoir of flood water) is primarily characterized by barotropy, typical of tropical low latitudes, precipitation is released by synoptic convective activity (e.g., tropical thunder storms), local convective activity (thunderstorms), and rising circumequatorial air at the Intertropical Convergence Zone (ITCZ). Flood hydrology deals with functionality of time- and space-dependent processes of waters and is focused on hydraulic and engineering dimensions of floods. Flood hydrology includes hydrometeorological aspects of flood, flood dynamics of channel, flood routing, and identification of flood stage, flood frequency analysis (i.e., design flood), flood risk estimation, and flood management [2, 7].

Nature has been kind to India in bestowing a season like monsoon (June–October) during which almost 75 to 80 percent of annual rainfall is received. Naturally, when almost the entire year's rainfall occurs in a period of few months, floods are bound to occur in its river [8]. Floods in relation to climate change and meteorological variability in recent past in India have already been analyzed and discussed by many researchers [8–17]. Large synoptic systems ranging in force from tropical lows to cyclones are the main cause of unusually large floods on the Indian rivers [18]. It is now found that the rainfall spreading over a short time may cause flood but the same spread over a longer span may result in a drought. There is considerable temporal variation in the monsoon rainfall related climatic variability in India. For example, 1917 and 1961 were excess monsoon years (>1000 m or +18 percent) and 2002, 1987, 1972, 1918, 1899, and 1877 were major deficient monsoon years (<700 mm or −18 percent) [16]. Also, twentieth century development in tropical meteorology reveals that poor monsoon is found to be associated with warm El nino events. The global warming and short-term climate change of Indian subcontinent simultaneously and forcefully influence the monsoon system affecting spatial trend of rainfall and

Citation: Sandipan Ghosh and Biswaranjan Mistri, "Geographic Concerns on Flood Climate and Flood Hydrology in Monsoon," Geography Journal, vol. 2015, Article ID 486740, 16 pages, 2015. doi:10.1155/2015/486740.

occurrences of extreme floods. On the basis of marine sediment cores and isotope dating in India, the monsoon intensity shows a steady increase from ~1700 AD to present [16]. Some recent studies [17, 19] have shown contrasting results of a decreasing trend in the frequency of tropical cyclones and monsoon depressions and significant rising trends in the frequency and magnitude of extreme rainfall events over India between 1951 and 2010. Wet extremes are projected to become more severe in many areas of India where mean precipitation is expected to increase and dry extremes are projected to become more intense in areas where mean precipitation is projected to decrease [17]. Now, it has been found that the rainfall in 6-hour and 3-hour duration can be about 75 and 60 percent of the 24 hour rainfall in the Himalayan foothills and central plateaus and it is intense rainfall which gives rise to flash floods in the rivers [11]. The phenomena of Mumbai flood in July 2005, Uttarakhand flood disaster in June 2013, and, most recently, flood of Jammu and Kashmir in September 2014 is compelled to think about predicting flood dynamics, changing climatic pattern, and improvement of flood forecasting. The flood generating rainstorms are usually of 3-day duration and are confined in the whole Ganga Basin where the Damodar River Basin is a subcatchment of it [12]. Entire Damodar River Basin is situated in the high-moderate flash flood zone (80 to 120 mm 3 hour rainfall). So, observing the flood potentiality of this basin, it is selected as a research area to assess the temporal trend of flood climate and flood hydrology. In this regard, when we consider the floodplains of West Bengal, the flood is regarded as extreme hydrometeorological annual phenomenon of monsoon season when the excessive surface runoff and over flow of alluvial rivers submerge the surrounding low lying areas of Barddhaman, Hooghly, and Howrah districts. At that time, the surface drainage systems of lower Damodar Basin failed to cope up with it. That is the main concern of inhabitants of southern West Bengal. The flood history of Damodar River is very much popular to West Bengal, as well as India, because this River is identified as "Sorrow of Bengal" due to havoc destruction in annual floods, particularly in the lower reach [20]. If we consider floods as extreme hydrological phenomenon and as quasidynamic equilibrium fluvial process, then it is not surprising to us that, to attain the present form of lower Damodar Floodplain, the flood aggradation and degradation are considered as the prime processes. But, in reality, settlements beside a river, like Damodar River, can be a mixed blessing; for once in a while, the river may overflow its banks and generate a heavy toll of property losses, income losses, and sometimes losses of life as well. In some cases, man has learned to live with such periodic inundations of the fertile floodplain [21]. The causes of floods can be natural, but human interference intensifies many floods. The decision to live in a floodplain, for a variety of perceived benefits, is one that is fraught with difficulties. The increase in flood damage is related to the increasing number of people living in floodplain regions [22]. Remembering and analyzing the past flood events of lower Damodar River, the very common but unsolved problem of Damodar River Basin is to manage the huge inflow water of runoff from upper catchment through the existing engineering structures of Damodar Valley Corporation (DVC) flood regulation system. It is a ground reality that incapable flood storage system, low canal consumption, declining carrying capacity of river, decayed palaeochannels, mounting unscientific embankments, and so forth are found to be the chief causes of recent floods and these are not capable of coping up with huge volume of runoff. The essential part of this study is the precise estimation and forecasting of runoff in relation to flood climate and flood risk of lower Damodar River. On the basis of hydrogeomorphic perspective of flood, we have considered the following five prime objectives of study:(1)understanding the geohydrological set-up of Damodar River Basin,(2)analyzing the influence of monsoon rainfall on flood risk,(3)estimating potential runoff to understand the magnitude of flood discharge,(4)quantifying the potential carrying capacity of Damodar River and current threshold level of bankfull discharge,(5)identifying the dominant key factors of flood risk in the lower part of Damodar River Basin.

PREVIOUS WORKS ON DAMODAR FLOODS

Floods, a popular theme of research as far as the monsoon-dominated rivers (namely, Narmada, Tapti, Kosi, Ganga, Kaveri, Godabari, Krishna, Brahmaputra, Luni, Pennar, Ajay, and Damodar) of India are concerned, are investigated with the perspectives of flood hydrology, flood geomorphology, palaeohydrology, and hydrometeorology. Large floods of Indian rivers basins and their temporal variations have been studied by many researchers [8, 12, 13, 15, 23–32]. In West Bengal, the lower segment of Damodar River is taken as main spatial unit of study by many researchers and scholars of different disciplines to analyze the flood ferocity and propensity. The flood history of Damodar during the period 1817–1917 can be traced from the Glass [33] report submitted to the then Bengal government as observed at Raniganj, West Bengal [21, 33, 34]. The pre-dam and post-dam temporal character of floods and results of dam construction are well analyzed by many workers [33, 35–42]. The causes, factors, extent, and trend of floods and phases of changing courses of Damodar River are studied, investigated, and assessed minutely by few Indian geographers and geologists [21,34, 43, 44].

METHODOLOGY

Uncertainty of physical event is very much allied with high monsoon rainfall, runoff generation, and flash floods [46] but if we estimate the current status of flood flow or bankfull discharge or water accommodation capacity of channel reach, then an idea of emergent flood risk can be found with some quantitative judgments. The long continuous, up-to-date, and reliable rainfall data, hydrological records, and geomorphic data are an indispensable part in the study of changing flood climate, fluvial dynamics, and flood risk assessment. Following fluvial hydrologic approach, the drainage basin is selected as the main spatial unit of study. Total workflow of methodology is confined within selection of research problem, study area choice, data collection, data processing through Geographic Information System (GIS) and statistical software, data analysis through quantitative judgments and thematic maps, and finally data interpretation including significance of this study.

Data Collection

The study about floods always incorporates an interdisciplinary approach of earth science which includes the method and techniques of hydrology, hydroclimatology, and fluvial geomorphology. We have collected secondary data and information mostly from Damodar Planning Atlas [45], book entitled "Lower Damodar River, India: Understanding the Human Role in Changing Fluvial Environment" [21], websites of Irrigation and Waterways Department of West Bengal, West Bengal State Marketing Board, Indian Meteorological Department of India, Geological Survey of India, Google Earth, and different research articles relating to Damodar River. We have also employed the maps of National Atlas Thematic Mapping Organization (NATMO), Survey of India (SOI) toposheets (73 M/7, M/11, M/12, M/15, M/16, N/13, and 79 A/4) and Geological Survey of India (GSI). To get elevation data, we have collected ASTER (Advanced Spaceborne Thermal Emission and Reflection Radiometer) data (2013) from the website of United State Geological Survey.

Data Analysis

MapInfo Professional 9.0 and ESRI ArcGis 9.2 GIS software are used to represent spatial data into organized format and thematic maps. The delimitation and subsetting of study area (i.e., Damodar River Basin) is completed with the help of Leica Erdas 9.1 imagine software using the topographical sheets (projected in UTM WGS 84 projection). After collecting secondary data, we have employed few statistical analyses, namely, mean, standard deviation, skewness, correlation, regression, probability distribution, and - and chi square test, for the hydrological interpretation of stream flow data in Microsoft Excel 2007 and SPSS 14.0. Finally, gathering the results, we have made significance of adopted methods, interpretation of outcomes, and conclusion to chalk out the major findings of this analysis.

Quantitative Hydrologic Techniques

To derive significant information and to assess the trend of Damodar floods, few quantitative techniques of fluvial hydrology are used here. The employed quantitative expressions are summarized as follows.

Runoff Volume Estimation

To measure the annual runoff volume of Damodar River at Rhondia (West Bengal), Dhir et al. (1958) had developed an empirical equation [48]. Consider

$$\text{Annual Runoff Volume } \left(\text{in million m}^3\right) = 13400 \text{ Annual Precipitation (in cm)} - 5.75\ 10^5. \tag{1}$$

Runoff-Rainfall Relation

Employing linear regression and t-test, we have established a significant rainfall-runoff equation to predict the trend of runoff on the basis of actual rainfall and runoff data (1934–1950) at Rhondia. The equation is expressed as follows:

$$\text{Runoff } (Y) = 0.819 \text{ Rainfall } (X) - 530.21. \tag{2}$$

Potential Runoff Estimation

Converting annual peak discharge ($m^3\ s^{-1}$) of Damodar to one-$m_3\ s^{-1}$ day in m^3 (i.e., maximum volume of water flow passing through the measuring station) and dividing it by drainage area (km^2) above Rhondia, we have obtained potential contributing runoff (mm) after satisfying the storage of dams. The expression is as follows [7]:

$$\text{Potential Contributing Runoff} = \frac{(86400 \text{ Peak Discharge})\, 10^4}{\text{Drainage Area } 10^6}. \tag{3}$$

Gumbel Flood Frequency

Flood frequency analysis is a widely used technique to derive flood discharge with certain return period. According to Gumbel probability distribution (GPD) the exceedance probability (P) is given by [7, 49]

$$P = 1 - e^{-e-y}, \tag{4}$$

where y is called reduced variates given by

$$y = \left[\left(X - X_{\text{mean}}\right) + \frac{0.45\sigma}{0.7797\sigma}\right], \tag{5}$$

where X_{mean} is the mean of sample, X is the observed data, and σ is standard deviation of sample. If X_T denotes the magnitude of the flood with return period of T years $[T = 1/(1 - e^{-e-y})]$,

$$X_T = X_{\text{mean}} + K_T \cdot \sigma, \tag{6}$$

where K_T is the frequency factor ($y_T - y_{\text{mean}}/\sigma_n$), y_T is the reduced variate, $y$mean is the mean of reduced variate, and σn is standard deviation of reduced variates (calculating these from table value of Gumbel distribution). In this case, we have used annual peak flow data of Damodar River (measured at Anderson Weir, Rhondia)—(1) pre-dam phase (1933–57) and (2) post-dam phase (1958–2007).

Carrying Capacity and Threshold Discharge

To estimate potential carrying capacity (m^3) of a particular reach and threshold level of flood discharge (m^3/s), we have calculated mean cross-sectional area (m^2) and mean maximum length of the reach (m) of three selected channel sections of Damodar, namely, (1) Rhondia to Jujuti reach, (2) Jujuti to Chanchai reach, and (3) Chanchai to Paikpara reach, using ASTER data and Global Mapper 13.0 software. The developed equations are as follows [7]:

$$\begin{aligned}&\text{Carrying Capacity of Reach} \\ &\quad = \text{Mean Cross} - \text{Sectional Area} \\ &\qquad * \text{Mean Maximum Length of Reach}, \\ &\text{Threshold Discharge or Bankfull Discharge} \\ &\quad = \frac{\text{Carrying Capacity of Reach}}{86400}.\end{aligned} \tag{7}$$

GEOGRAPHICAL OUTLINE OF STUDY AREA

The Damodar River, or the Deonad Nadi as it is known in its upstream sector, is a subsystem of the Ganges River system of India. The local meaning of the word Damodar is fire in "womb" or Udar, which is "full of fire." This implies that the Damodar flows through a coal-rich area of Gondwana Formation. The river rises in the Chotanagpur Plateau approximately at 23°37′N and 84°41′E [21]. The main tributaries are the Barakar, Tilaiya, and Konar. Below the confluence of the Barakar and Damodar, there are a few insignificant tributaries such as the Nunia and Sali. Once the main distributaries were the Khari, Banka, Behula, and Gangur, but now they look more like independent rivers. Near Palla, the River takes a sharp southerly bend. Below Jamalpur, the river bifurcates into the Kanki-Mundeswari and the Amta Channel-Damodar and joins the Hooghly River (also spelled Hugli) at Falta some 48.3 km south of Kolkata [21]. The Damodar River Basin is a subbasin and part of the Ganges River spreading over an area of about 23,370 km^2 in the states of Jharkhand and West Bengal. The sample study area (Figure 1) includes the lower reach of Damodar River in between Rhondia (Galsi I Block) and Paikpara (Jamalpur Block). Its latitudinal extension ranges from 23°00′ to 23°22′10″N and longitudinal extension ranges from 87°28′23″ to 88°01′00″E. At Anderson Weir (near Rhondia Village) the estimated basin area coverage of Damodar River is 19,920 km^2. The total stretch of Damodar between Rhondia and Paikpara is approximately 82 km. This reach crosses Sonamukhi and Patrasayer blocks (Bankura), Kanksa, Galsi I, Galsi II, Barddhaman I, Khandaghosh, Rayna, Memari I, and Jamalpur blocks (Barddhaman). The slope category for 5–30 metres of relief of lower Damodar Valley is flat lands of 0°–0°30′, having 7.74% of Basin area and slope category for upper 30–60 metres of relief is gentle sloping plain of 0°30′–1°00′, having 3.57% of Basin area [50]. Sixteen large magnitudes floods occurred in its basin during 1823–1943, though, after constructions of dams and reservoirs, the ferocity of flood has been reduced but the number of floods has been increased [44]. The floods of 1823, 1840, 1913, 1935, 1941, 1958, 1959, and 1978 had peaks of more than 16,992 m^3 s^{-1}. A peak flow of about 18,678 m^3 s^{-1} has been recorded three times: August 1913 and 1935 and October 1941 [21]. To mitigate flood, the Damodar Valley Corporation (DVC) came into existence on July 7, 1948, by an act of the constituent assembly [42]. In the first phase, only four dams, namely, Tilaiya (1953), Konar (1955), Maithon (1957), and Panchet (1959), were constructed by DVC. Then one more reservoir, Tenughat (1978) on the upper Damodar River (under the control of Jharkahnd Government) and one barrage, Durgapur Barrage (1955) on the middle Damodar River (under the control of West Bengal Government) are constructed sequentially [42].

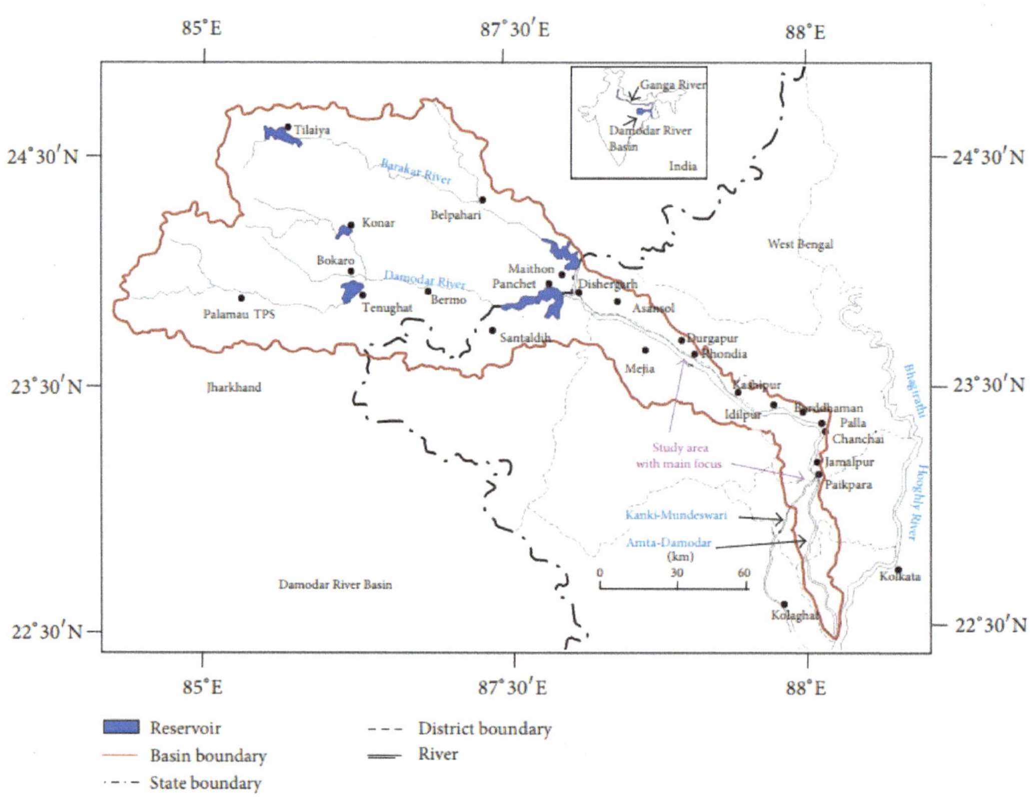

(a)

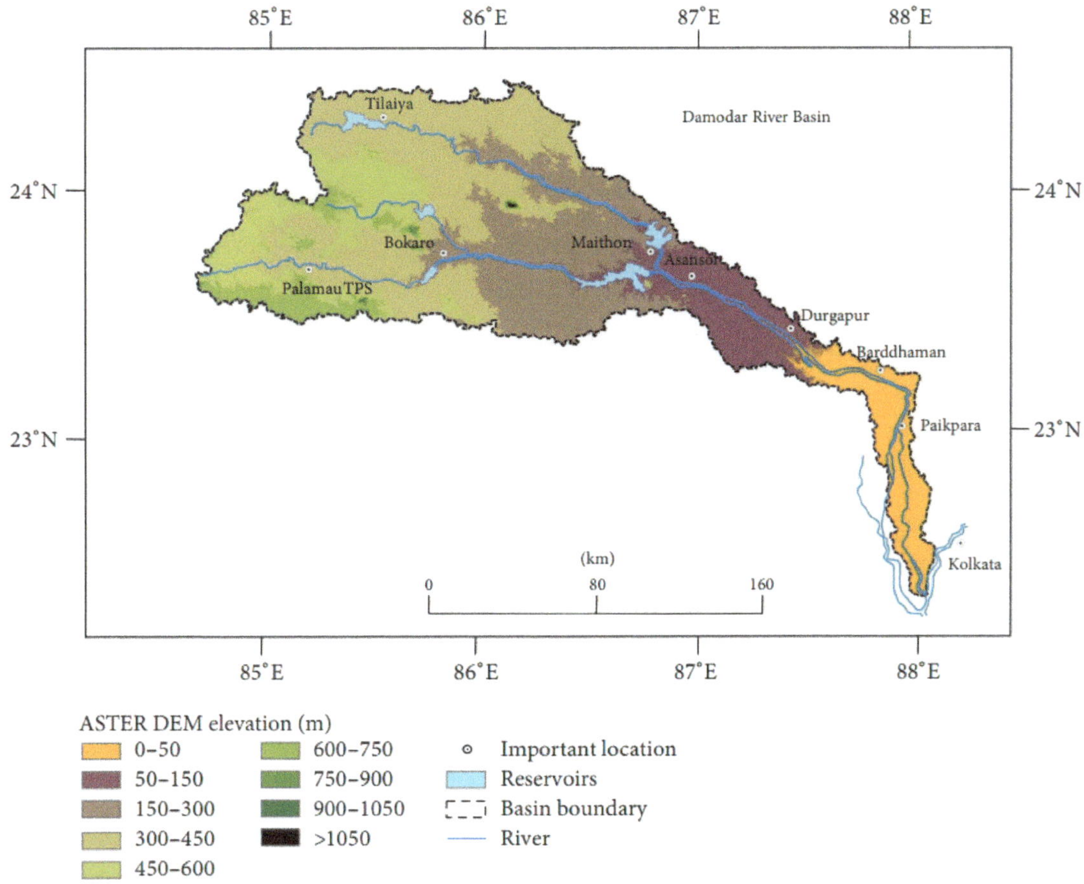

(b)

Figure 1: (a) Spatial extent of Damodar River Basin and (b) ASTER elevation map of the basin.

RESULTS AND DISCUSSION

Influence of Monsoonal Rainfall on Flood Propensity of Damodar River

The flow of any stream is determined by two entirely different sets of factors, the one depending upon the climate with special reference to the precipitation and the other depending upon the physical characteristics of the drainage basin. Rivers do not, however, remain at a high stage throughout the monsoon season. It is only after a spell of heavy rains, which may last for a period of several days, that large volume of runoff is generated in the catchments and the rivers experience floods annually [12]. Analysis of one-day extreme rainfall series shows that the intensity of extreme rainfall has increased over Orissa and West Bengal. The flood risk was more in the decades of 1981–1990, 1971–1980, and 1991–2000 [17]. The degree of flood risk has increased during the last two decades mostly over the eastern coast and West Bengal [17].

Monsoon depressions are the important synoptic systems that cause floods in the Peninsular Rivers, during the southwest monsoon season (June–October). On average, six depressions are mostly formed in the vicinity of Bay of Bengal during the three months of July to September [51]. These disturbances generally move west-north-westwards along the Lower Gangetic Plain to Chotanagpur Plateau after their formation at the head of the Bay of Bengal [37, 51]. The tracks of depressions are following the lower segment to upper catchment of Damodar River Basin. It is well know that heavy rainfall occurs in the southwestern sector of the monsoon depression due to strong convergence in that sector [51]. It has been found that the high flash-flood magnitudes of 1950–57 and 1958–69 are associated with 118 and 217 numbers of cyclones (which includes cyclonic disturbances: wind speed 17 knots or more, cyclones: 34 knots of more, and severe cyclones: 48 knots or more) of Bay of Bengal, respectively, but low flood magnitude of 1988–95 is related to only 95 numbers of cyclones [21].

Lower segment of the basin is affected by the floods due to upstream heavy rainfall and huge runoff volume generated in the upper catchment (i.e., western part of Maithon and Panchet dams) which consists of two drainage systems: (a) Damodar drainage system and (b) Barakar drainage system. It is estimated that Damodar catchment receives monsoon rainfall (June–October) of 855.57–1043.55 mm and Barakar catchment receives monsoon rainfall of 840.54–1079.81 mm annually [52]. At present, significant interannual and intraseasonal variabilities in the observed monsoon rainfall are displayed over this so-called sub-humid region, resulting in recurrent droughts and floods [53]. It is projected that 10–15 percent of rainfall will increase in area of average monsoon rainfall over the Indian subcontinent but the date of onset of summer monsoon over central India could become more variable in future. More intense rainfall spells are also projected in a warmer atmosphere, increasing the probability of extreme rainfall events and flash-floods in the sun-mountainous or plateau scarp region [11, 53].

The main hydroclimatic factors which influence runoff and flood of Damodar can be categorized as follows [54]: (1) type of rainfall, (2) rainfall intensity, (3) duration of rainfall on basin (intense short period or prolong duration), (4) direction of storm, depression and cyclone movement, and (5) antecedent rainfall and soil moisture. Heavy incessant and prolonged rainfall for long period is the basic cause of floods because enormous amount of water gets collected on the surface flowing as runoff. Higher magnitude of rainfall coupled with a larger catchment area, like Damodar Basin (23,370 km^2), leads to a greater volume of runoff. We have emphasized on the following influences of monsoonal rainfall and catchment characteristics on flood risk of lower Damodar Basin.

1. More or less same monsoonal rainfall pattern is observed both in the districts of upper catchment and lower catchment (Table 1). Within the Damodar Command Area of DVC, the upper and middle parts of Basin receive 1,209 mm rainfall annually and lower parts of Basin receive 1,329 mm [42]. Therefore, large fan-shaped contributing area of Damodar annually receives huge rainfall and due to low infiltration capacity and high runoff coefficient of rocky-deforested upper catchment, 70 to 80 percent of rainfall converted into surface runoff. Due to high drainage density of upper catchment, huge volume of runoff appears as excessive stream flow which finally contributes to the main Damodar River, below Panchet and Maithon Dams.

2. Rains of long duration and direction of monsoonal depressions (generally south to north-west from Gangetic West Bengal) annually aggravates huge volume of inflow water into the reservoirs of DVC. During late monsoon, the reservoirs including the rivers remain bankfull and groundwater of alluvial West Bengal is recharged to the fullest extent. A sudden cloudburst or strong depressional rainfall at this juncture frequently causes a disastrous flood. The month of September-early October is now considered as cruelest one for lower Damodar River, because all the major floods of 1956, 1958, 1959, 1978, 1995, and 2000 appeared in this session (Table 2).

3. The eastern parts of Jharkhand and Gangetic West Bengal are recommended as the one of major rainstorm zones and flood prone areas of India [12, 13]. It has been found that, on an average of 25 years, the average monsoonal rainfall in 6 hours ranges between 14 and 16 cm in the parts of eastern Jharkhand and western West Bengal [37, 39].

4. When Jharkhand Division (73.70 percent share in the basin) experiences intense rainfall of wide coverage, at that time, Gangetic West Bengal (26.30 percent share in the basin) is already oversaturated by continuing monsoonal depressions and lower catchment receives more than 1400 mm rainfall (Figure 2). Then, uncontrolled inflow runoff of upper catchment (released at last by Durgapur Barrage) integrates with further additional rainfall of Gangetic West Bengal and enormous volume of water passes through congested, silted, and oversaturated lower Damodar River (Figure 2) which is further narrowing down next to downstream of Barddhaman town. So, the overflow of river and additional runoff of lower catchment turns the inundated situation into misery.

Table 1: Subdivisional monsoonal rainfall covering the Damodar River Basin.

Year	Rainfall (mm) in Jharkhand subdivision	Rainfall (mm) in Gangetic West Bengal subdivision
2005	1105.0	1136.0
2006	1092.0	1126.0
2007	1093.0	1127.0
2008	1092.5	1291.7
2009	1002.6	971.5
2010	1084.5	1140.6
2011	1101.5	1394.7
1951–2000 (mean normal rainfall)	1091.9	1167.9

Table 2: Rainfall recorded during monsoonal major foods in the Damodar Valley (DV) and Barddhaman District (BD).

Year	Month	Date	Total storm rainfall (mm)
1913 (DV)	August	5 to 11	314
1935 (DV)	August	10 to 15	288
1958 (DV)	September	14 to 16	166
1959 (DV)	September-October	September 30 to October 7	231
1978 (DV)	September	26 to 29	184
1995 (BD)	September	26 to 28	369
2000 (BD)	September	18 to 21	800

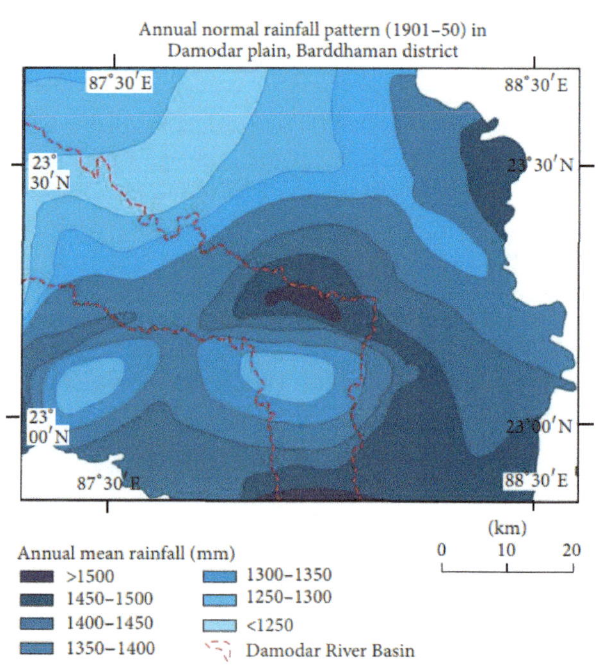

Figure 2: Spatial distribution of mean annual rainfall in lower segment of Damodar River Basin. Source: Chatterjee [45].

Estimating Surface Runoff and Flood Risk

Flood risk estimation includes the task of evaluating the probability of a flood arriving which is greater than that which can be safely accommodated [55]. Usually, the risk is expressed as the probability of such an occurrence or phenomenon within the intended design life of a scheme [55] or the risk involves quantification of the probability that a hazard (e.g., flood) will be harmful and the tolerable degree of risk depends upon what is being risked [46]. The uncertainty of event is very much allied with high monsoon rainfall, runoff generation, and flash floods but if we estimate the current status of flood discharge or runoff volume, then an idea of emergent flood risk can be found with some quantitative judgment.

Runoff is a portion of part of the hydrological cycle connecting precipitation and channel flow. The basin hydrological cycle can be viewed simply as inputs of precipitation being distributed through a number of storages by a series of transfers, leading to outputs of basin channel runoff, evapotranspiration, and deep outflow of ground water [56]. The watershed or catchment is the area of land draining into a stream at a given location [1]. To describe the surface water yield of basin (after soil moisture storage), the estimation of surface runoff obtained prime importance. It plays dual role in basin hydrogeomorphology—(1) fluvial erosional agent and gross sediment yield and (2) accumulation of all water and final discharge to main channel [13].

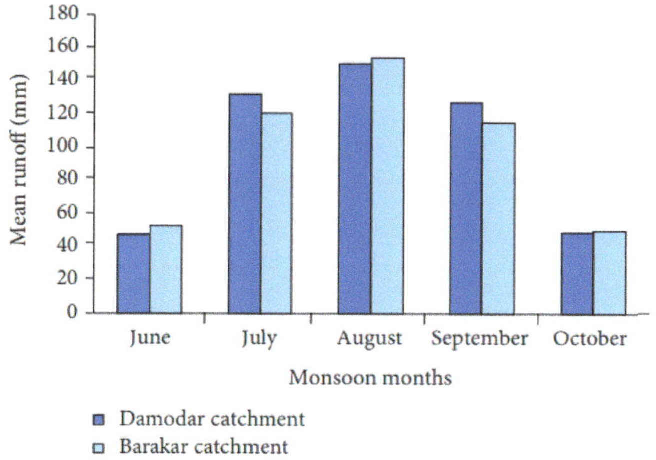

Figure 3: Estimated mean monsoon (June to October) runoff in the catchments of Damodar and Barakar.

It is calculated that Damodar catchment yields monsoon runoff (June–October) of 385.17–560.71 mm and Barakar catchment yields monsoon runoff of 416.59–565.84 mm annually (Figure 3) [52]. For estimation of runoff yield of Damodar Basin at Anderson Weir (Rhondia, West Bengal), Dhir et al. [48] had developed an empirical equation on the basis of a range of rainfall-runoff and stream flow data. Actually, they developed numerous individual equations for Kangsabati (West Bengal), Tapti (Gujrat), Chambal (Rajasthan), Tawa (Madhya Pradesh), Ghataprabha (Andhra Pradesh), and so forth river basins of India. These equations are well used by Central Water Commission (Government of India) and International Water Management Institute for ungauged sites of Indian catchments. It has been assumed that, from Rhondia, the remaining upper catchment of 19,920 km^2 has contained a considerable portion of rainfalls and produced runoff into the main channel, though it varies from region to region on the basis of land use, land cover, geology, soil, rainfall intensity, and so forth. At Rhondia, for the upper catchment area (19,920 km^2) of Damodar River, a direct rainfall-runoff equation is developed to estimate potential runoff volume at upper catchment [30].

From the above analysis (Table 3), we can say that annual runoff yield of Damodar Basin is much greater, ranging from 16961580 million m^3 (Barhi) to 19900200 million m^3 (Burdwan), for the decade of 1901–50. It is easily seen that the potentiality of runoff generation is more or less the same in the upper catchment area of Damodar (850–900 mm per year). Based on mean annual Basin rainfall of 1337.46 mm, we have been calculated potentially mean annual basin runoff of 871.71 mm which is generated, having positively skewed distribution. After getting idea of past condition, we have employed the recent districtwise monsoonal rainfall data (2006–10) which is regularly provided by website of Indian Meteorological Department (IMD, Pune). In between June and September, rainfall, soil moisture, and runoff reach up to maximum levels which contributed to greatest flood flows of lower Damodar Basin It have been found. Hazaribagh, Bokaro, and Dhanbad districts yield approximate mean runoff of 661, 640, and 669 mm, respectively, in monsoon (2006–10). This runoff of upper catchment reaches downstream of Burdwan Plain through the main channel and it is further cumulated with the monsoonal runoff of Burdwan and Bankura districts which are 758 and 756 mm, respectively, for 2006–10 (Figure 3). The important thing is that the basin area is reduced at downstream and mean annual runoff volume of 13882046 million m^3 passes through this narrow channel. It aggravates the risk of inundation and flood in the lower portion of Damodar Basin in late Monsoon.

Table 3: Calculated runoff of Damodar Basin from annual normal rainfall data (1901–50) of selected stations.

Stations	[1]Mean annual rainfall (mm)	Annual runoff volume (million m^3)	Annual depth of runoff (mm)	Runoff/rainfall ratio
Barhi	1308.7	16961580	852.34	0.651
Hazaribagh	1338.7	17363580	872.54	0.651
Ramgarh	1306.0	16925400	850.52	0.651
Jamtara	1376.7	17872780	898.13	0.652
Dhanbad	1309.0	16965600	852.54	0.651
Barjora	1469.0	19109600	967.84	0.658
Sonamukhi	1204.0	15558600	863.25	0.717
Asansol	1324.9	17178660	960.28	0.725
Mankar	1209.6	15633640	785.61	0.649
Burdwan	1528.0	19900200	999.00	0.653
Standard deviation	101.16	1355498	68.65	0.023
Mean	1337.46	17346964	871.71	0.665
Skewness	0.63	0.63	0.37	1.770

The above analysis (Table 4 and Figure 4) provides us with only a generalized idea about potentiality of runoff to be generated in the periods of 1901–1950 and 2006–2010. We have now established a linear regression equation dealing with seventeen years of estimated actual rainfall and runoff data (1934–1950) at Rhondia and a statistical significant test is employed to make that equation reliable and applicable to Damodar Basin. Here, we have assumed that the calculated runoff is generated within the basin area depending on the land use, land cover conditions, soil moisture, channel storage, rainfall intensity, evapotranspiration, and geological character of surface. So this rainfall-runoff relation reflects the hydrogeomorphic characteristics of Damodar Basin and flood risk of downstream region in particular. Again, establishing this significant regression equation, we can predict potential runoff depth of uncalculated years which will forecast the flood flow at Rhondia.

Table 4: Districtwise monsoonal rainfall-runoff estimation (2006–10) for Damodar River Basin.

(a) *Monsoonal rainfall in mm

District	2006	2007	2008	2009	2010
Hazaribagh	977.3	1113.5	1498.5	965.2	576.1
Bokaro	938.9	1027.1	1184.8	945.9	871.3
Dhanbad	1079.7	1488.4	977.8	896.0	748.9
Burdwan	1280.1	1483.0	1361.8	977.0	745.9
Bankura	1201.9	1555.2	1371.6	1001.7	704.5

(b) Monsoonal runoff volume in million m^3

District	2006	2007	2008	2009	2010
Hazaribagh	12520820	14345900	19504900	12358680	7144740
Bokaro	12006260	13188140	15301320	12100060	11100420
Dhanbad	13892980	19369560	12527520	11431400	9460260
Burdwan	16578340	19297200	17673120	12516800	9420060
Bankura	15530460	20264680	17804440	12847780	8865300

(c) Predicted runoff depth in mm

District	2006	2007	2008	2009	2010
Hazaribagh	628.55	720.17	979.16	620.41	358.67
Bokaro	602.72	662.05	768.13	607.43	557.25
Dhanbad	697.44	972.36	628.89	573.86	474.91
Burdwan	832.24	968.73	887.20	628.35	472.89
Bankura	779.64	1017.30	893.79	644.96	445.04

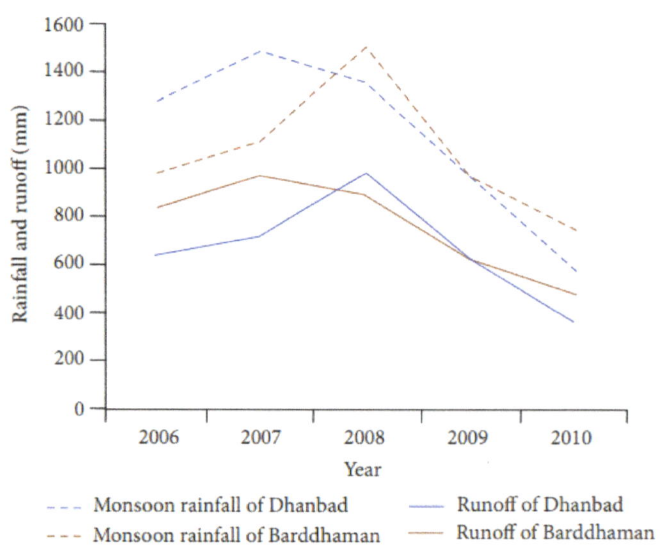

Figure 4: Monsoonal rainfall and runoff contrast (2006–10) of Dhanbad and Barddhaman districts.

Observing the above calculations (Table 5) and two graphs (Figure 5), we have been able to establish a meaningful rainfall-runoff relation of Damodar Basin at Rhondia. From "t-test" of b value and r value, we have obtained that both slope of positive increasing trend line and correlation between rainfall and runoff are very much significant, rejecting the null hypothesis. Also, the distribution is positively skewed (0.63). With the fluctuation of rainfall, the runoff was also deviated but from 1934 to 1950, the runoff has an increasing positive annual trend ($Y = 0.8194X - 530.21$).. On an average, estimated runoff is 548.89 mm, having standard deviation of 153.67 mm. From the trend of 1934–50, we have found that average annual runoff-rainfall ratio is 0.41 (0.25 to 0.54) which means that 41 percent of annual rainfall has possibility to convert into runoff in this area depending on different geographical conditions. It is a real fact that the largest part of runoff is generated at the time of peak monsoon (June to September) alone (when excess rainfall becomes direct runoff to the streams after fulfillment of surface storage). Therefore, using the districtwise normal monsoonal IMD rainfall data (1951–2003) of Bokaro, Dhanbad, Hazaribagh, Jamtara, Ramgarh, Bankura, and Burdwan, we have obtained the following status of runoff (Table 6).

Table 5: Establishing annual rainfall-runoff relation for Damodar River Basin at Rhondia (1934–50).

Year	[1]Mean annual rainfall (mm)	[2]Mean annual runoff (mm)	Calculations
1934	1088	274	$a = 530.21$
1935	1113	320	$b = 0.819$
1936	1512	543	Runoff (Y) = **0.819** rainfall (X) − **530.21**
1937	1343	437	Coefficient of determination (R^2) = **0.63**
1938	1103	352	
1939	1490	617	Standard deviation of Y = 158 mm
1940	1100	328	Standard error of estimate of Y = 96 mm
1941	1433	582	
1942	1475	763	Calculated t value of b = 3.25
1943	1380	558	Tabulated t value (0.01% confidence) = 2.95 Degree of freedom (df) = (n − 2) = 15
1944	1178	492	
1945	1223	478	Product moment correlation coefficient (r) = **0.794**
1946	1440	783	Calculated t value of r = 4.01
1947	1165	551	Tabulated t value (0.01% confidence) = 2.95
1948	1271	565	
1949	1443	720	Null hypothesis is rejected and the equation is very much significant
1950	1340	730	

Table 6: Normal monsoonal rainfall and calculated runoff amounts of selected districts (1951–2003).

Districts	[1]Normal rainfall in mm				Total monsoonal rainfall (mm)	Estimated monsoonal runoff (mm)
	June	July	August	September		
(1) Jharkhand Division						
Bokaro	178.5	318.8	286.0	256.9	1040.2	321.7
Hazaribagh	180.4	327.8	297.6	234.7	1040.5	321.9
Jamtara	216.8	347.6	339.9	278.2	1182.5	438.2
Dhanbad	192.3	336.1	300.8	245.5	1074.7	349.9
(2) Gangetic West Bengal						
Bankura	206.9	312.1	311.5	230.0	1060.5	338.3
Burdwan	229.4	324.8	303.6	240.3	1098.1	369.1

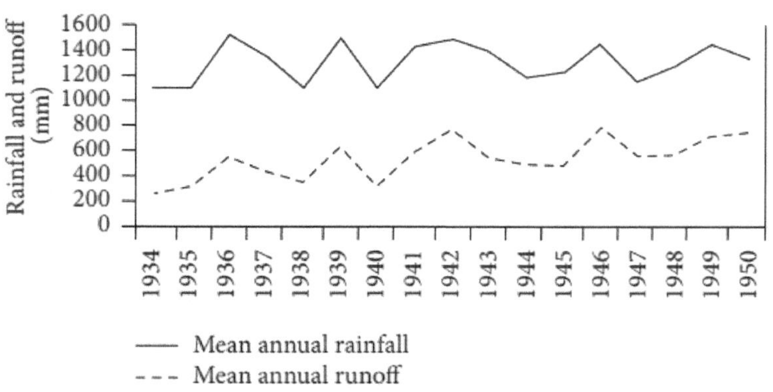

(a)

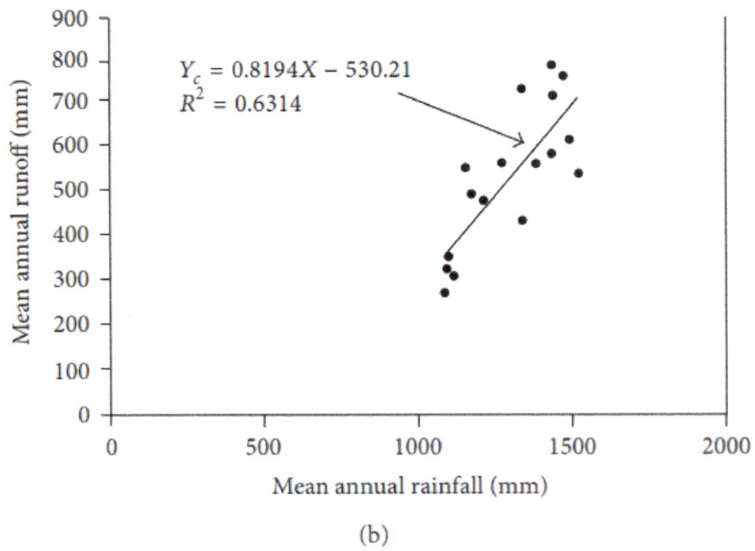

Figure 5: (a) Yearwise mean rainfall and runoff pattern at Rhondia from 1934 to 1950 and (b) increasing regression trend line between rainfall and runoff of Damodar Basin.

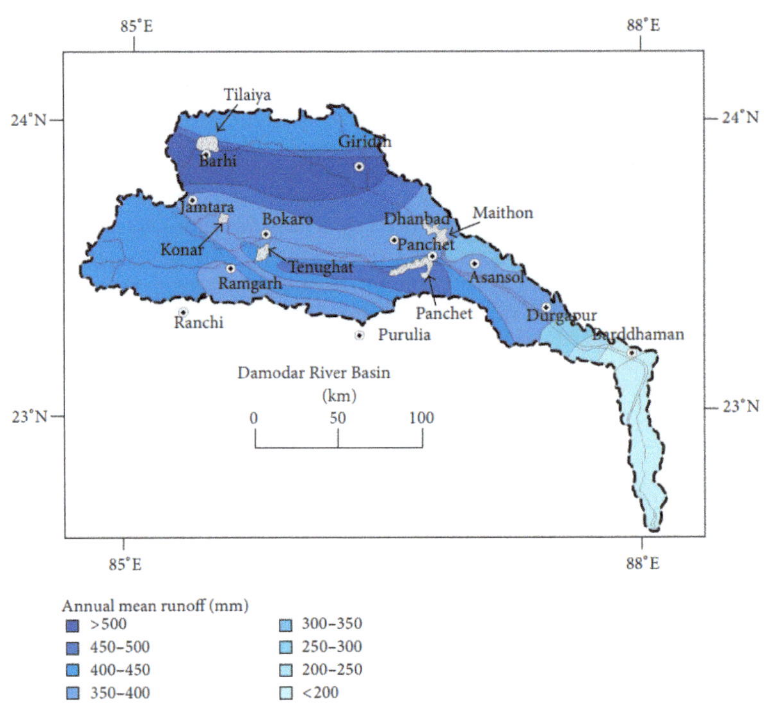

Figure 6: Average annual runoff pattern in the Damodar River Basin (2010).

On an average the normal monsoonal rainfall ranges from 1040.2 to 1182.5 mm per year and calculated runoff ranges from 321.7 to 438.2 mm per year. Both regions on an average yield have same runoff (357.9 to 353.7 mm) in monsoon. It signifies that now both upper and lower portions of Damodar Basin yield the same runoff which has a cumulative effect as increasing runoff progresses towards Gangetic West Bengal. As funnel shape upper catchment carries 321.7–438.2 mm of runoff towards West Bengal and further narrow elongated alluvial lower basin adds 338.3 to 369.1 mm of runoff on the same season or months, therefore, in late August to September, when soil moisture and ground water recharges up to maximum level, the agglomerative effect of this runoff within the same period creates huge volume of stream flow or peak discharge in Maithon, Panchet Dams, and Durgapur Barrage. According the Sen [34, 43], now DVC system is providing flood benefit of only 162.56 mm of surface runoff to the lower catchment, in place of 452.0 mm as advocated by Mr. Voorduin [43]. But our result shows that, now, estimated mean monsoonal runoff (1951–2003) is 355.0 mm, having 172.44 mm of unregulated runoff. This amount of runoff is not controlled by DVC storage system and is finally released by Durgapur Barrage into main River and canals. So there are ample chances of peak flood flow and inundation of Barddhaman Plains, when huge outflow of Durgapur Barrage will overtop its banks as the downstream cross-section of channel is reducing to a considerable extent from Barsul to Paikpara. In general, the annual runoff pattern of Damodar Basin is depicted as follows (Figure 6).

Predicting Potential Runoff in relation to Peak Discharge of Damodar

The channel flow or stream flow is the main form of surface water flow, and all the other surface and subsurface flow processes contribute to it. The precipitation which becomes stream flow may reach the stream by overland flow, subsurface flow, or both [3]. The observed or measured stream flow of Damodar River at Anderson Weir, Rhondia, is the flow collected from upper catchment area of Damodar River (19,920 km^2) and appeared at an outlet of Weir. But that measured flow does not reflect the actual stream flow because the reservoirs and dams store much water in them, though the measured stream flow can be regarded as uncontrolled inflow of upper catchment. It has been assumed that observed extreme level of steam flow rate of a day is the cumulative effect of total collective runoff of previous days in the upper catchment area of a river. Now, we have been attempting to estimate pre-dam (1933–1957) and post-dam (1958–2007) major contribution of runoff of upper catchment which is responsible for the peak stream flow at Rhondia.

Converting $m^3 s^{-1}$ (metric system unit of stream flow rate) into volume of flow (m^3) and dividing it by contributing catchment area (upper catchment of Damodar River, measured from Anderson Weir, Rhondia), we have been able to find the probable of potential depth of runoff responsible for that stream flow. The volume of water discharged in a period of 24 hours at a peak rate of 1 $m^3 s^{-1}$ is called "one $m^3 s^{-1}$ day" (1 $m^3 s^{-1}$ day = 86,400 m^3 = 8.64 ha-m) [7].

We have categorized the calculation (Table 7 and Figure 7) into two important phases, (1) pre-dam period (1934–57) and (2) post-dam period (1958–2007), to understand runoff amount in past natural condition and present controlled condition. The calculated potential runoff depths of both periods reaffirm that, to reach the observed peak discharge level and volume of water passing through Anderson Weir, the upper catchment of Damodar River should collect the maximum amount of runoff water. In pre-dam period, when a single dam was constructed, the stream flow of Damodar River was much more extreme in all years, having mean peak discharge of 8378 $m^3 s^{-1}$. In that natural condition, the estimated runoff ranges from 20.78 to 78.56 mm, but due to establishment of DVC Project, it is radically reduced and ranges from 1.79 to 47.36 mm. Now, much of runoff water is stored behind the dams of upper catchment. In spite of low runoff supply and storage, lower Basin of Damodar (Barddhaman, Hooghly, and Howrah districts) is still facing annual flood. The estimated runoff of post-dam period is the unregulated outflow of DVC, though the main portion of stream flow is coming from surface runoff. The above calculation can forecast the peak discharge or volume of flow at Rhondia if we can precisely estimate rainfall and runoff of upper catchment. When we look at the line graph of runoff (Figure 6), we have found that, after dam construction, the supply of runoff is decreased below 30 mm which is recommended as critical level for 7075 $m^3 s^{-1}$ peak discharge at Rhondia (threshold level of downstream flood, as recommended by DVC). But if we look at the trend of 1995–2007, we have found that after a sharp calm session, the contribution potential runoff amount is gradually increased. It signifies three facts: (1) declining storage capacity of reservoirs and unregulated outflow, (2) loss of carrying capacity of Damodar River to accommodate runoff, and (3) future risk of extreme flood at downstream section. We have mentioned the cumulative rainfall of past major flood events in relation to estimated runoff depth (Table 8). Here, the runoff coefficient to depict actual runoff of given days responsible for extreme peak discharge is estimated. Only 1 to 34 percent of rainfall is converted into runoff which appears as abnormal flood flow at Rhondia. So we can say that small amount of potential runoff appeared as gigantic flood flow due to wide areal coverage (19,920 km^2) of upper catchment (fan shaped basin) to accommodate huge volume of water within two to three days of prolonged rainfall.

Table 7: Estimating potential runoff of upper catchment of Damodar River on the basis of selective annual peak discharges at Rhondia for pre-dam and post-dam period.

Date of occurrence	[1]Peak discharge in $m^3 s^{-1}$	One-$m^3 s^{-1}$ day in m^3	Contributing potential runoff in mm
Pre-dam phase (1933–1957)			
12/7/1935	18,112	1564876800	78.56
6/9/1938	12,002	1036972800	52.05
4/8/1939	7,989	690249600	34.65
29/8/1940	8,773	757987200	38.05
10/10/1941	17,942	1550188800	77.82
10/8/1942	10,811	934070400	46.89
5/8/1943	8,384	724377600	36.36
18/9/1946	9,133	789091200	39.61
18/8/1950	9,561	826070400	41.47
11/9/1951	11,012	951436800	47.76
26/9/1956	8,579	741225600	37.21
Post-dam phase (1958–2007)			
2/10/1959	8,792	759628800	38.13
3/10/1961	4,371	377654400	18.96
19/9/1967	4,138	357523200	17.95
17/7/1971	4,556	393638400	19.76
13/10/1973	5,726	494726400	24.83
19/9/1976	5,297	457660800	22.98
27/9/1978	10,919	943401600	47.36
28/8/1980	4,210	363744000	18.26
14/9/1987	4,567	394588800	19.80
29/9/1995	6,522	563500800	28.28
25/9/1999	5,690	491616000	24.68
23/9/2000	6,387	551836800	27.70
24/9/2006	7,035	607824000	30.51
25/9/2007	8,883	767491200	38.53

Table 8: Relating rainfall, peak discharge, and estimated runoff of Damodar Basin in selected major flood events.

Consecutive date of rainfall	Cumulative average rainfall in mm	Date of peak discharge	Magnitude of peak discharge in $m^3 s^{-1}$	Estimated runoff in mm	Runoff coefficient
August 5–8, 1913	234	8/8/1913	18,406	79.83	0.34
August 10–12, 1935	196	12/8/1935	18,112	78.56	0.40
September 14–16, 1958	168	16/9/1958	4,682	20.30	0.12
September 30–Oct 2, 1959	152	2/11/1959	8,792	38.13	0.25
September 26-27, 1978	160	27/9/1978	10,919	47.36	0.29
September 18–21, 2000	359	23/9/2000	6,387	27.70	0.08

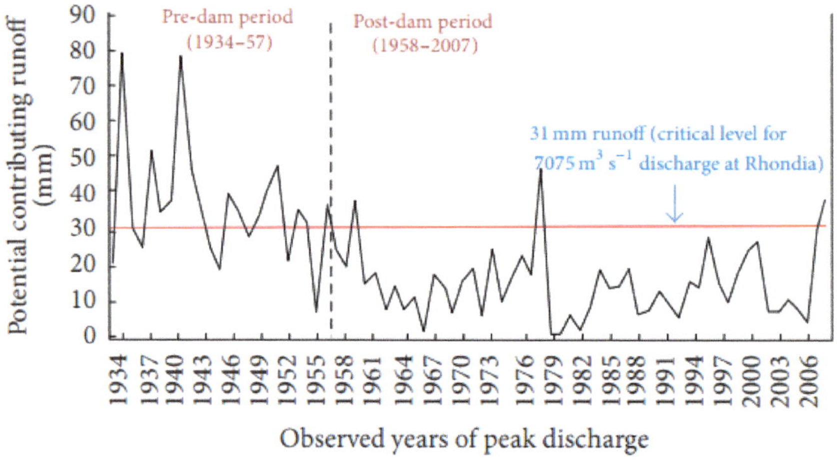

Figure 7: Pre-dam and post-dam change of potential contributing runoff in respect of peak discharge of Damodar River at Rhondia.

Now, we have employed Gumbel's distribution of extreme values to identify the standard project flood discharge and runoff of given return period (namely, 5, 10, 20, 50, 100, and 150 years flood) in Damodar Basin. In post-dam condition, 5 years return period of flood has possible peak discharge of 5,352 $m^3 s^{-1}$ (11,728 $m^3 s^{-1}$ in pre-dam condition) and predicted runoff of 23.21 mm (50.87 mm in pre-dam condition). Again, 100 years of return period of standard flood has potential peak discharge of 12,670 $m^3 s^{-1}$ (10,919 $m^3 s^{-1}$ in 1978) and predicted runoff of 54.95 mm (47.36 mm in 1978). When we observe the trend of peak discharge (1934–2007), we have found dominance 5000–7000 $m^3 s^{-1}$ of flood discharge in some years. So we should focus on 5, 10, and 20 years of flood to predict extreme flood discharge and contributing runoff (Table 9).

Table 9: Pre-dam and post-dam comparison of standard peak flood discharge and estimated runoff of variable return period using Gumbel distribution.

T return period (years)	Reduced variate	Pre-dam period			Estimated runoff (mm)	Post-dam period			Estimated Runoff (mm)
			K_T	Q ($m^3 s^{-1}$)			K_T	Q ($m^3 s^{-1}$)	
5	1.5		0.893	11,728	50.87		0.819	5,352	23.21
10	2.25	y_{mean} = 0.5296	1.583	14,317	62.10	y_{mean} = 0.54854	1.465	6,796	29.48
20	2.97	σ_n = 1.0864	2.246	16,805	72.89	σ_n = 1.16066	2.086	8,184	35.50
50	3.9	n = 24	3.102	20,016	86.81	n = 50	2.887	9,974	43.26
100	4.6		3.746	22,433	97.30		3.49	11,322	49.10
150	5.3		4.391	24,853	107.80		4.093	12,670	54.95

Note: K_T: frequency factor of Gumbel distribution, Q: estimated or projected peak discharge, y_{mean}: mean of peak discharges, σ_n: standard deviation of distribution, and n: number of variables.

Declining Carrying Capacity of Channel and Chances of Floods

It is now considered that lower Damodar River has constantly lost its carrying capacity to store or accommodate flood water within its confined valley in the time of peak monsoon [47, 57]. To verify it, we have calculated the potential volume water of a particular reach which may be accommodated limiting its carrying capacity. To estimate the maximum bankfull capacity to carry flood flow and threshold level of flood discharge, we have subdivided the river from Anderson Weir to downstream into three reaches, namely, (1) Rhondia to Jujuti reach, (2) Jujuti to Chanchai reach, and (3) Chanchai to Paikpara reach (Table 10). Using SRTM elevation data, we have drawn six to eight cross profiles of three reaches across Damodar River, taking an equal interval and mean cross-sectional area, mean depth, mean length of each reach, and mean bankfull volume of each reach are calculated using those profiles. It was estimated by Voodruin and DVC that if all the proposed dams were established, we could moderate the peak discharge up to 7075 $m^3 s^{-1}$ (611,280,000 $m^3 s^{-1}$/day) at Rhondia and any discharge above it would cause flood at downstream section [58, 59]. But, unfortunately, total DVC project is not completed till date and, after spending sixty years (from establishment of DVC), the dams, barrages, and river itself are much more silted due to engineering obstructions, annual regulation of stream flow, and embankments. So it is clear that, now, the threshold level is much less than 7075 $m^3 s^{-1}$, but question is how much it is decreased. For that reason, we have estimated the present threshold level of peak discharge to predict the overflow condition and flood risk of lower Damodar River (Table 10).

Table 10: Estimation of present threshold level of flood discharge in lower Damodar River.

Reaches of Damodar	Mean cross-sectional area (m^2)	Range of channel depth (m)	Bankfull volume (m^3) of total reach ($m^3 s^{-1}$/day)	Present estimated level of discharge ($m^3 s^{-1}$) (threshold level)	Discharge ($m^3 s^{-1}$) as recommended earlier by D.V.C.	Excess uncontrolled discharge ($m^3 s^{-1}$)
(1) Rhondia to Jujuti	12,290	3.89–7.84	346568177	4011		3064
(2) Jujuti to Chanchai	7082	5.08–6.85	204380800	2366	7075	4709
(3) Chanchai to Paikpara	7077	4.47–9.75	133259910	1542		5533

As the mean cross-sectional area of river is decreased downstream (12,290 to 7077 m^2), the bankfull volume of reach also is declined (346568177 to 133259910 m^3). It is estimated that Damodar River has lost its former carrying capacity (up to 611280000 m^3 for the discharge of 7075 $m^3 s^{-1}$), having a loss of 264711823 to 478020090 m^3 in one $m^3 s^{-1}$/day. Now, we have estimated that threshold levels of peak discharge are 4011, 2366, and 1542 $m^3 s^{-1}$, respectively, for the selected reaches of lower Damodar River (Rhondia to Paikpara). So, now, any bankfull discharge above 4011 $m^3 s^{-1}$ at Rhondia is considered to be flood discharge for lower catchment. That is why, in post-dam period, Barddhaman, Hooghly, and Howrah Districts had experienced high magnitude of floods in 1958, 1961, 1976, 1978, 1995, 1999, 1987, 2000, 2006, 2007, 2009, and 2013. The line graph also denotes that, from 1995, frequently the peak discharge have crossed the threshold limit and its magnitude is rising after the devastating floods of 1978 and 2000 (Figure 8).

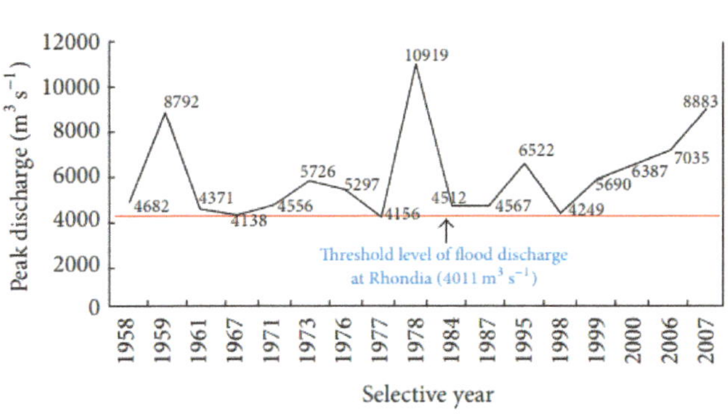

Figure 8: Post-dam trend of peak discharge at Rhondia above estimated threshold level of discharge (note: cumec: $m^3 s^{-1}$).

With increasing distance from the Anderson Weir, Rhondia, the channel dimensions of channel cross-sectional area (*Ab*), volume of that cross-section (V_c), bankfull width (W_b), maximum depth (thalweg depth) of that profile (*D*max), and width-depth ratio (*W*/*D*) are gradually decreased with low competence of accommodating peak discharge (Figure 9). The estimated maximum and minimum bankfull discharges are 5848 $m^3 s^{-1}$ and 529 $m^3 s^{-1}$, respectively, with the crosssectional channel area of 6360 m^2 and 812 m^2 being at those sections. The *W*/*D* is significantly decreased at downstream from 206.59 to 12.08, reflecting the narrowness of channel but maximum channel depth (D_{max}) is increased from 8.83 to 21.53 metres. The average monsoonal discharge is estimated to be near about 6071 m3 s −1 up to Bardhhaman town and 2781 $m^3 s^{-1}$ up to Paikpara. The annual peak flow is projected approximately 10,213 $m^3 s^{-1}$ and 12,393 $m^3 s^{-1}$ in between Ramgopalpur and Gohagram. This is reduced up to 2106–3048 $m^3 s^{-1}$ in between Jamalpur and Paikpara. This information again reconfirms the risk of overflow at the time of peak monsoon discharge in the lower segment of Damodar River.

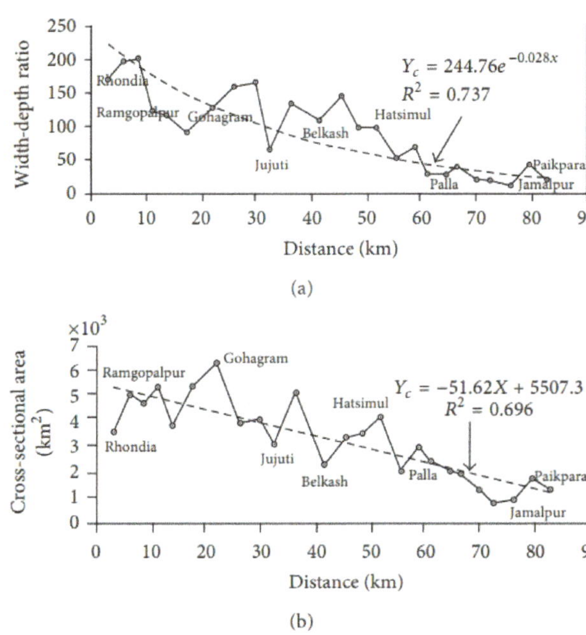

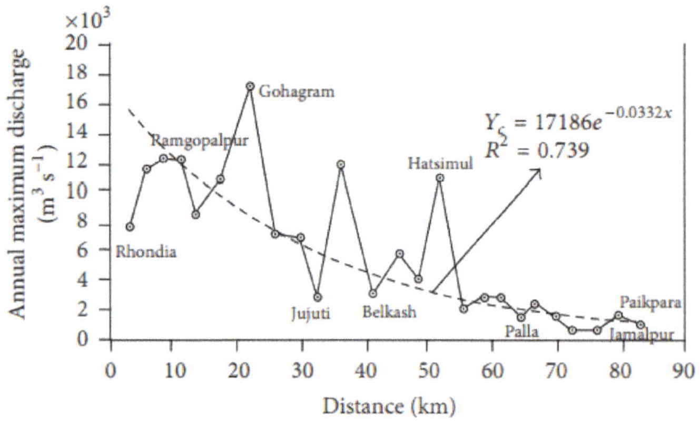

(c)

Figure 9: Downstream site-specific hydrogeomorphic changes of Damodar River, with increasing downstream distance from Rhondia: (a) $W_b/D_{\max}$ is decreased, (b) A_b is decreased, (c) level of Qmax is significantly decreased, signifying high flood risk.

CONCLUSION

Damodar River of India, its history of flood discharges, wide coverage of monsoonal rainfall, excessive volume of runoff, declining carrying capacity of River, and increasing human actions on its active domain are effectively explained here from a hydrogeomorphic perspective to analyze its dynamic fluvial phenomena in relation to flood climate and risk assessment. This study extracts significant information regarding past, present, and future trend of flood discharge and continuous river's responses to rainfall, runoff, and flood control system of DVC. Employing Gumbel's distribution of extreme annual discharge values, we have found that 100 years of flood discharge will reach up to 11,322 $m^3 s^{-1}$ in post-dam period, but the flood of that magnitude is associated with 5 years of recurrence interval in pre-dam period. It suggests that DVC flood regulation system manages to increase the recurrence interval of peak annual discharge up to 50 percent. In every 14 years, there will be 7.15 percent probability of occurrence to reach peak discharge up to 7075 $m^3 s^{-1}$. This discharge is contributed by increasing monsoonal runoff of upper catchment of Damodar which ranges from 320 to 440 mm per year and, in post-dam period, there is requirement of only 30 mm runoff to reach peak discharge of 7035 $m^3 s^{-1}$ at Rhondia. But, now, we have estimated that only 4011 $m^3 s^{-1}$ of discharge (return period of 3 years) is considered as threshold limit of flood discharge at Rhondia and, when we go downstream section, it is reduced up to only 1542 $m^3 s^{-1}$. For that reason, lower Damodar River has annually experienced overflow condition and adjoining parts of Barddhaman, Hooghly, and Howrah are flooded in peak monsoon. Now, it is said that present carrying capacity of lower Damodar is reduced to 1415 $m^3 s^{-1}$ only. Remembering the emerging problems of riverine flood, we have mentioned that the current factors of flood risk of lower Damodar River are as follows: (1) bottle-neck and physically handicapped location of lower Damodar Basin in the Gangetic West Bengal (huge volume of channel flow collected from funnel-shaped rocky upper catchment passing through narrow and shallow reach of lower Damodar) including the tidal effect of lower reach, (2) three to four days continuous heavy rainfall due to SW-NE directional monsoonal depressions, (3) uncontrolled runoff of upper catchment, (4) increasing siltation of dams, barrages, canals, and river beds, (5) only four large dams (i.e., Tilaiya, Konar, Maithon, and Panchet), Tenughat Reservoir, and Durgapur Barrage serving the purpose of all proposed eight large dams and combined flood moderation capacity of Maithon and Panchet dams reducing up to only 32 per cent, (6) the dams compelling to release excess water in the month of late September because of already storing water in the previous months of monsoon, (7) annual peak discharge of short duration occurring in between late September and October at the time of over-saturation of alluvial soils, ground water and existing streams, and (8) drainage congestion and encroachment of active river bed and floodplain.

From the perspective of flood climate, the recurrent floods of Damodar River is directly influenced by rainstorms of 3- to 4-day duration, path of cyclone, extreme rainfall event of 3 to 6 hours, runoff yield, and discharging of excess water from the upstream dams and Durgapur barrage. From the standpoint of flood hydrology, the stream flow during high magnitude floods in our study area is primarily confined within bankfull level, with occasional overtopping of the levees. The floodplain flow, whenever it took place, is intermittent in nature. To manage floods, we should focus on the travel time of flood waves from Durgapur barrage to the downstream end and on the up-to-date accurate estimation of critical bankfull discharge at the ungauged sites of lower Damodar River.

At last, it can be said that human and technology solely cannot control the river dynamics in our favour but we can adopt or adjust ourselves to the dynamicity of river discharges through a complete understanding of hydrogeomorphic behaviours of an active alluvial river, like Damodar. From the above analysis, it is understood that observing the drawbacks of large scale Damodar Valley Planning, we can only predict or manage the flood discharge to a certain level, not stopping it completely. So scrutinizing the exiting framework of basin planning, it is the exact time to rethink the increasing flood risk of lower Damodar River and renovation of Damodar Valley Planning in West Bengal in the frame of global warming and climate change.

ACKNOWLEDGMENTS

The authors sincerely thank Achim A. Beylich (Editor) for giving them suggestions to enrich the work in a scientific way. They are very much thankful to the reviewers for providing valuable comments and suggestions on the paper. This study was supported by the Department of Geography, The University of Burdwan (West Bengal, India).

REFERENCES

1. M. Allaby, Floods, Viva Books Private Limited, New Delhi, India, 2006.
2. H. M. Raghunath, Hydrology—Principles, Analysis and Design, New Age International Publishers, New Delhi, India, 2011.
3. V. T. Chow, D. R. Maidment, and L. M. Mays, Applied Hydrology, McGraw Hill Book, New York, NY, USA, 1988.
4. S. S. Sharma, "Floods in West Bengal—nature, analysis and solution," in Changing Environmental Scenario, S.

Basu, Ed., pp. 315–322, acb Publication, Kolkata, India, 2002.

5. B. P. Hayden, "Flood climate," in Flood Geomorphology, V. R. Baker, R. C. Kochel, and P. C. Patton, Eds., pp. 13–27, John Wiley & Sons, New York, NY, USA, 1988.

6. K. K. Hirschboeck, "Flood hydroclimatology," in Flood Geomorpholog, V. R. Baker, R. C. Kochel, and P. C. Patton, Eds., pp. 27–50, John Wiley & Sons, New York, NY, USA, 1988.

7. P. J. R. Reddy, A Textbook of Hydrology, University Science Press, Bangalore, India, 2011.

8. S. Nandargi and O. N. Dhar, "High frequency floods and their magnitudes in the Indian rivers," Journal of the Geological Society of India, vol. 61, no. 1, pp. 90–96, 2003. ·

9. K. Parthasarathy, "Weather in relation to floods in Bihar, Bengal and Assam during July and August 1954," Journal of Scientific and Industrial Research, vol. 14, no. A, pp. 115–119, 1954.

10. U. K. Bose, "Preliminary meteorological study associated with West Bengal rivers," Indian Journal of Power and River Valley Development, vol. 7, no. 9, pp. 23–24, 1957.

11. P. Singh, A. S. Ramanathan, and V. G. Ghanekar, "Flash floods in India," pp. 114–118, 1974, http://ks360352.kimsufi.com/redbooks/a112/iahs_112_0114.pdf.

12. V. S. Kale, "Geomorphic effects of monsoon floods on Indian rivers," Natural Hazards, vol. 28, no. 1, pp. 65–84, 2003. · ·

13. V. S. Kale, "Fluvial hydrology and geomorphology of monsoon-dominated Indian rivers," Revista Brasileira de Geomorfologia, vol. 6, no. 1, pp. 63–73, 2005.

14. V. S. Kale, "Research on floods in India," 2009, http://scholar.google.co.in/citations?view_op=view_citation&hl=en&user=eyv0fA8AAAAJ&cstart=60&citft=1&citft=2&email_for_op=sandipanghosh19%40gmail.com&citation_for_view=eyv0fA8AAAAJ:M3NEmzRMIkIC.

15. O. N. Dhar and S. Nandargi, "Hydrometeorological aspects of floods in India," Natural Hazards, vol. 28, no. 1, pp. 1–33, 2003. · ·

16. A. K. Singhvi and V. S. Kale, "Paleoclimate studies in India: last Ice Age to the Present," GBP-WCRP-SCOPE-Report series 4, Indian National Science Academy, 2009.

17. P. Guhathakurta, O. P. Sreejith, and P. A. Menon, "Impact of climate change on extreme rainfall events and flood risk in India," Journal of Earth System Science, vol. 120, no. 3, pp. 359–373, 2011. · ·

18. C. Ramaswamy, "Meteorological aspects of severe floods in India, 1923–1979," Meteorological Monograph Hydrology no. 10, Indian Meteorological Department, 1987.

19. World Bank, "Climate change impacts in droughts and floods affected areas: case studies in India," Report No. 43946-IN, Development of the World Bank, 2008.

20. M. K. Saha, "River flood forecasting and preparedness—Lower Damodar Valley in West Bengal," in River Floods—A Socio-Technical Approach, K. M. B. Rahim, N. Mukhopadhyay, and D. Sarkar, Eds., pp. 244–249, ACB Publications, Kolkata, India, 2005.

21. K. Bhattacharyya, The Lower Damodar River, India—Understanding the Human Role in Changing Fluvial Environment, Springer, New York, NY, USA, 2011.

22. G. Nagle, River and Water Management, Hodder and Stoughton, London, UK, 2003.

23. R. J. Garde and U. C. Kothyari, "Flood estimation in Indian catchments," Journal of Hydrology, vol. 113, no. 1–4, pp. 135–146, 1990. · ·

24. V. S. Kale, S. Mishra, Y. Enzel, L. L. Ely, S. N. Rajaguru, and V. R. Baker, "Flood geomorphology of the Indian peninsular rivers," Journal of Applied Hydrology, vol. 6, no. 1–4, pp. 40–55, 1995.

25. S. N. Rajaguru, A. Gupta, V. S. Kale, et al., "Channel form and processes of the flood-dominated Narmada River, India," Earth Surface Processes and Landforms, vol. 20, no. 5, pp. 407–421, 1995. ·

26. V. S. Kale, "Monsoon floods in India: a hydro-geomorphic perspective," in Flood Studies in India, V. S. Kale, Ed., pp. 229–256, Geological Society of India, Bengaluru, India, 1998.

27. V. S. Kale, "Long-period fluctuations in monsoon floods in the Deccan Peninsula, India," Journal of the Geological Society of India, vol. 53, no. 1, pp. 5–15, 1999. ·

28. D. C. Goswami, "Fluvial regime and flood hydrology of the Brahmaputra River, Assam," in Flood Studies in India, V. S. Kale, Ed., pp. 53–76, Geological Society of India, Bangalore, India, 1998.

29. P. R. Rakhecha, "Highest floods in India," in Proceedings of the Symposium on the Extreme of the Extremes: Extraordinary Floods, pp. 167–172, IAHS Publication, no. 271, Reykjvik, Icleand, 2002.

30. R. Jha and V. Smakthin, "Review of methods of hydrological estimation at ungauged sites in India,"International Water Management Institute Working Paper 130, 2008.

31. S. Mukhopadhyay, "A geo-environmental assessment of flood dynamics in lower Ajoy River inducing sand splay problem in eastern India," Ethiopian Journal of Environmental Studies and Management, vol. 3, no. 2, pp. 96–110, 2010. ·

32. V. C. Jha and H. Bairagya, "Floodplain planning based on statistical analysis of Tilpara Barrage discharge: a case study on Mayurakshi River Basin," Caminhos de Geografia, vol. 13, no. 43, pp. 326–346, 2012.

33. E. L. Glass, "Floods of the Damodar River and rainstorms producing them," Minutes of the Proceedings, vol. 217, pp. 333–346, 1924. ·

34. P. K. Sen, "The genesis of floods in the lower Damodar catchment," in The Concepts and Methods in Geography, P. K. Sen, Ed., pp. 71–85, University of Burdwan, Bardhaman, India, 1985.

35. M. R. Goodall, "River valley planning in India: the Damodar," The Journal of Land and Public Utility Economics, vol. 21, no. 4, pp. 371–375, 1945. ·

36. W. Krik, "The Damodar valley—valley

opima," Geographical Review, vol. 40, no. 3, pp. 415–443, 1950. ·

37. S. K. Pramanik and K. N. Rao, Hydrometeorology of the Damodar Catchment, 2012, http://iahs.info/redebooks/a036/036060.
38. K. Bagchi, "The Damodar Valley development and its impact on the region," in Indian Urbanization and Planning: Vehicle of Modernization, G. N. Allen and A. K. Dutt, Eds., pp. 232–241, Tata McGraw-Hill, New Delhi, India, 1977.
39. D. Roy, S. Mukherjee, and B. Bose, "Regulation of a multipurpose reservoir system: Damodar Valley, India," in Proceedings on Man's Influences on Freshwater Ecosystems and Water Use, IAHS Publication no. 230, pp. 95–99, 1995.
40. D. K. Mishra, "Living with floods: people's perspective," Economic and Political Weekly, vol. 36, no. 2, pp. 2756–2761, 2001.
41. S. Sengupta, "River and floods," Breakthrough, vol. 9, no. 2, pp. 2–8, 2001.
42. S. Chandra, India Flood Management—Damodar River Basin, 2012, http://www.apfm.info/publications/casestudies/cs_india_full.
43. P. K. Sen, "Flood hazards and river bank erosion in the Lower Damodar Basin," in Indian Geomorphology, H. S. Sharma, Ed., pp. 95–108, Concept Publishing Company, New Delhi, India, 1991.
44. M. Majumder, P. Roy, and A. Mazumdar, "An introduction and current trends of Damodar and Rupnarayan River Network," in Impact of Climate Change on Natural Resource Management, B. K. Jana and M. Majumder, Eds., pp. 461–480, Springer, New York, NY, USA, 2010.
45. S. P. Chatterjee, Damodar Planning Atlas, NATMO, Calcutta, India, 1969.
46. F. G. Bell, Geological Hazards: Their Assessment, Avoidance and Mitigation, E. & F.N. Spon, London, UK, 1999.
47. K. Rudra, "Floods in West Bengal, 2000—causes and consequences," in Changing Environmental Scenario, S. Basu, Ed., pp. 326–347, ACB Publications, Kolkata, India, 2002.
48. R. D. Dhir, P. R. Ahuja, and K. G. Majumdar, "A study on the success of reservoir based on the actual and estimated runoff," Research Session of Central Board of Irrigation and Power, vol. 68, 1958.
49. E. J. Gumbel, "The return period of flood flows," The Annals of Mathematical Statistics, vol. 12, no. 2, pp. 163–190, 1941. · ·
50. H. R. Betal, "Identification of slope categories in Damodar Valley, India," in Selected Papers on Physical Geography, S. P. Chatterjee and S. P. Dasgupta, Eds., vol. 1, pp. 4–6, National Committee for Geography, Calcutta, India, 1970.
51. G. N. Rao, "Occurrence of heavy rainfall around the confluence line in monsoon disturbances and its importance in causing floods," Journal of Earth System Science, vol. 110, no. 1, pp. 87–94, 2001. · ·
52. M. K. Goyal and C. S. P. Ojha, "Analysis of mean monthly rainfall runoff data of Indian Catchments using dimensionless variables by neutral networks," Journal Environment Protection, vol. 1, pp. 135–146, 2010.
53. M. Lal, T. Nozawa, S. Emori et al., "Future climate change: implications for Indian summer monsoon and its variability," Current Science, vol. 81, no. 9, pp. 1196–1207, 2001. ·
54. C. O. Wisler and E. F. Brater, Hydrology, John Wiley & Sons, New York, NY, USA, 1959.
55. D. W. Reed, "Reinforcing flood-risk estimation," Philosophical Transactions: Mathematical, Physical and Engineering Sciences, vol. 360, no. 1796, pp. 1371–1387, 2002.
56. R. J. More, "The basin hydrological cycle," in Water, Man and Earth, R. J. Chorley, Ed., pp. 67–75, Methuen and Co., London, UK, 1969.
57. P. K. Roy and A. Mazumder, "Hydrological impacts of climatic variability on water resources o the Damodar River Basin, India," in Regional Hydrological Impacts of Climate Change: Impact, Assessment and Decision Making, T. Wagener, S. Franks, H. V. Gupta, et al., Eds., pp. 147–156, IAHS Publication, London, UK, 2005.
58. S. C. Mukhopadhyay, "Comtemporary issues in geography with particular emphasis on the flood hazards of West Bengal," in Contemporary Issues and Techniques in Geography, R. Basu and S. Bhaduri, Eds., pp. 77–110, Progressive Publishers, Kolkata, India, 2007.
59. S. Ghosh, Flood Hydrology and Risk Assessment—Flood Study in a Dam-Controlled River of India, Lambert Academic Publishing, Saarbrücken, Germany, 2013.

CHAPTER 12 THE USE OF H-SAF SOIL MOISTURE PRODUCTS FOR OPERATIONAL HYDROLOGY

Christian Massari, Luca Brocca, Luca Ciabatta, Tommaso Moramarco1, Simone Gabellani, Clement Albergel, Patricia De Rosnay, Silvia Puca and Wolfgang Wagner

Keywords: Soil moisture, floods, remote sensing, hydrological modelling

INTRODUCTION

Flooding has become one of the events producing the most fatalities annually [1]. Whereas much progress has been made in meteo-forecasts and warnings and also in public preparedness, a comparable system for "quantitatively" predicting floods has experienced less progress, especially in floods occurring in medium–small catchment sizes (100–1000 km22), as demonstrated by recent events occurring in Italy (i.e., Liguria, Tuscany and Sicily at the end of 2011, Umbria in 2012 and Sardinia in 2014).

Indeed, the anticipation of the magnitude of an event is crucial for performing the correct actions within civil protection activities, but this is not a simple task. First, the amount of precipitation that transforms an otherwise ordinary rainfall event into an extraordinary one is led by complex interactions between meteorology and hydrology, such as, among other important factors, the soil moisture (SM) conditions prior to the flood event [2,3,4,5]. Second, predicting a flood event is not only a matter of being able to correctly describe such factors, but it is strongly related to the capability of the early warning system in terms of the data and tools upon which it can rely. That is: (i) an appropriate rainfall-runoff (RR) hydrological model able to infer, with a certain degree of accuracy, the discharge hydrograph; and (ii) a dense network of sensors able to provide good quality observations in near real time.

These two requirements are not independent of each other. Indeed, the choice of the most appropriate RR model (e.g., continuous versus event-based models) often stems from the availability of certain types of data (e.g., evapotranspiration or temperature data) and from the number of sensors available in the catchment, which leads the hydrologist to the choice of a continuous, event-based, distributed or lumped hydrological model.

Generally, continuous models try to describe the different hydrological processes able to generate runoff, but require long-term and uninterrupted time series of rainfall and evapotranspiration data. This could be a strong limitation in poorly-gauged areas, mainly if hourly observations are needed [6]. On the other hand, event-based RR models are very appealing and frequently employed within operational flood forecasting systems [3], because of their simplicity, the need for reduced parametrization and data records (i.e., only rainfall recorded during the event is needed) and last, but not least, the much lower computational demand. However, a major limitation of event-based models lies in the definition of the initial SM conditions of the catchment, which may strongly vary from one storm event to another [3,5,7], especially in regions characterized by strong seasonality, such as Mediterranean countries [8].

In the last few decades, many studies [4,5,9,10,11,12,13] have mentioned the high importance of SM, because of its ability to determine the partitioning of rainfall into runoff and infiltration [14]. In particular, the authors have demonstrated that when SM is characterized by high variability throughout the year (e.g., in Mediterranean regions), it represents, more than others quantities, a good proxy of the antecedent wetness conditions of the catchment, thus allowing better prediction of the runoff response to the rainfall input.

At the same time, the ever-increasing availability of SM measurements from in situ stations (International Soil Moisture Network (ISMN) [15]), satellite sensors (e.g., the Advanced Scatterometer (ASCAT), [16], the Advanced Microwave Scanning Radiometer for Earth observation (AMSR-E) and its successor, AMSR-2 [17], and the Soil Moisture and Ocean Salinity Mission (SMOS) [18]) and land surface models

Citation: Massari, C.; Brocca, L.; Ciabatta, L.; Moramarco, T.; Gabellani, S.; Albergel, C.; De Rosnay, P.; Puca, S.; Wagner, W. The Use of H-SAF Soil Moisture Products for Operational Hydrology. Hydrology 2015, 2, 2-22. doi:10.3390/hydrology2010002.

[19], at increasing temporal and spatial resolutions [20], has opened new possibilities for integrating such measurements into hydrological models, even at an operational level. By way of example, the "EUMETSAT Satellite Application Facility in Support of Operational Hydrology and Water Management" (H-SAF, http://hsaf.meteoam.it/), established by the EUMETSAT Council in 2005, generates and archives high-quality rainfall, soil moisture and snow products for operational hydrological applications, starting from the acquisition and processing of data from Earth observation satellites in geostationary and polar orbits operated both by EUMETSAT and other satellite organizations.

Such an overabundance of products has highly increased the number of studies concerning the assimilation of SM into hydrological models (the reader is referred to [21] for a complete review of these studies). Although many of them show very contrasting results in terms of how the assimilation of SM can improve the accuracy in flood forecasting, all seem to agree that such observations have a high potential to reduce uncertainty in flood prediction. The problem is more related to the preprocessing steps prior to the inclusion of the observations into the hydrological model [22] and in the correct use of the assimilation technique [23], rather than in the value that the observations themselves can bring to flood forecasting. At an operational level, especially in small–medium catchments, the problem is even more exacerbated, due to: (i) the high level of expertise required for implementing and setting up an appropriate assimilation scheme with the risk of not even exploiting these new source of data; and (ii) the lack of uninterrupted and good quality rainfall and evapotranspiration data for running continuous hydrological models, which are needed for current data assimilation approaches.

Recently, [24] proposed a "simplified continuous rainfall runoff model" (SCRRM) that, instead of modelling SM from continuous precipitation and evapotranspiration data, like in classical continuous RR models [12,25,26,27], directly uses SM recorded from external sources (e.g., ground observations, satellite sensors and land surface models) to set the initial conditions of an event-based model. SCRRM explicitly embeds the relationship existing between SM and the model's initial conditions [2,3,4,5] into an event-based RR model to simulate discharge hydrographs. The model was successfully applied in a small catchment of the Attica Region in Greece using different globally-available SM and yielding performances similar to a continuous model [24]. The main advantages of this approach are that SCRRM: (i) does not require continuously-recorded datasets; hence, it can be used also in poorly-gauged areas; (ii) is simple and parsimonious, which is an advantage for users with low hydrological expertise; (iii) requires low computational demand to be run, which makes it appropriate for early warning system applications operating in near real time; and (iv) can be used to "hydro-validate" satellite soil moisture observations. For these reasons, the model is very attractive for civil protection activities, especially in areas where a flood forecasting system is totally absent, and for testing the information content related to the satellite SM dataset.

Based on that, the objectives of this study are two-fold. The first goal is to investigate the performance of two of the satellite SM products of the H-SAF project (i.e., SM-OBS-1 and SM-DAS-2) in order to highlight the information content that they retain in flood modelling. The second goal is to test the performance of SCRRM against the "Modello Idrologico Semi-Distribuito in continuo" (MISDc, [12]) over the Italian territory in order to gain some understanding regarding under what conditions the information derived from the external source of SM can be used for flood forecasting at an operational level.

A total of 593 flood events are used, extracted from a dataset of rainfall and discharges recorded from 2010 to 2013 of 35 Italian catchments, representing a range of sizes, micro-climates, precipitation, streamflows and SM conditions. Through the analysis of such a large number of catchments, some understanding may be gained regarding under what conditions the information derived from the external source of SM can be used for flood forecasting at an operational level.

The paper is organized as follows: In Section 2, we describe the hydrological models used in this study and the characteristics of the selected catchments. In Section 3, we provide a comprehensive analysis of SCRRM performances running with SM-OBS-1 and SM-DAS-2, comparing it against the "Modello Idrologico Semi-Distribuito in continuo" (MISDc, [12]) in terms of reproducing the discharge hydrograph, the peak discharge and the total runoff volume. Finally, in Section 4, we provide the conclusions.

DATA AND METHODS

Catchment Selection and Hydro-Meteorological Data

An available dataset (2010–2013) of 35 Italian catchments having hourly streamflow, precipitation and temperature observations provided by the Italian Civil Protection Department (DPC) was used in the present analysis (see Table 1for more details). The catchments range in size from 800 to 7400 km22 (mean size 2507 km22). Most catchments are located in the northern part of Italy (see Figure 1). This region is characterized by a temperate Mediterranean climate with a wet winter and moderate summer rainfall, whereas the south part of Italy has a drier climate with semi-arid summers and temperate winters. The number of catchments is the result of an appropriate selection for excluding catchments with streamflow subject to regulation or diversion and characterized by a number of events less than three. The mean areal rainfall for each catchment was calculated by the GRISO model [28]. GRISO is an improved Kriging-based technique that preserves the values observed at the rain gauge location, allowing for a dynamical definition of the covariance structure associated with each rain gauge by the interpolation procedure. Each correlation structure may depend both on the rain gauge location and on the accumulation time considered. Rainfall events were extracted by selecting those with a continuous rainfall characterized by a total cumulated precipitation larger than 10 mm and no rainfall in the preceding day for a total of 593 flood events. Direct runoff was evaluated as in [29] by using an appropriate base flow separation technique. Mean temperature data for the catchments were obtained by averaging the temperature recorded by the thermometers inside the catchment boundaries.

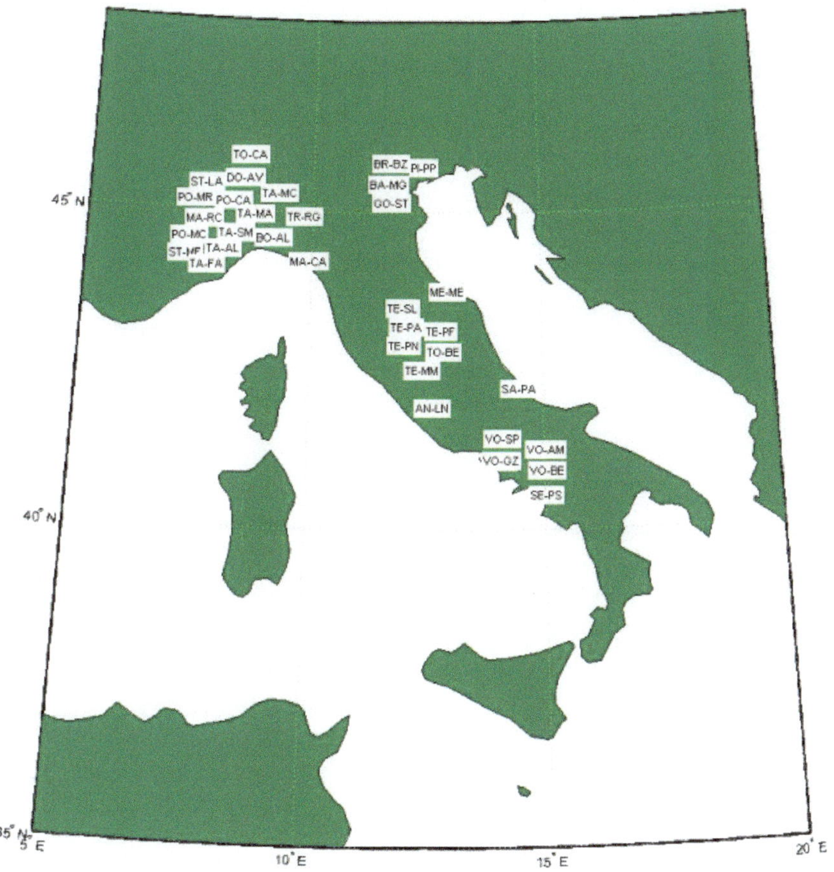

Figure 1: Approximative catchment position.

Table 1: Characteristics of the selected catchments

#	Code	Catchment	Longitude E	Latitude N	Area (km²)	No. Events	Mean Annual Rainfall (mm)	Temperature (°C)
1	AN-LN	Aniene at Lunghezza	12.66	41.93	984.6	5	1525.9	13
2	BA-MG	Bacchiglione at Montegalda	11.67	45.44	1321.3	18	2759.5	10.2
3	BO-AL	Bormida at Alessandria	8.65	44.91	2355.8	18	1157.6	11.6
4	BR-BZ	Brenta at Berzizza	11.73	45.78	1506.3	8	2255.1	7
5	DO-AV	Dorabaltea at Verolengo	8.04	45.19	3640.8	11	1089.7	4.4
6	GO-ST	Gorzone at Stanghella	11.76	45.15	1205.8	5	1901.5	13.6
7	MA-CA	Magra at Calamazza	9.95	44.2	857.8	26	2807.7	11.3
8	MA-RC	Maira at Raconigi	7.67	44.77	967.8	8	998.2	7.9
9	ME-ME	Metauro at Metauro	12.97	43.76	1206.4	13	1617.9	12.6
10	PI-PP	Piave at Pontedipiave	12.45	45.71	3902.7	11	2693.3	6.8
11	PO-CA	Po at Carignano	7.69	44.91	3569.5	9	993.6	8.7
12	PO-MC	Po at Moncalieri	7.68	45	4624.1	9	978.9	9.7
13	PO-MR	Po at Torino Murazzi	7.7	45.06	4899.9	9	986.3	9.7
14	SA-PA	Sangro at Paglieta	14.51	42.21	1522.8	4	1203.6	10.1
15	SE-PS	Sele at Persano Sele	15.03	40.54	2057.9	18	1556.9	12.5
16	ST-LA	Stura di Lanzo at Torino	7.71	45.11	799.9	26	1318.9	7.3
17	ST-MF	Stura di Demonte at Fossano	7.72	44.52	1129.7	8	1300.7	6.3
18	TA-AL	Tanaro at Alba	8.03	44.71	3070.3	20	1170.1	8.8
19	TA-FA	Tanaro at Farigliano	7.9	44.52	1364.5	19	1190.2	9.3
20	TA-MA	Tanaro at Masio	8.41	44.87	4157.4	15	1085.9	9.8
21	TA-MC	Tanaro at Montecastello	8.68	44.95	7400	14	1072.5	10.7
22	TA-SM	Tanaro at Asti San Martino	8.21	44.88	3229.7	14	1154.6	9
23	TE-MM	Tevere at Montemolino	12.39	42.79	4815.4	26	1341.7	12.8
24	TE-PA	Tevere at Pierantonio	12.38	43.26	1694.3	12	1397.7	12.2
25	TE-PF	Tevere at Pontefelcino	12.43	43.13	1879	28	1395.8	12.3
26	TE-PN	Tevere at Pontenuovo	12.43	43.01	3695.3	22	1379.8	12.5
27	TE-SL	Tevere at Santa Lucia	12.24	43.42	837.9	29	1456.9	11.8
28	TO-BE	Topino at Bettona	12.54	43.02	1054.9	24	1333.5	12.6
29	TO-CA	Toce at Candoglia	8.42	45.97	1264	17	1687.9	5.6
30	TR-RG	Trebbia at Rivergaro	9.58	44.9	839.5	27	1735.3	10
31	VO-AM	Volturno at Amorosi	14.45	41.2	1766.8	32	1574.4	13
32	VO-BE	Volturno at Benevento	14.77	41.13	1776.3	34	1173	13.3
33	VO-CA	Volturno at Cancello Arnone	14.02	41.07	4877.9	9	1430.4	13.4
34	VO-GZ	Volturno at Grazzanise	14.11	41.09	4871.1	16	1429.8	13.4
35	VO-SP	Volturno at Solopaca	14.57	41.21	2578.8	29	1307.2	13.3

Satellite and Modelled Soil Moisture Data

In this study, two different SM products, distributed by the H-SAF project, were used covering the period 2010–2013. The products are SM-OBS-1 (large-scale surface soil moisture by radar scatterometer) and SM-DAS-2 (scatterometer data assimilation in the European Centre for Medium-Range Weather Forecasts, ECMWF, Land Data Assimilation System). Given the different catchment extensions and spatial resolution of the two products, only data falling inside the catchment boundaries were selected (when present). When no pixel was contained inside the boundaries, the closest point to the centroid of the catchment was selected. Pixels near the sea or in very high mountain areas were not considered. Table 2 summarizes the main characteristics of the selected SM indicators, which are described in detail next.

Table 2: The main characteristics of the soil moisture products used in in this study. SM, soil moisture; ASCAT, Advanced Scatterometer.

Product	Code	Spatial Resolution (km)	Temporal Resolution (days)	Depth (cm)	Source
SM-OBS-1	H07	25	≃1	0–2	ASCAT
SM-DAS-2	H14	25	1	0–289	Assimilation of SM-OBS-1

SM-OBS-1

Product SM-OBS-1 is based on the radar scatterometer ASCAT onboard the MetOpsatellites. The instrument scans the scene in a push-broom mode by six side-looking antennas, three left-handed and three right-handed. ASCAT measures radar backscatter at the C-band (5.255 GHz) in VV polarization. The basic instrument sampling distance is 12.5 km. The primary ASCAT observation, sea-surface wind, is processed at 50-km resolution. For SM, processing is performed at 50-km (operational) and 25-km (research) resolution. Global coverage over Europe is achieved in ~1.5 days, while in Italy, measurements are available about once a day. The surface SM product (equivalent to a depth of 0–2 cm of soil) is calculated from the backscatter measurements through a time series-based change detection approach previously used for the ERS-1/2 by [30]. The SM is derived by selecting the historical lowest and highest backscatter measurement to which a 0% (dry) and a 100% (wet) reference is assigned, respectively.

SM-OBS-1 provides knowledge of SM for a very thin surface layer (about 0–2 cm); however, in RR transformation processes, this information may not be sufficient, since the root-zone SM has been shown to be much more important in determining the catchment response to a given storm event [31]. To obtain the root-zone SM product (SWI, Soil Water Index) from the satellite-based surface observations, the recursive formulation [32] of the exponential filter of [30] was adopted:

$$SWI(t_n) = SWI(t_{n-1}) + K_n\,[ms(t_n) - SWI(t_n - 1)] \quad (1)$$

Where $ms(t_n)$ is the surface SM observed by the satellite sensor SM-OBS-1, SWI_{tn} is the Soil Wetness Index representing the profile averaged saturation degree and time t_n is the acquisition time of $ms(t_n)$. The gain K_n at time t_n is given by (in a recursive form):

$$K_n = \frac{K_{n-1}}{K_{n-1} + e^{-\left(\frac{t_n - t_{n-1}}{T}\right)}} \quad (2)$$

where T is the characteristic time length expressed in days and represents the time scale of SM variation to obtain the SWI. For the initialization of this filter, K_1 and SWI_1 were set to 1 and $ms(t_1)$, respectively.

SM-DAS-2

SM-DAS-2 is the first global product of consistent surface and root zone SM that is available in near real time for the numerical weather prediction and climate and hydrological communities. It is based on ASCAT surface SM data assimilation in the ECMWF Land Data Assimilation System. Overall, SM-DAS-2 relies on an advanced land data assimilation system, which is based on an extended Kalman filter (EKF, [33]) able to ingest information contained in observations close to the surface (2-m temperature and relative humidity synoptic report), as well as new types of data, such as remotely-sensed surface SM. Within the EKF, the surface observation from ASCAT is propagated towards the root region down to 289 cm below the surface, providing estimates for 4 layers (thicknesses of 7, 21, 72 and 189 cm). The ECMWF model generates SM profile information according to the Hydrology Tiled ECMWF Scheme for Surface Exchanges over Land (HTESSEL, [19]). SM-DAS-2 is available at a 24-hour time step with a spatial resolution of 25 km, with a global daily coverage at 00:00 UTC. SM-DAS-2 is run continuously in order to ensure the time series consistency of the product (and also to provide values when there is no satellite data, from the model propagation). SM-DAS-2 SM has been validated against in situ SM from many locations in Africa, Australia and Europe, providing satisfactory results [34]. It has to be noted that SM-DAS-2 data prior to 2012 are experimental data (i.e., the full 2010–2013 dataset might not be completely consistent), so it is highly expected that it can provide better performances from 2012 onwards.

In this study, the information of the SM for any soil depth between 0 and 289 cm, θSM−DAS2θSM-DAS2, was obtained by the weighed mean of the SM provided by the related layer, according to:

$$\begin{aligned}
&\theta_{SM-DAS2} = \theta_1 && : z \leq 7\text{cm}\\
&\theta_{SM-DAS2} = \frac{(\theta_1 7 + \theta_2(z-7))}{z} && : 7 < z \leq 28\text{cm}\\
&\theta_{SM-DAS2} = \frac{(\theta_1 7 + \theta_2 21 + \theta_3(z-28))}{z} && : 28 < z \leq 100\text{cm}\\
&\theta_{SM-DAS2} = \frac{(\theta_1 7 + \theta_2 21 + \theta_2 72 + \theta_3(z-100))}{z} && : 100 < z \leq 289\text{cm}
\end{aligned} \quad (3)$$

where θ_1, θ_2, θ_3, θ_4 are the SM for each of the four layers of the SM-DAS-2 product and z is the parameter representing the depth.

Hydrological Models

In the following, we present a description of all of the hydrological models used for this study. A continuous model (MISDc, "Modello Idrologico Semiditribuito in Continuo", [12]) is used as a benchmark for evaluating the reliability of the SCRRM model presented in Section 2.3.2 running with the two SM products of the H-SAF project described inSection 2.2.

Continuous Model: MISDc

MISDc [12] is a continuous rainfall runoff model successfully applied to the Tiber River for flood prediction and operational purposes. The lumped version of the MISDc model used in this study couples a soil water balance model (SWB, [2]) to simulate the SM temporal pattern and a routing module [35] for transferring the rainfall excess to the outlet section of the catchment. The two models are linked through an experimentally-derived linear relationship between the potential maximum soil moisture retention S of the Soil Conservation Service-Curve Number (SCS-CN, [36]) and the relative SM at the beginning of the event.

The SWB model is a simple water balance model representing the main processes needed for SM simulation: infiltration, percolation and evapotranspiration. The processes are represented for infiltration through the Green–Ampt equation for drainage by a gravity-driven non-linear relationship and for actual evapotranspiration by a linear relationship with the potential evapotranspiration, calculated through a modified Blaney and Criddle method. The reader is referred to [2] and [12] for a detailed description of the model. The SM simulated by the SWB is used to calculate the parameter S method by means of an experimentally-derived relationship between S and SM [10]:

$$L = \eta 1.19 A^{0.33} \tag{4}$$

where θ is the modelled relative SM at the beginning of the event and a is a parameter to be estimated. Once the Sparameter is estimated, the routing to the outlet of the catchment is obtained from the convolution of the rainfall excess and the geomorphological instantaneous unit hydrograph (GIUH), such as proposed by [37]. In the model, the lag time is evaluated through the relationship proposed by [29]:

$$L = \eta 1.19 A^{0.33} \tag{5}$$

with L being the lag time in hours, A the area of the catchment (km22), and η the parameter to be calibrated [38].

MISDc requires as input data the rainfall and air temperature. The model outputs are both the direct runoff in correspondence to selected flood events and the catchment average relative SM.

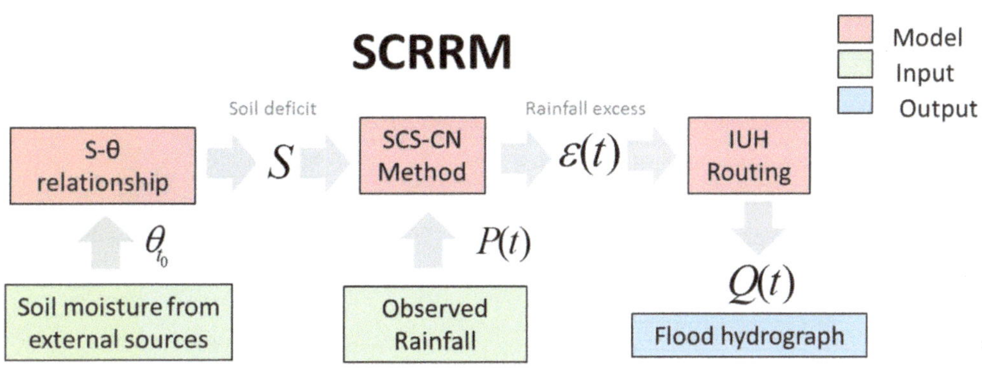

Figure 2: Structure of the simplified continuous rainfall-runoff model (SCRRM).

SCRRM

Unlike MISDc, where the SM at the beginning of the event is simulated by the SWB model, in SCRRM (Figure 2), it is provided by an external indicator, i.e., from satellite SM observations or model-based reanalysis products. SCRRM reflects the structure of MISDc, but has some significant differences. Indeed, the temporal evolution of the soil wetness conditions of the catchment is not modelled from rainfall and temperature data, as in MISDc, but it is integrated directly into the model from SM observations (i.e., SWB is replaced in SCRRM by SM observations immediately before the rainfall event) considering the SM products as proxies for the assessment of the wetness state of the catchment. The model can be also seen as the result of an assimilation technique in which the SM observations from the external source are assumed to be error free (i.e., a direct insertion technique).

Like in MISDc, the model exploits the observed linear behaviour between the wetness state of the soil and the parameter S of the SCS method by Equation (4). Once S is known, the rainfall excess is calculated by:

$$\epsilon = \frac{(P - F_a)^2}{P - F_a + S} \text{ if } \epsilon \geq F_a \tag{6}$$

Where F_α is the initial abstraction, ϵ is the rainfall excess and P is the rainfall depth. As in the SCS method, the quantity F_α is considered linearly dependent on S by:

$$F_a = \lambda S \tag{7}$$

and then is transferred to the outlet section with the same routing module used by MISDc. In Equation (7), λ is the initial abstraction coefficient.

In synthesis, SCRRM uses the SM and the event rainfall data as the sole inputs to simulate hourly flood hydrographs. Since the SM is provided by an external indicator, Equation (4) becomes a model relation embedded in the model structure. The calibration of the model involves the following three parameters: the coefficient of initial abstractions λ, the parameter α of the S-θ relationship and the parameter η of Equation (5).

As remarked in Section 2.2, the soil layer depth that controls the RR transformation is usually larger than a few centimetres. As a result, the application of SCRRM with the H-SAF products is taken into account by including the soil depth as an additional parameter of SCRRM. For SM-OBS-1, such a parameter is controlled by the characteristic time length T of the exponential filter described in Section 2.2.1, while for SM-DAS-2, the parameter z of Equation (3) was considered, which can vary from 0 to 289 cm (see Section 2.2.2). For this study, a lumped model was employed, even though the same concept can be easily applied to spatially-distributed models.

Performance Indexes

The performances of SCRRM and MISDc were evaluated by considering different indexes. The first one, commonly used for assessing the agreement between simulated and observed hydrographs, is the Nash–Sutcliffe efficiency coefficient, NSNS [39]:

$$NS = 1 - \frac{\sum_{t=1}^{T_{ev}} (Q_{obs} - Q_{sim})^2}{\sum_{t=1}^{T_{ev}} (Q_{obs} - \bar{Q}_{obs})^2} \qquad (8)$$

Where Q_{obs} and Q_{sim} are the observed and simulated discharges at time t, respectively, $\bar{Q}_{obs}$ is the mean value of the observed discharge during the event and T_{ev} is the event duration. NSNS was calculated for each of the selected events for every catchment considered in the analysis. In particular, the mean NSNS calculated over the selected events for each catchment was used as an objective function for calibrating the parameters of the models:

$$\overline{NS} = \frac{\sum_{j=1}^{N_{ev}} NS_j}{N_{ev}} \qquad (9)$$

Where N_{ev} is the number of events considered, whereas index j varies between 1 and the number of selected events for each catchment. For model calibration, a standard gradient-based automatic optimisation method (the "fmincon" function in MATLAB ®,[40]) was used. In addition, to evaluate the performance of the model in reproducing flood events, the percentage error on peak discharge:

$$E_{Q_p} = \frac{\max(Q_{obs}) - \max(Q_{sim})}{\max(Q_{obs})} \qquad (10)$$

and the percentage error on direct runoff volume:

$$E_V = \frac{\sum_{t}^{T_{ev}} Q_{obs} - \sum_{t}^{T_{ev}} Q_{sim}}{\sum_{t}^{T_{ev}} Q_{obs}} \qquad (11)$$

were both evaluated for each single event and as the mean of all of the selected events for every catchment considered in this study.

RESULTS AND DISCUSSION

In the following, we show the performances obtained by SCRRM using SM-OBS-1 and SM-DAS-2 compared with the results obtained by MISDc. Both MISDc and SCRRM were calibrated using Equation (9) as the objective function and tuning the four parameters of SCRRM and the seven parameters of MISDc. The calibration was carried out separately for each catchment yielding $\overline{NS}$ equal to 0.61, 0.58 and 0.66 for SM-OBS-1, SM-DAS-2 and MISDc, respectively.

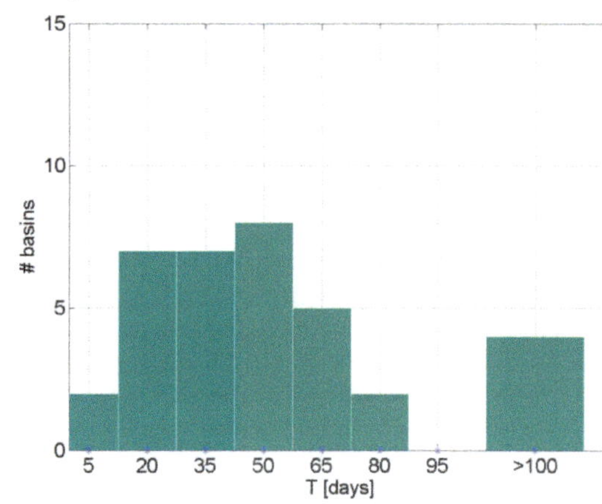

(a) Characteristic time length T

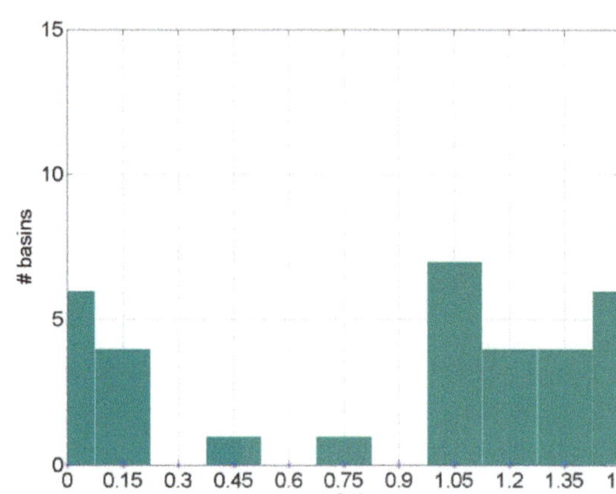

(b) Soil depth z

Figure 3: Histogram of the parameters of SCRRM representing the soil depth involved in the rainfall-runoff transformation. T = characteristic time length used in SCRRM with SM-OBS-1 (**a**); z = soil depth used in SCRRM with SM-DAS-2 (**b**).

In Figure 3(a) and Figure 3(b), the histograms of z and T (i.e., the parameters representing the influence of the soil depth in the RR transformation) show higher frequencies around 1–90 days for T and two main clusters for z: one below 20 cm and the other above 105 cm. The median T and z are about 48 days and 1 m, respectively, which are quite acceptable values in Italy (e.g., [31]). Values of T above 100 days are obtained for catchments PI-PP (Piave at Pontedipiave), AN-LN (Aniene at Lunghezza), BR-BZ (Brenta at Berzizza) and DO-AV (Dorabaltea at Verolengo). These values could be due to several reasons: (i) those related to the influence of a deeper soil layer in the RR transformation (catchments PI-PP and DO-AV have both T = 300 days and z = 80 cm and 130 cm, which are quite consistent values); (ii) and those related to the accuracy of the observations (e.g., observed discharge datasets), which may affect the consistency of the calibration procedure. Finally, another important reason could be the accuracy of the satellite observations, which results in contrasting T values if compared with z (z = 6 cm and T = 256 days for catchment BR-BZ).

Figure 4(a) shows the median NS_{50} obtained by SM-OBS-1, SM-DAS-2 and MISDc for the selected catchments. As can be seen, there is a general agreement between the three configurations, with MISDc being generally better than SM-OBS-1 and SM-DAS-2. The results for E_{Qp50} and E_{V50} in Figure 4(b) and Figure 4(c) reflect those for NS, with the line representing MISDc generally lower than the lines of SM-OBS-1 and SM-DAS-2. The reasons for this could be attributed to: (i) the better description of the water balance that MISDc can provide with respect to the SCRRM (indeed, MISDc uses the water balance model to "adjust" the initial condition prior to the RR event); (ii) the underlying errors present in the SM estimates and in the coarser resolution of the H-SAF SM products compared to the processes happening at the catchment scale; and (iii) the larger number of parameters of MISDc with respect to SCRRM. It is worth noting that the sensitivity of the selected catchments to the initial conditions was tested by running the SCRRM model with a constant value of soil moisture (i.e., equal to the mean of the soil moisture values associated with the events selected for each catchment). For both products, a sensible reduction of 15% was found for mean NS and an increase of 6% for both E_{Qp} and E_V.

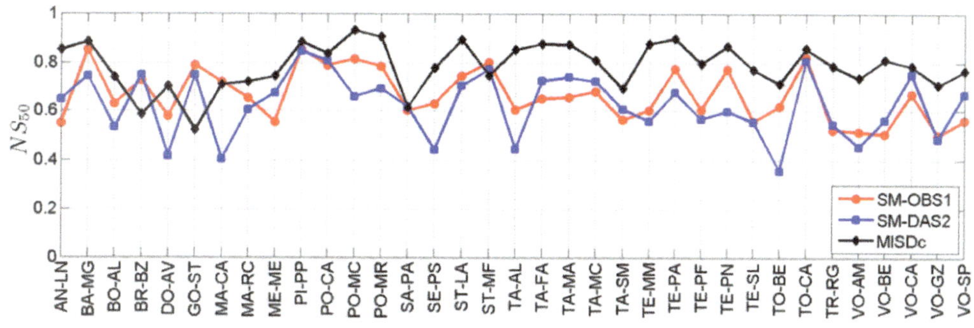

(a) NS_50

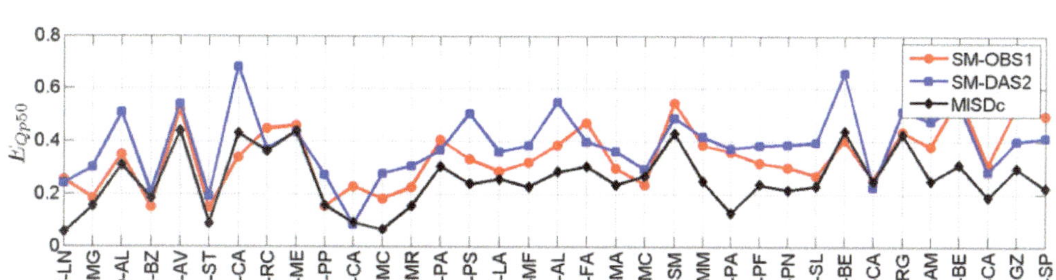

(b) E_{Qp50}

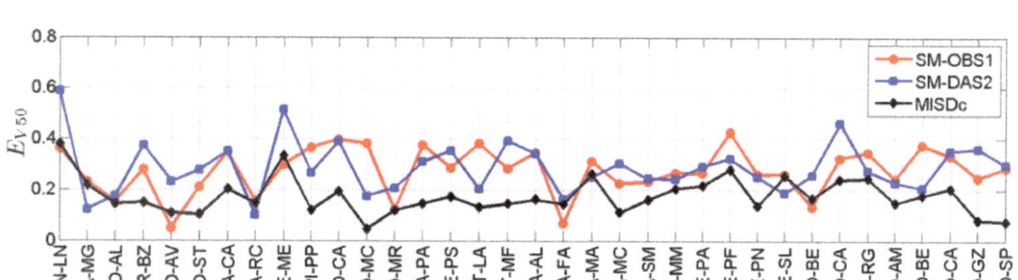

(c) E_{V50}

Figure 4: Median performance indexes for the selected catchments, NS_{50}=median Nash–Sutcliffe (**a**); E_{Qp50} = median relative error in peak discharge (**b**); E_{V50} = median relative error in volume (**c**).

The performance of SCRRM with respect to MISDc can be better visualized by the scatter plots shown in Figure 5(a) and 5(b). The figures plot the performance indexes NS_{50}, E_{Qp50} and E_{V50} of MISDc against those of SM-OBS-1 and SM-DAS-2, respectively. Overall, MISDc outperforms SCRRM, but it can be recognized that when MISDc provides a good performance, also SCRRM works well; thus, we may conclude that the results are somehow dependent on the quality of the input data and on the model structure. Overall, the median of NS, calculated on all selected events, yields 0.65, 0.60 and 0.74, for SM-OBS-1, SM-DAS-2 and MISDc, respectively, while 0.35, 0.40 and 0.25 are obtained for E_{Qp} Similarly, EV provides 0.29, 0.30 and 0.16. Note that MISDc generally outperforms SCRRM in both cases of SM-OBS-1 and SM-DAS-2, except for some catchments (e.g., for NS catchments MA-CA (Magra at Calamazza), GO-ST (Gorzone at Stanghella), ST-MF (Stura di Demonte at Fossano) and BR-BZ; see 4(a); for E_V , also MA-RC (Maira at Raconigi), TA-FA (Tanaro at Farigliano) and TO-BE (Topino at Bettona); see Figure 4(c)). Since the structures of MISDc and SCRRM are very similar (i.e., they rely on the same runoff production and routing models), the only reason for explaining the higher performance is the better estimation of SM at the beginning of the flood event. Indeed, in this study, MISDc is run in lumped mode, providing an estimate of SM for the entire catchment. On the contrary, the SM time series used in SCRRM rely on SM observations from different parts of the basin area; thus, it is likely that the spatial variability of SM is better taken into account by SCRRM. Moreover, SM estimates by MISDc rely on rainfall and temperature measurements, which, other than possibly being affected by many error sources [41], are averaged throughout the basin area, producing possible additional uncertainties [13].

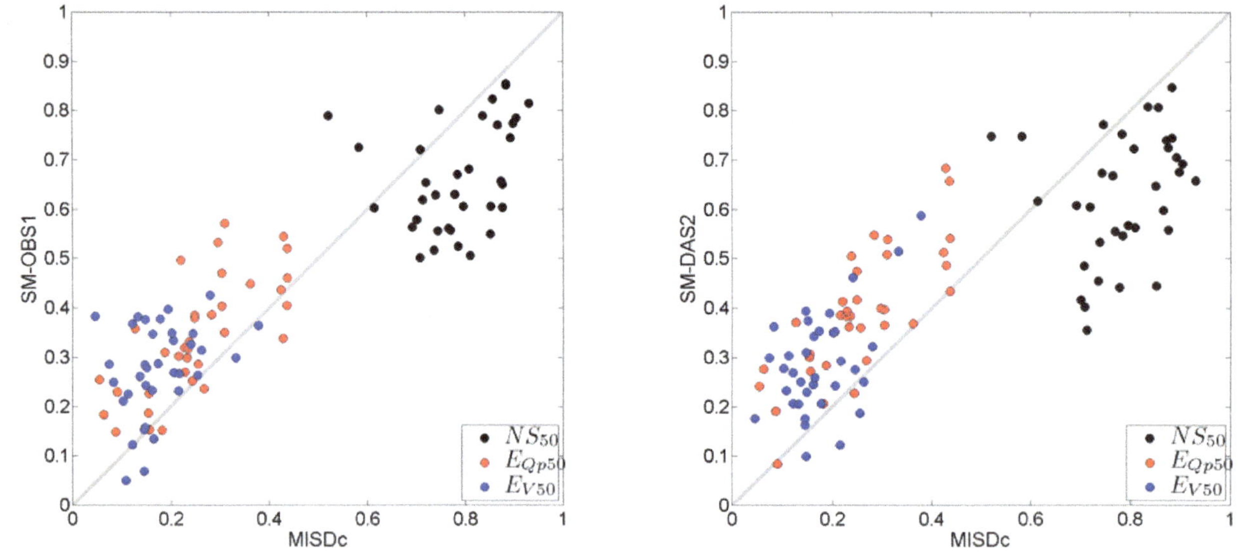

Figure 5: Comparison between MISDc (Modello Idrologico Semi-Distribuito in continuo) and SCRRM run with SM-OBS-1 (**a**) and SM-DAS-2 (**b**) in terms of median Nash–Sutcliffe, NS_{50}, median relative error on peak discharge, E_{Qp50}, and median relative error in total runoff volume, E_{V50}, for the selected catchments.

To assess if SCRRM results could be affected by catchment area extension, climatic conditions, average temperature and latitude, we calculated $\overline{NS}$ against these variables (not shown for the sake of brevity), but we did not find any strong evidence of any existing trend, except a slight increment of the performances as the area of the catchment is increased. Although we cannot draw any conclusions, because of the relatively small number of catchments analyzed (and the possible errors contained in the observed discharge), this increment is highly expected due to the coarse resolution of SM-OBS-1 and SM-DAS-2. Future investigations will consider a longer dataset and stronger climatic differences for the catchments in order to gain additional information on the SCRRM performances.

Finally, in Figures 6(a) and 7(a), we plotted the simulated and the observed discharge for two representative catchments (TE-SL (Tevere at Santa Lucia) and PO-CA (Po at Carignano)). The respective SM time series in terms of SW I, $\theta_{SM-DAS-2}$ and SM simulated by MISDc are plotted in Figures 6(b) and 7(b). Table 3 summarizes the results obtained for the two catchments.

For TE-SL (with an area of about 840 km^2), the models behave very similar, with MISDc slightly superior to SCRRM for all of the considered performance scores. In particular, SM-OBS-1 tends to overestimate the direct runoff volume and the peak discharges, both in the mean and in the median, especially for the largest events (see Figure 6(a)), providing the largest values of both E_{Qp} and E_V (see 3). The SM time series show very similar patterns for MISDc and $\theta_{SM-DAS-2}$ (z = 166 cm), while SW I seems characterized by a lower variability (T = 48 days).

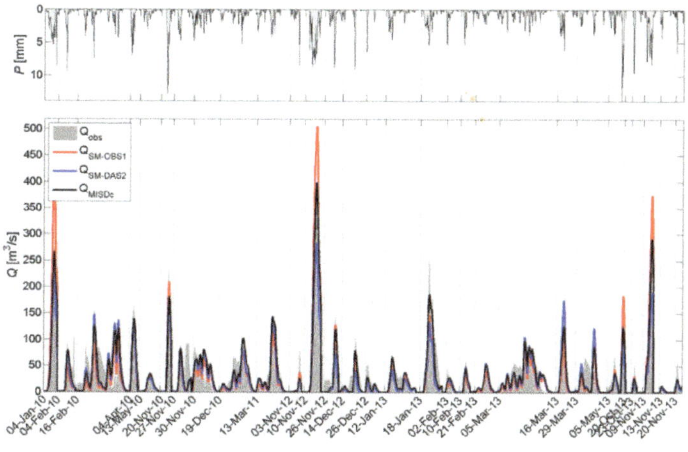

(a) Simulated *vs.* observed discharge for the selected events.

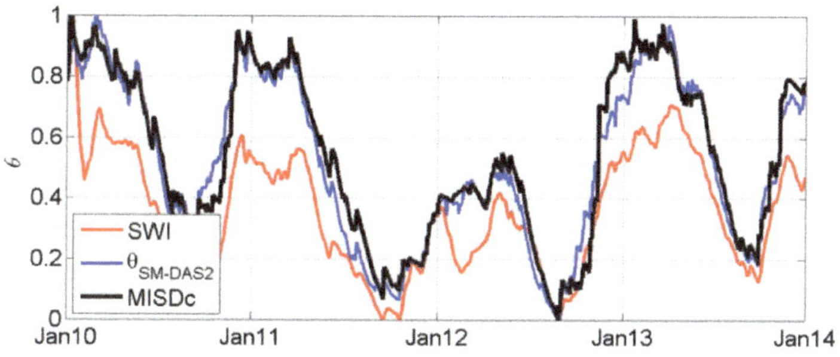

(b) Averaged SWI, $\theta_{SM-DAS-2}$ and MISDc soil moisture.

Figure 6: Observed rainfall (P) and simulated versus observed discharge (Q) for the selected events (**a**). Averaged SWISWI (Soil Water Index)(Equation (1)), θSM−DAS−2θSM-DAS-2 (Equation (3)) and MISDc soil moisture, θ, from 2010 to 2013 (**b**). Catchment: Tevere River at Santa Lucia (TE-SL).

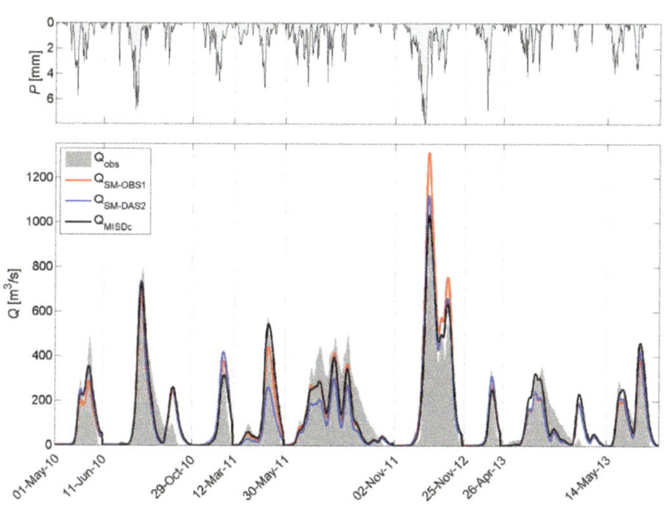

(a) Simulated *vs.* observed discharge for the selected events.

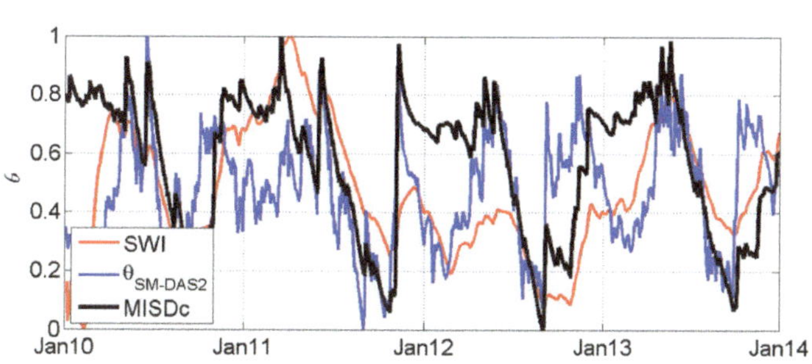

(b) Averaged SWI, $\theta_{SM-DAS-2}$ and MISDc soil moisture.

Figure 7: Observed rainfall (P) and simulated versus observed discharge (Q) (a). Averaged SW I (Equation (1)), $\theta_{\text{SM-DAS-2}}$

(Equation (3)) and MISDc soil moisture, θ, from 2010 to 2013 (b). Catchment: Po River at Carignano (PO-CA).

For PO-CA (with an area of about 3570 km^2, four-times the size of TE-SL), all of the models provide good results, both in the mean and in the median scores. A general agreement between the models can be also seen in Figure 6(a). It is interesting to note the good performance obtained by SM-DAS-2, which provides the lowest value of the median E_V. In contrast with TE-SL, the SM time series for PO-CA in 7(b) appear very different with respect to MISDc. Indeed, SW I (T = 82 days) and $\theta_{SM-DAS-2}$ (z = 168 cm) are the results of the average of multiple pixels falling inside the catchment boundaries of PO-CA; thus, the spatial variability and the inherent errors present in the observed time series may significantly affect the value of the calibrated parameters, as well as the averaged SM temporal pattern throughout the basin.

Table 3: Summary of the results obtained for the catchments Tevere River at Santa Lucia (TE-SL) and Po River at Carignano (PO-CA).

Catchment	Model	Mean			Median		
		NS	E_{Qp}	E_V	NS	E_{Qp}	E_V
	SM-OBS-1	0.452	0.298	0.401	0.558	0.269	0.367
TE-SL	SM-DAS-2	0.427	0.368	0.354	0.555	0.394	0.268
	MISDc	0.556	0.259	0.180	0.770	0.228	0.121
	SM-OBS-1	0.744	0.223	0.253	0.790	0.229	0.210
PO-CA	SM-DAS-2	0.692	0.213	0.305	0.808	0.084	0.277
	MISDc	0.805	0.135	0.146	0.836	0.090	0.102

CONCLUSIONS

In this paper, a simplified continuous rainfall-runoff model using SM from external sources for initialization has been used for testing the performance of two operational products (SM-OBS-1 and SM-DAS-2) of the "EUMETSAT Satellite Application Facility in Support of Operational Hydrology and Water Management (H-SAF)" and to gain some knowledge regarding under what conditions SCRRM can be used for operational purposes. The model was applied in 35 catchments of the Italian territory covering different sizes and climate conditions. The performances obtained by SCRRM in flood modelling have been compared against those of a classical continuous model (MISDc, [12]) showing satisfactory results (mean NS equal to 0.61, 0.58 for SM-OBS-1 and SM-DAS-2, respectively), but generally lower than MISDc (mean NS = 0.66).

In particular, it was found that:

- In 35 Italian catchments (800 to 7400 km22), satisfactory results can be obtained by the use SM-OBS-1 and SM-DAS-2 within SCRRM providing a median NSNS calculated on all of the selected events equal to 0.65 and 0.60, respectively. Similarly, the relative errors in median peak discharge and in runoff volume are 0.35 and 0.29 (SM-OBS-1) and 0.40 and 0.30 (SM-DAS-2). This means that the two products: (i) provide very similar performance and may both be satisfactorily used within SCRRM; and (ii) offer similar information content in flood modelling, which can be efficiently exploited in the context of soil moisture data assimilation in continuous models.
- MISDc generally outperforms SCRRM, except in a few cases. Although this aspect needs further investigation, the reason could be due to the fact that MISDc was run in lumped mode, while SCRRM SM was obtained by averaging the value of SM of different pixels falling inside the catchment boundaries, thus taking more into account for the SM spatial variability. In any case, although the performances of SCRRM are generally lower than those of MISDc (but not by far), they highlight two main interesting issues. First, for operational purposes, SCRRM is expected to be a valuable alternative to a continuous model in poorly gauged areas, since its structure is less sensitive to problems (not rare) of rain-gauge malfunctions and breakage. Second, the satisfactory results obtained indicate that H-SAF soil moisture products have the potential to improve flood modelling if used with more complex data assimilation schemes with continuous models (i.e., by assimilating SM-OBS-1 and SM-DAS-2 products into the MISDc model).
- Median T and z values, (i.e., the parameters representing the influence of the soil depth in the RR transformation) are 48 days and 100 cm, respectively, which are quite reasonable values in Italy.
- SCRRM can be used as a "hydro-validation tool" to assess the performance of different soil moisture products in terms of the ability to reproduce flood hydrographs. This is a new method for validating soil moisture data that has not been used before.

Overall, we can conclude that both products, SM-OBS-1 and SM-DAS-2, contain sufficient information for satisfactorily reproducing floods. The SCRRM benefits are the possibility to be used in poorly gauged areas (e.g., in areas characterized by discontinuous measures of rainfall and temperature), its simplicity and parsimony, which facilitate setup and operational use. On the other hand, it must be said that in areas characterized by dense and robust hydro-meteorological networks, like the ones used in this paper, a continuous modelling approach is preferable, since it allows one to obtain better performance. In this context, the satellite soil moisture products could be optimally integrated into the continuous model for obtaining a better modelling chain for flood forecasting. Future applications of the model will consider the use of satellite rainfall observations in order to rely completely on satellite data.

ACKNOWLEDGMENTS

The authors wish to thank EUMETESAT in the framework of the EUMETSAT Satellite Application Facility in Support of Operational Hydrology and Water Management (H-SAF) project and the Italian Civil Protection Department for providing the relevant datasets used in the paper.

REFERENCES

1. Jongman, B.; Hochrainer-Stigler, S.; Feyen, L.; Aerts, J.C.J.H.; Mechler, R.; Botzen, W.J.W.; Bouwer, L.M.; Pflug, G.; Rojas, R.; Ward, P.J. Increasing stress on disaster-risk finance due to large floods. Nat. Clim. Chang. 2014, 4, 264–268.
2. Brocca, L.; Melone, F.; Moramarco, T. On the estimation of antecedent wetness conditions in rainfall-runoff modelling. Hydrol. Process. 2008, 642, 629–642.
3. Coustau, M.; Bouvier, C.; Borrell-Estupina, V.; Jourde, H. Flood modelling with a distributed event-based parsimonious rainfall-runoff model: Case of the karstic Lez river catchment. Nat. Hazard. Earth Syst. Sci. 2012,12, 1119–1133.
4. Beck, H.E.; Jeu, R.A.M.D.; Schellekens, J.; Dijk, A.I.J.M.V.; Bruijnzeel, L.A. Improving curve number based storm runoff estimates using soil moisture proxies. IEEE J. Sel. Top. Appl. Earth Obs. Remote Sens. 2010, 2, 250–259.
5. Tramblay, Y.; Bouaicha, R.; Brocca, L.; Dorigo, W.; Bouvier, C.; Camici, S.; Servat, E. Estimation of antecedent wetness conditions for flood modelling in northern Morocco. Hydrol. Earth Syst. Sci. 2012, 16, 4375–4386.
6. Viviroli, D.; Mittelbach, H.; Gurtz, J.; Weingartner, R. Continuous simulation for flood estimation in ungauged mesoscale catchments of Switzerland–Part II: Parameter regionalisation and flood estimation results. J. Hydrol.2009, 377, 208–225.
7. Van Steenbergen, N.; Willems, P. Increasing river flood preparedness by real-time warning based on wetness state conditions. J. Hydrol. 2013, 489, 227–237.
8. Aronica, G.; Candela, A. A regional methodology for deriving flood frequency curves (FFC) in partially gauged catchments with uncertain knowledge of soil moisture conditions. In Proceedings of iEMSs, Osnabru, Germany, 14–17 June 2004; pp. 1147–1183.
9. Brocca, L.; Melone, F.; Moramarco, T.; Morbidelli, R. Antecedent wetness conditions based on ERS scatterometer data. J. Hydrol. 2009, 364, 73–87.
10. Brocca, L.; Melone, F.; Moramarco, T.; Singh, V.P. Assimilation of observed soil moisture data in storm rainfall-runoff modeling. J. Hydrol. Eng. 2009, 14, 153–165.
11. Tramblay, Y.; Bouvier, C.; Martin, C.; Didon-Lescot, J.F.; Todorovik, D.; Domergue, J.M. Assessment of initial soil moisture conditions for event-based rainfall-runoff modelling. J. Hydrol. 2010, 387, 176–187.
12. Brocca, L.; Melone, F.; Moramarco, T. Distributed rainfall-runoff modelling for flood frequency estimation and flood forecasting. Hydrol. Process. 2011, 25, 2801–2813.
13. Tramblay, Y.; Bouvier, C.; Ayral, P.A.; Marchandise, A. Impact of rainfall spatial distribution on rainfall-runoff modelling efficiency and initial soil moisture conditions estimation. Nat. Hazard. Earth Syst. Sci. 2011, 11, 157–170.
14. Aubert, D.; Loumagne, C.; Oudin, L. Sequential assimilation of soil moisture and streamflow data in a conceptual rainfall-runoff model. J. Hydrol. 2003, 280, 145–161.
15. Dorigo, W.A.; Wagner, W.; Hohensinn, R.; Hahn, S.; Paulik, C.; Xaver, A.; Gruber, A.; Drusch, M.; Mecklenburg, S.; van Oevelen, P.; et al. The International Soil Moisture Network: A data hosting facility for global in situ soil moisture measurements. Hydrol. Earth Syst. Sci. 2011, 15, 1675–1698.
16. Bartalis, Z.; Wagner, W.; Naeimi, V.; Hasenauer, S.; Scipal, K.; Bonekamp, H.; Figa, J.; Anderson, C. Initial soil moisture retrievals from the METOP-A Advanced Scatterometer (ASCAT). Geophys. Res. Lett. 2007, 34, L20401.
17. Owe, M.; de Jeu, R.; Holmes, T. Multisensor historical climatology of satellite-derived global land surface moisture. J. Geophys. Res. 2008, 113, F01002.
18. Kerr, Y.H.; Waldteufel, P.; Wigneron, J.P.; Delwart, S.; Cabot, F.; Boutin, J.; Escorihuela, M.J.; Font, J.; Reul, N.; Gruhier, C.; et al. The SMOS mission: New tool for monitoring key elements of the global water cycle. Proc. IEEE 2010, 98, 666–687.
19. Balsamo, G.; Albergel, C.; Beljaars, A.; Boussetta, S.; Brun, E.; Cloke, H.; Dee, D.; Dutra, E.; Pappenberger, F.; de Rosnay, P.; et al. ERA-Interim/Land: A global land-surface reanalysis based on ERA-Interim meteorological forcing. Era Rep. Series 2012, 13, 1–25.
20. Wagner, W.; Bloschl, G.; Pampaloni, P.; Calvet, J.; Bizzarri, B.; Wigneron, J.; Kerr, Y. Operational readiness of microwave remote sensing of soil moisture for hydrologic applications. Nordic Hydrol. 2007, 38, 1–20.
21. Liu, Y.; Weerts, A.H.; Clark, M.; Hendricks Franssen, H.J.; Kumar, S.; Moradkhani, H.; Seo, D.J.; Schwanenberg, D.; Smith, P.; van Dijk, A.I.J.M.; et al. Advancing data assimilation in operational hydrologic forecasting: Progresses, challenges, and emerging opportunities. Hydrol. Earth Syst. Sci. 2012, 16, 3863–3887.

22. Reichle, R.H. Bias reduction in short records of satellite soil moisture. Geophys. Res. Lett. 2004, 31, L19501.

23. Crow, W.T.; Van Loon, E. Impact of incorrect model error assumptions on the sequential assimilation of remotely sensed surface soil moisture. J. Hydrometeorol. 2006, 7, 421–432.

24. Massari, C.; Brocca, L.; Barbetta, S.; Papathanasiou, C.; Mimikou, M.; Moramarco, T. Using globally available soil moisture indicators for flood modelling in Mediterranean catchments. Hydrol. Earth Syst. Sci. 2014, 18, 839–853.

25. Todini, E. Rainfall-runoff modeling—Past, present and future. J. Hydrol. 1988, 100, 341–352.

26. Beven, K.; Freer, J. A dynamic TOPMODEL. Hydrol. Process. 2001, 15, 1993–2011.

27. Famiglietti, J.S.; Wood, E.F. Multiscale modeling of spatially variable water and energy balance processes. Water Resour. Res. 1994, 30, 3061–3078.

28. Pignone, F.; Rebora, N.; Silvestro, F.; Castelli, F. GRISO (Generatore Random di Interpolazioni Spaziali da Osservazioni incerte). Relazione delle attività del I anno inerente la Convenzione 778/2009 tra Dipartimento di Protezione Civile e Fondazione CIMA (Centro Internazionale in Monitoraggio Ambientale); Technical Report No. 272; Savona, Italy; p. 353. 2010.

29. Melone, F.; Corradini, C.; Singh, V.P. Lag prediction in ungauged basins: An investigation through actual data of the upper Tiber River valley. Hydrol. Process. 2002, 16, 1085–1094.

30. Wagner, W.; Lemoine, G.; Rott, H. A method for estimating soil moisture from ERS scatterometer and soil data.Remote Sens. Environ. 1999, 70, 191–207.

31. Brocca, L.; Melone, F.; Moramarco, T.; Wagner, W.; Naeimi, V.; Bartalis, Z.; Hasenauer, S. Improving runoff prediction through the assimilation of the ASCAT soil moisture product. Hydrol. Earth Syst. Sci. 2010, 14, 1881–1893.

32. Albergel, C.; Rüdiger, C.; Carrer, D.; Calvet, J.C.; Fritz, N.; Naeimi, V.; Bartalis, Z.; Hasenauer, S. An evaluation of ASCAT surface soil moisture products with in-situ observations in Southwestern France. Hydrol. Earth Syst. Sci. 2009, 13, 115–124.

33. De Rosnay, P.; Drusch, M.; Vasiljevic, D.; Balsamo, G.; Albergel, C.; Isaksen, L. A simplified Extended Kalman Filter for the global operational soil moisture analysis at ECMWF. Q. J. R. Meteorol. Soc. 2013, 139, 1199–1213.

34. Albergel, C.; de Rosnay, P.; Gruhier, C.; Muñoz Sabater, J.; Hasenauer, S.; Isaksen, L.; Kerr, Y.; Wagner, W. Evaluation of remotely sensed and modelled soil moisture products using global ground-based in situobservations. Remote Sens. Environ. 2012, 118, 215–226.

35. Melone, F.; Neri, N.; Morbidelli, R.; Saltalippi, C. A conceptual model for flood prediction in basins of moderate size. In Applied simulation and Modeling; Hamza, M.H., Ed.; IASTED Acta Press: Anaheim, CA, USA, 2001; pp. 461–466.

36. Kim, N.; Lee, J. Temporally weighted average curve number method for daily runoff simulation. Hydrol. Process. 2008, 4948, 4936–4948.

37. Gupta,V.;Waymire,C.Arepresentationofaninstantaneous unit hydrograph from geomorphology. Water Resour. Res. 1980, 16, 855–862.

38. Moramarco, T.; Melone, F.; Singh, V.P. Assessment of flooding in urbanized ungauged basins: A case study in the Upper Tiber area, Italy. Hydrol. Process. 2005, 19, 1909–1924.

39. Nash, J.; Sutcliffe, J. River flow forecasting through conceptual models part I—A discussion of principles. J. Hydrol. 1970, 10, 282–290.

40. Bober, W. Introduction to Numerical and Analytical Methods with MATLAB for Engineers and Scientists; CRC Press, Inc.: Boca Raton, FL, USA, 2013.

41. Crow, W.T.; van den Berg, M.J.; Huffman, G.J.; Pellarin, T. Correcting rainfall using satellite-based surface soil moisture retrievals: The Soil Moisture Analysis Rainfall Tool (SMART). Water Resour. Res. 2011, 47, W08521.

CHAPTER 13

FLOOD RISK MANAGEMENT IN REMOTE AND IMPOVERISHED AREAS

Valentin Heimhuber, Johann-Christian Hannemann and Wolfgang Rieger

Keywords: DEM, TanDEM-X, HEC-HMS, HEC-RAS, mapping drone, hydrologic and hydraulic modeling, flood risk analysis, flood risk management, flash-flood

INTRODUCTION

According to the World Disaster Report [1], seismic events killed the greatest number of people of all natural hazards in recent years, averaging 50,184 people per year from 2000 to 2008, whereas flood events affected the largest number of people, averaging 99 million people per year between 2000 and 2008. Haiti, currently rated as the poorest country in the Western hemisphere, is vulnerable to both of these natural hazards. In January 2010, a 7.0 magnitude earthquake struck Port-au-Prince, the capital of Haiti. It caused more than 300,000 deaths and left more than 1.5 million people homeless [2]. Only two years later, Hurricane Isaac crossed the southern peninsula of Haiti bringing flooding and storms to areas affected by the earthquake of 2010. As many as 5000 people were evacuated due to the flooding and 180 earthquake refugee camps were affected [3]. Due to the lack of disaster preparedness, the 2010 earthquake also led to an extensive resettlement process in which an estimated 500,000 people spontaneously left the capital and sought refuge in other provinces [4]. This process led to the formation of numerous informal settlements in previously uninhabited areas.

The area of Canaan north east of Port-au-Prince is a good example for this resettlement process. In order to address overcrowded refugee camps in acute risk of flooding in the capital, internally displaced people (IDP) camps were constructed in the previously uninhabited area of Canaan (Figure 1). Since then, Canaan has grown in an uncontrolled way and meanwhile, transformed into an increasingly urban settlement with likely more than 100,000 residents [5].

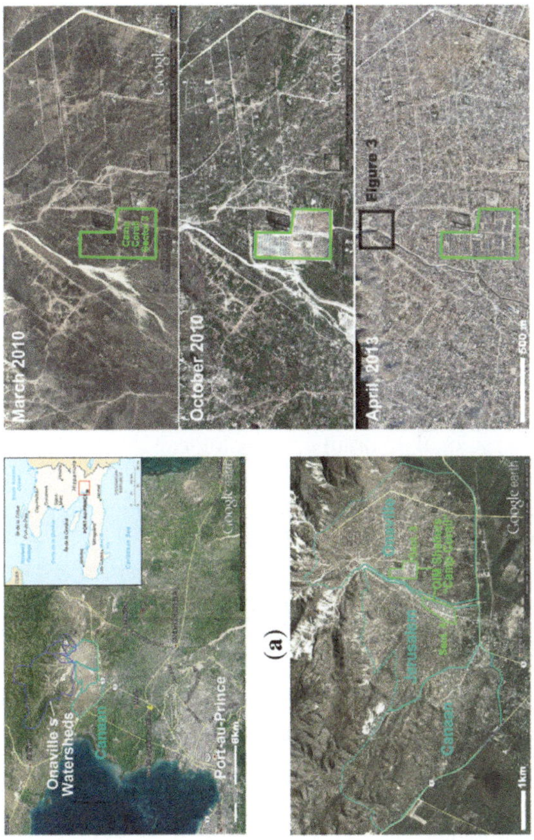

Figure 1: (**a**) Location of Canaan in Haiti; (**b**) Location of Camp Corail and Onaville in Canaan; (**c**) settlement process of Onaville between 2010 and 2013 [6].

The overall aim of this study was to perform a state-of-the-art flood risk analysis for Onaville by addressing the difficult socio-economic situation and the lack of readily available data through a variety of innovative approaches. The study was conducted in the frame of the Technical University of Munich (TUM) research project Urban Strategies for Onaville (TUM-USO) [7], in which researchers from different backgrounds as

Citation: Heimhuber, V.; Hannemann, J.-C.; Rieger, W. Flood Risk Management in Remote and Impoverished Areas. Water 2015, 7, 3832-3860. doi:10.3390/w7073832.

well as the non-governmental Organization (NGO) TECHO, currently active in Haiti, collaborate with local stakeholders in planning and implementing strategies for sustainable development of the settlement.

Previous studies indicate that parts of Onaville are currently at high risk of being affected by flooding [8,9]. The state-of-the-art to handle such natural hazards is to establish a defined procedure for handling risks like Flood Risk Management (Figure 2), which is the process of managing an existing flood risk situation which, beside the assessment and mitigation of flood risk, includes a continuous and holistic societal analysis [10]. The technical part for a functioning flood risk management and for preparing a decision basis is risk assessment, which comprises understanding, evaluating and interpreting the perceptions of risk and societal tolerances of risk [11]. Risk assessment for floods is based on hazard maps, which are part of the risk analysis process, a methodology to objectively determine risk by analyzing and combining probabilities and consequences [10].

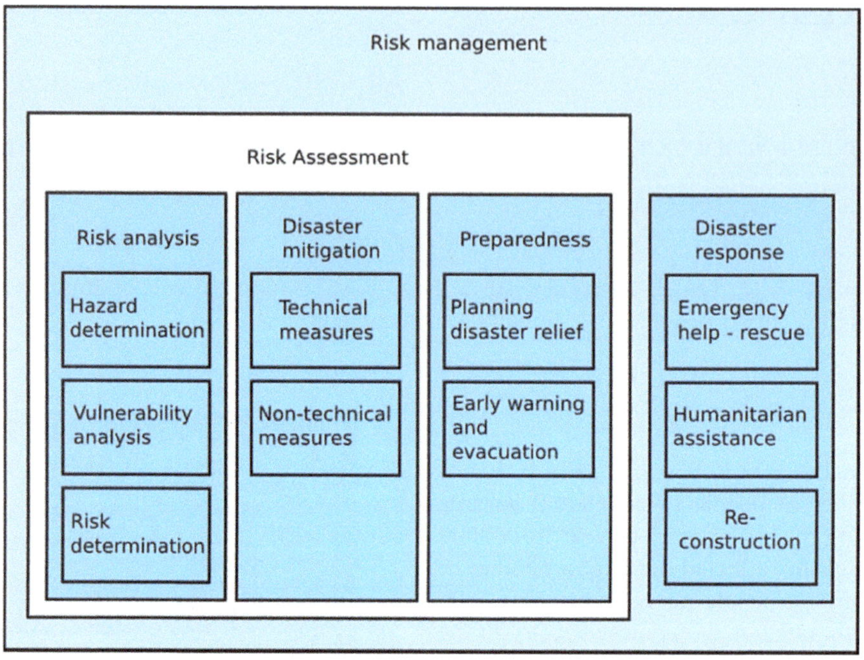

Figure 2: Elements of operational (flood) risk management, adapted from Plate [11].

In the case of remote and impoverished settlements, one of the main challenges is commonly the acquisition of sufficient regional, hydraulic, hydrological and meteorological data for the risk assessment. Additionally, it is crucial to assess the socio-economic situation of the study area in detail (Vulnerability analysis-risk determination) and to establish information channels to the residents of the settlement in order to communicate the risk situation and to improve disaster preparedness. Therefore, this study is focused on the materials and methods that are necessary to establish a functioning risk assessment for informal settlements located in remote, impoverished and data sparse areas like Onaville.

Based on the available data, a suitable modeling approach had to be found, which would allow performing the risk analysis while simultaneously serving as a flood risk management and planning tool. Since no records about the runoff characteristics of the rivers and watersheds that affect the Onaville settlement exist, rainfall-runoff modeling was used in combination with subsequent hydraulic modeling to estimate the flood hazard of Onaville and to create a flood hazard map as a basis for flood risk management. In such a coupled modeling approach, the annual exceedance probability (AEP) related to the magnitude of flood events is determined based on the AEP of the design storms used as the input to the rainfall-runoff model. The design storms were generated based on existing IDF curves, which are derived from several decades of rainfall measurements and thus represent the local long-term rainfall characteristics. Based on the encountered flood hazard situation, a grassroots approach for improving risk awareness and disaster preparedness in the complex socio-economic setting of Onaville is presented that is expected to be applicable to other informal settlements with similar risk situations. The main challenge for implementing a functioning risk management for the study area was the poor socio-economic state of the unplanned settlement, its political status and the uncertainties related to future urban development.

STUDY AREA AND DATA

Onaville and Canaan

Onaville is located about 15 km north-east of the Haitian capital and covers an area of approximately 7 km^2. Together with simultaneously developing informal settlements, it forms the rapidly urbanizing area commonly known as Canaan, situated in the transitional area between the Chaine-de-Matthieux mountain chain in the north and the Cul-de-Sac plain in the south (Figure 1).

Canaan is the result of a large resettlement process that was triggered by an expropriation decree that the Haitian government issued for this area in response to the 2010 earthquake. The decree was declared on 22 March 2010 along with the opening of the IDP camps Corail-Cesseless, which the international community had set up in the area. As interviews with local plot holders by the TUM research group revealed, these actions were widely understood as a land offer to the people. The location of Sector 3 and 4 of these IDP camps is outlined in Figure 1. Since the government and the so-called international community did not provide any further assistance to the area, Canaan has grown in a rapid and uncontrolled way ever since (see sequence of images in Figure 1). Public services such as water supply, sanitation and electricity are absent in most areas.

The area is currently expected to host 64,378 people (11,477 in Onaville; 31,156 in Canaan; 21,745 in Jerusalem) [5]. In comparison to that, Haitian government officials estimate about 300,000 people in the area [12]. The relative closeness of the area to the capital as well as the possibility of obtaining a piece of land informally can be seen as main drivers of this exceptional rise of a new city. A comprehensive overview of the geographic and socio-economic characteristics of Onaville and Canaan is given by Hannemann *et al.* [13].

All drainage courses (in Haiti commonly called "ravines") in the area are ephemeral, meaning that they are usually dry all year, except after strong rain events. The alluvial cone on which Onaville is situated is the result of accumulated sediment that the ravines deposit in this transitional zone. Due to the steep hills and the availability of debris as a result of erosive processes in the watersheds, mountain torrents often collect and carry significant amounts of sediment during intensive rainfalls. After passing a transit section, the streams enter the deposit region where sedimentation increases drastically due to the reduction in slope and the resulting decrease in flow velocity. The watersheds of Ravine Lan Couline in Onaville (also known as Ravine Madan-El) show many of these characteristics with high levels of erosion as well as numerous debris bodies next to the river channels resulting from landslides. Due to the abundance of debris in the watersheds, flood runoffs in Onaville can thus be expected to have high solid concentrations and bed load transport.

The Ravine Lan Couline marks the border between Onaville and Jerusalem (see Figure 1) and poses the largest flood related risk to the surrounding settlements. Figure 3 gives an impression of the characteristics of the study area, the size of the watercourse (2) and the dense settlement on its overbank areas (1). The location of the section of the river channel that is shown on the second image is referenced in Figure 1.

Shortly after the construction of Camp Corail, an artificial flow channel was constructed to guide flood runoffs in the Ravine Lan Couline safely through the lower part of the settlement, where flows originally used to spread out over the alluvial cone. Figure 4 shows a comparison of the area's topography before and after the construction of the channel based on hillshades derived from a light detection and ranging (LIDAR) DEM (2010) and a Drone DEM (2013). The severe changes in the flow path and channel geometry of the ravine that came along with the construction of the artificial channel can be seen clearly.

(a) (b)

Figure 3: Ravine Lan Couline flow channel from far (**a**) and close (**b**).

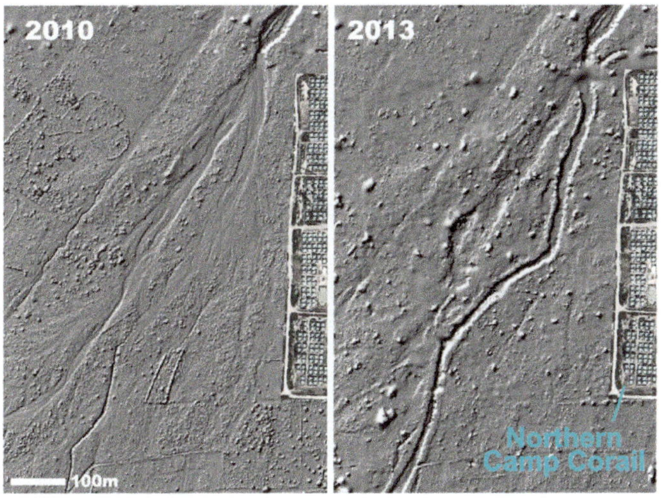

(a) (b)

Figure 4: Comparison of the Ravine Lan Couline in the area of northern Camp Corail (Sector 3) before and after channel constructions based on the LIDAR DEM (**a**) and the Drone DEM (**b**).

Digital Elevation Models

Accurate terrain data in form of Digital Elevation Models is crucial input for both hydrological and hydraulic modeling. The topographic datasets that are available for the study area are a 12 m spatial resolution DEM provided by the TanDEM-X mission from the German Aerospace Center (DLR) (TanDEM-X DEM), a LIDAR DEM with 1 m spatial resolution initiated by the World Bank (LIDAR DEM) [14] and a mapping drone DEM with 30 cm resolution provided by Drone Adventures (Drone DEM). A comparison of the three available DEMs based on hillshades is given in Figure 5.

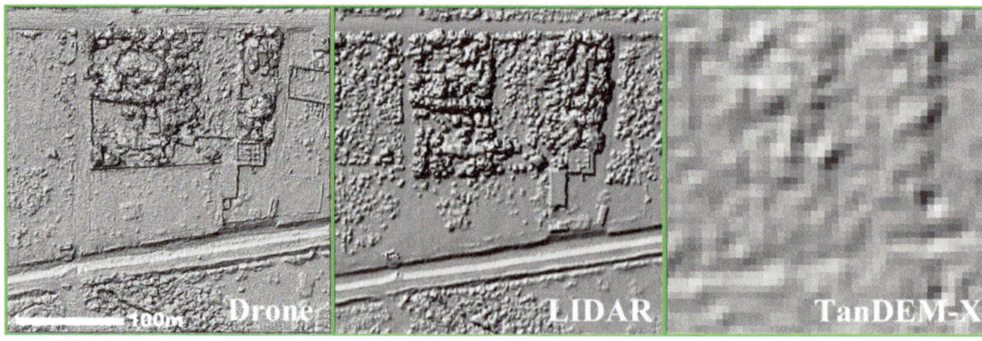

Figure 5: Comparison of available DEMs based on hillshades from an area south of Onaville, including Route Corail in the upper part and Boukanbrou Canal in the lower part.

The TanDEM-X DEM is an intermediate DEM provided by the DLR's TanDEM-X satellite mission. It is derived from the first coverage of Haiti taken on 26 June 2011. Compared to the final TanDEM-X DEM product, the intermediate DEM can be expected to have a lower vertical accuracy as well as phase unwrapping errors in mountainous regions [15]. Nevertheless, a rough analysis of the DEM together with the DLR's TanDEM-X ground segment showed that the DEM was suitable for the intended application [16].

The LIDAR DEM was initiated by the World Bank in response to the earthquake of 2010 in order to provide high resolution imagery and elevation data for the coordination of the disaster management. The Drone DEM provided for this study was created in 2013 by Drone Adventures, the Humanitarian OpenStreetMap Team (HOT) and the International Organization for Migration (IOM), using state-of-the-art fixed-wing mapping drones from the Swiss company senseFly. The DEM used in this study was provided by Drone Adventures and has a spatial resolution of 30 cm.

The steep and mountainous watersheds, which are imposing flood hazards for Onaville, are covered only by the TanDEM-X DEM, which was thus used to derive the hydrological structure of the catchments and to parameterize the rainfall-runoff model (Figure 6). The accuracy and resolution of this DEM was sufficient to derive the location of the flow channels and the boundaries of the sub-catchments in detail. For the hydraulic modeling of floods in the largest and most dangerous watercourse affecting Onaville, the Ravine Lan Couline (also known as Ravine Madan-Èl), the high-resolution Drone DEM was used in partial addition with the TanDEM-X DEM. This was done because of the limited extent of the Drone DEM in the upper part of the watercourse (see Figure 6). Nevertheless, the data availability for hydraulic modeling can be seen as sufficient since most parts of the flow channel, where overbank flow is to be expected within the settlement are covered by the high resolution DEM, which allowed the extraction of highly accurate channel cross sections. The methodology for the development of the hydraulic model structure based on these two DEMs is explained in detail in the hydraulic model application section.

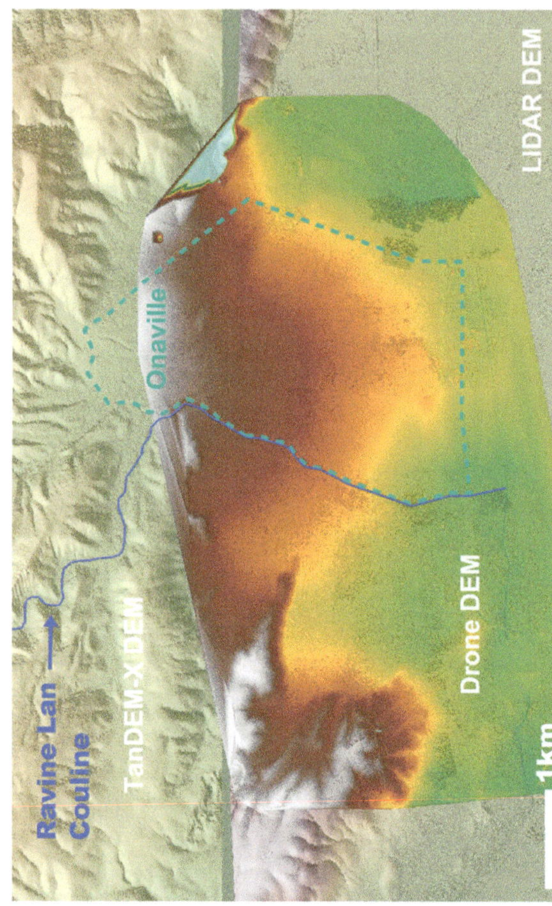

Figure 6: Spatial extent of the three DEMs used in this study.

Climate and Meteorological Data

Haiti has a tropical marine climate with two distinct rainy seasons occurring from March to May and from August to October. Average temperatures range from 25 to 28 °C. The closest weather station to the Canaan area is located at the Haitian Ministry of Agriculture, Natural Resources and Rural Development at Damien (18°35′44.49″ N; 72°17′21.07″ W), which is about 7 km south-east of Onaville. The average rainfall at Damien is based on 68 complete years of monthly rain data from 1927 to 2002 (Figure 7a). Due to its location in the center of the North Atlantic hurricane belt, Haiti is frequently exposed to strong tropical storms and hurricanes. The northern Atlantic hurricane season expands from 1 June to 30 November, September being the most active month [17].

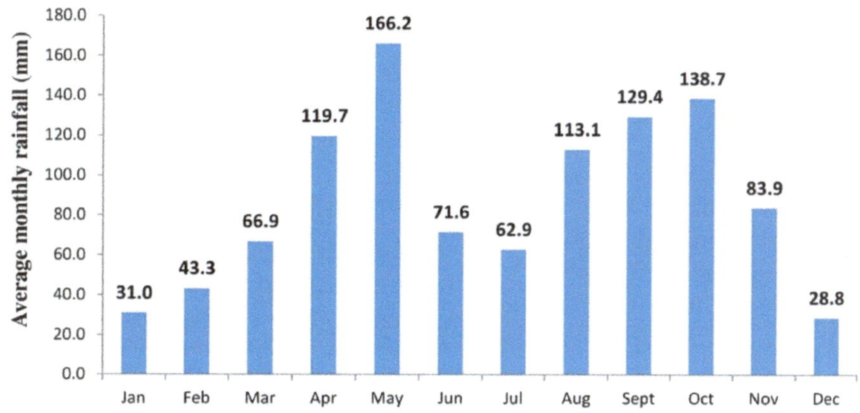

(a)

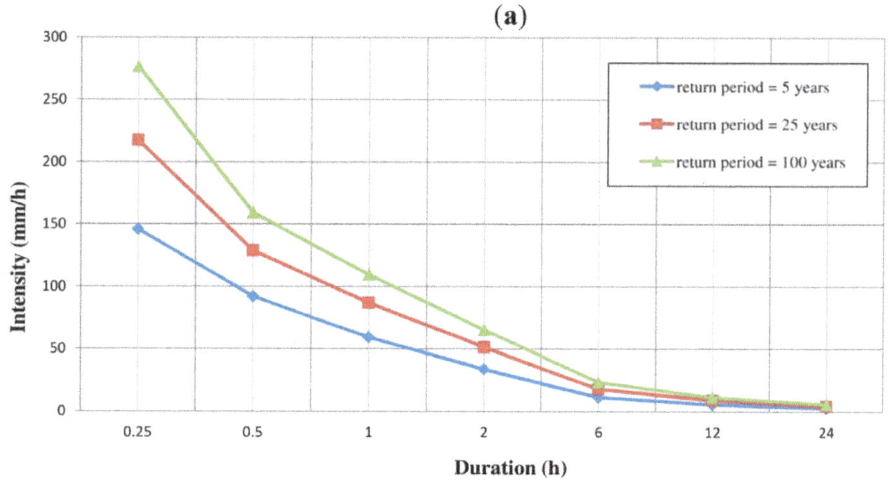

(b)

Figure 7: (**a**) Average monthly rainfall at Damien based on 68 years of record between 1927 and 2002 and (**b**) IDF-curves for Damien station [18].

Since the goal of this study was to assess the impact of flood runoffs with specified ARIs, statistical rain data in the form of IDF curves (Figure 7b) from the Damien weather station were used in the modeling process instead of specific historical storm events. These curves are provided by the Analysis of Multiple Natural hazards in Haiti (NATHAT) report [9] and were originally created in 1980 by SCET International for a Port-au-Prince drainage project [18]. The usage of IDF curves is common practice in flood risk assessment with the curves allowing the generation of design storms with a variety of relevant ARIs. Since the weather station is not located inside the study area, the generated design storms are not fully representative of the local rainfall characteristics so that the modeling results should be interpreted as estimations of the actual hazard situation.

Land Use and Soil Data

For the preparation of the rainfall-runoff model, the hydrologic and hydro-geological characteristics of the watersheds were analyzed in detail. A land use map [19] was provided by the Haitian National Center for Geospatial Information (CNIGS), in which the majority of the analyzed watershed is described as savanna. Savannas are defined as grass-covered areas of tropical or subtropical regions, which are nearly treeless in some places, but generally have a mix of widely spaced trees and bushes [20]. They also have distinct wet and dry seasons, which is in accordance with the local climate of the Onaville area. It is important to note that the currently existing vegetation is the result of a recovering process that started after the area had almost entirely been deforested in the past. According to the Food and Agriculture Organization of the United Nations (FAO)'s Global Forests Resource Assessment [21], the overall forest cover of Haiti is less than 4% compared to 30% in the Dominican Republic.

The geologic and hydrogeological maps of Haiti [19,22] gave rough information about the homogeneous fissured Limestone subsoil but were found to be insufficient for estimating the infiltration characteristics as a basis for the soil parameterization in the hydrological model. Thus, a field analysis of the top-soil, which is described in Section 3.2, was conducted at two representative locations in the watershed.

MATERIALS AND METHODS

Modeling Framework

Due to the limited data availability, hydrological (HEC-HMS) and hydraulic (HEC-RAS) modeling was applied in combination with a detailed analysis of the local hydro-geomorphology for the flood risk assessment (Figure 8). ArcMap was used for the management and manipulation of all spatial data. Together with the software extensions HEC-GeoHMS and HEC-GeoRAS, it was used for the preparation of the model input files and the visualization of the modeling results. These software extensions provide a powerful and interactive tool for preparing model input files and calculating the majority of model parameters from the available spatial datasets.

After this data preprocessing design, storms derived from the IDF-curves were used as model input for HEC-HMS which was set up for the watershed of the Ravine Lan Couline to generate flood hydrographs with different ARIs. These hydrographs were used as inflow boundaries for the Ravine Lan Couline HEC-RAS model at the northern edge of the settlements of Onaville to simulate the water levels with different ARIs along the river channel thereby forming the basis for a flood hazard zoning.

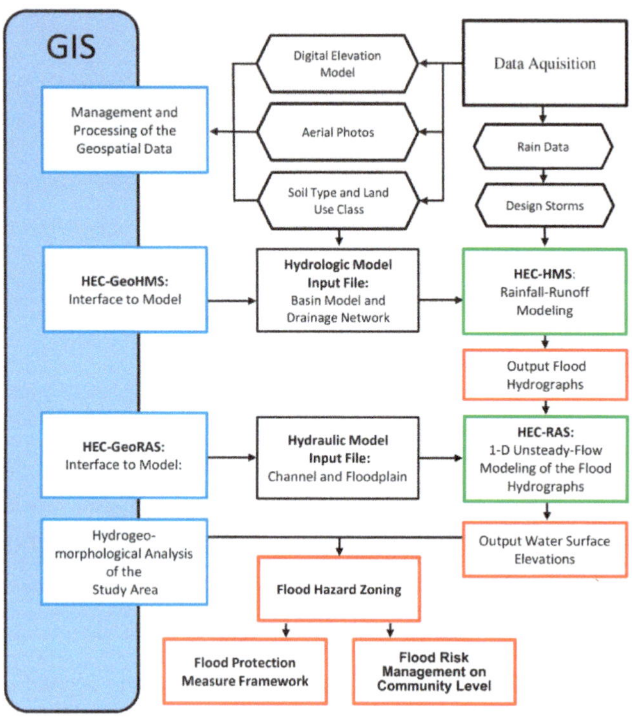

Figure 8: Flowchart of the GIS-based hydrologic and hydraulic modeling framework and flood risk assessment.

Hydrologic Modeling

The hydrologic modeling software HEC-HMS was used for modeling the rainfall-runoff process based on subsequent application of a suitable loss, transform and routing method. The analyzed rivers and watersheds are not gauged and due to their location in a remote and impoverished area, complex field surveys for a detailed estimation of hydrological parameters were not possible. Based on this background and the guidelines for model selection provided by the Hydrologic Engineering Center [23], the empirical Natural Resources Conservation Service (NRCS) Curve Number (CN) loss method, the empirical NRCS Unit Hydrograph transform method [24] and the Muskingum-Cunge routing method [25] were selected. The latter is quasi conceptual, since it is based on simplified equations of shallow water flow. Since baseflow does not occur in the analyzed watercourses, it was neglected in the modeling process. The chosen models represent a suitable way to simulate the rainfall-runoff process in the study area, since they are designed to model single storm events rather than continuous precipitation data [23], and can be calibrated under limited availability of detailed hydrological data. Using the HEC-HMS software environment, they were applied to transform single design storm events with different AEPs into flood hydrographs, for which the AEP is then assumed to be equal to the AEP of the design storms.

Generation of the Basin Model

The main input data for rainfall-runoff modeling were the TanDEM-X DEM, the IDF curves from the Damien rain gauging station, a land-use map and the results of the soil analyses. The watersheds were derived by applying the HEC-GeoHMS software functionality within the ArcMap GIS environment. The first major step in creating the basin model was to delineate the stream network and watershed boundaries of the area of interest by applying a number of subsequent processing steps to the input DEM commonly referred to as terrain preprocessing. In this process, the definition of streams is based on a threshold number of cells that drain into a given cell, which was set to 20,000 cells or 0.74 km^2, respectively. This threshold was chosen because then, the resulting streams begin where the beginning of clear flow channels is observable on aerial imagery. After the completion of the terrain preprocessing, project points were placed on the drainage lines to define the outlet of the watersheds to be modeled. For the Ravine Lan Couline, the resulting watershed had a total area of 25.8 km^2 (see watershed (WS) 1 in Figure 9) and originally consisted of 17 subbasins. In order to avoid very small subbasins and to equalize the size of the remaining subbasins, basins smaller than 0.3 km^2 were merged together with the up- or downstream basin. This resulted in a total of 13 subbasins with a minimum area of 0.3 km^2. Besides the watershed of Ravine Lan Couline, rainfall-runoff modelling was performed for four more relatively small watersheds (0.8 to 1.8 km^2) that are draining through the settlement (see WS 2, 3, 4 and 5 inFigure 9).

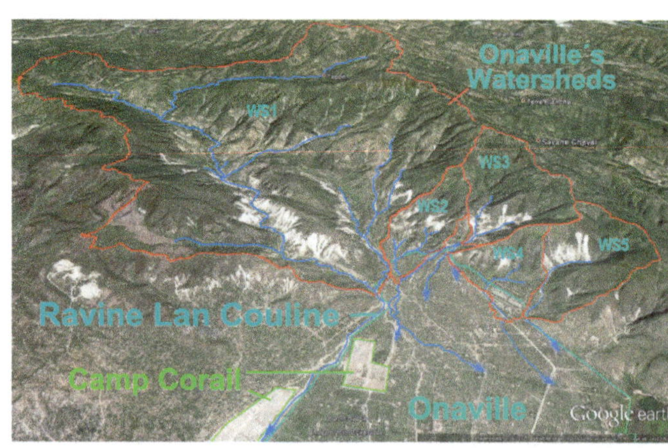

Figure 9: 3D view of the watersheds and drainage courses of Onaville including the watershed of Ravine Lan Couline [6].

Curve Number Loss Method

Based on the land-use classification and the semi-arid climatic characteristics of the area, the CN table for arid and semi-arid rangelands [26] was used for the determination of the CN. The cover type was defined on the basis of comparing major plants as listed in the CN table to observed vegetation cover on site. The definition of desert shrub in the CN table was found to most accurately describe the vegetation cover as observed on site (Figure 10a).

For the definition of the Hydrologic Soil Group (HSG), an on-site analysis of the top-soil was performed. Due to the uncertain security situation in the area of investigation and limited transportation possibilities, the field survey was limited to two sample points in the catchment area. These analyses were mainly based on the "estimation of the soil texture by feel" [27]. This method allows for a rough classification of the soil texture and type and is thus sufficient for the estimation of the HSG.

Excavations of the top soil layers revealed a large percentage of medium sized gravels surrounded by the actual soil at both sample points (Figure 10b). The sample included many small pebbles, but it was still plastic enough to form a ribbon with a length of almost 2 cm. Therefore, the texture of the soil was defined to be between sandy loam and loam. Since it shows a consistent soil matrix with loamy characteristics surrounding the gravel, the soil was assumed to have moderate infiltration rates. The available land use and hydro-geomorphology maps indicate that the area covered by the relevant watersheds has uniform soil and hydrologic characteristics. Because of that, the lack of more detailed soil data and the result of the soil analyses from both sample points, HSG B was assumed to be valid for all analyzed watersheds.

(a) (b)

Figure 10: Vegetation cover (**a**) and soil texture (**b**) at one sample point.

NRCS Unit Hydrograph Method

The NRCS Unit Hydrograph method was used to transform the excess precipitation into a flow hydrograph at the outlet of each subbasin. The main input parameter for this method is the basin lag time of each subbasin, which is defined as the time between the center of mass of excess precipitation and the peak of the hydrograph. In this method, the basin lag is approximated to be 0.6 times that of concentration. The time of concentration was determined based on the methodology described in the NRCS Technical Release 55 (TR-55) [26], which is integrated into the HEC-GeoHMS software environment. The TR-55 includes a table for the estimation of the surface roughness for sheet flow. The coefficient ranges from 0.011 for smooth surfaces, such as concrete, to 0.8 for forests with dense underbrush. Field observations revealed that the vegetation cover up to a height of 0.03 m, was best comparable to short grass prairie, as stated in the TR-55 table. Therefore, the coefficient was defined as 0.15.

Based on a table for the determination of roughness coefficients for mountain streams [25], the main flow channel was assigned a Manning coefficient of 0.045. Since channel observations further upstream into the watershed were not conducted, the value of 0.045 was assigned to all flow channels in the analyzed watershed. The resulting basin model shown in Figure 11 consists of 13 subbasins, eight junctions, seven reaches and the main outlet and includes all previously defined basin, reach and model parameters. Table 1 gives an overview of hydrological key parameters of the 13 subbasins. Subbasins W190, W250, W290 and W340 together contain the longest flow path of the watershed indicated by the grey marked columns in the table.

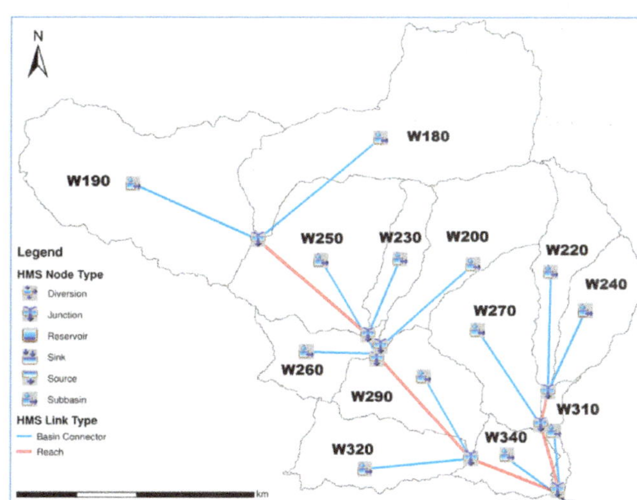

Figure 11: Model representation of the Ravine Lan Couline watershed in HEC-HMS.

The design storms for different statistical return periods were derived from the IDF curves using the frequency storm method as implemented in HEC-HMS. HEC-HMS uses the alternating block method [25] to compute design hyetographs with ARIs of five, 25 and 100 years that were used as input for rainfall-runoff modeling (Figure 12).

Table 1: Overview of subbasin features.

Subbasin	W180	W190	W200	W220	W230	W240	W250	W260	W270	W290	W310	W320	W340	Total
Area (km^2)	5.0	4.7	2.2	1.4	0.8	1.3	3.0	1.0	2.1	1.7	0.3	1.6	0.8	25.8
Average basin slope (%)	61	39	60	62	58	59	51	48	46	49	29	34	34	48
Longest flowpath (m)	6509	4975	4003	3855	2619	2746	3681	2284	3619	3365	2054	3543	2336	14357
Average slope of longest flowpath (%)	10.6	5.7	18.5	19.9	20.4	25.5	10.5	14.7	13.4	8.6	8.4	7.2	9.9	8.3
Time of concentration (min)	37	48	21	22	19	16	25	21	32	26	18	36	22	121

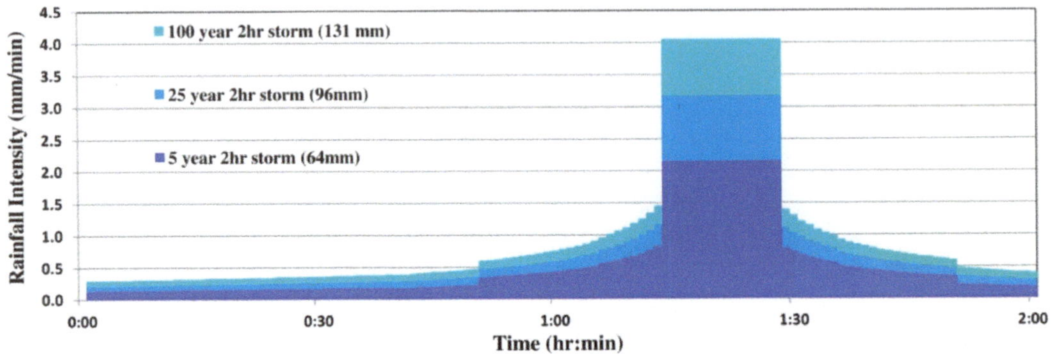

Figure 12: Design storms used as input for rainfall-runoff modeling.

Hydraulic Modeling

Hydraulic modelling was primarily used to define the locations of the Ravine Lan Couline channel in the settlement, where bank overtopping can be expected to occur first after a certain threshold flood discharge is exceeded. Due to the topography of the channel and the alluvial cone, it was assumed that the location of the cross sections where bank overtopping begins during the modeling of a flood wave, roughly defines the "bottleneck" fraction of the channel. From this point on, only the threshold discharge of the bottleneck is conveyed further downstream in the channel whereas the rest of the discharge is leaving the channel and spreads out laterally on the alluvial cone. For the analysis of the threshold flow carrying capacity, the definition of levees on both sides of the flow channel was found to be crucial for the flow simulations. As shown in Figure 13, levee points limit the flow to the main channel until the calculated water surface level in the channel exceeds the elevation of any of the two levee points on each side of the channel. Thus, the model can be expected to produce realistic flow simulations as long as the modeled discharge does not exceed the flow carrying capacity of the main channel. Nevertheless, the basic assumptions for 1d flow routing are violated as soon as the water level exceeds the levee elevation. Figure 13 shows an exemplary cross section from the Ravine Lan Couline channel along with the water surface and energy line just before and after levee overtopping occurs during an unsteady flow simulation.

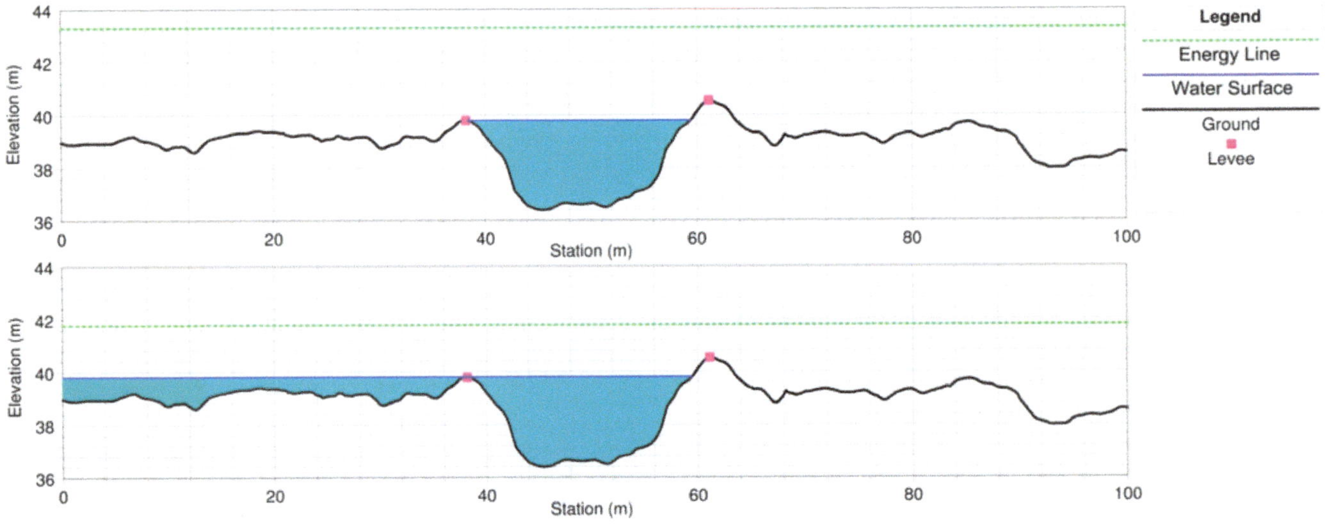

Figure 13: Water surface elevation and energy head before and after levee overtopping.

Once levee overtopping begins, the active flow area increases drastically due to the low elevation of the overbank areas behind the levees. Since the model only calculates a single horizontal water surface for each cross section, the water surface extends until the end of the cross section, where a vertical wall is assumed that defines the border of the model area. This sudden drastic increase in active flow area leads to an abrupt decrease in flow velocity observable as a drop in the elevation of the energy line (Figure 13). Furthermore, the entire flow is conveyed to the next downstream cross section which potentially could have enough flow carrying capacity to convey the entire flow in the main channel resulting in another jump in the energy line. Besides the resulting unrealistic velocity and flow pattern along the direction of flow in the ravine, another key problem is that the active flow area for overbank flow increases with increasing length of the defined cross sections. While the extent of the cross sections is predefined for larger streams with a classical channel and floodplain geometry, the definition was found to be rather arbitrary for the given situation. To address this problem, the extent of all cross sections was chosen as 100 m, which limited their extent to the main channel, its levees and a small and constant stripe of its left and right overbank areas. In the flow computations, each cross section is assumed to be representative for the reach geometry half way to the next up- and downstream cross section. The average bank full depth of the Ravine Lan Couline was estimated to be 4.5 m, which, together with an average invert slope of 0.03 (m/m), impedes that cross sections should be placed no more than 22.5 m apart from neighboring cross sections [28]. Based on these conditions, the cross section configuration shown inFigure 14 was defined for the extraction of the channel geometry for hydraulic modelling. It consisted of 374 cross sections that were placed at regularly spaced intervals along a 3150 m long section of the Ravine Lan Couline. The river stations (red numbers in Figure 14) are used to locate cross sections along the channel and are defined as the distance to the downstream end of the modeled reach in meters.

The creation of the model geometry based on two different DEMs (TanDEM-X and Drone DEM) was done in HEC-GeoRAS. The two DEMs were first referenced vertically by adding a constant value to each elevation value so that both DEMs had similar elevations in the area of the overlap. Based on a manually defined border between the two overlapping DEMs, HEC-GeoRAS automatically extracts the topography of cross sections depending on their location in respect to the border. Due to the insufficient spatial resolution of the TanDEM-X DEM, as well as the presence of inaccuracies resulting from interpolation in small areas of the Drone DEM, some of the cross sections of the geometric model had to be adjusted manually in order to accurately represent the geometry of the channel. In the same way as the parameterization of the flood routing method in the rainfall-runoff model, the main channel and overbanks were assigned a roughness coefficient of 0.045.

The flood hydrographs resulting from hydrologic modeling were used as the upstream boundary condition. In order to avoid instability issues in the flow computations, a base flow of approximately 10% of the peak flow of the flood wave was applied before and after the flood wave. Since the flood hydrograph at the downstream end of the modeled reach was unknown, the normal depth (approximated as the local channel bed slope) was used as the downstream boundary condition. This approach involves a high level of uncertainty so that the last cross section was placed far enough downstream from the area of main interest within the settlement (Figure 14).

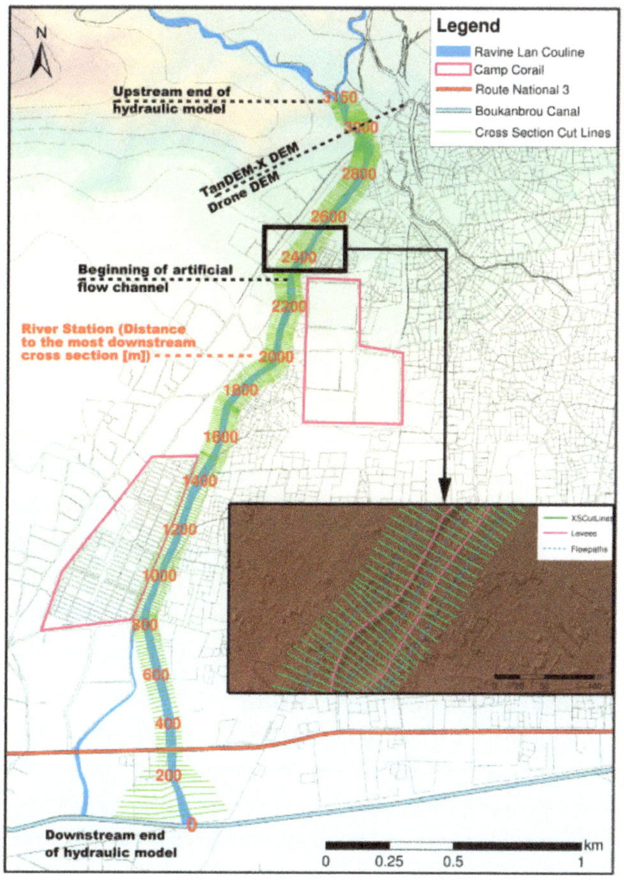

Figure 14: Geometric model configuration and development in HEC-GeoRAS.

RESULTS AND DISCUSSION

Hydrologic Modeling Results

Figure 15 shows the design floods resulting from rainfall-runoff modeling of the design storms. It can be seen that a design storm with an ARI of two years only leads to a peak discharge of 2 m^3/s in the Ravine Lan Couline whereas a storm with an ARI of five years already leads to a peak discharge of 112 m^3/s. This is in accordance with the fact that no significant discharges have been seen in the ravine since the beginning of urban development in the area in 2010 (Interviews with local residents by members of TUM-USO). Even though this observation could not be verified due to the lack of historic data, it strengthens the prediction of the model that the Ravine Lan Couline watershed only produces runoff in response to larger precipitation events. Furthermore, it is seen that exceptional storms with short durations and very high intensities can be expected to produce large peak runoffs of up to more than 400 m^3/s (ARI = 100 years) and more than 300 m^3/s (ARI = 50 years). Considering the relatively small size of the analyzed watershed (25.8 km^2), discharges of more than 400 m^3/s seem very high at first. The shape of the hydrographs, however, is typical for flash flood events as the result of short and intense rainfalls in mountainous watersheds. The high rainfall intensities that the study area experiences in combination with the short time of concentration of the analyzed watershed leads to the steep and short rising limb of the flood hydrographs as well as a short overall duration of the flood.

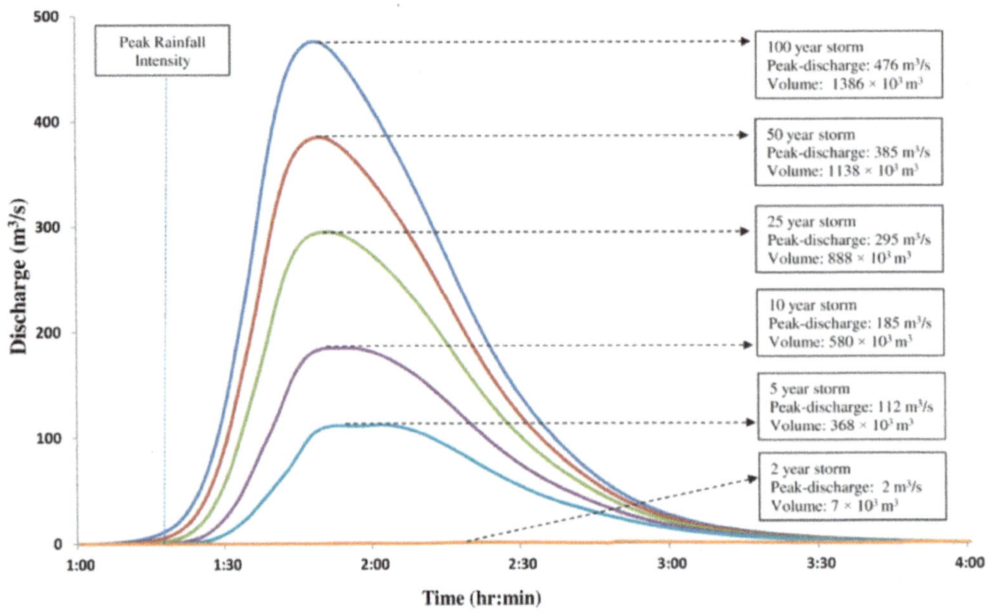

Figure 15: Flood hydrographs with ARIs of two to 100 years (Ravine Lan Couline).

Hydraulic Modeling Results

Based on the hydrographs resulting from rainfall-runoff modeling (Figure 15), unsteady flow simulations were performed for ARIs of five, 25 and 100 years. Figure 16 illustrates the hydraulic modeling results in form of a three-dimensional view of the maximum water surface resulting from a 25 year ARI design flood, which was chosen as this is the lowest ARI where levee overtopping occurs. Additionally, the modeling results for ARIs of five, 25 and 100 years are shown in the form of cross-section views for two locations along the channel (river station 2258 and 1972). The river station is used to locate cross sections along the modeled reach and is defined as the distance to the downstream end of the modelled reach in meters. An overview of the entire reach that was modelled along with the river stations was given in Figure 14.

The artificial flow channel constructed along with Camp Corail roughly begins at river station 2350. The maximum water surface resulting from the peak discharge of the five year ARI flood runoff shows that the bottleneck fraction of the channel is roughly located between river station 1800 and 2000. In this area, levee overtopping occurs first during the rise of the flood wave. The cross section at river station 2285 (Figure 16) is representative for the beginning section of the artificial flow channel. The stability of the levees in this section is questionable, since they are entirely constructed from unconsolidated material excavated from the channel invert and clearly deviate from state-of-the-art levee design standards [29]. It is therefore expected that the levee cannot resist the 25 and 100 year ARI flood runoffs so that lateral outflow is likely to occur in this area at the right overbank in downstream direction. It is further estimated that the levees in this area are capable of resisting the five year ARI flood runoff because in this case the maximum water level is in the area of the foot of the levee (Figure 16). The cross section at river station 1972 represents the beginning of the bottleneck fraction of the channel. It can be observed that the peak discharge during the five year ARI flood almost fills up the main channel so that larger discharges are likely to cause lateral outflow in this section. The cross sections located downstream represent the channel in the area of the northern extent of southern Camp Corail. It is obvious that the flow carrying capacity of the channel is significantly larger here than at the bottleneck section, since the 25 year ARI flood peak discharge is conveyed within the main channel.

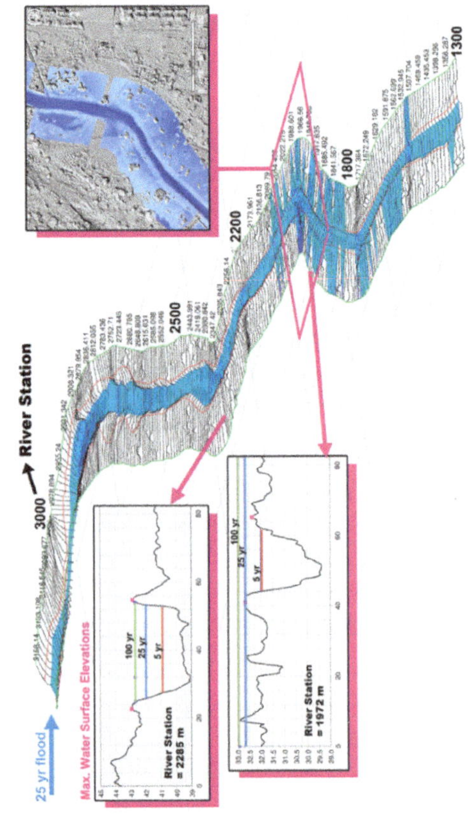

Figure 16: Maximum water surface for a 25 year ARI flood in 3D view and for a five, 25 and 100 year ARI flood for two channel cross sections (river station 2285 and 1972).

Flood Hazard Zoning

In order to provide a practical overview of the outcome of the hydrologic and hydraulic modeling as a basis for the flood risk assessment, the results were incorporated into a flood hazard zoning (Figure 17). It is important to note that a hazard zoning in comparison to a flood risk zoning does not account for the potential economic and social damages resulting from flooding. The hazard map is intended to provide an easy to understand overview of the flood hazard situation in Onaville. Following the basic structure of European flood hazard maps, the zoning outlines areas based on the ARI of flood events. The zoning can further be separated into areas potentially affected by large flood runoffs or torrential debris flows in Ravine Lan Couline and areas affected by smaller scale, torrential debris flows generated in the other four watersheds in the north of Onaville (see WS 2–5 in Figure 11). For the latter, hydraulic modeling was not performed as part of this study, since the Drone DEM did not provide sufficient coverage for the network of channels through which these smaller watersheds are draining.

To realize a more understandable zoning especially for non-professionals, the flood discharges with ARIs of five, 25 and 100 years in this study were defined as frequent, rare and exceptional hazard. In case of the Ravine Lan Couline, areas of all of these discharges were outlined. For the rest of the settlement, hazard zones were only categorized into frequent and exceptional hazards. Since hydraulic modeling was limited to locating hydraulic deficiencies of the Ravine Lan Couline channel and did not provide information about the propagation of the flood wave on the alluvial cone, a hydro-geomorphologic analysis of the area was used for the completion of the hazard zoning. Such a combined approach is common practice for the development of flood hazard maps, and was based on the following datasets:

- The digital elevation models presented in Chapter 2
- Historical aerial imagery in Google Earth
- On-site photographs

The main steps in generating the hazard zonation were as follows. Before the construction of the artificial flow channel, the propagation of floods over the alluvial cone was easily traceable on historic aerial imagery and the DEMs. The hydraulic modeling showed clearly, where overbank flow and levee breaching is to be expected first during large flood events. Floods in a five year ARI range can be conveyed safely by Ravine Lan Couline through the entire settlement. Floods in a 25 year ARI range exceed the flow carrying capacity of the artificial flow channel at the bottleneck section and are expected to lead to levee breaching right at the beginning of the artificial channel (see cross section at river station 2285 in Figure 16). Therefore, the zone of rare hazard essentially represents the propagation of floods before the construction of the artificial flow channel along two distinct and relatively wide longitudinal depressions in the alluvial cone in the area of southern Camp Corail. Since a levee breach will lead to a large increase of the flow area and a resulting decrease in flow velocity, bed load deposits are expected in the area of the breach as indicated in the map. The zone of exceptional hazard was created by gradually extending the rare zone towards more elevated areas using the DEMs.

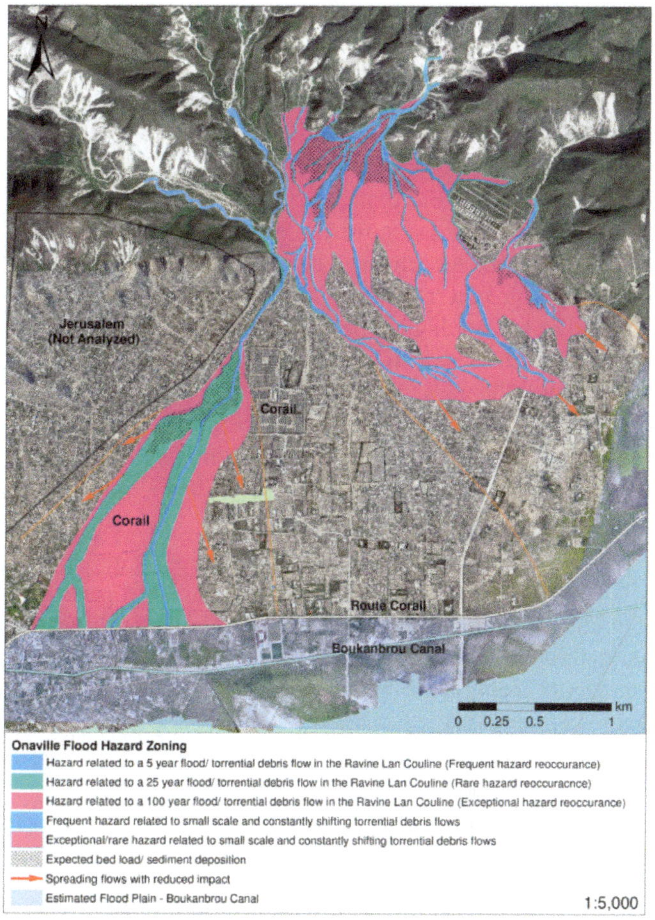

Figure 17: Flood hazard map of Onaville and the eastern part of Jerusalem (Canaan).

Interviews with residents of northern Onaville revealed that the additional four smaller watersheds frequently produce smaller scale flash flood events carrying large amounts of solids. All four watersheds drain through a clearly visible network of relatively small drainage channels that are carved into the alluvial cone, thereby allowing a good estimation of the propagation of flood discharges through the settlement (Figure 17). The frequently affected zone was defined based on the currently visible drainage channel network. The rare zone was defined by using the DEM and by considering that the existing drainage channel network is likely to change over time with new channels emerging potentially in areas with similar but not higher elevation than the current channel network. Due to a reduction of the channel slopes along the various flow channels, these concentrated flows eventually spread out on the alluvial cone so that their impact is expected to decrease drastically from this point on. Therefore, areas that are expected to experience spreading flows are outlined separately in the map.

Overall, the flood hazard map shows that large areas in Onaville are exposed to flood hazards. The artificial part of the Ravine Lan Couline channel was found to provide protection only to flood events in the frequent hazard range. Flood events of rare and exceptional hazard range exceed the capacity of the channel and are thus expected to cause severe flooding in the outlined urbanizing areas.

Interviews with residents of northern Onaville revealed that the additional four smaller watersheds frequently produce smaller scale flash flood events carrying large amounts of solids. All four watersheds drain through a clearly visible network of relatively small drainage channels that are carved into the alluvial cone, thereby allowing a good estimation of the propagation of flood discharges through the settlement (Figure 17). The frequently affected zone was defined based on the currently visible drainage channel network. The rare zone was defined by using the DEM and by considering that the existing drainage channel network is likely to change over time with new channels emerging potentially in areas with similar but not higher elevation than the current channel network. Due to a reduction of the channel slopes along the various flow channels, these concentrated flows eventually spread out on the alluvial cone so that their impact is expected to decrease drastically from this point on. Therefore, areas that are expected to experience spreading flows are outlined separately in the map.

Overall, the flood hazard map shows that large areas in Onaville are exposed to flood hazards. The artificial part of the Ravine Lan Couline channel was found to provide protection only to flood events in the frequent hazard range. Flood events of rare and exceptional hazard range exceed the capacity of the channel and are thus expected to cause severe flooding in the outlined urbanizing areas.

Preventive Flood Risk Management on Community Level

In Onaville, ongoing land ownership issues, uncontrolled and rapid urbanization and the lack of functioning political administration of the settlement lead to difficult and complex circumstances for implementing any kind of management. Especially, due to unresolved land tenure and ownership issues, currently no major interventions are executed by the Haitian government, so that the neighborhoods mostly rely on work executed by NGOs and self-organized initiatives.

Through preceding projects within the framework of TUM-USO research group, a relatively strong connection and communication channel has been established to residents of Onaville. The NGO TECHO has been active in Canaan since the early days of the settlement in 2010 and provides a major link to the population of Onaville. TECHO assists the community by providing temporary housing and basic infrastructure as well as social guidance and skills training imparted in weekly meetings and frequent workshops. During field research, these networks and communication channels were used to raise awareness of the encountered hazard situation among affected residents and community leaders. A small focus group on environmental hazards was established, consisting of four TECHO volunteers and a few active and influential members of the local community of Onaville. This group was familiarized in detail with the methodology and outcomes of the flood risk analysis serving as a continuing link between the research group and an ongoing flood risk management on site.

In April 2014, the planning phase for a campaign aiming at community-wide increase of awareness started with a kickoff-event in Onaville, including a thematic workshop on flood hazards (Figure 18). Together with TECHO volunteers, local leaders and a group of representatives from Camp Corail a basic emergency management plan was elaborated.

Figure 18: In a flood risk workshop, the hazard situation was presented and discussed with a group of community leaders from Onaville and nearby Camp Corail.

The core of the Neighborhood Contingency Plan is a network consisting of six local risk management representatives identified and elected during that same workshop. In order to take over emergency information coordination, each leader will be a contact person for residents, coordinating danger alarms within their specific neighborhood. Further, they will communicate with the Contingency Plan leader network, which would be in charge of coordination with local and national authorities and administration. Identified communication methods vary from (mass) text messages, "*tele-dyol*" (word of mouth) and megaphone announcements to local radio station broadcasts carried out at the same time in order to reach the maximum of people as fast as possible. When looking for potential emergency shelters, appropriate buildings and structures were discussed, as well as their fundamental requirements such as good construction quality, location in "safe" zones, good accessibility, proper size and use.

During a joint site-visit, workshop participants were able to experience and reflect on previously discussed construction issues in risk prone areas. Upcoming activities proved that the workshop has contributed to increased collaboration among leaders from Onaville and Corail camps. The spreading of information in both sectors of Camp Corail has been facilitated, since representatives and the director were involved in the collective coordination and planning by the leader network.

Additionally, a first phase of an awareness-raising campaign in risk zones was carried out on three weekends in June 2014 with about 10 volunteers, consisting of community leaders and TECHO volunteers, distributing about 200 flood risk awareness flyers to households in hazard-exposed areas of the ravines and steep slopes of Ti Sous and Fon Dyab (eastern and upper Onaville). These flyers (Figure 19) were based on both the outcomes of this study and the flood risk workshop. They were designed to communicate the current risk situation considering local circumstances in a way that is easy to understand. In personal talks, risk exposure and prevention measures such as the Neighborhood Contingency Plan and personal emergency kits were explained to the residents. Telephone numbers of local risk management representatives were handed out to allow fast communication in case of hazard observation or warning of tropical storms and heavy rains. During this awareness-raising campaign it turned out that most families living in hazard zones are well conscious about their exposure to flood and landslides, but simply have no choice, as they cannot afford living in safer zones due to the high parcel and rental prices. This fact as well as the related real estate speculations are major drivers for the invasion of risk zones by underprivileged families in Onaville.

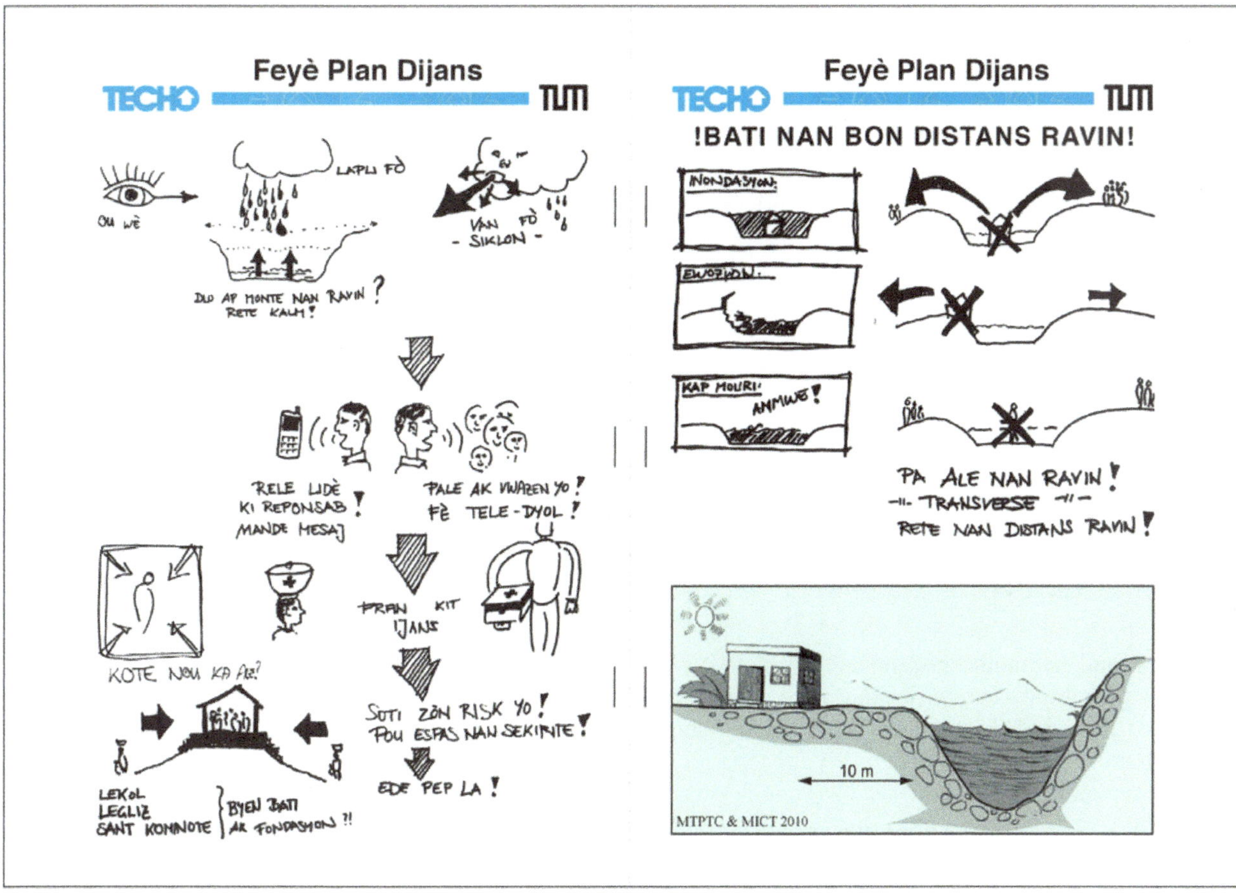

Figure 19: Excerpt from the flood risk awareness information flyers.

Beside the community-based work, the results of the flood hazard analysis were presented to key actors within the Haitian government (Unit for Housing Construction and Public Buildings (UCLBP), National Center for Geospatial Information (CNIGS), Inter-ministerial Territorial Planning and Development Committee (CIAT), Department of Civil Protection (DPC), Ministry of Environment (MDE)) same as to international aid and development agencies (Oxfam, ONU-Habitat, CordAid, American Red Cross). The flood hazard map (Figure 17) was shared with CNIGS to realize future integration into the national geo-information platform for housing and city quarters (SILQ).

In combination with the participatory field work (Contingency Plan), this research was presented by members of the focus group on environmental hazards at the exhibition of the international experts' and practitioners' forum Understanding Risk in Haiti: Innovate to Prevent in Port-au-Prince, 2014. Independent meetings of the Contingency Plan leader network were held together with (inter-)national disaster experts as well as with local and national authorities in order to evaluate potential support. Both showed interest in the grassroots initiative since its volunteer methodology of disaster prevention is considered to have potential for upscaling on the regional or national level.

Discussion

Considering the insufficient database, the predominantly empirical rainfall-runoff models can be seen as highly suitable for the generation of the design flood hydrographs. Although mainly physically-based water balance models like WaSiM or MIKE SHE are state-of-the-art in simulating the rainfall-runoff process, the suitability of such models for the given situation is limited due to the lack of data. To realize an accurate parameterization of such models regional distributed land use and soil data in high resolution are necessary as well as measured rainfall and runoff data for model calibration. Furthermore, the additional features of such physically based models (e.g., simulation of infiltration rates, evapotranspiration or interception) were not needed in this study and their validation would not have been feasible because of the limited data. In order to increase the reliability of the modeling results, measured precipitation and runoff in at least hourly resolution would have been necessary. For this study, information about the runoff behavior of the analyzed watercourses was based on the statements of the inhabitants only, so that a validation of the modeling results with measured data was not possible. Therefore, the modeling results should be interpreted as an approximation of the flood events that may occur in this region.

In contrast to this, the acquired data can be considered as suitable for generating the river channel through the settlements in HEC-RAS in a widely realistic way. Only in the northern part of the modeled reach, the TanDEM-X DEM had to be adjusted manually, because of the lower resolution of this DEM compared to the Drone DEM. Since the model is able to simulate the relevant water levels and locate the overflowing sections of the Ravine Lan Couline, it is definitely suitable to accurately estimate the flood hazard situation in the study region.

Despite the good suitability of the topographic data, the applied one-dimensional routing model shows obvious deficits regarding the simulation of lateral outflow of water from the main channel and the flow on the overbank areas, because the model is limited to simulating flow perpendicular to each channel cross section. For accurate modeling of lateral overbank flow, a two-dimensional modeling approach (e.g., HYDRO_AS-2D, MIKE-FLOOD) would be preferable. However, the development and manual correction of the model geometry based on a combination of the Drone DEM and the TanDEM-X DEM, would have been difficult for the 2D modeling approach. The Drone DEM has a number of small areas with poor representation of the terrain, which was easily adjustable for the cross sections in the 1D flow model. These difficulties would have made the development of a 2D flow model complex so that it would have exceeded the scope of this analysis.

Hydrologic and hydraulic modeling revealed that large areas of the settlement are currently exposed to a high level of flood related hazard. The artificial channel was found to mark the beginning of the high-risk zones. Before the artificial channel, the Ravine Lan Couline channel has enough capacity to convey a 100 year ARI flood runoff safely. Lateral outflow of flood runoff in the area of the market place of Onaville (see location of Figure 3 as referenced inFigure 1) as predicted by previous studies, is unlikely. The constructed channel, however, only provides protection against flood events in the frequent hazard range. Hydraulic modeling showed that flood events in the rare and exceptional hazard range are likely to have a devastating impact on the settlement in this area. Large amounts of flood runoff leaving the main channel as a result of levee overtopping and breaching have to be expected along with high flow velocities and bed load depositions in the settlement. In general, Sector 4 of Camp Corail is likely to be more vulnerable to flooding than Sector 3. Besides flood runoffs in the Ravine Lan Couline, runoffs generated in the other four examined watersheds also pose flood hazards to certain areas of the settlement. These smaller scale flow events have an unpredictable nature regarding their flow path through the settlement along with high destructive potential as a result of the high solid loads.

Due to a variety of uncertainties in the rainfall-runoff modelling and the incorporation of the hydraulic modelling results into the flood hazard map, the hazard zones have to be seen as a first approximation of the actual hazard situation. The ARIs of the flood events that are outlined in the hazard map have to be proven in more detail by acquiring additional data and applying state-of-the-art process based models. Developing a 2D hydraulic model is recommended since such a model could potentially lead to better estimates of the areas that will be affected by floods in the 25 and 100 year ARI range, if solid transport and deposition can be accurately accounted for.

Based on the GIS and simulation results, the methodology of bringing academic knowledge and technical aspects of flood risk assessment in the form of "applied science" back to local neighborhoods has turned out to be successful. The participants of community workshops were trained with technical, site-specific, methodological and important background knowledge as well as with the skills to come up with low-tech, low-cost solutions, which are applicable on the individual, family or neighborhood level. Yet, until this point in time, only a relatively small group of individuals has been able to benefit from that due to the fact that the undertaken activities were not up-scaled and the leader network was not reinforced.

CONCLUSIONS AND OUTLOOK

Key Issues and Solutions

One of the key issues related to the fulfillment of the study goals was found to be the acquisition of adequate hydrologic and topographic data. Hereby, the lack of readily available data related to the poor socio-economic state of Haiti was overcome by acquiring data generated by state-of-the-art remote sensing technologies and on-site analyses. The extensive data acquisition process lead to a database that was sufficient to set up a combined-hydrological and hydraulic-conceptual model approach. This leads to the next key issue which was found to be the limited suitability of the one-dimensional HEC-RAS model for the given situation. Even though the model was capable of evaluating the hydraulic capacity and deficiencies of the Ravine Lan Couline channel, a detailed simulation of the propagation of flows outside the main channel was not possible. In addition to that, the geometry of the hydraulic model had to be created based on two separate DEMs. Due to the insufficient spatial resolution of TanDEM-X DEM the cross sections that were derived from this DEM were manually adjusted based on the examination of the channel geometry on high-resolution aerial images.

Flood Risk Management and Future Work

Considering the available data sets and the selected modeling approach, the flood hazard zonation can be seen as a rough determination of the actual hazard situation that the settlement is facing. Since the presented zonation is at the moment by far the most accurate estimation of the flood hazard situation for the study area, it can be seen as a suitable basis for establishing a functioning flood risk management for Onaville. More complex risk assessments from governmental side are, as a result of the ongoing informal character of the settlement, rather unlikely. To realize a substantially complete operational flood risk management system, a flood protection frame work including designed technical and non-technical flood mitigation measures has to be generated in the next steps. For this, more data have to be collected concerning e.g., runoff data as well as the characteristics, the stability and erosion tendency of soils. Furthermore, the existing circumstances in this region have to be checked to ensure that only feasible measures are taken into account.

The third pillar of modern risk management strategies is prevention and preparedness, which comprises behavioral-, constructional- and risk-based prevention. The flood hazard map presented in this study provides a solid basis for preventive management, which was undertaken in the course of this study on the community as follows:

- Skills and technical training of local community leaders
- Initiation of an awareness-raising campaign and conceptualization of a local contingency plan
- Network-building among local leaders and between these leaders and external national and
- International organizations and entities
- Development of community-driven risk mitigation measures

This strategy for flood risk management on the community level is mainly based on an established focus group on environmental hazards and a network of local flood risk representatives. It is a reliable, low-cost and highly autonomous basis for risk prevention and mitigation, as it can incorporate resources and information dissemination networks of the community. It could additionally help the (local) authorities to become better prepared and react faster to unforeseeable disasters on a local scale.

Even though the flood hazard map along with the ongoing community work is a big step towards a functioning flood risk management for Onaville, this study has shown that there is an urgent need for further planning and implementation of flood protection measures and a more centralized and integrated flood risk management strategy. In this context, an Onaville-wide upscaling and reinforcement of the Neighborhood Contingency Plan activities is highly recommended, as different stakeholders have already been involved in the first pilot phase, and very positive feedback for the community-driven working methodology of the leader-led focus group has been reported by international and national entities and the community. The organizing focus group of the Neighborhood Contingency Plan should be immediately supported in its grassroots approach to bundle energies and capacities of this volunteer initiative before they are replaced by disillusion. Additionally, the campaign should be backed by technological prevention infrastructure such as the construction of simple weather stations in the hills of Onaville, and discharge gauge instruments at crucial points of the major water course Ravine Lan Couline. As for the emergency plan in general, the leader network needs to be reinforced by involving other established community leaders such as local and national experts, local teachers, pastors and so on. The network has to be supported in collecting and regularly evaluating experiences in the execution of alerts. Additionally, a site-specific assessment of available shelter structures and their surroundings ought to be required in a further step to assure the safety of the location.

ACKNOWLEDGMENTS

This work was kindly supported by the German Research Foundation (DFG) and the Technical University of Munich within the funding scheme Open Access Publishing.

We would like to thank the Technical University of Munich and the German Academic Exchange Service (DAAD) for funding several research trips to the study area.

We also want to thank TECHO-Haiti for hosting us during our field trips and the successful and fruitful collaboration on the Onaville project.

The TanDEM-X DEM was kindly provided by the German Aerospace Center (DLR). We would like to especially thank Thomas Fritz and Christian Rossi from the TanDEM-X ground segment for their effort, support and advice.

The Drone DEM was developed and provided by Drone Adventures, the Humanitarian Open Street Map Team (H.O.T.) and the International Organization of Migration (IOM). We want to especially thank Alexandre Habersaat from Drone Adventures who invested a lot of effort and time in the development of a DEM that was suitable for this study.

REFERENCES

1. International Federation of Red Cross and Red Crescent Societies. World Disasters Report Focus on Urban Risk; Geneva, 2010. Available online: https://www.ifrc.org/en/publications-and-reports/world-disasters-report/wdr2010/ (accessed on 20 October 2013).
2. U.S. Central Intelligence Agency (CIA). The World Factbook. Available online: https://www.cia.gov/library/publications/the-world-factbook/index.html (accessed on 5 September 2013).
3. Office for the Coordination of Humanitarian Affairs (OCHA). Tropical Storm Isaac in Haiti: Damage Needs and Responses. Analysis Report. Available online: http://www.humanitarianresponse.info/en/operations/haiti/document/tropical-storm-isaac-haiti-damage-needs-and-responses (accessed on 2 September 2013).
4. Office for the Coordination of Humanitarian Affairs (OCHA). Haitiearthquake Situation Report#23. Available online: http://reliefweb.int/sites/reliefweb.int/files/resources/CEB1B7C987DB2763492576D3001ECF55-Full_Report.pdf (accessed on 25 June 2015).
5. International Organization for Migration (IOM). Displacement Tracking Matrix (DTM)-Report Version 2.0, 2013. Available online: http://iomhaitidataportal.info/dtm/downloads.aspx?file=downloads/16.DTM%20Report_Sep %2713/DTM_V2_Report_Sep_2013_English.pdf (accessed on 15 October 2013).
6. Aerial Imagery ©2013 Google, Digital Globe. Imagery obtained with Google Earth. Available online: http://www.google.com/earth/ (accessed on 6 July 2015).
7. TUM Research Platform—Urban Strategies for Onaville (Haiti). Available online: http://openarchitecturenetwork.org/projects/urban_strategies_ona (accessed on 10 May 2015).
8. Kervin, S. Drainage, Water and Sanitation Infrastructure Improvement in Impoverished Areas—A Case Study of Onaville, Haiti. Master's Thesis, Technical University Munich, München, Germany, 28 March 2013.
9. Mora, S.; Roumagnac, A.; Asté, J.-P.; Calais, E. *Analysis of Multiple Natural Hazardsin Haiti (NATHAT)*; Government of Haiti: Port-au-Prince, Haiti, 2010. Available online: https://www.iris.edu/hq/files/workshops/2010/03/docs/Report-MULTIHAZARDS-HA-English-SergioMora-Final-Red.pdf (accessed on 31 October 2013).
10. FLOODsite. *Language of Risk—Project Definitions*, Report T32-04-01. 2005. Available online: http://www.

adbarno.it/rep/biblio/FLOODsite_Language_of_Risk_v4_0_P1.pdf (accessed on 11 March 2015).

11. Plate, E.J. Flood risk and flood management. *J. Hydrol.* **2002**, *267*, 2–11.
12. Lamothe, L. Press Statement about Canaan, Haitian Prime Minister, 2013. Available online: https://soundcloud.com/laurentlamothe/point-de-presse-du-premier-7 (accessed on 15 December 2013).
13. Hannemann, J.-C.; Werthmann, C.; Hauck, T. Designing for uncertainty: The case of Canaan, Haiti. In *Revising Green Infrastructure: Concepts between Nature and Design*; Czechowski, D., Hauck, T., Hausladen, G., Eds.; CRC Press: Boca Raton, FL, USA, 2014; pp. 323–351.
14. Information Products Lab for Emergency Response (IPLER). LIDAR DEM of Port-au-Prince Region, 2010. Available online: http://ipler.cis.rit.edu/projects/haiti (accessed on 10 May 2013).
15. Deutsches Amt für Luft- und Raumfahrt (DLR). TanDEM-X DEM Product Specification, 2013. Available online:https://tandemx-science.dlr.de (accessed on 25 September 2013).
16. Rossi, C.; (German Aerospace Center (DLR), Oberpfaffenhofen, Germany). TanDEM-X Ground Segment. Personal communication, 2013.
17. U.S. National Weather Service-National Hurricane Center. Tropical Cyclone Climatology. Available online: http://www.nhc.noaa.gov/climo/ (accessed on 5 November 2013).
18. *Hydrology and Hydraulic Analysis for Ravine Millet Petion-Ville, Haiti*; The United States Agency for International Development: Port-au-Prince, Haiti, 2011.
19. Haiti Landcover (Occupation du Sol), SPOT-CNIGS [04.1998]-polygon, Port-au-Prince, September 2008. Available online: http://www.haitidata.org/layers/cnigs.spatialdata:hti_biota_landcover_spot_cnigs_041998_polygon (accessed on 14 June 2015).
20. American Heritage Dictionary. *The American Heritage Science Dictionary*, 1st ed.; Houghton Mifflin Company: Boston, MA, USA, 2005.
21. Food and Agriculture Organization of the United Nations (FAO). Global Forests Resource Assessment, Progress towards Sustainable Forest Management, 2005. Available online: ftp://ftp.fao.org/docrep/fao/008/A0400E/A0400E00.pdf (accessed 22 October 2013).
22. Bureau des Mines et de l''Énergie (BME). Haiti Geology (géologie), BME-polygon, Haiti, Port-au-Prince, 2005. Available online: http://www.haitidata.org/layers/cnigs.spatialdata:hti_geology_geology_polygon_082005(accessed on 14 June 2015).
23. *HEC-HMS Technical Reference Manual*; US Army Corps of Engineers Hydrologic Engineering Center (HEC): Davis, CA, USA, 2010.
24. Natural Resources Conservation Service (NRCS). *National Engineering Handbook: Part 630—Hydrology*; United States Department of Agriculture: Washington, DC, USA, 2004; Chapters 1, 4, 6.
25. Chow, V.T.; Maidment, D.R.; Mays, L.W. *Applied Hydrology*; McGraw-Hill: New York, NY, USA, 1988.
26. *Urban Hydrology for Small Watersheds*; Technical Release Number 55 (TR-55). United States Department of Agriculture: Washington, DC, USA, 1986.
27. *Soil Mechanics Level 1, Module 3, USDA Textural Soil Classification, Study Guide*; United States Department of Agriculture: Washington, DC, USA, 1987.
28. *HEC-RAS Hydraulic Reference Manual, Version 4.1*; US Army Corps of Engineers Hydrologic Engineering Center (HEC): Davis, CA, USA, 2010.
29. *Hochwasserschutzanlagen an Fließgewässern*; DIN 19712; Deutsches Institut für Normung e.V. (DIN): Berlin, Germany, 2013. (In German)

CHAPTER 14

REMOTE SENSING-BASED ASSESSMENT OF THE VARIABILITY OF WINTER AND SUMMER PRECIPITATION IN THE PAMIRS AND THEIR EFFECTS ON HYDROLOGY AND HAZARDS USING HARMONIC TIME SERIES ANALYSIS

Eric Pohl, Richard Gloaguen and Ralf Seiler

Keywords: Westerlies, Indian summer monsoon, weather/climate variability, hydrology, hazards

INTRODUCTION

Central Asia's mountain ranges act as water towers for downstream regions by intercepting moisture transported by the eastward Westerlies and the northward Indian summer monsoon [1,2,3]. This general characterization has been addressed to precisely point out underlying circulation mechanisms [4], but the interaction of summer and winter circulations on moisture transport are still relatively unconstrained. The largest portion of moisture is provided as snow [1,5], which, when melted during summer, feed the Central Asian rivers. Moreover, snow precipitation provide moisture to glaciers, which represent the most important source of river discharge during late summer [1,6,7,8]. Despite such relevance, quantifying the actual moisture supply and obtaining systematic precipitation distributions remains a major challenge. Various studies dealt with this issue by assessing rain/snow gauges, remote sensing and climate model data (e.g., [5,9,10]). However, the lack of a dense gauging network, the inability of remote sensing instruments to detect snowfall over persisting snow cover [11,12] and the huge variances in climate model data result in substantial qualitative uncertainties in where and how much precipitation is actually provided [1,10].

Studies based on hydrological modeling are able to validate precipitation quantities by comparing modeled snow and glacier melt amounts with observed discharge. This attempt is prone to errors due to the high impact of glacier melts in the hydrological budget [1,6,8], because adapting glacier melt [1], a quantity that is even less restricted than precipitation amounts, due to an even sparser measuringnetwork [8,13], may allow the compensation of wrongly-assessed precipitation amounts. The reconstruction of the paleoclimate based on past glacier retreats and advances [14,15,16] is even more challenging, as boundary conditions, such as insolation, temperature, weakening of synoptic scale circulations and timing of precipitation, are less constrained than they are today. These high uncertainties often lead to simplified interpretations, such as generally intensified Westerlies or Indian summer monsoon (ISM) [14,15,16].

These uncertainties demand for alternative approaches to constrain the sphere of influence of the Westerlies and the ISM in the Pamirs or, more generally speaking, the influences of summer and winter precipitation. While there is consensus of the transition from more summer to winter precipitation-

Citation: Pohl, E.; Gloaguen, R.; Seiler, R. Remote Sensing-Based Assessment of the Variability of Winter and Summer Precipitation in the Pamirs and Their Effects on Hydrology and Hazards Using Harmonic Time Series Analysis. Remote Sens. 2015, 7, 9727-9752. doi:10.3390/rs70809727.

dominated regions from the Himalayas to the Pamirs [5,9,10,17], it remains unclear if this transition has a sharp boundary or if there is a smooth gradient. Maussion *et al.* [17] have delineated seasonal figures that already indicated such regions, showing that winter precipitation dominates far into the Karakoram region. This is especially interesting with respect to contrasting reports of positive and negative glacier mass balances across the Pamirs (Gardelle *et al.* [18] and Gardelle *et al.* [19]; *cf.* Sorg *et al.* [8] and Unger-Shayesteh *et al.* [13]), which might be related to yet unresolved local-scale climate variations or even unresolved precipitation occurrence in general. The impact of onset timing of the ISM on Tibetan glacier evolution [20] raises the question if a likewise dependency is effecting glacier evolution in the Pamirs. However, Mölg *et al.* [20] also point out that local- or regional-scale circulations rather than synoptic-scale climate is the important key aspect to describe glacier dynamics. This demands for a sufficiently spatially-resolved analysis.

Global climate models (GCM) often have resolutions in several tens to hundreds of km22 spatially and, hence, are unemployable to resolve local- and regional-scale characteristics, thus leading to weak correlation with *in situ* data. Regional climate models (RCM) and remote sensing data provide much better resolutions. The RCM HAR10 (10-km resolution High Asia Reanalysis) [17], for example, shows a high correlation with *in situ* precipitation measurements in the Gunt catchment in the Central Pamirs [1], although overestimating actual precipitation amounts. A major drawback of such high-resolution datasets is the often limited time span for which these data are available (12 years for HAR10). The analysis of precipitation time series is even more challenging if there is no pronounced seasonality and high intra-annual variation. A promising and steadily improving method to overcome such issues is the harmonic time series analysis (HANTS) [21,22] that is used to analyze a variety of natural phenomenon, such as phenology(e.g., [21,22]), diurnal precipitation distributions [23], airtemperatures [24] or land surface temperatures [25]. Several studies already explored the application of harmonic analysis, showing its potential to reveal precipitation seasonalities [26,27,28,29]. The previously developed, open access toolbox for HANTS analysis "phenotemp" [30,31] provides a collection of different HANTS methods for such analyses.

We propose to use remote sensing data products to estimate the spatiotemporal variability of some climatic indicators in the region. As water availability seems one of the major parameters affecting the climatic conditions in the orogen, we focus on HANTS to explore the inter-annual variability of precipitation in the Pamir region. Specifically, this paper addresses the following issues:(1) determination of the spatiotemporal influence of winter and summer precipitation in the Pamirs; (2) to support the analysis of precipitation time series, we incorporate snow cover, temperature and water storage data to reveal interrelations with precipitation and to independently validate the inferences based on the HANTS of precipitation; (3) we relate the obtained results to recent glacier evolution, to point out possible key mechanisms related to weather variability; and (4) we compare exceptional meteorological/weather conditions with the occurrence of regional-scale hazard events, including floods and droughts in the first decade of the 21st century in the greater Pamir region to point outpossible causalities.

STUDY AREA

The Pamirs mark the transition zone between the Westerlies and the Indian summer monsoon, but several additional mesoscale circulation systems [4] increase the complexity of this binomial delineation. Complex terrain with strongly incised geomorphology on the Pamir margins and a low relief, high altitude plateau in the central Pamirs further highlight that the Pamirs include a more diverse meteorological setting. Westerly winds are deflected to the north and south during winter and provide precipitation mainly as snow to the Pamir margins [5,32]. Precipitation records of few existing meteorological stations show a minor, but noticeable amount of summer precipitation in the eastern central Pamir region [1]. Precise information is rare about the actual precipitation distribution due to the lack of a dense gauging network. Different studies comfort the assumption that the region receives variable amounts of precipitation with a decreasing W-E trend [1,8,33]. We hence divide the greater Pamir region (Figure 1) into seven subregions bounded by notable orographic barriers and name them based on dominant mountain systems.

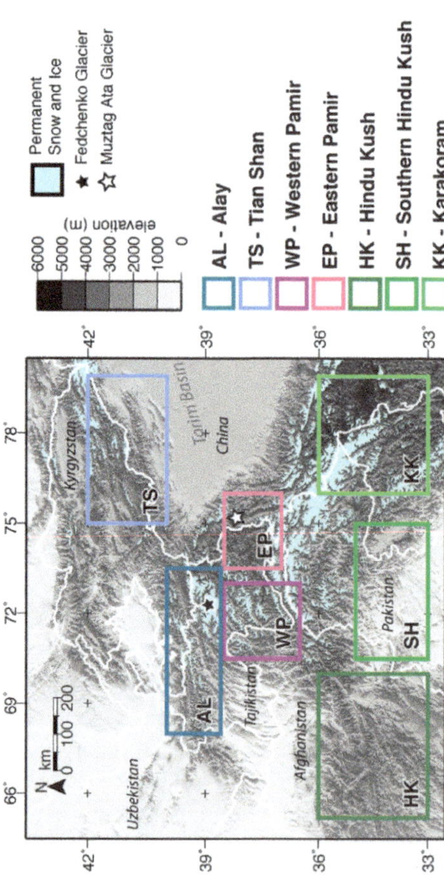

Figure 1: Study area with selected subdivided study sites. Glacier extent is approximated from MODIS MCD12Q1 land cover. Subdivided study sites were defined based on expected precipitation regimes and are named based on the dominant surrounding mountain system.

The subregions Alay Mountains (AL) and Tian Shan (TS) in the northern part mark the outermost orographic barrier that would intercept southward moisture supply. In the central part, the Western Pamir (WP) would intercept direct westward moisture supply, while the Eastern Pamir (EP) lies in the orographic precipitation shadow and is assumed to generally receive very little precipitation, but might have a dominant summer influence [1]. The Western Hindu Kush (HK) is assumed to intercept the southward deflection of westerly winds mainly during winter and might also act as a barrier for far-reaching ISM extensions that are assumed to be most dominant in the lowlands in the Southern Hindu Kush (SH). The Karakoram Range (KK) receives much snowfall, probably related to westerly extensions [5], but since it is located even more eastwards than SH, a noticeable ISM influence is assumed.

DATA

In order to relate the precipitation analysis to a broader context in terms of hydrology, meteorology and hazards, we incorporate temperature, snow cover and liquid water equivalent thickness (LWET) data. Ideally, we would have selected only remote sensing products for this study. Pohl *et al.* [1] and Palazzi *et al.* [10] have shown that state-of-the-art precipitation datasets for the Pamir region show big differences, both in intensity and spatial distribution. In a previous work [1], we tested different precipitation products for their application in a hydrological modeling approach in the Pamir Mountains. We found that the frequently-used TRMM (Tropical Rainfall Measuring Mission) 3B42 V7 dataset [34,35,36] most likely suffers from issues with snowfall determination [11,37,38,39]. Even though the used microwave imagers can recognize snowfall, the quality relies on the discrimination between frozen precipitation and antecedent snow cover [11]. Kamal-Heikman *et al.* [38] also reported an underestimation of precipitation in cases of intense snowfall in the Himalayas. We found that the RCM dataset HAR (High Asia Reanalysis) [17] provides the most reliable estimates for precipitation in the Central Pamirs. However, good correlation with *in situ* measurements contrasts an overestimation of actual precipitation amounts. The reported overestimation is consistent for low and high elevation pixels, hence the required spatial patterns, and event timings are not affected by a lack of calibration.

The relatively short time spans of the used meteorological data (≈12 years) does not allow a full analysis in a climatological context. However, comparison of obtained HANTS results for the timing of winter and summer precipitation is in strong agreement with ECMWF (European Centre for Medium-Range Weather Forecasts) ERA-Interim data [40] that span a 35-year period (Figure 8). For the meteorological parameters snow cover, temperature and LWET, we do not apply the HANTS algorithm, because these parameters already show very continuous signals that allow analysis based on long-term means and corresponding anomalies for individual periods (monthly, winter, summer).

HAR results from a dynamical downscaling of global analysis data (final analysis data from the Global Forecasting System [41]; dataset ds083.2) using the Weather Research and Forecasting (WRF) model [42]. HAR data are available from 2001 to 2012 at different spatial extents and spatial resolutions. We used the 30-km resolution HAR30 dataset that fully covers the study area. We further use HAR30 2-m air temperature data that result from the same modeling. Pohl *et al.* [1] have validated the finer resolution HAR10 (10-km resolution) datasets in the central Pamirs. Because HAR10 does not spatially cover the study region, we use the HAR30 dataset, which is consistent with the 10-km resolution version. Although HAR is a modeled and not a directly-derived remote sensing dataset, it comprises the same structure as other remote sensing data products. The applied method in this study, therefore, suits the application to other precipitation datasets, where those data provide reliable estimates.

To evaluate whether precipitation falls as snow or rain, we use the MODIS (Moderate-Resolution Imaging Spectroradiometer) MOD10CM [43] monthly snow cover dataset. We use the time span from January 2001 until December 2012 to match the period available for HAR. This dataset comes originally in a spatial resolution of 0.05°° and was resampled using mean values to ≈0.3°° to allow for faster processing and to match the resolution of the precipitation and temperature datasets. MODIS snow cover data are very frequently used for hydrological and climate change scenario studies [3,5,44]. MOD10CM is an averaged monthly product based on the daily MODIS MOD10C1 [45] product.

To evaluate meteorological influences on the water balance of the regions, we use the GRACE (Gravity Recovery and Climate Experiment) [46,47] dataset GFZ (German Research Centre for Geosciences) release RL05 [48]. GFZ RL05 is available from 2003 to 2012 in monthly temporal and 1 × 1°° spatial resolution. GRACE measures the Earth's gravity field and its time variability. The major influence on time variability is the amount of stored water, and hence, variations in observed gravity can be used as a proxy for liquid water equivalent thickness (LWET). We use the GFZ RL05 dataset in this study to highlight the influences of meteorological configurations on the regional hydrological water budget. Several hydrological studies make use of GRACE data to access monthly changes in the water balance [49,50,51], but GRACE data have also been utilized to assess regional and global glacier mass balance changes [51,52]. We further use the Emergency Events Database (EM-DAT) of the Centre for Research on the Epidemiology of Disasters (CRED) [53] to qualitatively relate the meteorological interplay with hazard events. Assessing hazards is always coupled to the effect an event, such as a flood, mudflow or avalanche, has on local communities. It is hence difficult to localize the exact extent or area of influence of such events if very few people or little infrastructure is affected. The needed detailed analysis of appropriate imagery data would go beyond the scope of this work. Hence, we use EM-DAT events that have at least 100 people affected (see http://www.emdat.be/explanatory-notes for the full list of criteria). EM-DAT does not claim completeness, but lists the major events and, thus, provides event records for the entire study period and lists major events by country. Selected events are summarized in Table 1. The fact that events are stated by country prevents a local-scale analysis. However, in the case of droughts and severe floods, large proportions of the river catchments can be assumed to play an important role. This is because the main rivers in the study area originate in the high mountains and are affected by the regional meteorological conditions in the orogen. We want to point out that we do not intend to provide a local hazard assessment in this study. We rather want to demonstrate whether large-scale climatic events coincide with large-scale hazards and discuss a probable causal relationship.

Table 1: Selected hazard events from Emergency Events Database (EM-DAT) [53] in the greater Pamir region and adjacent regions for the years 2005, 2008 and 2010. Events are listed according to affected country, rather than exact locations. AF, Afghanistan; PK, Pakistan; TJ, Tajikistan; KY, Kyrgyzstan; UZ, Uzbekistan.

Country	Event Type	Date	People Affected
2005			
AF	Storm	January 2005	22,656
AF	Flood	05 March 2005	11003
AF	Flood	16 June 2005	5040
PK	Flood	09 February 2005	7,000,450
PK	Flood	21 June 2005	460,073
PK	Flood	05 July 2005	58,020
PK	Flood	02 March 2005	5000
PK	Flood	20 March 2005	3500
TJ	Flood	08 June 2005	3222
TJ	Mass movement wet	03 February 2005	1953
TJ	Flood	23 July 2005	1890
KY	Flood	10 June 2005	2050
UZ	Flood	24 February 2005	1500
2008			
AF	Drought	October 2008	280,000
AF	Flood	August 2008	1180
PK	Flood	02 August 2008	200,012
PK	Flood	09 August 2008	90,752
TJ	Drought	October 2008	800,000
2010			
AF	Flood	05 May 2010	40,000
AF	Flood	27 July 2010	5000
PK	Flood	28 July 2010	20,359,496
PK	Mass Movement Wet	04 January 2010	26,700
PK	Storm	06 June 2010	4000
PK	Flood	21 July 2010	4000
PK	Mass Movement Wet	18 February 2010	3705
TJ	Flood	06 May 2010	6708
TJ	Flood	11 April 2010	1914
KY	Mass Movement Wet	03 June 2010	8350

METHODS

Precipitation occurs as discrete events, *i.e.*, it takes place during a day or over a few days with intermediate gaps. Even in regions influenced by monsoon, there is no continuous signal of precipitation. In the Pamirs, most precipitation occurs during winter and early spring with a strong decreasing W-E gradient [1,2]. As a result, precipitation amounts and intensities are generally low, and it is difficult to derive seasonal patterns based on ordinary moving averages (Figure 2a). Few precipitation events further complicate the determination of the timing of when an individual season peaks. To overcome this issue, we apply the HANTS algorithm [21,22].

HANTS

The aim of HANTS as presented by Roerink *et al.* [22] and DeJong *et al.* [21] is to remove pixel-wise cloud-affected observations and to temporally interpolate a given time series of remote sensing vegetation data. HANTS is a multi-step approach that in the end represents a given time series by means of a limited amount of cosine functions with different phases, frequencies and amplitudes [21]. In the first step, the Fourier series are calculated. Depending on the assumed number of harmonic signals, e.g., annual, seasonal, *etc.*, a corresponding number of Fourier series with according frequencies are calculated. Each data point of the original time series is replaced by the (smoothed) value of the combined Fourier series, if its residual is above a manually set threshold. Once this replacement has been done for all points of the time series, the algorithm checks whether (1) a predefined number of iterations have been calculated or (2) an also *a priori*-specified max percentage of data points has been changed from the original to a smoothed value. If one of the two criteria is met, the algorithm terminates. Otherwise, it starts a new iteration with calculating the set of Fourier series from the now partially-smoothed data. Seiler [30], and Seiler and Gloaguen [31] developed a standalone, open access Java application "phenotemp", which provides a compilation of different modifications of HANTS, of which the version after de Jong *et al.*[21] was used in this study. We used phenotemp to obtain (1) the average annual pattern (based on the full time series) and (2) the individual annual patterns (Figure 2b).

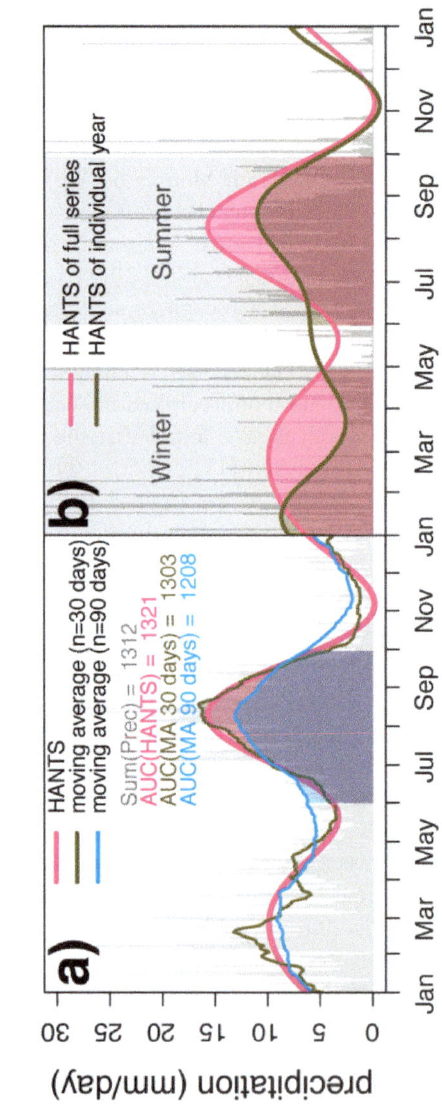

Figure 2: Harmonic time series analysis (HANTS) principals. (**a**) HANTS *vs*. running means with different window widths applied to a precipitation time series (grey) of a randomly-selected pixel; (**b**) HANTS is applied to find the maximum of the full time series and of individual years to determine the peak precipitation timing for each pixel (see Section 5). Anomalies are calculated as the difference of the area under the curve (AUC) of either the winter or summer season of an individual year and the full time series. Note that "HANTS of full series" is calculated based on the full length of the time series and

shows the same shape in every year. "HANTS of individual year" is calculated individually for each year of the full time series, and thus, its curves show different shapes from year to year. Here, only one year is displayed for the sake of brevity.

Differences between the integrals of the average and individual years are calculated as winter (JFMA; January–April) and summer (JJAS; June–September) anomalies. The algorithm needs three parameters: (1) the maximum number of iterations*maxIt*; (2) the maximum tolerance of original data points from the smoothed time series *maxTol*; and (3) the minimum number of preserved data points *minPres*. HANTS lacks an objective rule to set these parameters [21]. Because we intend to unambiguously determine HANTS maxima, we set *maxTol* = 0, which preserves a continuous HANTS signal (*maxTol* > 0 can result in several maxima, as is the case for moving averages; see Figure 2a). *maxIt* and *minPres* were set to 10 times and 67%, respectively. We have tested the HANTS algorithm on three *in situ* precipitation time series in the Central Pamirs that show different intensities and seasonalities. In either case, HANTS was able to precisely represent the long-term average annual cyclicality and intensities. This ability has been utilized by other studies to identify precipitation seasonalities and the influences of synoptic and regional-scale atmospheric circulations on the distribution of precipitation [28,29].

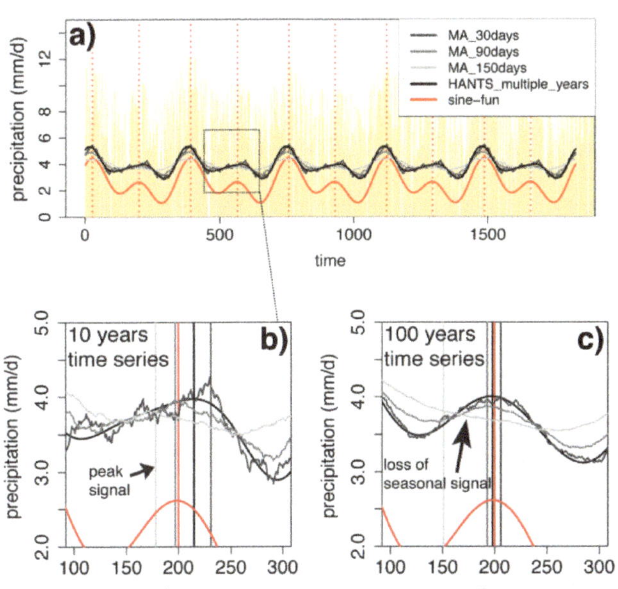

Figure 3: Comparison of HANTS and moving averages (MA). (**a**) Synthetic time series of precipitation data (yellow) based on two added sine functions with frequencies of 2ϖ and 4ϖ per year and a phase offset to model a signal similar to possible bi-modal curves with a winter and a summer peak (red curve). Noise added by (1) means of a uniform distribution with values between 0 and 8 and (2) additional random removal of individual values. (**b**) Different MA with varying window sizes applied to mean daily values of the generated time series with a 10-year period. Displayed HANTS ("full series") is based on the same time series. (**c**) Same as for (b), but based on time series of 100 years. Vertical bars indicate the maximum of individual curves within a 100-day window around the actual maxima of the original overlain sine functions.

The application of HANTS comprises noticeable additional effort in comparison to the application of, e.g., moving averages (MA). Therefore, we summarize some major issues that can result from the application of moving averages and why HANTS provides an advantage in particular cases. The example provided in Figure 3 shows a synthetic precipitation time series with a bi-modal peak that can be expected in the transition zone between winter and summer precipitation-dominated regions. The basis of the time series is a combined sine function with distinct frequencies of $2\pi 2\pi$ and $4\pi 4\pi$, respectively. Noise was added by means of uniformly-distributed values and by randomly removing data points. This was done to test both MA's and HANTS' ability to reveal the underlying periodic signal. The calculation of mean daily values, followed by an application of MA, is able to identify the timing of the peak amplitude of the underlying signal if the time series is long enough or the average signal is pronounced enough (Figure 3c).

Based on the choice of the moving window, the results are not as distinct. In particular, the choice of the moving window affects the determination of the real underlying seasonal patterns. Small window sizes are sensitive to fluctuations (Figure 3b; 30-day window size) and might introduce temporal bias, and windows that are too long might result in the diminution of the seasonal signal (Figure 3c; 150-day window size). The presented HANTS in this study, by definition, does not allow high frequency fluctuations, but preserves a continuous signal, nor does its application dampen (or even remove) the amplitude of the signal, which is the case for MA with a large window sizes. As shown in Figure 2, the dampening of the signal amplitude from MA with a large window size will lead to a biased assessment of AUC. This bias prevents meaningful quantitative analysis. In summary, HANTS combines the robustness of MA with large window sizes to reveal low-frequency seasonal signals and the ability of MA with small window sizes to preserve amplitudes of seasonal signals. HANTS series are better at preserving the real times of maxima and the original signal intensity.

RESULTS

The greater Pamir region encompasses complex topographic and climatic settings, which prevent a simple spatiotemporal classification. Instead of trying to describe all of the processes and their interactions, we present the general patterns of precipitation intensities and timing (Figure 4 and Figure 5). In order to display the temporal dynamics, we present regionally-averaged time series for individual regions, which we have described in Section 2. This step is required due to the large amount of spatiotemporal data that would exceed the scope of this article by far. However, the simplification by means of regional averaged time series hampers the identification of the associated spatial patterns. Therefore, we highlight three individual years that show the outstanding interplay of meteorological parameters and specific effects on LWET (Figure 6) in Section 5.2.

Mean Patterns and Inter-Annual Variability

There are strong differences in the average winter and summer precipitation distribution (Figure 4). Most significant amounts in winter are provided to the western sites AL and WP, but also to the northern Hindu Kush and to the northern Karakoram Range (Figure 4b). In contrast, the HK site and the lowlands in SH, as well as TS receive very little precipitation. In summer, significant amounts are provided to TS, EP and SH (Figure 4c). The WP and HK receive almost no precipitation in summer.

The maxima of the HANTS analysis for the full time series (Figure 5) show similar timings of peak precipitation for the HAR30 (Figure 5a) precipitation dataset (12 years) and ERA-Interim (Figure 5b) (35 years). The higher spatial resolution of HAR30 discriminates regions that are entirely winter-/summer-dominated and regions showing partial domination. Entirely winter-dominated are HK and WP, while TS and the southern part of SH, *i.e.*, the lowlands, are summer-dominated. Striking is a sharp discrimination of either winter- or summer-domination in the SH and KK, but also the EP site. In contrast to these sharp differences, there are several transitional regions that show successive temporal gradients of peak precipitation input. Most pronounced here is the AL region with a W-E gradient of later peak precipitation occurrence. Similarly, the entire Tarim Basin shows smooth transitional timings. While TS shows a distinct summer peak, northward and southward timings are successively earlier.

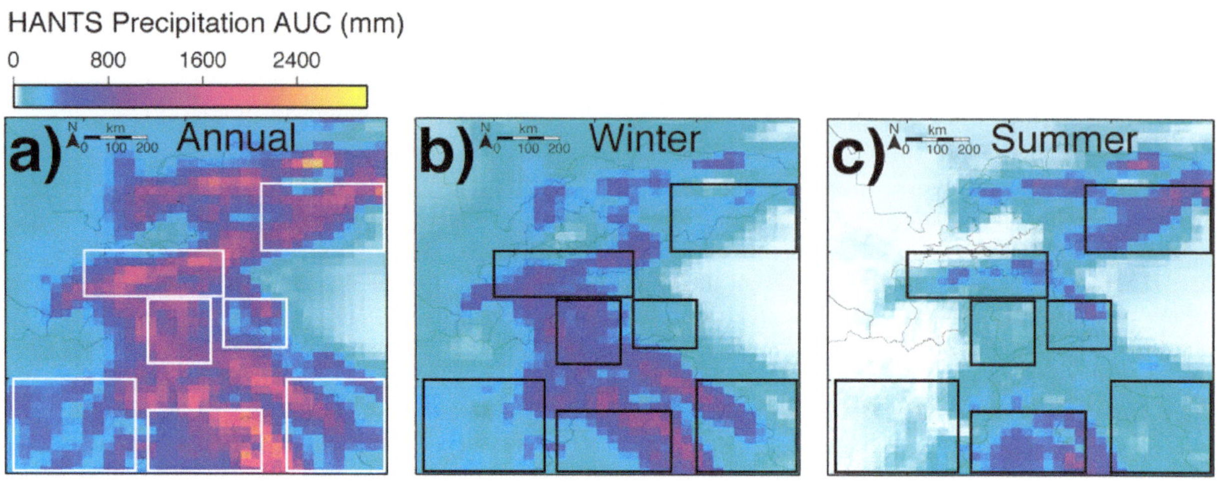

Figure 4: Precipitation distributions of the applied HANTS method for the (**a**) annual, (**b**) winter and (**c**) summer precipitation patterns.

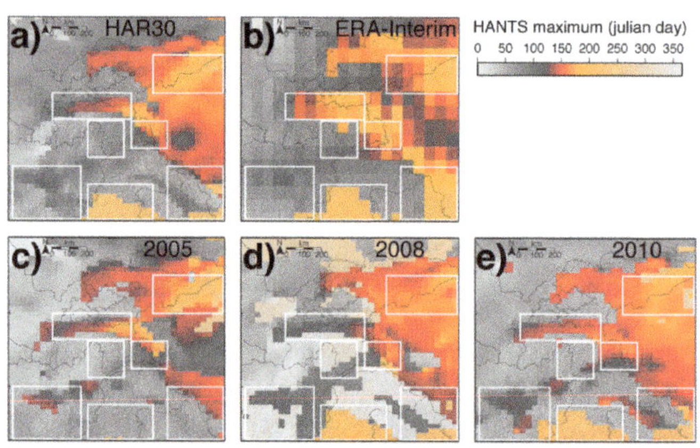

Figure 5: HANTS timing results for (**a**) the 12-year period of HAR30 and (**b**) the 35-year period of ERA-Interim for comparison. Timing results for the individually-analyzed years (**c**) 2005, (**d**) 2008 and (**e**) 2010. White bounding boxes correspond to subdivided study sites from Figure 1.

Figure 6 shows the combined meteorological parameters for the individual study sites. All but LWET show very high and rather unsystematic variabilities. LWET in contrast shows high variability between individual years, but intra-annual variabilities are rather weak compared to other parameters. Several study sites display similar temperature anomalies, but show huge differences in precipitation, snow cover and LWET. No causative relationship between individual parameters are immediately apparent, but individual years show combinations of different mechanisms that lead to specific LWET anomalies. To highlight the relationship of meteorological parameters and LWET anomalies, we exemplarily chose and analyze the three years 2005, 2008 and 2010. These years show very pronounced and different interactions of meteorological parameters. Therefore, they provide the opportunity to systematically describe dominant processes, point out regional differences and describe specific interrelations between meteorological conditions, hydrology and hazards.

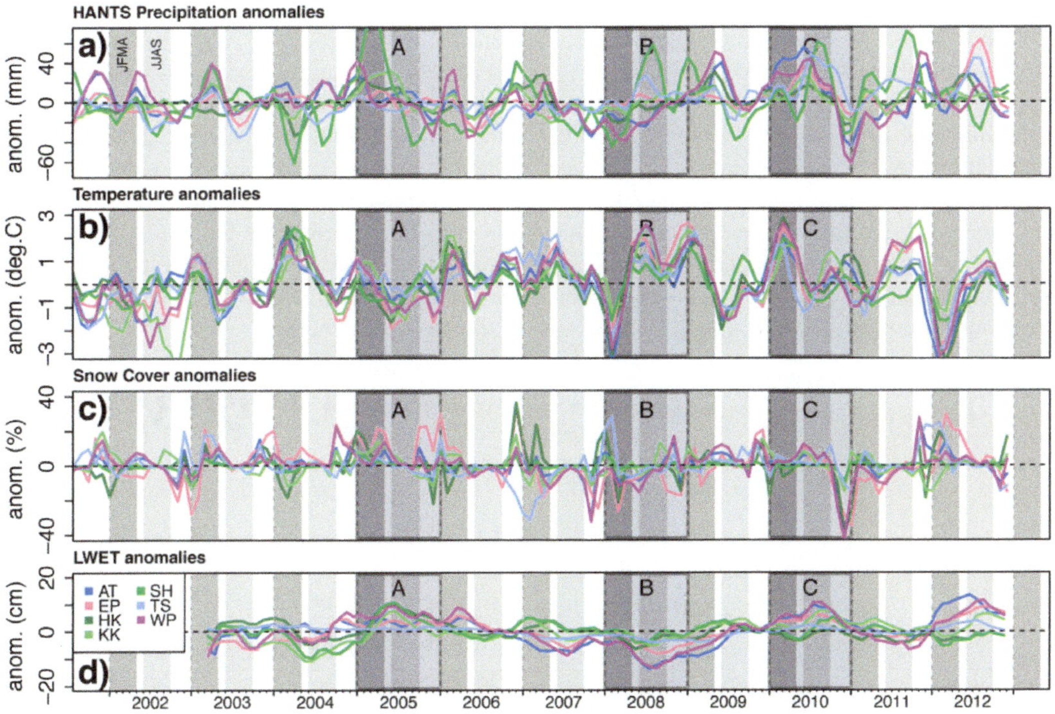

Figure 6: Monthly meteorological anomalies for (**a**) HANTS precipitation, (**b**) temperature, (**c**) snow cover and (**d**) liquid water equivalent thickness (LWET). The running mean of three months is applied for clearer visualization. Capital letters A, B and C are the three individual years that are analyzed in detail to point out the interplay between meteorological parameters. Vertical dark grey bars denote the winter (JFMA; January–April) and light grey bars the summer (JJAS; June–September) season.

Individual Years

Figure 7 displays that in 2005 the central and northern sites show only small anomalies for precipitation, while the southern sites, especially SH, show very high winter precipitation. In summer, SH shows reversed, negative precipitation anomalies. A very pronounced negative temperature anomaly in the snow melting season (AMJJA; April–August) (Figure 6b) for the central and southern sites is accompanied by a high snow cover (Figure 6c and Figure 7d). The prevented snow melt correlates with strong positive LWET anomalies (Figure 6d). By September, temperature and snow cover anomalies have declined. LWET anomalies for the southern sites and WP strongly decline after temperature anomalies become less negative. Reported floods in Tajikistan in June and July, in Afghanistan in June and in Pakistan in June and July (Table 1) coincide with the reduction in snow cover and temperature anomalies. Timings of peak precipitation input show that the usually summer-dominated parts of SH and KK are winter dominated. The central and northern Pamir sites show a picture similar to the average pattern, contrasting the southern Pamir sites (Figure 5c). Spots along the Hindu Kush syntaxis show slightly later peak inputs.

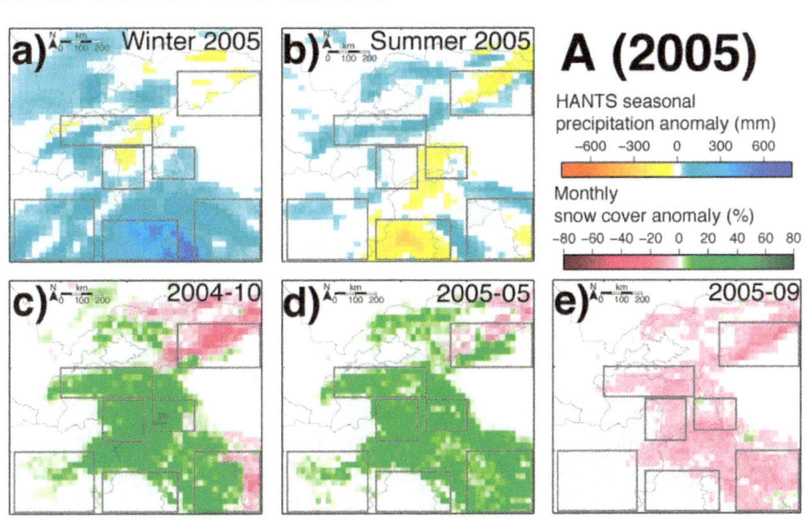

Figure 7: 2005 HANTS precipitation pattern for (**a**) winter and (**b**) summer. Very early snow cover in (**c**) October 2004 lasted till the end of (**d**) spring 2005 before diminution of snow cover to normal values occurred (see Figure 6), leading to slightly negative anomalies in (**e**) September 2005.

The year 2008 (Figure 8) is characterized by negative winter precipitation anomalies at all sites. In summer, SH solely shows strong positive anomalies. Western sites show negative anomalies and eastern sites positive ones. Positive temperature anomalies start in May and are most pronounced in July, concentrated in the central Pamirs and the Karakorum Range, where the most extensive glacier cover is present (see Figure 1). Negative anomalies for LWET and positive ones for temperature decline synchronously in summer. The strongest LWET anomalies are evident for AL and WP, which also show the most pronounced negative precipitation anomalies in both winter and summer. Heavy droughts are reported for October in Tajikistan and Afghanistan [53,54], along with two flood events in August for Pakistan (Table 1). Earlier timings of peak precipitation are apparent in the western sites, while summer-dominated regions in the southern parts of SH and KK match the timing of the average pattern (Figure 5d). Slightly earlier peak precipitation occurrence in the most western parts contrasts later occurrence along the Hindu Kush syntaxis and the EP site. The AL site, despite showing later occurrence than the average pattern, also stands out from the surrounding areas with later occurrences.

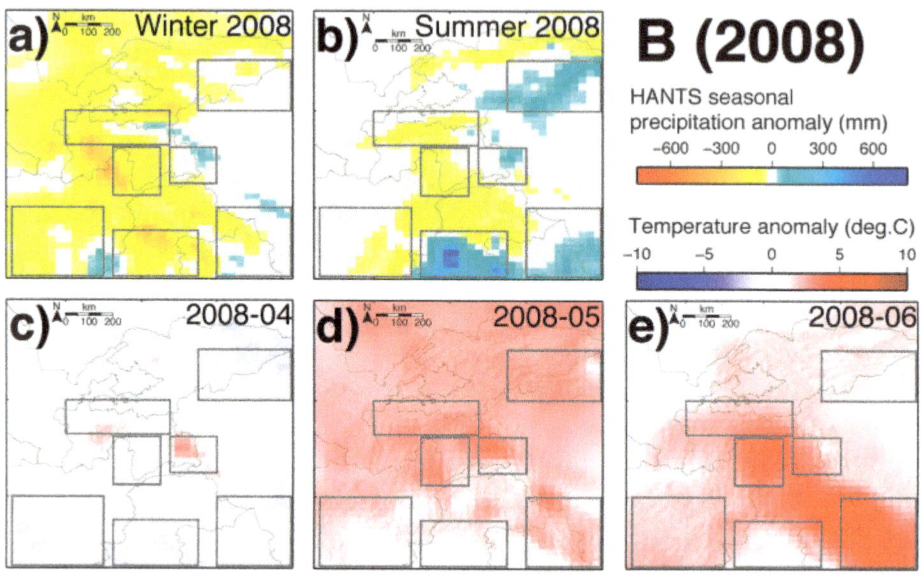

Figure 8: 2008 HANTS precipitation pattern for (**a**) winter and (**b**) summer. Gradually-evolving temperature anomalies (**c–e**) with a concentration of high temperatures at high altitude in the WP, EP and KK sites at the end.

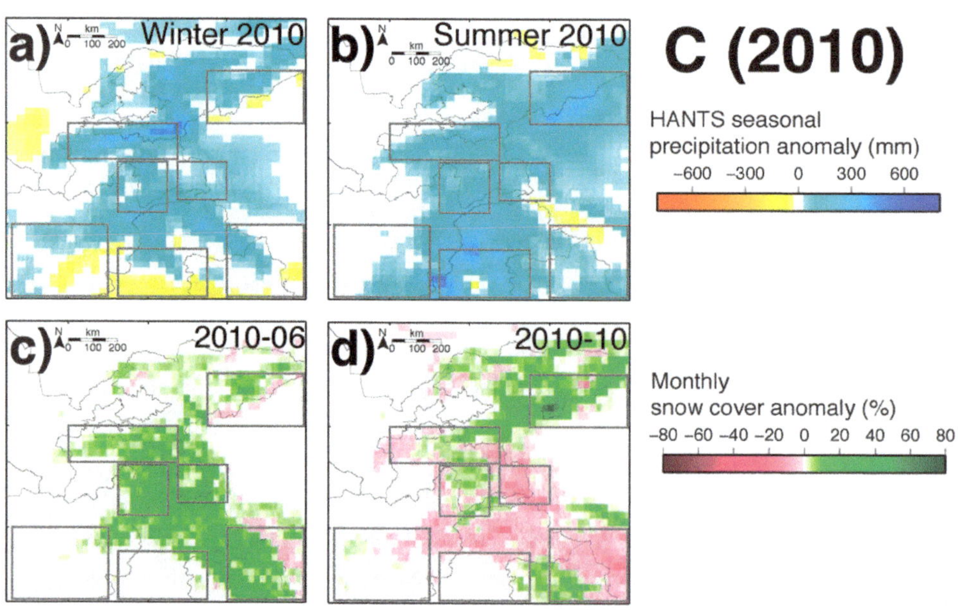

Figure 9: 2010 HANTS precipitation pattern for (**a**) winter and (**b**) summer. Strong positive summer precipitation anomalies accompanied by area-wide positive snow cover anomalies for high elevation sites in (**c**) June 2010. Diminution of snow cover except for TS until (**d**) the end of summer.

The year 2010 stands out because both summer and winter precipitation anomalies are strongly positive (Figure 9). It also marks the second wettest recorded monsoon year for Pakistan [55]. Especially SH and TS received more precipitation than usual in summer. In winter, the northern and central sites already show positive anomalies that are most pronounced in the AL and WP sites. Strong positive temperature anomalies in winter are not reflected by negative snow cover anomalies. Striking is that the strong June precipitation anomaly coincides with strong positive snow cover anomalies in the central Pamirs, northern Hindu Kush and Karakoram region. By October and later into the year, strong negative precipitation and snow cover anomalies are apparent. This coincides with a leveling out of high positive LWET anomalies. The timing of the Pakistan flood in July/August (e.g., [56]) (Table 1) correlates with the peak of precipitation anomalies for SH. Neither temperature nor snow cover show any anomalies that might be associated with this event. Tajikistan and Afghanistan also encountered flood events that were, however, much earlier in the beginning of May. These events coincide with the strong positive precipitation anomalies in the AL and WP sites starting in winter and continuing into mid-summer, as well as with the still positive temperature anomalies in April and May. Peak precipitation timings show similar timings to the average pattern, but transitional regions in the AL and TS sites show a more discriminated pattern. Along the Hindu Kush syntaxis into the EP site, local spots of later timings are apparent as in 2005, but their spatial extent is more pronounced, almost connecting the HK and EP sites. Summer-dominated regions in the SH and KK site are also slightly expanded compared to the average pattern.

Spatial Correlation of Winter and Summer Precipitation

Summer and winter anomalies for precipitation and snow cover were used to reveal similar behavior between individual sites and to highlight possible relations of strong/weak summer and winter precipitation anomalies (Figure 10). Significant correlations (p-value < 0.05) are primarily apparent between either winter or summer anomalies of individual sites. In winter, the northern and north-western sites WP, TS and AL show a similar behavior. In the south, SH shows the highest correlation with KK and HK. In summer, WP also correlates with EP and HK, while SH no longer shows significant correlation with any other site.

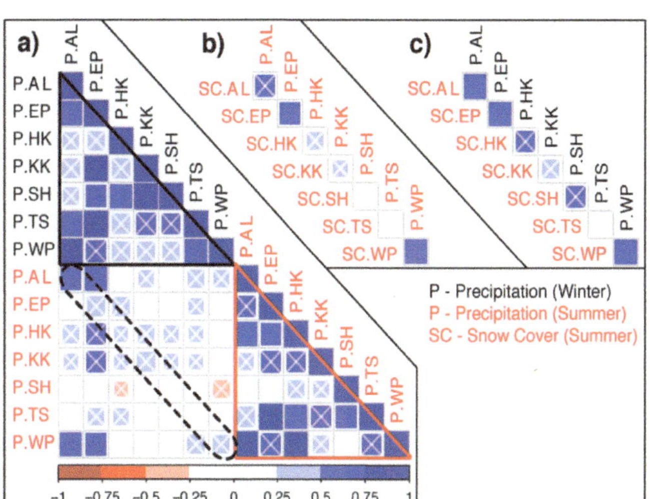

Figure 10: Spatiotemporal correlation of (**a**) HANTS summer and winter precipitation anomalies; (**b**) HANTS summer precipitation anomalies with snow cover anomalies; and (**c**) HANTS winter precipitation anomalies and summer snow cover anomalies. Color coding corresponds to the correlation coefficient multiplied by either one (positive) or −1 (negative correlation). Crosses mark non-significant correlations (p-value ≥ 0.05).

Comparison of winter and summer anomalies only shows a significant positive correlation for AL. The lowest correlations (≤ 0.25) can be seen for SH and TS. Although other correlations appear non-significant, no negative correlations are found that would suggest an antagonistic relationship between winter and summer precipitation. Comparison of winter or summer precipitation with summer snow cover suggests that AL, WP and EP winter anomalies are related to observed summer snow cover. This relation also holds true for summer precipitation, suggesting that precipitation in both summer and winter contributes to snow cover and partial water storage.

DISCUSSION

HANTS Precipitation and Interplay of Atmospheric Circulations

There are clearly winter and summer precipitation-dominated regions within the greater Pamirs. For both the average and for individual years, this discrimination is most pronounced in the SH and KK sites, as is revealed by the HANTS maxima of the respective time series. The resulting line in the southern Pamirs can be seen as the border of mainly Westerlies- or ISM-dominated regions. The border shows variable locations for individual years and seems (for ISM-dominated precipitation) to be limited by the orographic barrier of the Hindu Kush. Strong winter precipitation in combination with little summer precipitation as, e.g., in 2005 leads to a disappearance of this discrimination line in the study area. Opposite conditions, with stronger summer and weaker winter precipitation, only expand the summer-dominated region by a relatively small portion (2008, 2010), still showing winter dominance along the Hindu Kush syntaxis into the WP and EP sites.

The north-eastern sites TS, AL and EP show especially high variability in peak precipitation timing. The Alay Range shows later timings than the surrounding regions. Aizen *et al.* [4] have pointed out that the northern regions of the Pamirs, including the AL and TS sites, receive a high portion of recycled water during the warm season transferred by northern intrusions. Based on deuterium excess values of sampled snow/ice cores, Aizen *et al.* [4] discriminate the Pamirs (Fedchenko Glacier within the AL site) to receive either more slowly evaporated or less often evaporated water than the Tian Shan. This differentiation qualitatively follows the presented differences in peak precipitation timing under the assumption that summer precipitation undergoes more recycling than winter precipitation in this area. Although HAR30 precipitation is available for only 12 years, it shows comparable results to 35 years of ERA-Interim precipitation in terms of spatiotemporal distribution as presented by peak precipitation input. This is quite remarkable considering the high weather variability (*cf.* [57]). The obtained spatiotemporal precipitation patterns in the Central Pamirs corroborate a reported strong negative W-E precipitation gradient and later peak precipitation timings in the eastern Pamirs [1]. Because the spatial coverage of rain gauges is sparse in the greater Pamir region, there is no possibility to comprehensively validate the HANTS results. A meaningful comparison with *in situ* data is furthermore impeded by the point-wise character of precipitation gauges that are strongly biased compared to gridded data in mountainous regions. This is due to different orographic effects, such as precipitation shadows and precipitation-altitude gradients. No significant correlations of winter and summer precipitation are apparent for individual sites, except for AL (Figure 10). Most correlations, including correlations between different sites, are positive, but insignificant. Observed strong correlations between different sites in either season show a similarity that might be related to prevailing winds and associatedmoisture transport.

Inter-Annual Weather Variability

The three selected years 2005, 2008 and 2010 show very different meteorological settings, allowing one to investigate their effects on the water budget. Although timings of abrupt LWET anomaly changes coincide with major hazard events, the lack of precise hazard locations prevents a derivation of causalities with certainty. The year 2005 illustrates the effect of temperature anomalies on existing snow cover. Especially for high altitude regions (excluding SH), winter precipitation plays a crucial role in different ways. Low temperatures in the preceding year 2004 caused high amounts of precipitation to be stored as snow (Figure 6b–d and Figure 7c). This snow cover persisted late into 2005 due to low temperatures in late spring and summer 2005 (Figure 6b). Even though there are no especially high anomalies in precipitation, this situation favored positive LWET anomalies. The subsequent rapid decline in LWET anomalies suggests high amounts of snow melt. This highlights the strong impact of positive temperature anomalies on the mobilization of large quantities in water. The timing coincides with the occurrence of floods. The role of the high mountains in Central Asia as water towers for the downstreamregions [13,58] point to a possible relationship of such melting events and their impact on flooding. However, initially positive precipitation anomalies for the southern sites HK, SH and KK (Figure 6a) equivocally coincide with the timing of flood events. This points to a possible synergetic effect on runoff by (1) increasing temperatures and resulting snowmelt and (2) increased snowmelt from liquid precipitation on top of the snow layer [59].

Different from 2005, 2008 had strong negative winter precipitation anomalies. LWET anomalies reveal that there was slightly less water stored during winter 2008 than usual. Again, positive temperature anomalies in late spring and early summer caused a further decrease in LWET. Negative snow cover and winter precipitation anomalies indicate that snow stocks during early 2008 were already very low. Snow cover in the Central Pamirs disappears on average by June/July [1,5]. This might have caused strongly increased glacier melt. This assumption is further strengthened by the observed concentrated positive temperature anomalies at high altitudes where the glaciers are located (Figure 1). Pohl *et al.* [1] previously have shown that about 30% of the central Pamirs' annual discharge results from glacier melt. The remaining 70% of annual discharge result almost entirely from snowmelt that also replenishes the groundwater reservoirs. The reduced groundwater discharge and low total amounts of glacier melt, hence, suggest reduced overall river discharge. Again, the role of Central Asia's high mountains as water towers point to reduced river discharge in the mountain foreland regions in summer. The strong arid, continental climate in these regions make river discharge the essential source of irrigation water [13] (Figure 4c), and hence, the occurrence of droughts and reduced stream flow generation in the mountains in Tajikistan and Afghanistan seem to be closely related. At the same time, positive precipitation anomalies in the SH site have caused flood events in the Peshawar district, Pakistan (within the SH site) [60], highlighting the high spatial meteorological variability.

In 2010, the effects of the strong precipitation anomalies are more straight-forward. The high amount of water input is directly reflected by positive LWET anomalies. The southern sites KK and SH that show only small precipitation anomalies in winter show accordingly less LWET. Precipitation in Tajikistan is concentrated in winter and at a high elevation (Figure 4b). Hence, positive winter precipitation anomalies are likely to result in increased snow stocks, without a noticeable effect on winter river discharge [1,5]. The positive temperature anomalies in the winter season are unlikely to affect snowmelt in high mountains, because winter temperatures are well below the dew point [1]. For intermediate elevations, however, temperature anomalies of up to 3∘∘C in combination with high snow stocks and/or increased portions of rainfall over snowfall might result in increased runoff, similar to the effect described for 2005.

These conditions might have led to the reported flood events in Tajikistan and Afghanistan in late spring. Different from that, the severe flood in Pakistan at the end of July 2010 is well documented. Extensive precipitation with high intensities provided moisture to the entire norther regions of Pakistan [56]. Striking is the strong increase of snow cover in June in the high altitude regions all over the Pamirs and the Karakorum Range (Figure 9). This shows that summer precipitation accompanied by normal temperatures for this time can result in significant amounts of precipitation received as snow. Compared to a more immediate response of discharge to rainfall, the response to snowfall is more intermediate, and water is released successively, which weakens the intensity of floods. The delayed release of snowmelt allows for infiltration of water into shallow and subjacent deep groundwater aquifers that buffer regional discharge and provide sustainable groundwater discharge (see LWET) in winter. Such events might further preserve glaciers due to the increased albedo of fresh snow in the early ISM period in June [20,61].

Impacts on Glacier Evolution

There are only a few studies that have assessed glacier mass balances in the greater Pamir region. Two studies by Gardelle*et al.* [18] and Gardelle *et al.* [19] use a geodetic method that calculates the difference of digital elevation models obtained in two different years (e.g., 1998 and 2010). They show that the Pamir and Karakorum regions show slightly positive glacier mass balances for the first decade of the 21st century (slightly alternating time periods between the sites). Given the strong inter-annual and intra-annual variability of weather that is presented in this study, one could also assume different outcomes of glacier mass balance studies based on the actual time period that is considered. Moreover, Mölg *et al.* [20] have shown that the Indian monsoon onset timing has a high impact on glacier mass balance in the Tibetan Plateau, because the reflective properties of such precipitation received as snow early in the monsoon season attenuate melting. This highlights the two important aspects of precipitation: (1) providing moisture to accumulate on glaciers; and (2) protecting properties of fresh snow against melting. Furthermore, Liu *et al.* [62] and Khromova *et al.* [63] have pointed out that regional temperature variability plays a key role in glacier retreat at Muztag Ata, eastern Pamir. All of these findings combined highlight that years such as 2008, with negative winter and summer precipitation anomalies, strong positive summer temperature anomalies concentrated at high altitude glacier regions, might have the strongest impacts on glacier retreat. However, it also points out that if glacier evolution studies are based on only two points in time (e.g., Gardelle *et al.* [18], Gardelle *et al.* [19]), they might actually be strongly influenced by the inter-annual variability rather than a persisting trend. Glaciers as a whole feature a response to climate/weather integrated over a spatial area and over a much longer time than, e.g., river discharge or LWET. It would therefore be highly beneficial for the scientific community to investigate short-term responses of glaciers to weather, e.g., how special weather conditions (such as 2008 or 2010) quantitatively affect glacier evolution. The presented timing maps of peak precipitation input could be used as indicators, where such measurements would yield the most valuable insights. The strong spatiotemporal variability of glacial advances in the Quaternary across the greater Pamir region [2] does not provide a picture as consistent as the presented timing maps of peak precipitation input. This is to a large degree due to the very limited amount of available studies, which prevents any meaningful analysis at this moment. However, our results can help to develop sampling strategies for such studies and, thus, help to relate past and present-day climate and their effects on glacier evolution.

Impacts on Hazards

We demonstrated the qualitative agreement of meteorological and hydrological anomalies and their described interplay with hazard events. Hazard events are reported at the country scale and, thus, prevent deriving causalities with certitude. However, rivers of Central Asia originate in the high mountains [13,58], and therefore, strong floods and/or droughts are bound to hydrological conditions in the mountains. This gives reason to imply a causal link even at the country scale. The derived insights into mechanisms that can lead to hazardous events can help to complement more geomorphologic and static hazard potential studies (e.g., Gruber and Mergili [64]), which disregard quickly changing meteorological parameters. Integration of such studies with our approach could help to get a more complementary approach that transitions from general risk potential (based on geomorphology and long-term meteorological/climate conditions) into acute risks resulting from short-term weather patterns. A major limitation is the lack of real-time RCM data, allowing only post analysis with the selected datasets. The presented maps of precipitation anomalies, however, reveal similarities in precipitation for several regions that could provide feasible locations for future precipitation and temperature gauges in the scarcely-monitored Pamirs. Due to the similarity in temperature anomalies between the study sites (Figure 6b), the few available meteorological stations might already provide sufficient information to infer strong melting events that can result in floods. Remote sensing precipitation products are not feasible for this approach at the moment due to issues with detecting snowfall and/or very low precipitation intensities [11,12], especially in mountainous regions. The recently launched Global Precipitation Measurement mission (GPM) [65] might close this gap and allow analysis with little time delay. Even until then, LWET anomalies from GRACE data provide high explanatory power for quick hazard identification, especially for spring and summer flood assessments. This is due to the temporal decoupling of snow accumulation in winter and the release of water in spring and summer, allowing a certain time lag between observation and compilation of the dataset. Even one-month delayed data could provide the essential LWET values that can be used for the assessment of a potential water release. Remote sensing products to capture land surface temperatures, which have been shown to be a reliable proxy for near surface air temperatures [1], such as MODIS MOD11, can then be utilized to quickly issue warnings if positive temperature anomalies and, hence, strong meltings are expected.

CONCLUSIONS

We present a multi-source approach based on climate model data and remote sensing products to highlight the effects of winter and summer precipitation anomalies on the greater Pamirs. We use an in-house developed toolbox of HANTS that explores seasonal patterns in time series of gridded datasets, such as remote sensing or climate model data. We incorporate independent remote sensing-based datasets of snow cover (MODIS) and liquid water equivalent thicknesses (GRACE), along with RCM data for temperature and precipitation. Anomalies of precipitation, temperature, snow cover and LWET reveal different interplays of meteorological parameters. We select three characteristic climatic situations to exemplify the complex climatic feedbacks. This consistency is further highlighted by the qualitative agreement of water mobilization or lack thereof, with droughts and floods. Probably because of issues related to snowfall detection, we had to discard remotely-sensed precipitation data that are still not a reliable input parameter in mountainous areas. Hence, we rely on the best performing precipitation dataset for this region, HAR30. The near future will hopefully provide promising alternatives to RCM data, e.g., with the Global Precipitation Measurement (GPM) datasets, allowing one to directly incorporate remote sensing data with little time delay from acquisition to application. The GRACE and MODIS remote sensing data proved to be useful additional material to corroborate HANTS analysis and to provide supplementary information to infer the underlying processes.

By applying the HANTS algorithm, we were able to investigate precipitation time series that show high variability and are difficult to investigate using simple averaging or smoothing methods. We define distinct regions of summer and winter dominance surrounding intermediate regions. This quantitative approach allows us to refine the previous vague characterization of the Pamirs as a transition zone between Westerlies and the ISM to a now spatially more precise distinction of individual regions as winter-, summer- or transitionally-dominated. Anomalies inter-comparison of (1) precipitation, (2) temperature, (3) snow cover and (4) LWET display high consistency and highlight the complex processes that affect hydrology, glacier evolution and hazard potentials in the high mountains of Central Asia. We do not find any significant indications for causal summer and winter precipitation interactions. Instead, high variability of either meteorological component results in the specific scenarios that are presented and discussed. We find that temperature in combination with (without) snow cover has high impacts on the change in LWET. Consequently, this must correspond to varying streamflow and, thus, the occurrence of floods (droughts). Especially in the high mountain regions, positive precipitation anomalies result in large snow stocks that, depending on temperature anomalies, can quickly release high amounts of melt water, baring the risk of floods. However, the strongly heterogeneous distribution of precipitation might cause local floods, while only a few hundred kilometers away, severe droughts occur. The presented approach and delineated regions of summer, winter and transitional dominance might help to provide a valuable basis for glaciological, hydrological, but also paleoclimatological studies that often struggle to refer to certain moisture transport systems. The open access HANTS toolbox promises to be an efficient tool to explore the spatiotemporal variability of cyclic signals in remote sensing data.

ACKNOWLEDGMENTS

This work is part of the BMBF (Bundesministerium für Bildung und Forschung; German Federal Ministry of Education and Research) research program PAMIR (Impact of climate change on the water balance of a river basin in the Pamir) (funding code: FKZ 03G0815) within the CAME (Central Asia and Tibet: Monsoon dynamics and geo-ecosystems) project and funded by the BMBF. MODIS MCD12Q1 data are distributed by the Land Processes Distributed Active Archive Center (LP DAAC), located at the U.S. Geological Survey (USGS) Earth Resources Observation and Science (EROS) Center (lpdaac.usgs.gov). GRACE land data were processed by Sean Swenson, supported by the NASA MEaSUREs (Making Earth System Data Records for Use in Research Environments) Program, and are available at http://grace.jpl.nasa.gov. Generic Mapping Tools (GMT) http://gmt.soest.hawaii.edu/ were used to produce the maps. We would like to thank four anonymous reviewers for their comments, which helped to improve this manuscript. We further like to thank the editorial board for their guidance and decisions made in the review process. Processing and analyzing of the data were conducted with the aid of the statistical software R [66], including the packages "raster" [67] and "rgdal" [68]. The PAMIR team further includes Malte Knoche, Christiane Meier, Stephan Weise, Karsten Osenbrück, Stefan Geyer, Tino Rödiger, Christian Siebert and Wolfgang Busch.

REFERENCES

1. Pohl, E.; Knoche, M.; Gloaguen, R.; Andermann, C.; Krause, P. The hydrological cycle in the high Pamir Mountains: How temperature and seasonal precipitation distribution influence stream flow in the Gunt catchment, Tajikistan.*Earth Surf. Dyn. Discuss.* 2014, *2*, 1155–1215.
2. Fuchs, M.C.; Gloaguen, R.; Pohl, E. Tectonic and climatic forcing on the Panj river system during the Quaternary. *Int. J. Earth Sci.* 2013, *102*, 1985–2003.
3. Tahir, A.A.; Chevallier, P.; Arnaud, Y.; Ahmad, B. Snow cover dynamics and hydrological regime of the Hunza River basin, Karakoram Range, Northern Pakistan. *Hydrol. Earth Syst. Sci. Discuss.* 2011, *8*, 2821–2860.
4. Aizen, V.B.; Mayewski, P.A.; Aizen, E.M.; Joswiak, D.R.; Surazakov, A.B.; Kaspari, S.; Grigholm, B.; Krachler, M.; Handley, M.; Finaev, A. Stable-isotope and trace element time series from Fedchenko glacier (Pamirs) snow/firn cores.*J. Glaciol.* 2009, *55*, 275–291.
5. Immerzeel, W.; Droogers, P.; de Jong, S.; Bierkens, M. Large-scale monitoring of snow cover and runoff simulation in Himalayan river basins using remote sensing. *Remote Sens. Environ.* 2009, *113*, 40–49.
6. Lutz, A.F.; Immerzeel, W.W.; Shrestha, A.B.; Bierkens, M.F.P. Consistent increase in High Asia's runoff due to increasing glacier melt and precipitation. *Nat. Clim. Chang.* 2014, *4*, 587–592.
7. Hagg, W.; Braun, L.; Kuhn, M.; Nesgaard, T. Modelling of hydrological response to climate change in glacierized Central Asian catchments. *J. Hydrol.* 2007, *332*, 40–53.
8. Sorg, A.; Bolch, T.; Stoffel, M.; Solomina, O.; Beniston, M. Climate change impacts on glaciers and runoff in Tien Shan (Central Asia). *Nat. Clim. Chang.* 2012, *2*, 725–731.
9. Bookhagen, B.; Burbank, D.W. Toward a complete Himalayan hydrological budget: Spatiotemporal distribution of snowmelt and rainfall and their impact on river discharge. *J. Geophys. Res.* 2010, *115*, 1–25.
10. Palazzi, E.; Hardenberg, J.V.; Provenzale, A. Precipitation in the Hindu-Kush Karakoram Himalaya: Observations and future scenarios. *J. Geophys. Res.: Atmos.* 2013, *118*, 85–100.
11. Skofronick-Jackson, G.; Weinman, J. A physical model to determine snowfall over land by microwave radiometry. *IEEE Trans. Geosci. Remote Sens.* 2004, *42*, 1047–1058.
12. Prigent, C. Precipitation retrieval from space: An overview. *Comptes Rendus Geosci.* 2010, *342*, 380–389.
13. Unger-Shayesteh, K.; Vorogushyn, S.; Farinotti, D.; Gafurov, A.; Duethmann, D.; Mandychev, A.; Merz, B. What do we know about past changes in the water cycle of Central Asian headwaters? A review. *Glob. Planet. Chang.* 2013, *110*, 4–25.
14. Zech, R.; Abramowski, U.; Glaser, B.; Sosin, P.; Kubik, P.; Zech, W. Late Quaternary glacial and climate history of the Pamir Mountains derived from cosmogenic be exposure ages. *Quat. Res.* 2005, *64*, 212–220.
15. Röhringer, I.; Zech, R.; Abramowski, U.; Sosin, P.; Aldahan, A.; Kubik, P.W.; Zöller, L.; Zech, W. The late Pleistocene glaciation in the Bogchigir Valleys (Pamir, Tajikistan) based on 10Be surface exposure dating. *Quat. Res.* 2012, *78*, 590–597.
16. Seong, Y.B.; Owen, L.a.; Yi, C.; Finkel, R.C. Quaternary glaciation of Muztag Ata and Kongur Shan: Evidence for glacier response to rapid climate changes throughout the Late Glacial and Holocene in westernmost Tibet. *Geol. Soc. Am. Bull.* 2009, *121*, 348–365.
17. Maussion, F.; Scherer, D.; Mölg, T.; Collier, E.; Curio, J.; Finkelnburg, R. Precipitation seasonality and variability over the Tibetan Plateau as resolved by the High Asia reanalysis. *J. Clim.* 2014, *27*, 1910–1927.
18. Gardelle, J.; Berthier, E.; Arnaud, Y. Slight mass

gain of Karakoram glaciers in the early Twenty-First century. *Nat. Geosci.* 2012, *5*, 322–325.

19. Gardelle, J.; Berthier, E.; Arnaud, Y.; Kääb, A. Region-wide glacier mass balances over the Pamir-Karakoram-Himalaya during 1999–2011. *Cryosphere* 2013, *7*, 1263–1286.
20. Mölg, T.; Maussion, F.; Yang, W.; Scherer, D. The footprint of Asian monsoon dynamics in the mass and energy balance of a Tibetan glacier. *Cryosphere* 2012, *6*, 1445–1461.
21. de Jong, R.; de Bruin, S.; de Wit, A.; Schaepman, M.E.; Dent, D.L. Analysis of monotonic greening and browning trends from global NDVI time-series. *Remote Sens. Environ.* 2011, *115*, 692–702.
22. Roerink, G.J.; Menenti, M.; Verhoef, W. Reconstructing cloudfree NDVI composites using Fourier analysis of time series. *Int. J. Remote Sens.* 2000, *21*, 1911–1917.
23. Li, W.; Luo, C.; Wang, D.; Lei, T. Diurnal variations of precipitation over the South China Sea. *Meteorol. Atmos. Phys.*2010, *109*, 33–46.
24. Justino, F.; Setzer, A.; Bracegirdle, T.J.; Mendes, D.; Grimm, A.; Dechiche, G.; Schaefer, C.E.G.R. Harmonic analysis of climatological temperature over Antarctica: Present day and greenhouse warming perspectives. *Int. J. Climatol.* 2011,*31*, 514–530.
25. Xu, Y.; Shen, Y.; Wu, Z. Spatial and temporal variations of land surface temperature over the Tibetan Plateau based on Harmonic analysis. *Mount. Res. Dev.* 2013, *33*, 85–94.
26. Horn, L.H.; Bryson, R.A. Harmonic analysis of the annual March of precipitation over the United States 1. *Ann. Assoc. Am. Geogr.* 1960, *50*, 157–171.
27. Kirkyla, K.I.; Hameed, S. Harmonic analysis of the seasonal cycle in precipitation over the United States: A comparison between observations and a general circulation model. *J. Clim.* 1989, *2*, 1463–1475.
28. Kadioğlu, M.; Öztürk, N.; Erdun, H.; Şen, Z. On the precipitation climatology of Turkey by harmonic analysis. *Int. J. Climatol.* 1999, *19*, 1717–1728.
29. Tarawneh, Q. Harmonic analysis of precipitation climatology in Saudi Arabia. *Theor. Appl. Climatol.* 2015.
30. Seiler, R. Analyse linearer Trends in der Phänologie multimodaler Vegetation, Auswertung von NDVI Zeitreihen für das Niger Binnendelta (Rep. Mali / Westafrika) (in German). In Proceedings of the 32nd Scientific–Technical Annual Meeting of the DGPF (Deutsche Gesellschaft für Photogrammetrie, Fernerkundung und Geoinformation e.V.), Potsdam, Germany, 14–17 March 2012; Volume 21, pp. 65–74.
31. Seiler, R.; Gloaguen, R. (Non-)linear phenological trends in an ecosystem with multiple growing seasons derived from AVHRR-NDVI time series. In Proceedings of the 2012 IEEE International Geoscience and Remote Sensing Symposium (IGARSS), Munich, Germany, 22–27 July 2012; pp. 6789–6792.
32. Yao, T.; Thompson, L.; Yang, W.; Yu, W.; Gao, Y.; Guo, X.; Yang, X.; Duan, K.; Zhao, H.; Xu, B.; *et al.* Different glacier status with atmospheric circulations in Tibetan Plateau and surroundings. *Nat. Clim. Chang.* 2012, *2*, 663–667.
33. Fuchs, M.C.; Gloaguen, R.; Krbetschek, M.; Szulc, A. Rates of river incision across the main tectonic units of the Pamir identified using optically stimulated luminescence dating of fluvial terraces. *Geomorphology* 2014, *216*, 79–92.
34. Huffman, G.J. Estimates of Root-Mean-Square Random Error for finite samples of estimated precipitation. *J. Appl. Meteorol.* 1997, *36*, 1191–1201.
35. Huffman, G.J.; Adler, R.F.; Arkin, P.; Chang, A.; Ferraro, R.; Gruber, A.; Janowiak, J.; McNab, A.; Rudolf, B.; Schneider, U. The Global Precipitation Climatology Project (GPCP) combined precipitation dataset. *Bull. Am. Meteorol. Soc.* 1997,*78*, 5–20.
36. Huffman, G.J.; Adler, R.F.; Bolvin, D.T.; Gu, G.; Nelkin, E.J.; Bowman, K.P.; Hong, Y.; Stocker, E.F.; Wolff, D.B. The TRMM Multisatellite Precipitation Analysis (TMPA): Quasi-global, multiyear, combined-sensor precipitation estimates at fine scales. *J. Hydrometeorol.* 2007, *8*, 38.
37. Yin, Z.Y. Using a geographic information system to improve Special Sensor Microwave Imager precipitation estimates over the Tibetan Plateau. *J. Geophys. Res.* 2004, *109*, D03110.
38. Kamal-Heikman, S.; Derry, L.A.; Stedinger, J.R.; Duncan, C.C. A simple predictive tool for lower Brahmaputra River Basin monsoon flooding. *Earth Interact.* 2007, *11*, 1–11.
39. Prigent, C. Precipitation retrieval from space: An overview. *Comptes Rendus Geosci.* 2010, *342*, 380–389.
40. Dee, D.P.; Uppala, S.M.; Simmons, A.J.; Berrisford, P.; Poli, P.; Kobayashi, S.; Andrae, U.; Balmaseda, M.A.; Balsamo, G.; Bauer, P.; *et al.* The ERA-Interim reanalysis: Configuration and performance of the data assimilation system. *Q. J. R. Meteorol. Soc.* 2011, *137*, 553–597.
41. National Centers for Environmental Prediction–NOAA, U.S. Department of Commerce, National Weather Service (NWS). NCEP FNL Operational Model Global Tropospheric Analyses, Continuing from July 1999. Available online: http://rda.ucar.edu/datasets/ds083.2/#access (accessed on 10 July 2015).
42. Skamarock, W.C.; Klemp, J.B. A time-split nonhydrostatic atmospheric model for weather research and forecasting applications. *J. Comput. Phys.* 2008, *227*, 3465–3485.
43. Hall, D.K.; Salomonson, V.V.; Riggs, G.A. *MODIS/Terra Snow Cover Monthly L3 Global 0.05Deg CMG, Version 5. [MOD10CM]*; National Snow and Ice Data Center: Boulder, CO, USA, 2006.
44. Dietz, A.; Conrad, C.; Kuenzer, C.; Gesell, G.; Dech, S. Identifying changing snow cover characteristics in Central Asia between 1986 and 2014 from remote sensing data. *Remote Sens.* 2014, *6*, 12752–12775.
45. Hall, D.K.; Salomonson, V.V.; Riggs, G.A. *MODIS/Terra Snow Cover Daily L3 Global 0.05deg CMG V005. [MOD10C1]*; National Snow and Ice Data Center: Boulder, CO, USA, 2006.

46. Landerer, F.W.; Swenson, S.C. Accuracy of scaled GRACE terrestrial water storage estimates. *Water Resour. Res.* 2012,*48*, W04531.

47. Swenson, S.; Wahr, J. Post-processing removal of correlated errors in GRACE data. *Geophys. Res. Lett.* 2006, *33*, L08402.

48. Dahle, C.; Flechtner, F.; Gruber, C.; König, D.; König, R.; Michalak, G.; Neumayer, K.H. *GFZ GRACE Level-2 Processing Standards Document for Level-2 Product Release 0005: Revised Edition*; Technical Report; Deutsches GeoForschungsZentrum GFZ: Potsdam, Germany, 2013.

49. Lettenmaier, D.P.; Famiglietti, J.S. Hydrology: Water from on high. *Nature* 2006, *444*, 562–563.

50. Syed, T.H.; Famiglietti, J.S.; Rodell, M.; Chen, J.; Wilson, C.R. Analysis of terrestrial water storage changes from GRACE and GLDAS. *Water Resour. Res.* 2008.

51. Ramillien, G.; Famiglietti, J.S.; Wahr, J. Detection of continental hydrology and glaciology signals from GRACE: A review. *Surv. Geophys.* 2008, *29*, 361–374.

52. Jacob, T.; Wahr, J.; Pfeffer, W.T.; Swenson, S. Recent contributions of glaciers and ice caps to sea level rise. *Nature* 2012,*482*, 514–8.

53. Guha-Sapir, D.; Below, R.; Hoyois, P. *EM-DAT: International Disaster Database*; Technical Report; Université Catholique de Louvain: Brussels, Belgium, 2014.

54. Beekma, J.; Fiddes, J. Floods and droughts : The Afghan water paradox. In *Afghanistan Human Development Report 2011*; Kabul University (UNDP-Afghanistan): Kabul, Afghanistan, 2011; pp. 1–30.

55. Mahmood, A.; Faisal, N.; Jameel, A. *Special Report on Pakistan's Monsoon 2011 Rainfall*; Technical Report January; Pakistan Meteorological Department: Karachi, Pakistan, 2012.

56. Houze, R.A.; Rasmussen, K.L.; Medina, S.; Brodzik, S.R.; Romatschke, U. Anomalous atmospheric events leading to the summer 2010 floods in Pakistan. *Bull. Am. Meteorol. Soc.* 2011, *92*, 291–298.

57. Kapnick, S.B.; Delworth, T.L.; Ashfaq, M.; Malyshev, S.; Milly, P.C.D. Snowfall less sensitive to warming in Karakoram than in Himalayas due to a unique seasonal cycle. *Nat. Geosci.* 2014, *7*, 834–840.

58. Immerzeel, W.W.; van Beek, L.P.H.; Bierkens, M.F.P. Climate change will affect the Asian water towers. *Science* 2010,*328*, 1382–1385.

59. Harr, R.D. Some characteristics and consequences of snowmelt during rainfall in western Oregon. *J. Hydrol.* 1981, *53*, 277–304.

60. World Health Organization. *Pakistan Floods Situation Report#F-1-2008 (Aug 3–5, 2008)*; Technical Report; World Health Organization–Country Office, National Institute of Health: Islamabad, Pakistan, 2008.

61. Fujita, K.; Ageta, Y. Effect of summer accumulation on glacier mass balance on the Tibetan Plateau revealed by mass-balance model. *J. Glaciol.* 2000, *46*, 244–252.

62. Liu, X.; Herzschuh, U.; Wang, Y.; Kuhn, G.; Yu, Z. Glacier fluctuations of Muztagh Ata and temperature changes during the late Holocene in westernmost Tibetan Plateau, based on glaciolacustrine sediment records. *Geophys. Res. Lett.* 2014, *41*, 6265–6273.

63. Khromova, T.; Osipova, G.; Tsvetkov, D.; Dyurgerov, M.; Barry, R. Changes in glacier extent in the eastern Pamir, Central Asia, determined from historical data and ASTER imagery. *Remote Sens. Environ.* 2006, *102*, 24–32.

64. Gruber, F.E.; Mergili, M. Regional-scale analysis of high-mountain multi-hazard and risk indicators in the Pamir (Tajikistan) with GRASS GIS. *Nat. Hazards Earth Syst. Sci.* 2013, *13*, 2779–2796.

65. Hou, A.Y.; Kakar, R.K.; Neeck, S.; Azarbarzin, A.A.; Kummerow, C.D.; Kojima, M.; Oki, R.; Nakamura, K.; Iguchi, T. The Global Precipitation Measurement Mission. *Bull. Am. Meteorol. Soc.* 2014, *95*, 701–722.

66. R Core Team. *R: A Language and Environment for Statistical Computing*; R Foundation for Statistical Computing: Vienna, Austria, 2014.

67. Hijmans, R.J. raster: Geographic Data Analysis and Modeling, 2015; R Package Version 2.4-15. Available online: http://CRAN.R-project.org/package=raster (accessed on 29 July 2015).

68. Bivand, R.; Keitt, T.; Rowlingson, B. rgdal: Bindings for the Geospatial Data Abstraction Library, 2015; R Package Version 1.0-4. Available online: http://CRAN.R-project.org/package=rgdal (accessed on 29 July 2015).

CHAPTER 15

HYDROLOGICAL EXCITATION OF POLAR MOTION DERIVED FROM GRACE GRAVITY FIELD SOLUTIONS

L. Seoane, J. Nastula, C. Bizouard, and D. Gambis

INTRODUCTION

The excitation of polar motion is, to a large extent, related to the mass redistribution of geophysical fluids. The importance of atmospheric and oceanic angular momentum signals at monthly and seasonal periods is well known. The contribution of the continental hydrological signals, originating from land water, snow, and ice, is, however, less known. A number of previous studies have estimated hydrological excitation from climatological measurements, numerical climate models, and global hydrology models based upon the observed distribution of surface water, snow, ice, and soil moisture [1–5].

The hydrological part of polar motion excitation can also be obtained, as a residual series, by removing atmospheric and oceanic signals from the geodetic excitation of polar motion. The general conclusion of these studies is that the change in continental water storage plays a major role in the seasonal polar motion although the results of the hydrological models do not agree with each other and with observed polar motion [6, 7]. This is mainly due to the lack of global measurements of related hydrological parameters which are difficult to predict (like evaporation, run-off, groundwater, and snow/ice mass change). Other analyses show that hydrological signals seem to exert a substantial influence at interannual scales [7, 8].

Thanks to the gravity recovery and climate experiment (GRACE) mission, the mass redistributions are observed over whole Earth. Atmospheric mass variations and the corresponding oceanic response are largely removed during GRACE data processing, then the obtained time-variable gravity fields reflect hydrological phenomena (i.e., terrestrial water storage variations from the soil water, groundwater, and snow/ice sheets) plus other non-modeled effects like postglacial rebound, earthquakes, and baroclinic oceanic signal. Thus, considering gravity field variations over continents mainly due to water and ice mass change, we have a new approach to evaluate the effect of the continental Earth's hydrology on polar motion. Other studies [9] show that GRACE observations are useful to evaluate oceanic and hydrological polar motion excitations in order to close global mass budget.

Our main objective is to compare the hydrological excitations computed from different data sets to better understand the influence of water mass variations on polar motion. Additionally, GRACE can help to located significant hydrological mismodelling effects. The data sets used in our study are geodetic observations of polar motion (IERS C04 polar motion series), GRACE gravity field solutions provided by the Groupe de Recherche de Géodésie Spatiale (GRGS), the Climate Prediction Center (CPC) [10] hydrological model, and the newer surface modelling system global land data assimilation system (GLDAS) [11].

DATA SETS

Hydrological Models

Latitude-longitude grids providing the charge of continental water by surface unit are predicted by the CPC and the surface model GLDAS.

The CPC uses the Noah land surface model [10] forced by observed precipitation derived from CPC daily and hourly precipitation analyses, downward solar and long-wave radiation, surface pressure, humidity, temperature, and horizontal wind speed from the atmospheric model reanalysis

Citation: L. Seoane, J. Nastula, C. Bizouard, and D. Gambis, "Hydrological Excitation of Polar Motion Derived from GRACE Gravity Field Solutions," International Journal of Geophysics, vol. 2011, Article ID 174396, 10 pages, 2011. doi:10.1155/2011/174396.

of the National Centers for Environmental Prediction (NCEP). The net water storage Δ_q at a grid point is estimated using the conservation equation [12]

$$\Delta q = P - E - R, \tag{1}$$

where P, E, and R are precipitation, evapotranspiration, and run-off. CPC water storage grids are estimated from January 1948 to December 2007 with a monthly resolution. The GLDAS generates optimal fields of land surface states and fluxes by integrating ground- and satellite-based observational data products, using advanced land surface modelling and data assimilation techniques [11]. GLDAS has resulted in a massive archive of modelled and observed global surface meteorological data, parameter maps, and output which includes 1 and 0.25 degrees resolution of the 1979 to present simulations of the Noah, community land (CLM), mosaic, and variable infiltration capacity (VIC) land surface models [13]. In this case, water storage is the sum of soil moisture, snow water equivalent, and canopy surface water not counting changes in groundwater below the depth defined by the model [12], that is, variations below 2 meters depth. Hydrological excitation functions can be computed from the monthly fields of global water storage from January 1979 to February 2008.

Gravimetric Data

The GRACE time-variable gravity field solutions of the GRGS are provided at 10 day intervals from July 29 2002 to May 27 2008 in the last release RL02 [14]. They use a different data processing strategy than those applied by the other analysis centers (CSR, GFZ, and JPL). The GRGS gravity field solution is computed by a combination of GRACE and LAGEOS-1 and 2 observations. LAGEOS data provide over 90% of the information on the (2,0) degree coefficients [14]. GRACE data provide nearly 100% of the information on all other harmonic coefficients. A different background model is used mainly concerning the nontidal oceanic model. The GRGS uses a barotropic oceanic model Modele 2D d'Ondes de Gravit ` e (MOG2D). Lastly, it is ´ known that the geoid's height difference plots show a northsouth streaking due to systematic errors in higher degree coefficients. To avoid this problem, GRGS constrained the higher degrees coefficients gradually to those of the static solution [15]. The other centers (CSR, GFZ, and JPL) use instead a filtering procedure for avoiding this problem [16]. Usually, the excitation based on gravimetric observations is derived from the C_{21} and S_{21} coefficients of the gravity field solution using the classical methodology (see, e.g., [17]). However, these coefficients contain hydrological variations and also residuals over oceans, in our case, this is a baroclinic signal. In this paper, we propose to estimate the hydrological GRACE-based excitation in a different way employing equivalent water height maps (EWH) restricted to continents in order to exclude oceanic effects. In fact, GRGS provides converted maps into meters of EWH from which we estimate the terrestrial water storage grids by Δ_q = EWH $*\rho$ where $\rho = 1000$ kg m^{-3} is the water density.

Geodetic Residuals

The International Earth Rotation and Reference Systems Service (IERS) provides a combined time series C04 of the Earth Orientation Parameters (EOP) at daily intervals [18, 19], in particular the pole coordinates x and y. The geodetic polar motion excitation in polar motion can be written as

$$\chi = \chi_1 + i\chi_2 = p + i\frac{\dot{p}}{\sigma_c}, \tag{2}$$

where $p = x - iy$ is the complex pole coordinate obtained from C04 series and σ_c is the frequency of the Chandler pulsation ($2\pi/433$ rad/days) with an adopted quality factor of 175 [20]. We derived the geodetic excitation according to Wilson [21]. The hydrological influences are obtained by removing atmospheric and oceanic signals from the geodetic excitation. For computing these geodetic residuals, we use the atmospheric excitation series based on the NCEP/NCAR reanalysis [22] and the oceanic excitation series derived from the ECCO model [23]. In order to match the temporal spectral consistency with GRACE gravity field solutions, geodetic, atmospheric, and oceanic excitations are smoothed using Vondrak filter [24, 25] and interpolated every 10 days.

ANALYSIS AND RESULTS

Hydrological Polar Motion Excitations

According to Chen and Wilson [7] formulation, by integrating the grids of water thickness of the CPC and GLDAS models over continents (Greenland and Antarctica are excluded), we have reconstituted the hydrological excitation function reduced to its mass term as follows:

$$\begin{bmatrix}\chi_1\\ \chi_2\end{bmatrix} = \frac{-1.098R_e^4}{C-A}\iint \Delta q\,\sin(\theta)\cos^2(\theta)\begin{pmatrix}\cos\lambda\\ \sin\lambda\end{pmatrix}d\theta d\lambda, \tag{3}$$

where R is the Earth's mean radius, C and A are the Earth's principal moments of inertia, and (θ, λ) is the latitude and longitude.

The factor 1.098 accounts the combined effects of the yielding of the solid Earth to surface load, core-mantle decoupling, and rotational deformation [26]. The series are interpolated to the GRACE temporal resolution. In the same way, that is, integrating over continents the terrestrial water storage derived from GRACE (3), we obtain another approach of the hydrological excitation. From November 2002 to February 2003 are the main GRACE data gaps, and so, our study period is restricted to March 2003 until December 2007, which is the last common date with the hydrological models time span. Figure 1 shows the estimates of the hydrological excitation (GRACE grids were integrated over continents) and the oceanic residuals (GRACE grids were integrated over oceans only). We can notice significant oceanic variations that we will not consider in our study. Spatial spectral leakage due to using a [0 or 1] mask is expected to be negligible over large regions. Figure 2 shows the excitation functions obtained from residual geodetic observations, GRACE gravity field solutions, and models estimates. Means and linear trends are removed from all data sets. As expected, we observe that the hydrological signals are dominated by the annual fluctuations. GRACE-based excitations are in agreement with the residual geodetic excitations mainly at annual scales, and both show variations between ±20. Hydrological models underestimate variations, probably because some mismodeled process, except for the χ_2 component derived from CPC.

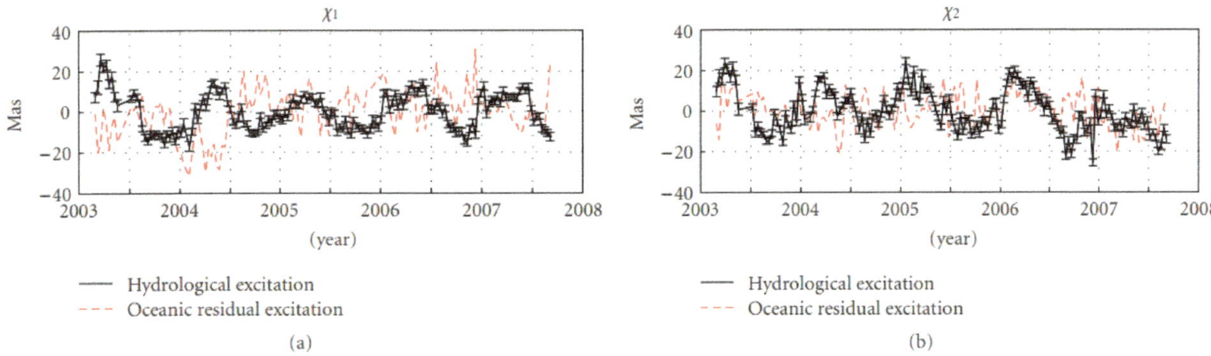

Figure 1: Oceanic residual excitation (mask over continents) and hydrological excitation (mask over oceans) estimated by integrating GRACE equivalent water height grids. We observed a significant oceanic residual signal. Bars represent the calibrated errors provided by the GRGS.

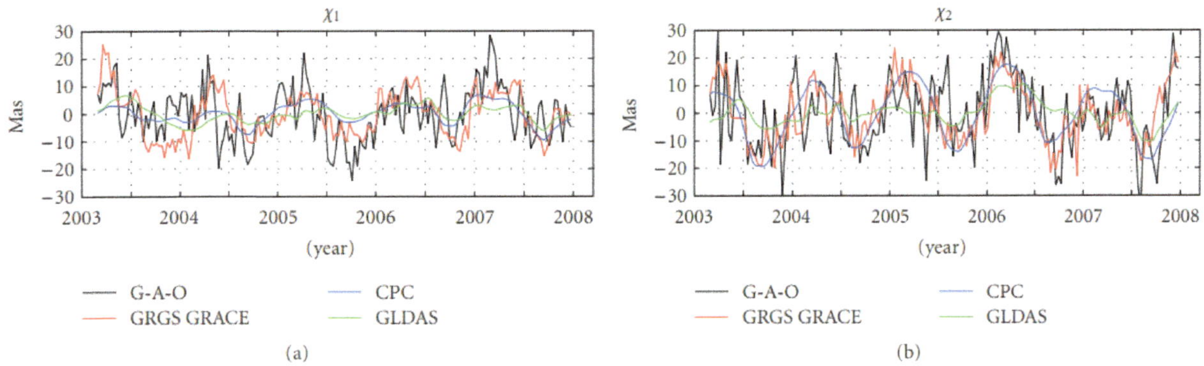

Figure 2: Hydrological excitation functions computed from geodetic residuals (G-A-O), GRACE solutions, and CPC and GLDAS hydrological models. We observe that GRACE-based excitations are well correlated with geodetic residual. We note the large differences between the hydrological models especially in the case of χ_2.

Seasonal Variations

Seasonal fluctuations are estimated by least-square fitting (Table 1 and corresponding phaser plots Figure 3). The annual variations of GRACE-based and the residual geodetic excitations are in good agreement for the both components χ_1 and χ_2. Phase differences are more important in the case of χ_1 but not large (23° versus 345°); however, CPC model is closer in phase (23° versus 3°). The χ_1 is more sensitive to the ocean mass redistributions and the uncertainties of the ocean models used for estimating residual geodetic series can probably be a cause of disagreement. To the contrary, χ_2 is affected by the Earth hydrology due to the location of main continents [7]. Whereas, considering the hydrological models predictions with respect to GRACE observations and geodetic residual, there are larger discrepancies in amplitude and the modeled annual signals are generally underestimated. Except for χ_2 computed from the CPC model, which has an amplitude of nearly 3 mas larger than the GRACE estimate.

Table 1: Amplitude and phase of seasonal variations estimated from geodetic residuals (G-A-O), GRACE, CPC, and GLDAS-based excitations. For the annual variations, we note that GRACE-based excitation reach a better agreement in amplitude with geodetic residuals than the models[a]

	Annual		Semiannual	
	Amplitude mas	Phase degree	Amplitude mas	Phase degree
χ_1 G-A-O	7.9 ± 1.1	23 ± 8	1.3 ± 1.2	319 ± 54
χ_1 CPC	3.9 ± 0.3	3 ± 4	1.2 ± 0.3	125 ± 16
χ_1 GLDAS	2.2 ± 0.3	311 ± 9	1.1 ± 0.3	120 ± 18
χ_1 GRACE	9.0 ± 0.7	345 ± 4	1.2 ± 0.6	180 ± 27
χ_2 G-A-O	8.6 ± 1.5	20 ± 10	4.2 ± 1.2	90 ± 16
χ_2 CPC	13.0 ± 0.4	18 ± 2	1.9 ± 0.5	156 ± 13
χ_2 GLDAS	2.6 ± 0.4	2 ± 10	1.7 ± 0.6	129 ± 19
χ_2 GRACE	10.0 ± 1.1	27 ± 6	1.6 ± 0.9	174 ± 32

Phases agree quite well. In conclusion, the annual hydrological budget estimated by GRACE is more consistent with the geodetic observations than models predictions. In the case of the semiannual oscillations, they have lower amplitudes (up to 4 mas). It seems that there is a reasonably agreement between models predictions and GRACE observations. Phasor diagram (Figure 3) confirms that GRACE-based excitation is in better agreement at geodetic residuals (phases are similar and amplitude difference is about 2 mas). However in the case of the semiannual signal, GRACE and models-based excitation are in agreement, but geodetic residuals are large different probably due to atmospheric or oceanic models errors at this frequency.

Regional Hydrology Contributions at Annual Scale

In order to find the Earth's regions, where models CPC and GLDAS do not predict consistent annual variations with respect to GRACE observations, the amplitudes and phases for every grid point are computed by least-square fitting (Figure 4). In Figure 4, we note that the prominent annual signals are due to the monsoonal climates [27] situated at latitudes smaller than 30∘N. The monsoon regions are located in the Amazon, Central Africa, South Africa, North Australia, India, and Indochina. We observe that CPC model provides stronger amplitudes than the GLDAS model. The groundwater storage not counting in the GLDAS model seems to be the cause of the lower amplitudes over major river basins [12]. In fact, GLDAS model is defined with a surface layer of 2 m depth, then groundwater variations below are not taken into account. So, what is the contribution of the monsoonal climate regions to the polar motion excitation? To answer this question, we compute the hydrological excitations restricting the integration of the water storage grids of CPC, GLDAS, and GRACE (see (3)) over those latitudes smaller than 30∘N. Results are shown in Figure 5. As expected, annual oscillations are clearly dominant, and there is a very good agreement between the GRACE observations and models. Generally, the contribution of this Earth domain does not explain the hydrological annual budget found in Table 1, thus influences of the north high latitudes regions must not be neglected. At these latitudes (latitudes ≥ 30°N), we find the main areas of snow cover and continental climates. We notice the large discrepancies between the hydrological excitations computed from models and GRACE (Figure 5), especially concerning χ_2, come from this region, where models snow water estimations seems to be inaccurate.

Interannual Polar Motion Excitations

The hydrology plays an important role at low frequencies. Chen and Wilson [7] have shown significant interannual variations which seem to be of hydrological nature. From wavelet analysis of the excitation functions derived from the geodetic residuals and the hydrological model predictions, Fernandez et al. [′ 8] have found strong energy values around 4 years in the retrograde band. For obtaining interannual variations, we remove seasonal signals (annual and semiannual) from geodetic, CPC, GLDAS, and GRACE-based hydrological excitations. Afterward, we applied a low pass digital filter of cutoff frequency 1 year to the residuals. The interannual signals are shown in Figure 6. For the χ_2 component, models and GRACE-based excitations are well correlated with the geodetic residuals (Table 2). Future longer series of GRACE observations and improvements in the data processing can help to gain a greater understanding of the hydrology influences on polar motion at interannual scales.

Intraseasonal Polar Motion Excitations

Various studies note the important influence of the atmosphere and oceans on polar motion excitations at interseasonal scales [22, 28– 31]. Gross et al. [32] have observed that a third of the geodetic excitation was not explained by atmospheric and oceanic phenomenon. They attribute this discrepancy to the uncertainties in the data sets and also to the effect of other excitation process not considered like the continental hydrology. However, models do not predict enough power at those frequencies. To obtain the intraseasonal variations we removed from all series the annual, semiannual signals, and low frequency fluctuations. The latter are estimated in the previous section.

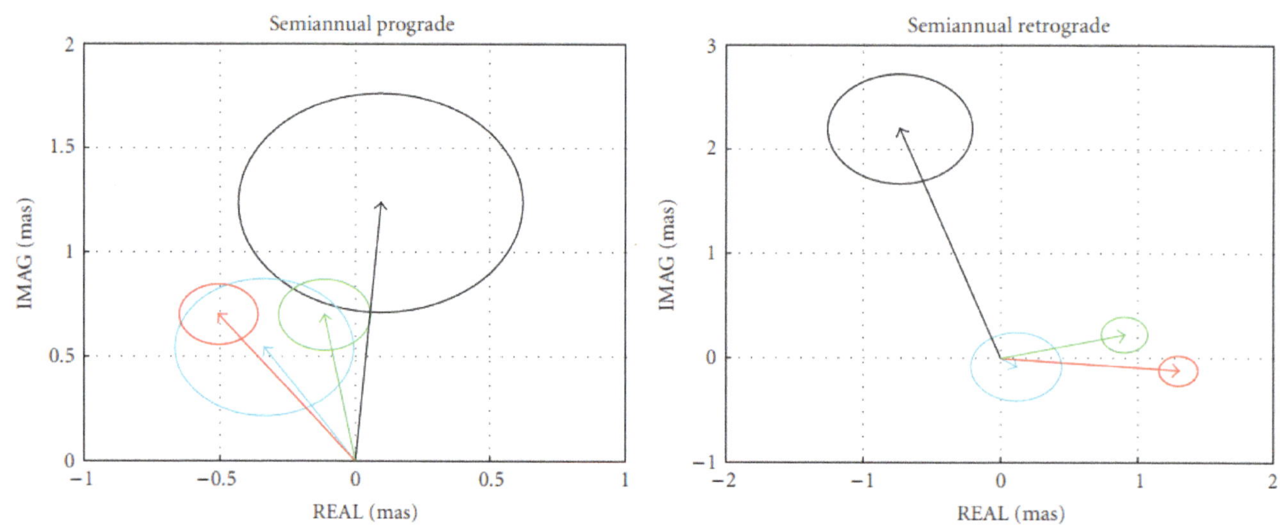

(a) (b)

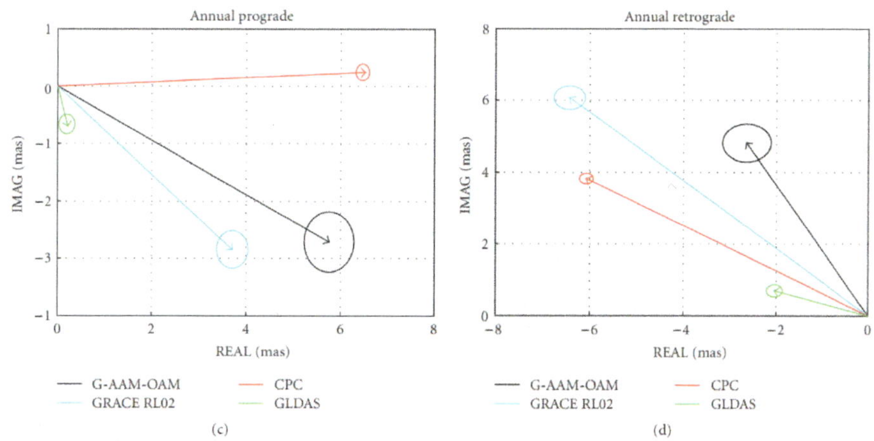

Figure 3: Phasor diagram of seasonal variations. Circles represent the formal errors.

Table 2: Correlations between the polar motion interannual variations estimated from geodetic residuals, GRACE solutions, and CPC and GLDAS hydrological models. Correlation values are better for χ_2 than χ_1. This discrepancy is possibly caused by errors of the oceanic model removed from the geodetic observations[a]

	Percentage of explained variance		Correlation			
					$\chi_1 + i\chi_2$	
	χ_1	χ_2	χ_1	χ_2	Magnitude	Phase degree
G-A-O/GRACE	−7	40	0.3	0.7	0.6	−34
G-A-O/CPC	9	37	0.3	0.6	0.6	−25
G-A-O/GLDAS	11	45	0.3	0.7	0.6	−22
GRACE/CPC	−22	34	0.2	0.6	0.6	27
GRACE/GLDAS	37	4	0.6	0.4	0.5	23

[a]According to the Student-*t* test, the 90% significance level of correlation is 0.17. The standard error of the difference between two values of correlation is 0.19. G-A-O: geodetic residuals (geodetic-atmospheric-oceanic excitations), GRACE: GRACE solutions of GRGS, CPC, and GLDAS: hydrological excitations.

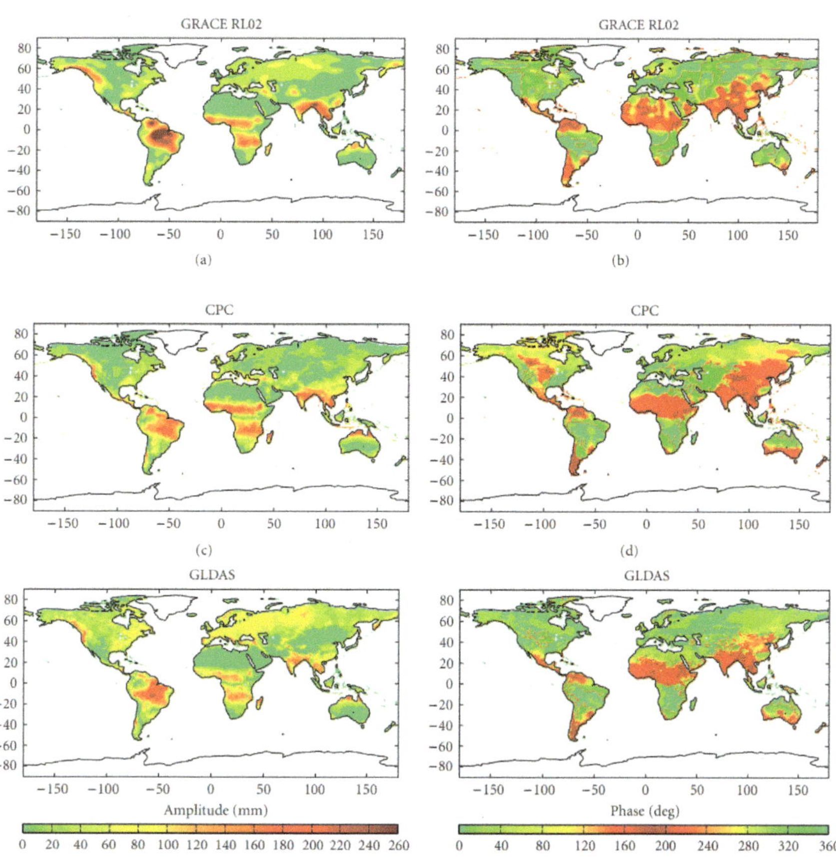

Figure 4: Amplitudes (left column) and phases (right column) of annual oscillations at every grid point computed from GRACE solutions and

CPC and GLDAS models. The reference date for the phase is 2004 January 1, 0h UTC. The prominent annual signal are located in the monsoon regions.

Figure 7 shows the spectrum of the different hydrological excitation data sets. GRACE observations provide more power than hydrological models, especially in the prograde and retrograde bands around the frequency of 3 cycles per year. At those frequencies, the power of the hydrological excitation derived from GRACE is comparable to those derived from geodetic residual and the hydrology seems to be a nonnegligible process of intraseasonal polar motion excitation.

CONCLUSION

This study explores the hydrological excitation of the polar motion computed from GRACE gravity field observations. So far, the excitation produced by the water storage variations was estimated using advanced hydrological models. These models include algorithms to assimilate observations of terrestrial hydrology. However, the main causes of errors are the lack of the global measurements of related hydrological parameters. The GRACE mission provides continental water change with unprecedented accuracy. After removing the atmospheric and oceanic effect from geodetic observations, we compare these residual signals to the excitations computed by GRACE estimates. At the annual timescale, we find that the GRACE-based excitations are in better agreement with geodetic residuals than the excitations predicted by the hydrological models. The prominent annual excitation is mainly due to the monsoon regions located at latitudes smaller than 30°N, where GRACE and models are in a very good agreement, in particular for CPC. In fact, we note that the groundwater variations unaccounted in the GLDAS model can have significant influences. Moreover, the annual signal observed by GRACE and obtained by the models at the high latitudes regions show important discrepancies probably due to the inaccurate snow water and ice estimations.

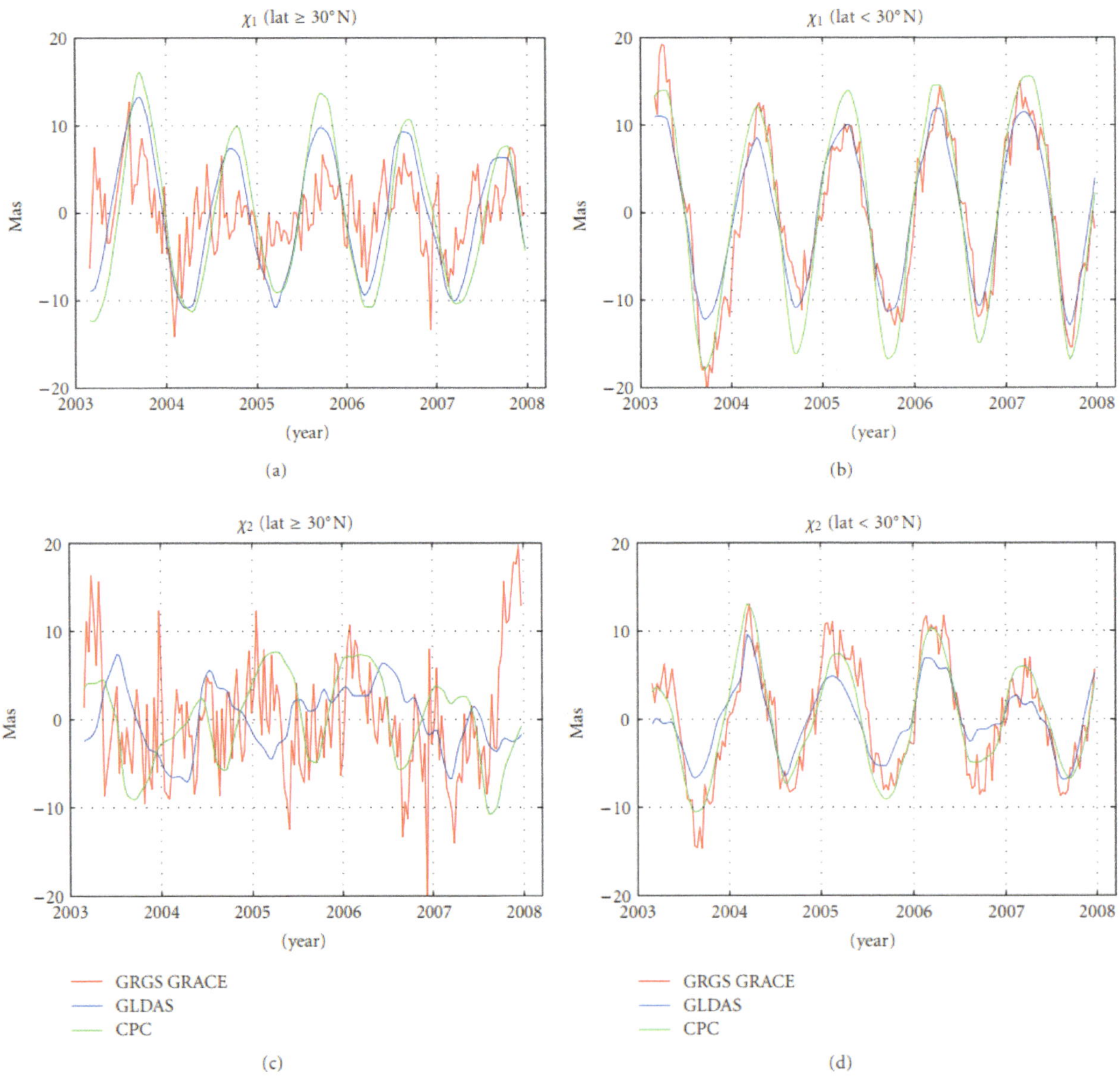

Figure 5: Polar motion excitations integrated over the Earth's north high latitudes (latitudes ≥ 30°N) and over the other latitudes (latitudes < 30°N) using GRACE solutions and CPC and GLDAS models. Larger discrepancies between the models and GRACE observations come from the north high latitudes estimations.

We find also considerable interannual variations which confirm that the continental hydrology exerts a significant influence on the variations at low frequencies. Even so, the models or the GRACE solutions cannot explain the fluctuations of the geodetic residuals especially concerning the χ_1 component. Finally, from the spectral analysis of interseasonal variations, GRACE observations show that continental hydrology can excite polar motion at frequencies of about 3 cpy. In conclusion, mass water variations of terrestrial water storage components (like groundwater or ice/snow sheets) are very difficult to model precisely. The mismodelling signals have a significant impact in the determination of hydrological influence on polar motion. Still, terrestrial water storage variations are observed by the GRACE mission, and it could be useful to constrain the models. We can expect that data processing improvements and longer series of GRACE observations will help in the understanding of polar motion excitation caused by the Earth's hydrology and closing the global mass budget.

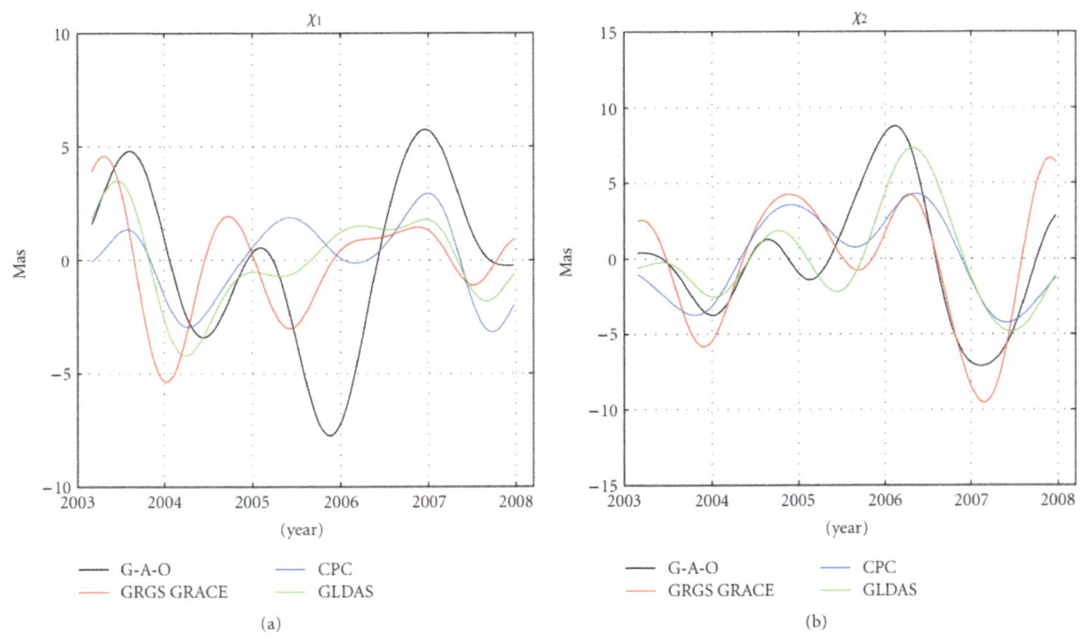

Figure 6: Interannual variations derived from the geodetic residual (G-A-O), GRACE, CPC, and GLDAS-based excitations. We observed a reasonably agreement for the χ_2 term.

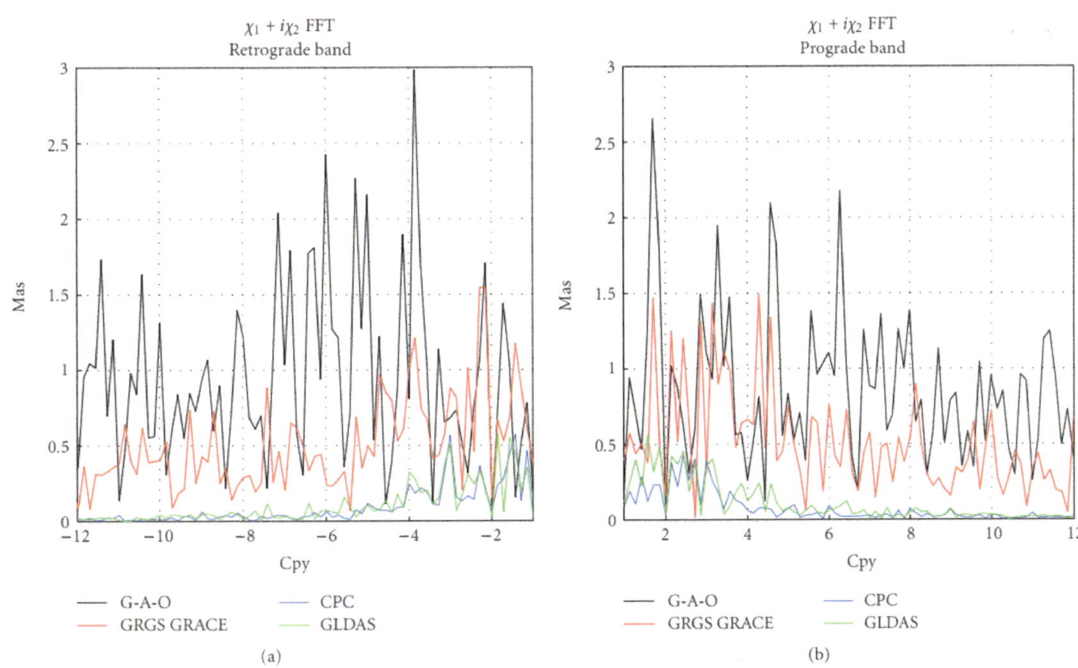

Figure 7: Intraseasonal variations spectrum computed from the geodetic residuals (G-A-O), GRACE, CPC, and GLDAS-based excitations. Data sets are interpolated with an uniform sampling of 10-days. GRACE observations provide more power than hydrological models, especially in the prograde and retrograde bands around the frequency of 3 cycles per year.

ACKNOWLEDGMENTS

The authors are grateful to Jean Michel Lemoine for providing different GRGS data sets and David Salstein for his suggestions. During her stay in Paris, J. Nastula was sponsored by the Observatoire de Paris. J. Nastula acknowledges the support of Project no. N526 140 735 from the Polish Ministry of Science and Higher Education.

REFERENCES

1. T. E. A. Van Hycklama, Water Balance and Earth Unbalance, vol. 92, IASH AISH, 1970.
2. L. A. Hinnov and C. R. Wilson, "An estimate of the water storage contribution to the excitation of polar motion," Geophysical Journal Royal Astronomical Society, vol. 88, no. 2, pp. 437–459, 1987. ·
3. B. F. Chao and W. P. O'Connor, "Global surface-water-induced seasonal variations in the Earth's rotation and gravitational field," Geophysical Journal, vol. 94, no. 2, pp. 263–270, 1988. ·
4. J. Kuehne and C. R. Wilson, "Terrestrial water storage and polar motion," Journal of Geophysical Research, vol. 96, no. 3, pp. 4337–4345, 1991. ·
5. J. L. Chen, C. R. Wilson, B. F. Chao, C. K. Shum, and B. D. Tapley, "Hydrological and oceanic excitations to polar motion and length-of-day variation," Geophysical Journal International, vol. 141, no. 1, pp. 149–156, 2000. · ·
6. J. Nastula, B. Kolaczek, and W. Popiński, "Comparisons of hydrological angular momentum (HAM) of different models," in Proceedings of the Journées, A. Brzeziński, N. Capitaine, and B. Kolaczek, Eds., pp. 207–210, 2005.
7. J. L. Chen and C. R. Wilson, "Hydrological excitations of polar motion, 1993–2002," Geophysical Journal International, vol. 160, no. 3, pp. 833–839, 2005. · ·
8. L. I. Fernández, H. Schuh, M. Schmidt, and F. Seitz, "Effects of inter-annual water storage variations on polar motion," Geophysical Journal International, vol. 169, no. 1, pp. 12–18, 2007. · ·
9. S. Jin, D. P. Chambers, and B. D. Tapley, "Hydrological and oceanic effects on polar motion from GRACE and models," Journal of Geophysical Research B, vol. 115, no. 2, Article ID B02403, 2010. · ·
10. Y. Fan, H. van den Dool, K. Mitchel, and D. Lohmann, "A 51-year reanalysis of the US land-surface hydrology," GEWEX News, vol. 13, no. 2, pp. 6–10, 2003.
11. M. Rodell, P. R. Houser, U. Jambor et al., "The global land data assimilation system," Bulletin of the American Meteorological Society, vol. 85, no. 3, pp. 381–394, 2004. · ·
12. J. L. Chen, M. Rodell, C. R. Wilson, and J. S. Famiglietti, "Low degree spherical harmonics influences on Gravity Field Recovery and Climate Experiment (GRACE) water storage," Geophysical Research Letters, vol. 32, p. L14405, 2005.
13. H. Fang, "Readme document for Global Data Assimilation System (GLDAS) products," 2008.
14. R. Biancale, J. M. Lemoine, G. Balmino, et al., 6 years of gravity variations from GRACE and LAGEOS data at 10-day intervals over the period from July 29th 2002 to November 2008, http://bgi.cnes.fr:8110/geoid-variations/README.html.
15. J. M. Lemoine, S. Bruinsma, S. Loyer et al., "Temporal gravity field models inferred from GRACE data,"Advances in Space Research, vol. 39, no. 10, pp. 1620–1629, 2007. · ·
16. S. Swenson and J. Wahr, "Post-processing removal of correlated errors in GRACE data," in Geophysical Research Letters, vol. 33, L08402, 2006.
17. J. L. Chen, C. R. Wilson, B. D. Tapley, and J. C. Ries, "Low degree gravitational changes from GRACE: validation and interpretation," Geophysical Research Letters, vol. 31, no. 22, Article ID L22607, 5 pages, 2004. · ·
18. D. Gambis, "Monitoring Earth orientation using space-geodetic techniques: state-of-the-art and prospective," Journal of Geodesy, vol. 78, no. 4-5, pp. 295–303, 2004. · ·
19. C. Bizouard and D. Gambis, "The combined solution C04 for Earth Orientation Parameters consistent with International Terrestrial Reference Frame 2005," Technical Note of IERS Earth Orientation Center, 2007.
20. C. R. Wilson and R. O. Vicente, "Maximum likelihood estimates of polar motion parameters," inVariations in Earth Rotation, D. D. McCarthy and W. Carter, Eds., vol. 53 of Geophysical Monograph Series, pp. 151–155, AGU, Washington, DC, USA, 1990.
21. C. R. Wilson, "Discrete polar motion equations," Geophysical Journal Royal Astronomical Society, vol. 80, no. 2, pp. 551–554, 1985. ·
22. D. A. Salstein, D. M. Kann, A. J. Miller, and R. D. Rosen, "The sub-bureau for atmospheric angular momentum of the International Earth Rotation Service: a meteorological data center with geodetic applications," Bulletin of the American Meteorological Society, vol. 74, no. 1, pp. 67–80, 1993. ·
23. R. S. Gross, "An improved empirical model for the effect of long-period ocean tides on polar motion,"Journal of Geodesy, vol. 83, no. 7, pp. 635–644, 2009. · ·
24. J. Vondrak, "A contribution to the problem of smoothing observational data," Bulletin of the Astronomical Institutes of Czechoslovakia, vol. 20, p. 349, 1969.
25. J. Vondrak, "Problem of smoothing observational data II," Bulletin of the Astronomical Institutes of Czechoslovakia, vol. 28, pp. 84–89, 1977.
26. T. M. Eubanks, "Variations in the orientation ofthe Earth," in Contributions of Space Geodesy to Geodynamic: Earth Dynamics, D. Smith and D. Turcotte, Eds., vol. 24 of Geodynamics Series, pp. 1–54, AGU, Washington, DC, USA, 1993.
27. Y. Fan and H. van den Dool, "Climate Prediction Center global monthly soil moisture data set at 0.5° resolution for 1948 to present," Journal of Geophysical Research D, vol. 109, no. 10, p. D10102, 2004. · ·
28. B. Kolaczek, W. Kosek, and H. Schuh, "Short-period

oscillations of Earth rotation," in Polar Motion: Historical and Scientific Problems, S. Dick, D. McCarthy, and B. Luzum, Eds., vol. 178 of IAU Colloquia, pp. 533–544, Astronomical Society of the Pacific, San Francisco, Calif, USA, 2000.

29. B. Kołaczek, M. Nuzhdina, J. Nastula, and W. Kosek, "El Niño impact on atmospheric polar motion excitation," Journal of Geophysical Research B, vol. 105, no. 2, pp. 3081–3087, 2000. ·

30. R. M. Ponte, D. Stammer, and J. Marshall, "Oceanic signals in observed motions of the Earth's pole of rotation," Nature, vol. 391, no. 6666, pp. 476–479, 1998. · ·

31. J. Nastula, R. M. Ponte, and D. A. Salstein, "Regional signals in atmospheric and oceanic excitation of polar motion," in Proceedings of the 178th IAU Colloquium, Polar Motion: Historical and Scientific Problems, pp. 463–472, Cagliari Sardinia, Italy, 2000.

32. R. S. Gross, I. Fukumori, and D. Menemenlis, "Atmospheric and oceanic excitation of the Earth's wobbles during 1980–2000," Journal of Geophysical Research B, vol. 108, no. 8, pp. 6–16, 2003.

CHAPTER 16

SPATIAL EVALUATION OF HEAVY METALS CONCENTRATIONS IN THE SURFACE SEDIMENT

Yong Niu, Wei Jiao, Hui Yu, Yuan Niu, Yong Pang, Xiangyang Xu and Xiaochun Guo

Keywords: Heavy metals, sediment, spatial distribution, toxic unit, Taihu Lake

INTRODUCTION

Heavy metals in the water reservoirs affect environmental quality by accumulating in reservoirs, resulting in serious human health hazards and significant ecological effects throughout the food chain [1,2]. Most heavy metals are initially absorbed by the suspended particles in the water, after which they undergo a series of processes and are then deposited into the sediment [3,4]. When external environmental factors such as pH, Eh, or temperature change, metals in the sediment can be released back into the water column causing secondary pollution [2]. Thus, it is essential to attach great importance to the contamination of heavy metals during the strategy making of water management.

Due to the fact that the levels of heavy metals in aquatic environment are not the same in different seasons or other environmental factors, the contamination status cannot be indicated by heavy metals all the time [3]. As the sediment cannot only carry but also absorb different kinds of pollutants in water, they can be regarded as a helpful indicator of contamination [4]. Nevertheless, because of some certain disturbances, it is still possible for the heavy metals to be released back to the upper water, which may lead to the serious destruction of the ecological systems [5]. Therefore, analyzing and evaluating the sediment is of great significance in assessing the contamination status of heavy metals in water ecological environment. At present, distinct organizations in both federal and provincial levels have put forward the numerical sediment quality guidelines (SQGs for short), which is conducive to the evaluation of environmental threats caused by heavy metals [6].

It was reported that among 27 lakes monitored in China, almost 60% of them have been polluted by anthropogenic activities and could not have met the requirements of daily life and agricultural irrigation [7]. Taihu Lake is one of the lakes to which great importance has been attached by the Chinese government in the control of water contamination, and its serious eutrophication is well known to the public [8,9]. Nevertheless, with the constant increase of population and rapid growth of economy in its surrounding regions, this lake is also faced with a serious problem of heavy metal contamination. It can be seen from the research of Fu *et al.* [10] that a much higher degree of accumulation of heavy metals has been seen in fishes from the Taihu Lake compared with those from the Yangtze River, which will exert negative influence on human health. Furthermore, in order to further supervise the contamination in Lake Taihu basin, an extensive analysis has been carried out in the sampling gathered from Taihu Lake as well as its significant tributaries [11,12,13]. Nevertheless, these investigations focused primarily on natural waters and were generally focused on the small scale. Fewer studies have paid attention to how heavy metals are distributed in sediment and the evaluation of threats in the scale of the entire lake.

Thus, following major targets are set for this research: firstly, it aims to examine how Pb along with Cd, Cu, Zn, Cr and Ni distributes in the overlying deposits of Lake Taihu; secondly, it aims to find out the source of heavy metal contamination through multivariate analysis; and thirdly, it aims to evaluate the environmental threats of these heavy metals based on the SQGs and the total amount of toxic units, thereby enabling ranking and prioritization of sites and metals of concern.

Citation: Niu, Y.; Jiao, W.; Yu, H.; Niu, Y.; Pang, Y.; Xu, X.; Guo, X. Spatial Evaluation of Heavy Metals Concentrations in the Surface Sediment. Int. J. Environ. Res. Public Health 2015, 12, 15028-15039. doi:10.3390/ijerph121214966.

MATERIALS AND METHODS

Description of Research Area

With regard to the size of China's freshwater lakes, Taihu Lake, situated in lower Yangtze River Delta between 30°56″–31°33′ N and 119°53′–120°36″ E as shown in Figure 1, ranks third. Despite of its total surface area of 2338 km^2, Taihu Lake has a mean water depth of only 1.9 m [14]. At the same time, this lake is regarded as a vital source of drinking water for a few of cities like Shanghai as well as Suzhou and in general the feature of Taihu Lake basin can be concluded as a great plain of river networks because about 200 rivers as well as tributaries flow into it. Due to the fact that the surrounding areas of Taihu Lake experience a rapid economic growth, serious water contamination has been experienced in this lake from the 1980s [8]. Nowadays, according to the Chinese Environmental Quality Standards for Surface Water (GB3838-2002), water in Taihu Lake is lower than the standard of Class V.

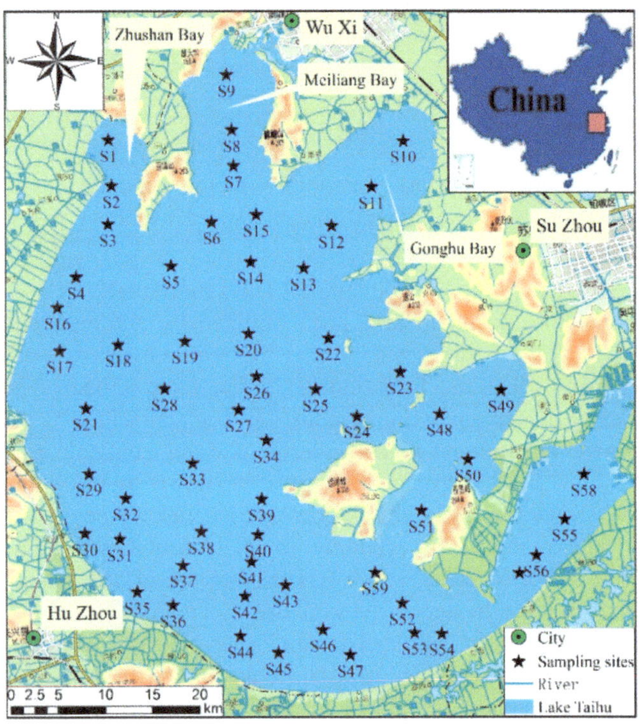

Figure 1: Location of sampling sites in the Lake Taihu, Jiangsu Province, China.

Collection and Analysis of Samples

In August of 2013, researchers gathered 59 samples of surface sediment at random from the entire Taihu Lake as shown inFigure 1. At every place where the samples were obtained, a Petersen grab of stainless steel material was used to take and intensively mix three sub-samples for purpose of acquiring the composite sample. Then the samples were taken to the laboratory where they experienced air drying at ambient temperature and grind with the application of petal and mortar so that a nylon filter of 0.149 mm could be passed through by them. In order to investigate the concentration of Pb, Cd, Cu, Zn, Cr as well as Ni, the approach of HNO_3–HF–$HClO_4$ as well as inductively coupled plasma-atomic emission spectroscopy (ICP-AES) was applied to the digestion and determination of the samples that had been dried. In addition, an automatic analyzer (Shimadzu TOC 5000) was adopted to identify the concentration of total organic carbon and the reference material GBW-07402 was also used to assess the quality of analytical data at the same time. Consequently, great consistency was achieved between the values that have been measured and certified, while the recoveries were in the range between 96.74% and 105.22%, Pb 96.74%, Cd 105.22%, Cu 99.21%, Zn 98.46%, Ni 103.16%, Cr 97.74% individually.

Sediment Quality Guidelines

Sediment quality guidelines (SQGs) have been developed by various federal and provincial agencies in North America for both freshwater and marine ecosystems, which are very useful to screen sediment contamination by comparing sediment contaminant concentration with the corresponding quality guidelines [10]. These guidelines assess the degree to which the sediment chemical status might adversely affect aquatic organisms and are designed to assist the interpretation of sediment quality. With regard to the assessment of environmental threats resulted from heavy metals in sediment, two sets of SQGs were adopted, namely the threshold effect level (TEL)/probable effect level (PEL) and the effect range low (ERL)/effect range median (ERM) values [15,16]. ERLs and TELs are categorized into the low range values referring to the concentrations below which adverse effects upon sediment dwelling fauna will be infrequently. Nevertheless, ERMs along with PELs stand for the concentrations above which negative influences are likely to occur [15]. Besides, toxic units (TU) were also calculated to normalize the toxicities caused by various heavy metals, which allow the comparison of their relative effects and can be defined as the ratio of the determined concentration (C_i) to PEL value (P_i), Equation (1). The sum toxic units ($\sum TU$) are sum of TU_i [17].

$$TU_i = C_i/P_i \quad (1)$$

Spatial and Statistical Analyses

When it comes to the areas where samples were not gathered, it is recognized that Kriging is an effective method of spatial interpolation and forecast [18]. Based on the results of 59 sampling sites, Ordinary Kriging integrated into a Geographical Information System (GIS) was used to obtain the spatial distributions of the potential ecological risk values of heavy metals. On contour maps, distinctly darker shades were used to indicate zones of higher values and lighter shades were used for lower values zones. At the same time, the connection among six heavy metals in the sediment was examined with the application of the Pearson correlation coefficient. In the end, in order to find out probable sources of contamination, factor analysis was carried out through the assessment of major components and calculation of eigenvectors. Besides, the SPSS 17.0 for Windows (Inc. Chicago: Chicago, IL, USA) was applied to all of the statistical analyses conducted in this research.

RESULTS AND DISCUSSION

Average Concentrations of Heavy Metals in Sediment

Table 1 presents the concentrations of six heavy metals including Pb, Cd, Cu, Zn, Cr as well as Ni in the sediment of Taihu Lake. Generally speaking, the concentrations of the above heavy metals fluctuate in a wide range, which can be seen from the fact that Pb, Cd, Cu, Zn, Cr and Ni have the maximum concentrations which are 3.40, 7.64, 3.31, 4.91, 11.61 and 7.25 times larger than their minimum concentrations, respectively. It can be concluded from this result that the concentrations of heavy metals are very different in terms of different areas in the sediment of lakes. The concentrations of heavy metal in our study was in order of Zn, Cr, Ni, Pb, Cu and Cd, which was in agreement with the related researches of heavy metals in sediment from Lake Taihu (Table 1). Moreover, all concentrations of Cd, Pb, Zn, Cu and Ni in the sediment were increased and their average concentrations were far greater than the values in basin background. Furthermore, it can be seen from the findings that the average concentration of Pb along with Cd and Cr were greater than the values of TEL. Nevertheless, they were not higher than the values of PEL or ERM.

Table 1: Statistical description of heavy metal concentrations in the surface sediments of Lake Taihu and sediment guideline values (mg/kg)

	Pb	Cd	Cu	Zn	Cr	Ni
Minimum	16.76	0.21	16.1	41.61	12.08	10.87
Maximum	57.09	1.61	53.24	204.37	140.22	78.83
Average	29.87	0.74	28.27	79.74	72.11	34.33
S.D.	8.27	0.23	6.11	26.89	29.48	13.93
CV (%)	28%	31%	22%	36%	42%	41%
TEL [a]	18	0.596	35.7	123	37.3	35
PEL [a]	36	3.53	197	315	90	91.3
ERL [a]	35	5	70	120	80	30
ERM [a]	110	9	390	270	145	50
Yin [19]	51.8	0.94	36.7	/	56.2	/
Yuan [20]	36.3	/	27.2	99.5	86.7	40.6
Fu [10]	33.55	0.14	34.14	105.55	68.09	36.23
background value [b]	15.7	0.27	18.9	59	79.3	19.5

S.D.: standard deviation; CV: coefficients of variation; [a] TEL: threshold effect level, PEL: probable effect level [15,16], ERL: effects range low, ERM: effects range median [15,16]; [b] Soil background concentrations of heavy metals in Taihu basin [21].

Spatial Distribution of Heavy Metals in Sediment

With contour maps, Figure 2a–f presents the pattern considering the spatial distribution of the above six heavy metals in sediment. On contour maps, distinctly darker shades were used to indicate zones of higher values and lighter shades were used for lower values zones.

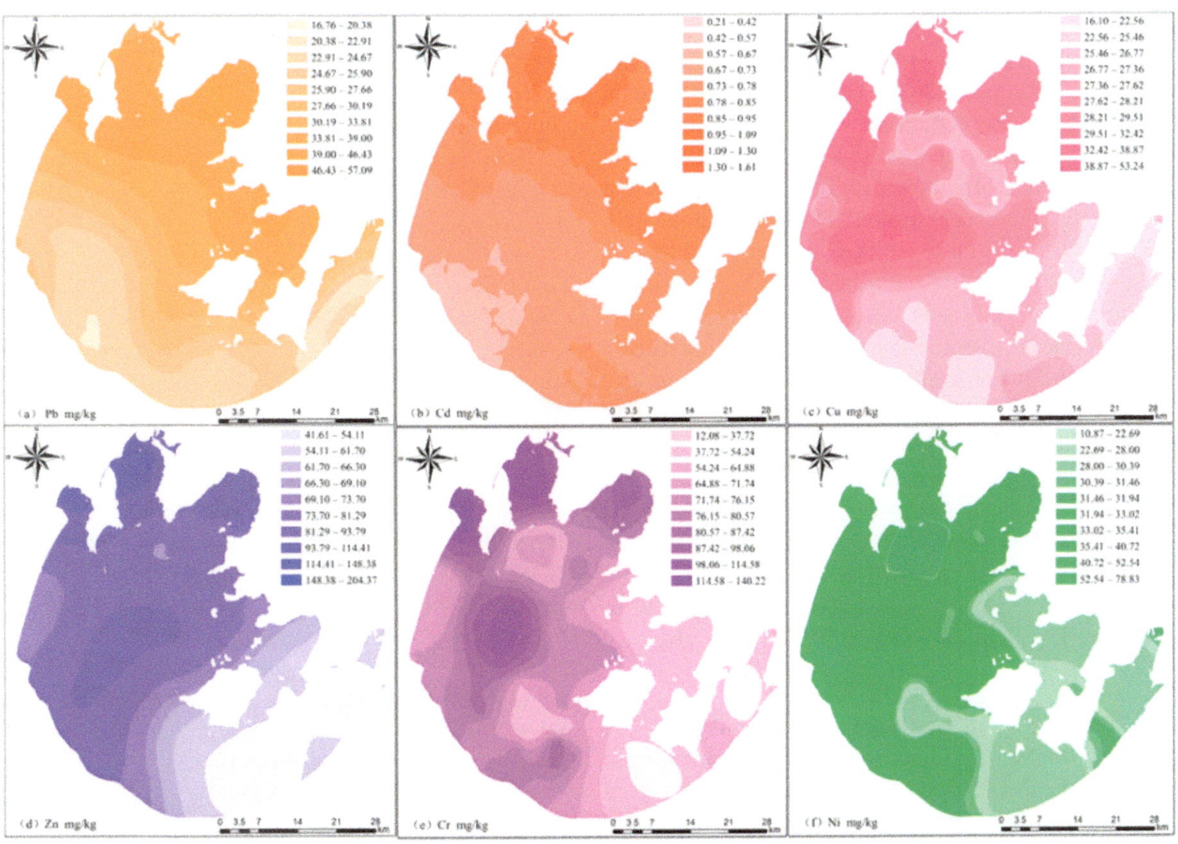

Figure 2: Spatial distributions of **(a)** Pb; **(b)** Cd; **(c)** Cu; **(d)** Zn; **(e)** Cr and **(f)** Ni in the surface sediments of Lake Taihu. Note: On contour maps, distinctly darker shades were used to indicate zones of higher values and lighter shades were used for lower values zones.

According to Figure 2a, the major distribution areas of Pb with higher concentrations were the north bays, namely the Gonghu, Zhuashan as well as Meiliang bays and the eastern lake. The distribution of Cd resembles that of Pb and it was found that the distributions of the above two metals were correlated positively with each other in a significant way, which was shown in Table 2. Despite the fact that not only Cu but also Zn belong to minor nutrients, they may be poisonous for the aquatic organisms if their concentrations are high [22]. Based on Figure 2c,d, the major distributions areas of Cu and Zn with higher concentrations were the north bays and the lake's central regions, while Figure 2e,f presented the spatial distribution and regional difference of Cr and Ni, which were similar with that of Cu and Zn. In addition, Cr and Ni are often connected with numerous rocks and thus their significant regional difference in soils has been widely recognized even if the scale is small. With regard to this research, all variation coefficients of the above two metals were larger than 40%. Besides, the major distribution areas of Cr and Ni with higher concentrations were the north bays and the lake's central and southwest regions.

Table 2: Correlation matrix between heavy metals in the surface sediments of Lake Taihu

Element	Pb	Cd	Cu	Zn	Cr	Ni	TOC
Pb	1						
Cd	0.516 **	1					
Cu	0.359 **	0.290 *	1				
Zn	0.344 **	0.319 *	0.760 **	1			
Cr	0.172	0.030	0.551 **	0.678 **	1		
Ni	0.250	0.146	0.630 **	0.753 **	0.932 **	1	
TOC	0.362 **	0.407 **	0.444 **	0.513 **	0.573 **	0.485 **	1

* Represents significant correlation at the level of $p < 0.05$; ** Represents significant correlation at the level of $p < 0.01$.

Based on lots of former researches, because of organic matter's excellent complexing ability for metallic pollutants, it could play the role of a key sink of heavy metals [7,16]. The total organic matter (TOC) as indicator of organic matter pollution of surface water or sediment was widely accepted and cited by related researches [13,14]. In our study, the TOC varied between 0.76% and 0.23%. The sampling sites with higher concentrations of TOC were mainly distributed in the north bays, the middle and southwest areas, suggesting anthropogenic influence. As shown in Table 2, it was found in this research that the concentrations of heavy metals were positively correlated with those of TOC in a significant way in the sediment, with $p < 0.01$. Furthermore, it was known from the findings of Shi and Sun [23] that considerable contaminants (nutrients, organic matter) could be assimilated by the water plants in relatively shallower water. Mishra *et al.* [24] also reported that aquatic macrophytes revealed high removal (>90%) of heavy metals (Zn, Cu, Cr and Cd) by bioaccumulation. Due to the abundance of plants in the southeastern lake, the concentrations of heavy metals in the sediment may be lower and reducing the internal contamination through plant harvest may be an effective method.

Source Identification of Heavy Metals in the Sediment

Important information about the source of contaminants can be offered by the correlation among different heavy metals. It can be seen from Table 2 that Pb and Cd were positively correlated with each other in a significant way. However, none of them had presented significant correlations with Cr and Ni, meaning that the input sources of Pb and Cd might be similar, but their sources might differ from those of Cr and Ni. In addition, both Cu and Zn showed a significant correlation with not only Pb and Cd but also Cr and Ni, which indicated that the sources of Cu and Zn might be different.

The correlations of heavy metals are also likely to be obtained through factor analysis in the factor which will provide some information on the sources of contamination [25]. Table 3 presented the factor analysis results regarding the concentrations of heavy metals in sediment of Taihu Lake. Additionally, a biplot of the PCA loadings is presented in Figure 3. The results of factor analysis for heavy metal concentrations in the surface sediments of Lake Taihu are listed in Table 3. Two factors with eigenvalues greater than 1 were extracted, which explained over 70% of the data variation. The first factor (PC1), with a variance of 51.122%, was strongly and positively correlated with Cu, Zn, Cr and Ni. The second factor (PC2) accounted for 19.606% of the total variance and showed highly positive factor loadings on Pb and Cd. These results suggested that Pb and Cd might come from the common sources, which was consistent with the correlation analysis. According to spatial distribution patterns, Pb and Cd arising from atmospheric deposition might mainly contribute to their accumulation in the north bays and east area, where the urban and traffic pollution was especially serious [26]. By analyzing the soil samples in different historical periods, Huang *et al.* [27] have also demonstrated the impact of atmospheric deposition on Pb and Cd accumulations in this area. However, river inputs most likely controlled the distribution of Cu, Zn, Cr and Ni in the lake sediments [28]. For example, the Cu, Zn, and Ni concentrations in a typical influent river nearby the north bays were 46.5, 149.9 and 58.7 mg/kg, respectively [29]. These values all exceeded their corresponding concentrations in the lake sediments. It should be noted that the patches of higher Cu, Zn, Cr and Ni concentrations also appeared in the middle area of Lake Taihu. This was most likely attributed to the subsequent distribution of river sediments after deposition [30]. In fact, owning to the effect of factors in physical aspect like flat slope or wave, undisturbed sediment are restricted in Taihu Lake. As a result, the sediment in most regions nearby the shores are soft and their thicknesses are different, while the sediment in the lake's middle area are relatively hard [30].

Table 3: Factor analysis of heavy metals in the surface sediments of Lake Taihu

	Initial Eigenvalues			Element	Rotated Component Matrix	
	Total	% of Variance	Cumulative %		PC1	PC2
	Explanation of Total Variance				Component Matrixes	
1	3.067	51.122	51.122	Pb	0.249	0.845
2	1.176	19.606	70.782	Cd	0.101	0.844
3	0.818	13.632	84.361	Cu	0.867	0.232
4	0.396	6.597	90.958	Zn	0.839	0.322
5	0.347	5.780	96.738	Cr	0.751	−0.032
6	0.196	3.262	100.000	Ni	0.690	0.407

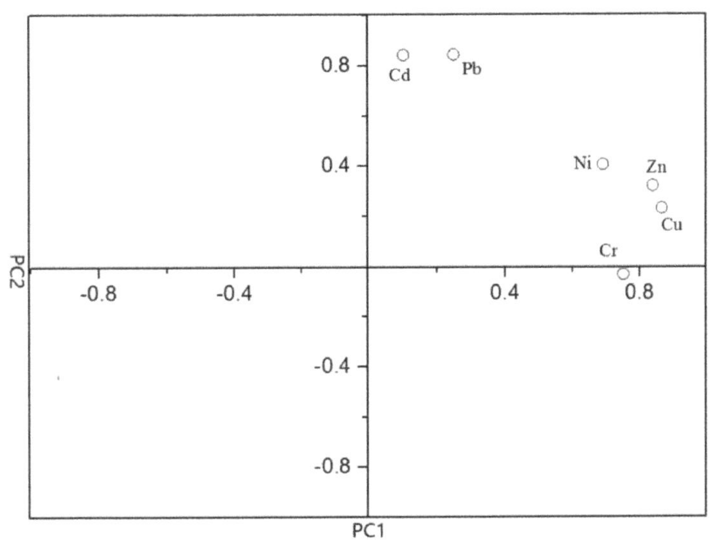

Figure 3: Plot of the two principal components.

Risk Appraisal of Heavy Metals in the Sediment

The above-mentioned SQGs have been extensively applied to the assessment of the extent to which negative influence may be produced by the chemical conditions related to the sediment and such chemical conditions can be adopted to provide help for the improvement of the deposit quality [31]. Meanwhile, the SQGs can also be applied to screening the pollution of sediment by making a comparison between the concentrations of deposit pollutants and relevant quality guidelines related to the ecosystem of freshwater. Based on the viewpoints of Harikumar and Nasir in 2010 [32], a range of minimal influence is represented by the contents lower than ERL which are designed for the evaluation of situations in which little attention is paid to the biological effects. With regard to the contents that are equal to or larger than ERL, but smaller than ERM, a range in which biological effects will be aroused at times is represented by them. A range of possible effect in which negative biological effects can often be aroused is represented by the contents equal to or larger than the values of ERM. Through the comparison between the values of ERL and those of ERM as shown in Table 1, it can be found that the toxicity of sediment from 54 sampling locations was lower because the concentrations of all metals were not larger than the ERMs. However, the toxicity of sediment from the other five sampling locations was higher because the concentration of Ni was larger than the value of ERM. Nickel is essential for animal metabolism, but the excessive ingestion of Ni generates serious toxicological responses, such as vomiting, cramps, convulsions, or even death [33]. Higher concentrations of metals were observed in the north bays (Meiliang bay, Zhushan bay and Gonghu bay), the lake's central and southwest regions. Therefore, this metal should be given special consideration. At all areas where samples were gathered, the sediment regarding Cd and Cu were considered lower than ERL. Comparatively, the sediment regarding Pb and Cr were categorized into ERL-ERM at over 22% of the areas where samples were obtained. Heavy metals were confirmed to pose potential threats to both aquatic animals [34] and mammals [35,36], as well as to humans [37]. A great deal of research shows that plumbum (Pb) has obvious toxicity effects on organisms, including nerve, blood, growth, gastrointestinal tract and immune disease though inducing oxidative stress and apoptosis [38]. Additionally, as another widespread environmental pollutant, chromium (Cr) was demonstrated to cause tissue injury, including testicular lesions, renal tubular necrosis, liver toxicity and even cancer [39,40].

Furthermore, for the purpose of assessing the extent to which the aquatic organisms may be influenced by sediment pollution, underlying acute toxicity of pollutants was sincerely suggested. Potential acute toxicity of contaminants in sediment samples can be estimated as the sum of the toxic units (ΣTU) defined as the ratio of the determined concentration to PEL value [17]. In Figure 4, a bar graph presenting the contribution rate of every heavy metal to the total amount of toxic units was shown. The result of sorting the heavy metals by the reduction of mean heavy metal toxic units in Taihu Lake would be: Pb,

Cr, Ni, Zn, Cd and Cu, which suggested that the contribution rate of Pb, Cr and Ni to the total amount of toxic units (33.02% ± 7.74%, 29.05% ± 8.53% and 14.44% ± 2.21%, respectively) was higher. Nevertheless, Cu made the lowest contribution to the toxic units, which was 5.54% ± 1.14%.

The total amount of toxic units (ΣTU) for the Taihu Lake was computed as per the total amount of the above six heavy metals that had been measured. Subsequently, the contour map indicating the toxic units for the Taihu Lake was drawn and illustrated in Figure 5. All samples with toxic units greater than 4 accounted for over 7% of the total deposit samples, meaning that the toxicity was moderate. According to the contour map, the major distribution areas of sediment with higher toxicity were the regions like the north bays as well as the lake's central region. Sediments are not only a sink for discharged metals but also a known source of pollutants in lake ecosystem [5]. Heavy metals can be released back to overlying waters from sediments in response to certain disturbances and have potential risks on drinking water source [4]. In addition, aquatic organisms assimilate metals by bio-assimilation and bioaccumulation, with potential long-term implications for human health and the ecosystem [41]. According to a previous study [10], the heavy metals have accumulated in fishes from the Lake Taihu, which were much higher than those from the Yangtze River and might pose risk of adverse health effects to humans. Therefore, it is necessary to control external sources by reducing heavy metal concentrations in industrial sewage discharged into inflowing rivers and remove internal pollution by aquatic plant uptake.

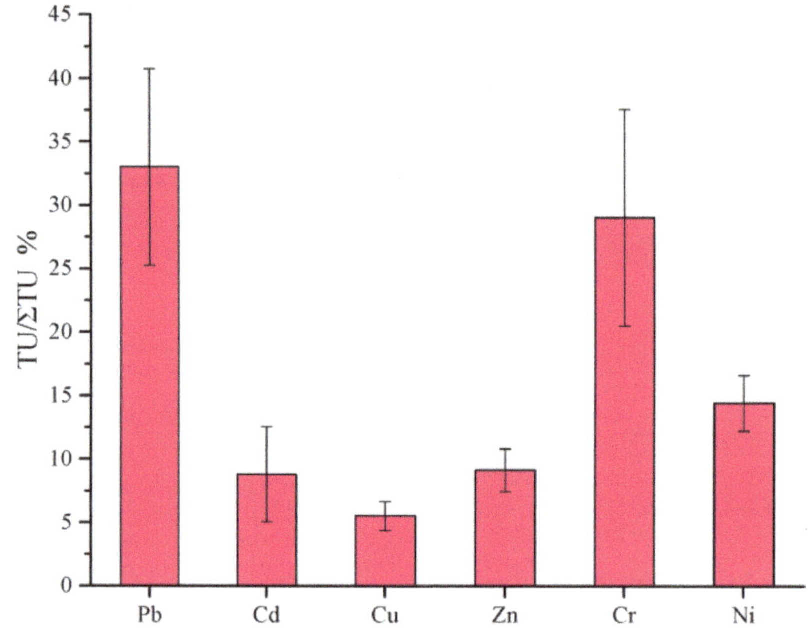

Figure 4: Contributions of respective heavy metal to the sum of toxic units.

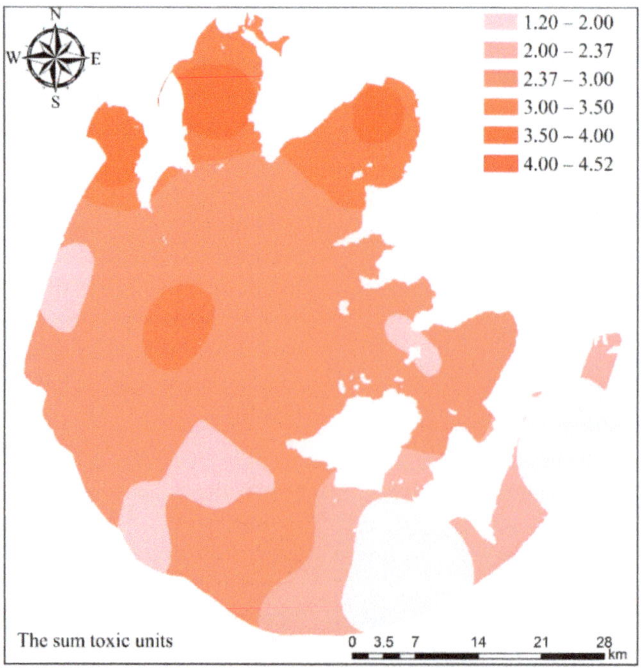

Figure 5: Spatial distribution of the sum of toxic units in the Lake Taihu.

CONCLUSIONS

The major distribution areas of Pb and Cd with higher concentrations were the north bays and east outlet region. At the same time, in the north bays, we observed Cu, Zn as well as Ni with higher concentrations. In comparison, the major distribution areas of the above heavy metals with lower concentrations were the southeast lake, where water plants were prosperous. Besides, it is possible for heavy metals like Cu, Zn, Cr as well as Ni to stem from influent streams; however, atmospheric precipitation caused by urban and traffic contamination played a decisive role in Pb and Cd's higher concentrations. At present, a threat, namely moderate human-caused contamination, is faced by the Taihu Lake. At all areas where samples were gathered, the sediment regarding Cd and Cu were considered lower than ERL. Comparatively, the sediment regarding Pb and Ni were categorized into ERL-ERM at over 22% of the areas where samples were obtained. Sorting the above heavy metals according to the decrease of their mean heavy metal toxic units in descending order gives: Pb, Cr, Ni, Zn, Cd and Cu. Among the above six heavy metals, Pb, Cr and Ni exerted greater influence on the total toxic units, while Cu exerted the least influence on the toxic units. Therefore, in order to protect the wetland and restore the Taihu Lake in the future, some strategies have to be adopted to decrease the amount of heavy metals in sediment resulting from river input and atmospheric precipitation. Moreover, during the recovery of the ecological system, the introduction of water plants that can grow better in heavy metals is of great significance for the purpose of reducing the ecological threat of Taihu Lake.

ACKNOWLEDGMENTS

This work was financially supported by the National Water Pollution Control and Management Technology Major Program of China (No. 2012ZX07101-001). We also thank the editors and anonymous reviewers for comments regarding the manuscript.

REFERENCES

1. Alkorta, I.; Hernández-Allica, J.; Becerril, J.; Amezaga, I.; Albizu, I.; Garbisu, C. Recent findings on the phytoremediation of soils contaminated with environmentally toxic heavy metals and metalloids such as zinc, cadmium, lead, and arsenic. *Rev. Environ. Sci. Biol.* 2004, *3*, 71–90.
2. Salomons, W.; Förstner, U. *Metals in the Hydrocycle*; Springer Science & Business Media: Berlin, Germany; Heidelberg, Germany, 2012.
3. Morillo, J.; Usero, J.; Gracia, I. Partitioning of metals in sediments from the Odiel River (Spain). *Environ. Int.* 2002, *28*, 263–271.
4. Suresh, G.; Sutharsan, P.; Ramasamy, V.; Venkatachalapathy, R. Assessment of spatial distribution and potential ecological risk of the heavy metals in relation to granulometric contents of Veeranam lake sediments, India. *Ecotox. Environ. Saf.* 2012, *84*, 117–124.
5. Varol, M.; Şen, B. Assessment of nutrient and heavy metal contamination in surface water and sediments of the upper Tigris River, Turkey. *Catena* 2012, *92*, 1–10.
6. Kwok, K.W.; Batley, G.E.; Wenning, R.J.; Zhu, L.; Vangheluwe, M.; Lee, S. Sediment quality guidelines: Challenges and opportunities for improving sediment management. *Environ. Sci. Pollut. Res.* 2014, *21*, 17–27.
7. Bai, J.; Cui, B.; Chen, B.; Zhang, K.; Deng, W.; Gao, H. Spatial distribution and ecological risk assessment of heavy metals in surface sediments from a typical plateau lake wetland, China. *Ecol. Model.* 2011, *222*, 301–306.
8. Paerl, H.W.; Xu, H.; McCarthy, M.J.; Zhu, G.; Qin, B.; Li, Y. Controlling harmful cyanobacterial blooms in a hyper-eutrophic lake (Lake Taihu, China): The need for a dual nutrient (N & P) management strategy. *Water Res.* 2011, *45*, 1973–1983.
9. Qin, B.; Zhu, G.; Gao, G.; Zhang, Y.; Li, W.; Paerl, H.W. A drinking water crisis in Lake Taihu, China: Linkage to climatic variability and lake management. *Environ. Manag.* 2010, *45*, 105–112.
10. Fu, J.; Hu, X.; Tao, X.; Yu, H.; Zhang, X. Risk and toxicity assessments of heavy metals in sediments and fishes from the Yangtze River and Taihu Lake, China. *Chemosphere* 2013, *93*, 1887–1895.
11. Zhang, R.; Jiang, D.; Zhang, L.; Cui, Y.; Li, M.; Xiao, L. Distribution of nutrients, heavy metals, and PAHs affected by sediment dredging in the Wujin.'gang River basin flowing into Meiliang Bay of Lake Taihu. *Environ. Sci. Pollut. Res.*2014, *21*, 2141–2153.
12. Bing, H.; Wu, Y.; Sun, Z.; Yao, S. Historical trends of heavy metal contamination and their sources in lacustrine sediment from Xijiu Lake, Taihu Lake Catchment, China. *J. Environ. Sci.* 2011, *23*, 1671–1678.
13. Bo, L.; Wang, D.; Li, T.; Li, Y.; Zhang, G.; Wang, C. Accumulation and risk assessment of heavy metals in water, sediments, and aquatic organisms in rural rivers in the Taihu Lake region, China. *Environ. Sci. Pollut. Res.* 2015, *22*, 6721–6731.
14. Yuan, H.-Z.; Shen, J.; Liu, E.-F.; Wang, J.-J.; Meng, X.-H. Assessment of nutrients and heavy metals enrichment in surface sediments from Taihu Lake, a eutrophic shallow lake in China. *Environ. Geochem. Health* 2011, *33*, 67–81.
15. MacDonald,D.D.;Ingersoll,C.G.;Berger,T.Development and evaluation of consensus-based sediment quality guidelines for freshwater ecosystems. *Arch. Environ. Contam. Tox.* 2000, *39*, 20–31.
16. Bai, J.; Cui, B.; Yang, Z.; Xu, X.; Ding, Q.; Gao, H. Heavy metal contamination of cultivated wetland soils along a typical plateau lake from southwest China. *Environ. Earth Sci.* 2010, *59*, 1781–1788.
17. Pedersen, F.; Bjørnestad, E.; Andersen, H.V.; Kjølholt, J.; Poll, C. Characterization of sediments from Copenhagen Harbour by use of biotests. *Water Sci. Technol.* 1998, *37*, 233–240.
18. Benndorf, J.; Menz, J. Improving the assessment of uncertainty and risk in the spatial prediction of environmental impacts: A new approach for fitting geostatistical model parameters based on dual Kringing in the presence of a trend. *Stoch. Environ. Res. Risk Assess.* 2013, *28*, 627–637.

19. Yin, H.; Gao, Y.; Fan, C. Distribution, sources and ecological risk assessment of heavy metals in surface sediments from Lake Taihu, China. *Environ. Res. Lett.* 2011, *6*, 044012.
20. Yuan, H.; Liu, E.; Shen, J. The accumulation and potential ecological risk of heavy metals in microalgae from a eutrophic lake (Taihu Lake, China). *Environ. Sci. Pollut. Res.* 2015, *22*, 17123–17134.
21. Wenchuan, Q.; Dickman, M.; Sumin, W. Multivariate analysis of heavy metal and nutrient concentrations in sediments of Taihu Lake, China. *Hydrobiologia* 2001, *450*, 83–89.
22. Miao, A.J.; Wang, W.X.; Juneau, P. Comparison of Cd, Cu, and Zn toxic effects on four marine phytoplankton by pulse-amplitude-modulated fluoromentry. *Environ. Toxicol. Chem.* 2005, *24*, 2603–2611.
23. Shi, L.P.; Sun, Q.Y. Brief talk on effect of aquatic plant purifying water body of Yilonghu Lake. *Yunnan Environ. Sci.*2005, *24*, 40–42. (In Chinese).
24. Mishra, S.; Mohanty, M.; Pradhan, C.; Patra, H.K.; Das, R.; Sahoo, S. Physico-chemical assessment of paper mill effluent and its heavy metal remediation using aquatic macrophytes—A case study at JK Paper mill, Rayagada, India.*Environ. Monit. Assess.* 2013, *185*, 4347–4359.
25. Yongming, H.; Peixuan, D.; Junji, C.; Posmentier, E.S. Multivariate analysis of heavy metal contamination in urban dusts of Xi'an, Central China. *Sci. Total Environ.* 2006, *355*, 176–186.
26. Cheng, H.; Hu, Y. Lead (Pb) isotopic fingerprinting and its applications in lead pollution studies in China: A review.*Environ. Pollut.* 2010, *158*, 1134–1146.
27. Huang, S.; Liao, Q.; Hua, M.; Wu, X.; Bi, K.; Yan, C. Survey of heavy metal pollution and assessment of agricultural soil in Yangzhong district, Jiangsu Province, China. *Chemosphere* 2007, *67*, 2148–2155.
28. Niu, Y.; Niu, Y.; Pang, Y.; Yu, H. Assessment of Heavy Metal Pollution in Sediments of Inflow Rivers to Lake Taihu, China. *Bull. Environ. Contam. Tox.* 2015, *95*, 618–623.
29. Fang, X.H.; Liu, X.W.; Wei, D.Y.; Xu, Z.C.; Yang, Z.; Huang, Z.S.; Liu, W. Heavy metal distribution characteristics in sediment of Caoqiao River, Taihu Basin. *Environ. Chem.* 2012, *31*, 771–776. (In Chinese).
30. Hu, K.; Pang, Y.; Wang, H.; Wang, X.; Wu, X.; Bao, K. Simulation study on water quality based on sediment release flume experiment in Lake Taihu, China. *Ecol. Eng.* 2011, *37*, 607–615.
31. Burnett-Seidel, C.; Liber, K. Evaluation of sediment quality guidelines derived using the screening-level concentration approach for application at uranium operations in Saskatchewan, Canada. *Environ. Monit. Assess.* 2012, *184*, 1593–1602.
32. Harikumar, P.; Nasir, U. Ecotoxicological impact assessment of heavy metals in core sediments of a tropical estuary.*Ecotox. Environ. Saf.* 2010, *73*, 1742–1747.
33. Dou, C.; Zhang, J. Effects of lead on neurogenesis during zebrafish embryonic brain development. *J. Hazard Mater.*2011, *194*, 277–282.
34. Domingo, J.L. Metal-induced developmental toxicity in mammals: A review. *J. Toxicol. Environ. Heal. A* 1994, *42*, 123–141.
35. Courtois, E.; Marques, M.; Barrientos, A.; Casado, S.; López-Farré, A. Lead-induced downregulation of soluble guanylate cyclase in isolated rat aortic segments mediated by reactive oxygen species and cyclooxygenase-2. *J. Am. Soc. Nephrol.* 2003, *14*, 1464–1470.
36. Gurer-Orhan, H.; Sabır, H.U.; Özgüneş, H. Correlation between clinical indicators of lead poisoning and oxidative stress parameters in controls and lead-exposed workers. *Toxicology* 2004, *195*, 147–154.
37. Patrick, L. Lead toxicity part II: The role of free radical damage and the use of antioxidants in the pathology and treatment of lead toxicity. *Altern. Med. Rev.* 2006, *11*, 114.
38. Davidson, T.; Kluz, T.; Burns, F.; Rossman, T.; Zhang, Q.; Uddin, A. Exposure to chromium (VI) in the drinking water increases susceptibility to UV-induced skin tumors in hairless mice. *Toxicol. Appl. Pharm.* 2004, *196*, 431–437.
39. Ueno, S.; Kashimoto, T.; Susa, N.; Furukawa, Y.; Ishii, M.; Yokoi, K. Detection of dichromate (VI)-induced DNA strand breaks and formation of paramagnetic chromium in multiple mouse organs. *Toxicol. Appl. Pharm.* 2001, *170*, 56–62.
40. Moore, J.W.; Ramamoorthy, S. *Heavy Metals in Natural Waters: Applied Monitoring and Impact Assessment*; Springer Science & Business Media: Berlin, Germany; Heidelberg, Germany, 2012.
41. Lavoie, R.A.; Jardine, T.D.; Chumchal, M.M.; Kidd, K.A.; Campbell, L.M. Biomagnification of mercury in aquatic food webs: A worldwide meta-analysis. *Environ. Sci. Technol.* 2013, *47*, 13385–13394.

CHAPTER 17

FROM CATCHMENT TO NATIONAL SCALE RAINFALL-RUNOFF MODELLING: DEMONSTRATION OF A HYDROLOGICAL MODELLING FRAMEWORK

Susan M. Crooks, Alison L. Kay , Helen N. Davies and Victoria A. Bell

Keywords: Rainfall-runoff, digital databases, generalized parameterization, gridded flow routing, modelling framework, HOST soils

INTRODUCTION

Hydrological models for simulating river flow from rainfall have evolved over many decades [1] into a wide range of representations in terms of structure, complexity, data requirements, and scale of application from field plot to global, with a similarly wide range of purposes from floods to droughts, past to future climate, water resources and water quality. Much hydrological research has been based on catchment studies using models calibrated to observed flow (e.g., [2,3,4]), as well as continued exploration and evolving understanding of hydrological processes to enable more reliable simulation in ungauged catchments [5]. Generalization of relationships between catchment properties, hydrological response and model parameters allows estimation of flow metrics in data-scarce areas [6], and flow simulation for gauged catchments with little or no direct calibration [7,8,9] as well as for ungauged locations within gauged catchments [10]; although use of such relationships implies that similarity between catchments can be defined [11]. Grid-based hydrological models (e.g., [12,13]) have exploited the advantages of digital databases and broadened the spatial scale of model application. Modelling systems have been designed to incorporate components from different models to customize a structure appropriate for individual catchments (e.g., [14,15]).

While the wealth of available hydrological models provides tailored solutions to a range of hydrological problems they often have very individual requirements, in terms of their supporting spatial and configurative datasets, drivers, data-format resolution and internal structure. Calibrated catchment models can capture the key behavior of the area to which they are applied, but often say little about how the model can be applied to other catchments without recalibration. The overall aim of the generic Hydrological Modelling Framework (HMF) introduced here is to support and enhance existing models used for a range of water-management applications, by providing a single GIS-based data and spatial configuration framework. Potentially, existing models can be adapted to run within this spatial configuration, allowing a previously catchment-based model to operate in a gridded framework over a wide area and at a range of spatial resolutions. Although, a pre-requisite for transfer of a model to the HMF is methodology to calculate model parameter values from physical properties. The concept recognises that hydrological (and land-surface) models increasingly share a common spatial representation of the landscape provided by digital datasets, and that some of this functionality can be shared without necessarily losing the unique characteristics that a specialised model can provide. The HMF enables models to be applied across Great Britain, or a wider area, and enables users to focus resources on a specialist science question, or model application, while relying on an existing framework for underpinning data and spatial information.

Citation: Crooks, S.M.; Kay, A.L.; Davies, H.N.; Bell, V.A. From Catchment to National Scale Rainfall-Runoff Modelling: Demonstration of a Hydrological Modelling Framework. Hydrology 2014, 1, 63-88. doi:10.3390/hydrology1010063.

The aim of this paper is to demonstrate, as an example, the transfer of one existing catchment-based model, CLASSIC [16], to operation in the HMF. Transfer of other models may require different changes, dependent on the model, but could follow a similar procedure and analysis of results. The paper gives a description of the structure of the HMF, background information on CLASSIC (for which a generalised methodology exists to determine parameter values from physical properties), followed by details of transfer of CLASSIC from simulation of individual catchments to simulation of gridded river flows across the whole of Britain. Final model performance is discussed for over 50 catchments and analysed to show the effect of spatial resolution and demonstrate how the HMF can be used to further model understanding, development and application.

BACKGROUND AND METHODOLOGY

Description of the Hydrological Modelling Framework

The HMF essentially comprises a set of modules for: model set-up; input of gridded driving data; production of gridded runoff (including optional use of a snow module); routing runoff along a network of flow paths to produce river flows; and output of simulated variables such as flow or soil-moisture (Figure 1). The modular nature of the HMF ensures flexibility, by allowing substitution or addition of modules. For example, the main driving data module provides observed climate data but could be replaced by a module providing projected climate data for a future period (taken from a regional climate model for instance). Similarly, alternative snowmelt, runoff-production or routing modules could be applied. Modules could be added to provide information on groundwater abstractions or to simulate water quality for example, thus more easily enabling sharing of existing model expertise.

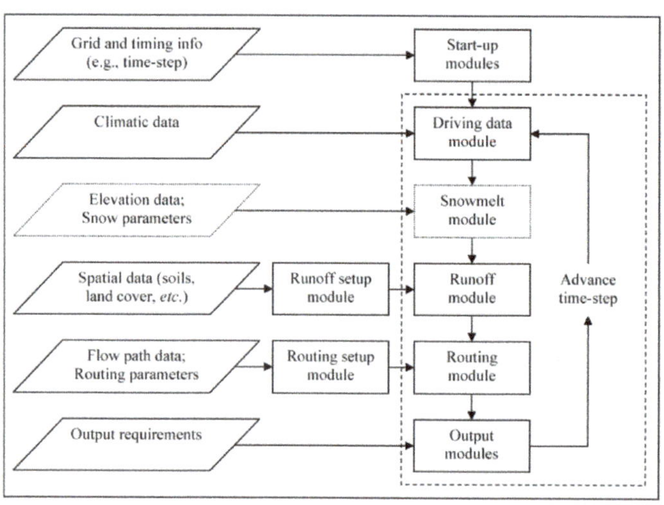

Figure 1: Schematic of the Hydrological Modelling Framework, showing modules (rectangles) and data requirements (parallelograms), with optional parts indicated by dotted lines. The parts within the dashed box are performed at each time-step.

There are currently two domain applications of HMF. The first (HMF-GB) was set up for use over Great Britain, with kilometre-scale grids aligned with the GB national grid. The second (HMF-GLOBAL) was set up for use over a user-defined region up to global scale, with degree-scale grids aligned with standard latitude and longitude. For each variant a key requirement is derivation of the network of flow paths along which runoff is routed, and definition of a land-sea mask; sea boxes are subsequently ignored (all data set to missing values). Flow path delineation for a range of spatial resolutions has been undertaken using the COTAT+ method for deriving lower-resolution river networks from higher-resolution Digital Terrain Model (DTM) data [17], which assumes that flows are routed in one of eight directions. Currently, HMF-GB can operate at four spatial resolutions (10, 5, 2.5 and 1 km) while HMF-GLOBAL can operate at three spatial resolutions (0.25, 0.125 and 0.05 deg). Hereafter, this paper focuses on HMF-GB; the representation of river flow at the four spatial resolutions is shown in Figure 2. The introduction and testing of the runoff-production scheme from a semi-distributed catchment-based model, CLASSIC [16], into HMF-GB (called CLASSIC-GB) are described later in this paper. Other key features of the HMF are briefly described below.

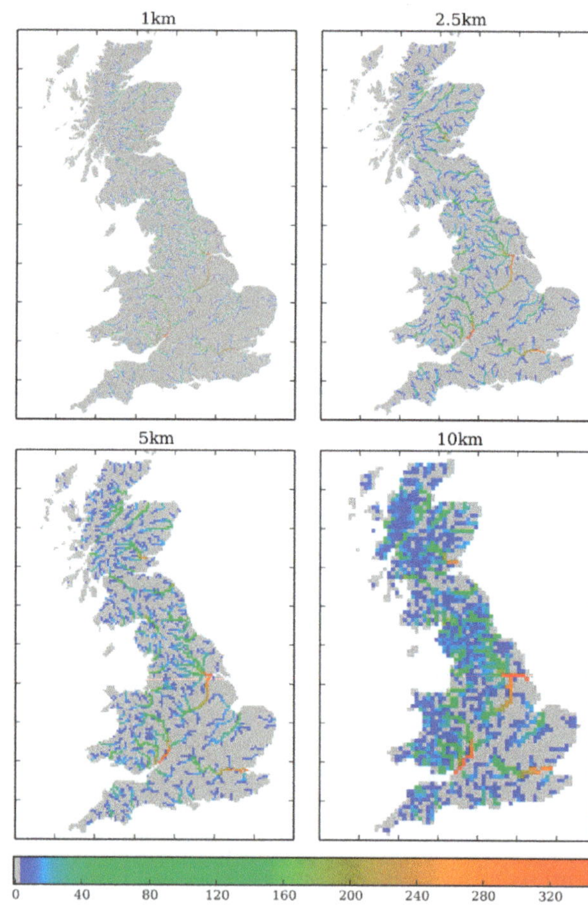

Figure 2: Representation of river flow in Britain at four spatial resolutions; mean monthly flow ($m^3 \cdot s^{-1}$) in January 1982.

The routing module currently in HMF-GB is based on a kinematic wave scheme [18], and allows parallel routing of surface and sub-surface runoff. In addition, each (non-sea) grid box is designated as either 'land' or 'river' using a cumulative catchment area threshold of 20 km^2 [18]; at lower resolutions (10 and 5 km) all grid boxes are designated as 'river', whereas for the higher resolutions (2.5 and 1 km) boxes can be either 'land' or 'river'. Thus four wave speeds are defined (for land surface, land sub-surface, river surface and river sub-surface runoff), with slower speeds for land boxes and sub-surface runoff. There is also a fractional return flow, similar to [18], from sub-surface to surface runoff, for land and river grid boxes. Lakes and reservoirs are not currently differentiated. For stability of the routing scheme, the routing time-step must be sufficiently small in relation to the spatial grid size, meaning that a sub-daily time-step is necessary for the routing module. The main HMF-GB time-step is generally the same as the routing time-step, although it can be a multiple of the routing time-step if faster run-times are required.

The optional snowmelt module in HMF-GB currently uses a simple temperature-related snow store and melt rate with eight parameters (based on [19]), and operates with separate accumulation and melt in different elevation zones. Apart from temperature the module also requires the altitude to which the temperature relates, and information on fractional grid-box area within the different elevation zones.

Spatial outputs from HMF-GB are generally produced by an output module as netcdf files. The HMF can also be used in validation mode, in which simulated flow time-series are output for grid boxes corresponding to the locations of river flow gauging stations, along with observed flow time-series for those stations. Other variables, like soil-moisture, can also be output, allowing extended model validation.

The climate data provided by HMF-GB are rainfall, potential evaporation (PE) and temperature. The GB application presented here uses observed data; daily rainfall on a 1 km grid [20], monthly total PE on a 40 km grid provided by the Met Office Rainfall and Evaporation Calculation System (MORECS [21]) and daily minimum and maximum temperature on a 5 km grid [22]. Rainfall and PE are averaged up or copied down to the HMF-GB grid, as required, and divided equally down to the HMF-GB time-step. The daily minimum and maximum temperatures are lapsed to the required HMF-GB elevations and interpolated through the day using a sine curve, assuming that the minimum occurs at 2am and the maximum at 2pm [23].

Other datasets provided by HMF-GB, as 500 m grids, include: percentages of each soil class in the Hydrology of Soil Types (HOST [24]), percentages of each land cover class in the Land Cover Map 2007 (LCM2007 [25]), and mean slope derived from the Integrated Hydrological Digital Terrain Model (IHDTM [26]) which has a 50 m horizontal resolution. Each of these is spatially averaged to the required HMF-GB grid size. Also provided is mean altitude on a 100 m grid, derived from the IHDTM, which can be averaged as required, or used to determine the proportion of each HMF grid square in different elevation zones for use with the snowmelt module.

Background on CLASSIC

The semi-distributed continuous simulation model CLASSIC (Climate and Land-use Scenario Simulation in Catchments) was originally developed in the mid-1990s for estimating the impacts of climate and land use change in three large catchments in Britain (~10,000 km^2). The model is described in detail in [16] with the main features outlined below.

CLASSIC is applied on a grid framework superimposed on a catchment, defined by a topographic boundary. The grid size is not fixed but selected so as to be compatible with the catchment area and heterogeneity of the physical and climatic characteristics of the catchment. Grid sizes of 40 km (original size), 20 km (suitable for large catchments e.g., > 5000 km^2), 10 km (normally used) and 5 km have all been used. The model is normally run at a daily time-step. The model's three component modules are soil-moisture accounting, soil-drainage and channel routing. Inputs of rainfall and PE to the soil-moisture accounting module determine effective rainfall, which provides the input to the soil-drainage module, the output from which enters the channel routing module. The soil-moisture accounting and soil-drainage modules operate in each grid square, while the channel routing module transfers runoff from each grid square directly to the catchment outlet. The river flow at the outlet (normally a river gauging station) is the summation of the routed flow from all contributing grid squares. An optional fourth module, a snowmelt module, can be used as a precursor to the soil-moisture accounting. This catchment-based version of CLASSIC is referred to in subsequent sections as CLASSIC-catchment.

A generalised methodology [16] has been developed to allow parameter values in the soil-moisture and soil-drainage modules to be determined from physical properties, with only the two routing parameters determined by calibration against gauged flows. Parameter values in the soil-moisture module are derived from land cover type, soil type, slope and altitude, while those in the soil-drainage module are derived from soil type. Land cover is from CEH digitised databases, of which three are currently available (for years 1990 [27], 2000 [28] and 2007 [25]). Land cover types are amalgamated into six groups for use in CLASSIC; grass, deciduous woodland, coniferous woodland, arable, upland and urban. The arable group contains a seasonal growth cycle from bare ground in the autumn to maximum growth in early summer. Information on soils is from the HOST soil classification system, which has been shown to provide a meaningful approach to hydrological generalisation [29,30]. HOST groups soils into 29 classes based on the dominant features controlling water movement through the soil. Determining characteristics are whether the soil is mineral based or peat, whether there is an underlying aquifer, and the presence, or not, of an impermeable layer within the top metre (Figure 3.3 from [24]) Parameter generalisation was initiated through the calibration of a range of catchments with predominantly one HOST class and natural flow (*i.e.*, without artificial influences through abstraction, regulation or augmentation). The channel routing module uses network width functions (the number of drainage channels at kilometre distances from the catchment outlet, derived from the 50 m IHDTM) to represent the drainage network, combined with calibrated parameters for wave velocity and attenuation.

The aim of the generalised method for determining grid square parameter values is to simulate the natural flow regime, rather than the gauged flow. Few catchments in Britain, apart from headwater catchments, have natural flow so the generalised methodology provides a more stationary approach to model parameterisation than is obtained with calibration against observed flow, which is often performed over a time period of only a few years and can be affected by time-varying changes in water usage. The method also ensures spatial continuity of flow simulation within and between river basins. When CLASSIC is applied, although the generalised methodology for setting grid square parameter values is used to start with, there is the option to adjust parameter values if appropriate. Catchments where this option has been implemented are those with soils overlying aquifers of chalk, limestone and sandstone (HOST classes 1, 2 and 3 respectively). One representative response time is assigned for each of these HOST classes (default value) but, as there is considerable regional variation in response rate within each class, more appropriate response times have been determined for defined groundwater regions to match recession curves and improve simulation of the dominant baseflow in relevant catchments.

CLASSIC-catchment has been set up for over 30 catchments across Britain with catchment areas ranging from 200 km^2 to 10,000 km^2. It has been used for a range of studies including: assessing the effect of land-use change on Thames floods [31]; investigating transient climate change and snowmelt in the Dee, north-east Scotland [23]; event attribution for the Autumn 2000 floods in England [32]; regionalising the impacts of climate change on flooding in Britain [33]; assessing impact uncertainty for larger catchments [34]. It is also one of a suite of models used to provide consistent sets of transient daily river flow projections across Britain for 1950–2098 [35].

Setting up CLASSIC-GB

Given the generally successful application of CLASSIC-catchment to catchments across Britain [33,35], the model was used to test the concept of the HMF and demonstrate the steps undertaken to reconfigure CLASSIC-catchment for use in HMF-GB (CLASSIC-GB). The generalised method for assigning parameter values from physical catchment properties in the soil-moisture and soil-drainage modules of CLASSIC-catchment is a key aspect in the transfer. These modules are used directly in the runoff module in HMF-GB and operate in the same way in both CLASSIC-catchment and CLASSIC-GB. However, the routing method is different. As the network width functions used in CLASSIC-catchment are not compatible with the structure of HMF-GB, CLASSIC-GB uses the kinematic wave routing scheme instead.

CLASSIC-GB uses the HMF-GB climate and physiographic datasets. Two further datasets have been created to implement other operations developed for CLASSIC-catchment; a PE station index and a response-time index. In CLASSIC-catchment, estimation of PE rates for different land cover types is derived from a set of 12 monthly regression equations between monthly MORECS PE for a short grass land cover and five other land cover types (deciduous woodland, coniferous woodland, upland, winter and spring sown grain crops). The regression equations were calculated for 10 climate stations, from daily data provided by the UK Met Office for each station derived for the six land cover types for eight years (1985–1992). The PE Index specifies which climate station to use for each grid square in HMF-GB. Boundaries between the stations have been mainly determined using appropriate catchment boundaries. The response-time index is described in Section 3.2.

To compare simulated discharge with observed, a grid square is selected as appropriate for a gauging station. The upstream contributing area and flow direction associated with each grid square are obtained from the network of flow paths; this information allows the contributing area for these grid squares to be compared with the required catchment area and drainage pattern to ensure that the most relevant square is selected for comparison of discharges. An advantage of CLASSIC-GB is that simulated river flow can be extracted for multiple grid cell locations (catchments) on the drainage network without setting up CLASSIC-catchment separately for each one. But, as catchment boundaries are not used, the upstream area contributing to a particular grid square is affected by the framework resolution—the finer the grid the more accurate the contributing area and potentially the more accurate the volume of simulated discharge.

Catchments

Details of the catchments used to assess the performance of CLASSIC-GB are given in Table 1, with their boundaries mapped in Figure 3 (for further catchment details see [36]). Table 1 is in two parts; the first 13 catchments are those used in the initial testing of CLASSIC-GB (with boundaries in cyan in Figure 3), the following 41 were used in further development and testing (boundaries in blue). Catchments were selected to give a range of areas, geographical location, climatic properties and HOST classes. The HOST classes listed in Table 1 each have at least 10% catchment coverage and are given in order of decreasing percentage. Also in the table are $SAAR_{61-90}$, the standard average annual rainfall for 1961–1990, and BFI, the baseflow index, described as an effective measure of indexing catchment geology [37] and used here as a combined measure of soil type.

All catchments in the initial selection have been previously modelled with CLASSIC-catchment, and most have an area of at least 1000 km^2, reflecting the fact that CLASSIC-catchment was developed to simulate flows in larger catchments. Two smaller catchments (27034 and 39081) were included to investigate the effect of coarse grid-resolution on simulated flows. The Thames at Kingston (39001) has a naturalised flow record, which has been used for comparison with modelled flow; in all other catchments modelled flow is compared with gauged flow, which may be affected by abstractions, augmentation or regulation. Of the 41 catchments selected for further assessment, 14 had been previously modelled with CLASSIC-catchment. Availability of other medium to large non-modelled catchments with reasonable quality flow data is limited and some were selected knowing that the observed flow data are affected by substantial alterations to flow. A group of 12 smaller catchments (107 km^2 to 570 km^2) was included to test more fully the capability of CLASSIC-GB at a finer spatial scale than that at which CLASSIC-catchment was developed.

Table 1: Details of catchments used to test CLASSIC-GB (the first 13, above the dashed line, are used for the initial assessment)

Station Number	River	Area (km^2)	$SAAR_{61-90}$ (mm)	Altitude range (m)	BFI	HOST
12002	Dee @ Park[1]	1844	1081	23–1309	0.53	17, 15, 29
15006	Tay @ Ballathie[1]	4587	1425	26–1210	0.64	17, 15
27009	Ouse @ Skelton[1]	3315	900	5–714	0.39	24, 29, 26
27034	Ure @ Kilgram[1]	510	1342	88–710	0.32	29, 26, 6
27041	Derwent @ Buttercrambe[1]	1586	765	10–452	0.69	24, 4, 2
39001	Thames @ Kingston[1]	9948	706	5–330	0.63	25, 1, 2
39081	Ock @ Abingdon[1]	234	639	51–260	0.64	25, 2, 3
43021	Avon @ Knapp Mill[1]	1706	810	1–294	0.86	1
47001	Tamar @ Gunnislake[1]	917	1216	8–580	0.46	17, 24, 21
54001	Severn @ Bewdley[1]	4325	913	17–826	0.53	24, 17, 18
54057	Severn @ Haw Bridge[1]	9895	792	11–826	0.56	24, 18
71001	Ribble @ Samlesbury[1]	1145	1353	10–688	0.33	24, 26, 29
84013	Clyde @ Daldowie[1]	1903	1129	8–745	0.46	24, 15, 29
04001	Conon @ Moy Bridge	962	1770	10–1100	0.57	15, 19, 29, 12
10002	Ugie @ Inverugie	325	812	9–234	0.63	17, 24
11001	Don @ Parkhill[1]	1273	885	32–874	0.68	17, 15
13007	North Esk @ Logie Mill	732	1074	11–929	0.51	15, 18, 17
21008	Teviot @ Ormiston Mill	1110	939	43–611	0.45	19, 6, 15, 24, 17
21009	Tweed @ Norham[1]	4390	955	4–838	0.52	17, 15, 24
22001	Coquet @ Morwick	570	850	5–775	0.44	24, 15
22009	Coquet @ Rothbury	346	905	71–775	0.48	24, 29, 15, 19
23004	South Tyne @ Haydon Bridge	751	1148	59–893	0.34	24, 29, 26
24009	Wear @ Chester le Street[1]	1008	885	6–745	0.46	24, 26
27007	Ure @ Westwick Lock[1]	915	1118	14–710	0.39	24, 26, 29, 6
27021	Don @ Doncaster	1256	799	4–543	0.56	24, 4
28022	Trent @ North Muskham[1]	8231	747	5–634	0.65	24, 21
28066	Cole @ Coleshill	130	722	79–202	0.42	24, 21
28085	Derwent @ St Mary's Bridge	1054	1012	44–634	0.63	4, 24, 15
33019	Thet @ Melford Bridge	316	620	11–71	0.78	1, 24, 5, 18
33026	Bedford Ouse @ Offord[1]	2570	609	11–247	0.50	21, 23, 25
36006	Stour @ Langham	578	580	6–128	0.52	21, 5
39006	Windrush @ Newbridge	363	743	63–317	0.86	2, 23
39016	Kennet @ Theale[1]	1038	759	43–296	0.88	1, 25, 18
39020	Coln @ Bibury	107	820	101–330	0.93	2, 23
39034	Evenlode @ Cassington Mill[1]	430	691	60–267	0.71	2, 25
40003	Medway @ Teston[1]	1256	744	7–268	0.40	25, 18, 24
40011	Great Stour @ Horton	345	747	13–196	0.69	1, 18, 25, 3
42004	Test @ Broadlands	1040	790	10–296	0.94	1, 6
45001	Exe @ Thorveton	601	1295	26–514	0.50	17
45005	Otter @ Dotton	202	976	15–302	0.53	21, 3, 24
50006	Mole @ Woodleigh	327	1307	48–490	0.47	17, 24
53018	Avon @ Bathford[1]	1552	817	18–304	0.57	2, 25, 23
54002	Avon @ Evesham	2210	654	20–317	0.52	25, 24, 21, 23
54029	Teme @ Knightsford Bridge	1480	818	21–545	0.55	18, 4
55023	Wye @ Redbrook[1]	4010	1011	9–749	0.54	18, 17, 24
56001	Usk @ Chain Bridge	912	1363	23–885	0.50	4, 17, 15, 5, 26
60003	Taf @ Clog-y-Fran	217	1420	9–392	0.55	17
62001	Teifi @ Glan Teifi[1]	894	1382	66–592	0.54	17, 24
67015	Dee @ Manley Hall	1013	1369	25–878	0.54	17, 15, 24, 29
68005	Weaver @ Audlem	207	719	45–221	0.54	24, 18, 5, 10
72004	Lune @ Caton[1]	983	1523	11–734	0.32	29, 24, 26, 15
73005	Kent @ Sedgwick	209	1732	19–812	0.41	17, 29
76007	Eden @ Sheepmount[1]	2286	1183	10–945	0.49	24, 5, 29
79003	Nith @ Hall Bridge	155	1505	173–607	0.27	24, 15, 29

[1] Catchments previously modelled with CLASSIC-catchment; $SAAR_{61-90}$ — standard average annual rainfall for 1961–1990; BFI — baseflow index; HOST — Hydrology of Soil Types [24].

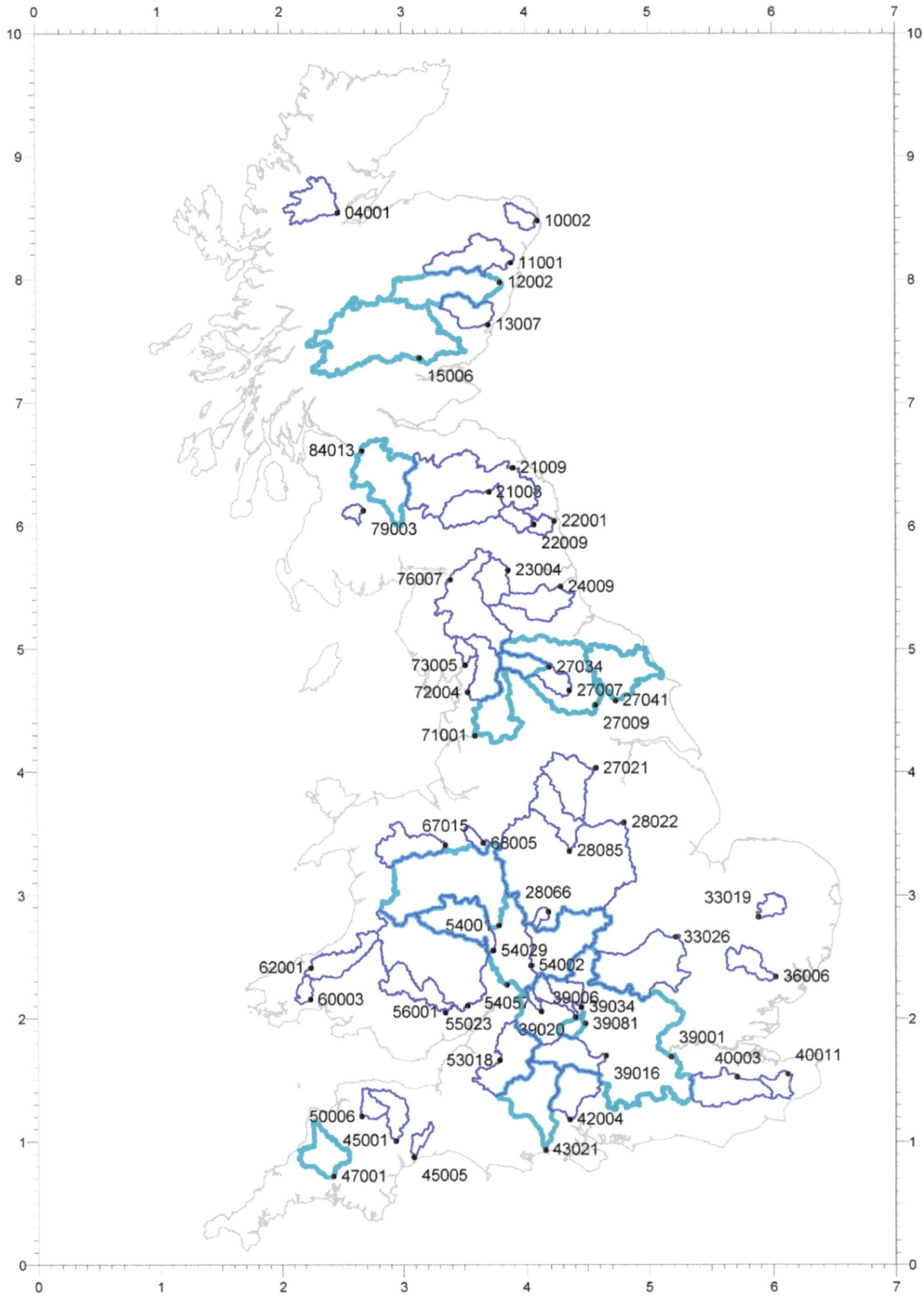

Figure 3: Map of Britain showing catchment boundaries and station numbers. Catchments used in initial tests have boundaries in cyan, others in blue. The frame is the GB national grid with numbers at 100 km intervals.

Model Assessment Methods

Three measures of fit between daily observed flows (Q_d) and simulated flows (q_d) have been used to assess performance of CLASSIC-GB; Nash-Sutcliffe efficiency (NS, Equation (1) [38]), water balance bias (Bias, Equation (2)), and a fit of flow statistics score (derived from monthly mean flows and flow duration curves; mmfd, Equation (3)). The first two of these are frequently used to characterise model performance, while the third is introduced here as a measure of how well the model simulates flow response at seasonal/monthly time scales and across the flow range. Note that over-bars in the equations below indicate overall mean flows, while Q_m (and q_m) indicates mean monthly flows and Q_n (and q_n) indicates daily flows exceeded n% of the time.

$$\mathrm{NS} = 1 - \frac{\Sigma(Q_d - q_d)^2}{\Sigma(Q_d - \overline{Q})^2} \quad (1)$$

$$\mathrm{Bias} = 100\left(\frac{\overline{q}}{\overline{Q}} - 1\right) \quad (2)$$

$$\mathrm{mmfd} = \mathrm{MM} + \mathrm{FD}$$

$$\mathrm{MM} = \frac{1}{12}\sum_{m=1}^{12} 100\frac{|Q_m - q_m|}{Q_m}$$

$$\mathrm{FD} = \frac{1}{5}\sum_{n\in\{1,10,25,50,75\}} 100\frac{|Q_n - q_n|}{Q_n} \quad (3)$$

NS is a dimensionless performance measure which expresses the proportion of variability in observed flows accounted for by the model simulation. A value of 1 indicates an exact fit between observed and modelled flow, a value of 0 that the model is only as good as using the mean flow while a negative value indicates that model simulations are worse than using the mean. NS is sensitive to timing differences of peak flows, which may not be important when just considering whether the model can reproduce the characteristics of the observed flow regime. The Bias (%) indicates how well the balance between rainfall and evaporation agrees with the observed volume of flow over the simulation period. This relates in part to differences between catchment and contributing HMF-GB areas and in part to how well losses from evaporation (adjustment from potential to actual evaporation (AE)) are simulated in the model (as well as measurement errors of all variables). The two component measures within mmfd are derived from percentage differences in mean monthly flow (MM) and flow duration (FD). Here, five points have been used to characterise the fit of the simulated flow duration curve, but more points or a wider range could be included. The two measures have been combined here for convenience of presentation but each provides more specific information than NS or Bias on how well the seasonal soil-moisture cycle (and snowmelt processes where appropriate) are simulated. While the overall Bias can mask considerable variation in water balance at an annual scale, mmfd uses absolute values of percentage differences so that positive and negative values are not cancelled out. For most catchments, apart from the Thames (which has naturalised flow), the Bias and mmfd are also affected by alterations to the natural flow regime.

As well as calculating measures of fit it is also useful to assess the quality of the values [39,40]. Three performance bands (Table 2) have been defined as additional indicators of how well CLASSIC-GB performs (no protocols or procedures for model testing are currently standard hydrological practice [5]).

Table 2: Model performance bands for each performance measure

Band	NS	Bias (%)	mmfd
1	$\mathrm{NS} \geq 0.8$	$-10 \leq \mathrm{Bias} \leq 10$	$\mathrm{mmfd} \leq 20$
2	$0.6 \leq \mathrm{NS} < 0.8$	$-20 \leq \mathrm{Bias} < -10$ or $10 < \mathrm{Bias} \leq 20$	$20 < \mathrm{mmfd} \leq 40$
3	$\mathrm{NS} < 0.6$	$\mathrm{Bias} < -20$ or $\mathrm{Bias} > 20$	$\mathrm{mmfd} > 40$

RESULTS AND DISCUSSION

Initial runs of CLASSIC-GB tested the overall operation of the modelling framework and determined how well the generalised methodology in the runoff-production scheme of CLASSIC-catchment combined with the routing module of HMF-GB. Results are given in Section 3.1. Analysis of these indicated where modifications were required, as described inSection 3.2. The enhanced version was tested on a much larger set of catchments, over a longer time period, with results given in Section 3.3 and discussed in Section 3.4. All testing of CLASSIC-GB used the snowmelt module.

Initial Results

The runoff setup module in HMF-GB was applied using the same parameter value to catchment property relationships as derived for CLASSIC-catchment, but taking the default parameter values for all areas of HOST classes 1, 2 and 3. CLASSIC-GB was run for four years, 1980–1983, with four grid resolutions—10, 5, 2.5 and 1 km—using 6, 12, 24 and 72 time-steps per day respectively (for compatibility with the grid resolution). Those model parameters that are time-step related, and developed for running at a daily time-step, are adjusted for the time-step length. Results are given in Table 3 for the 13 catchments selected for initial testing, for NS and Bias, and compared with results from CLASSIC-catchment.

For the smallest catchment (39081) performance improves as the grid size decreases. For the next two smallest catchments (27034 and 47001) there is significant improvement between 10 km and 5 km resolutions but similar performance for the two higher resolutions. For the remaining catchments the best performance is provided by either the 10 or 5 km resolutions, with no improvement in NS by modelling at a higher resolution, though the Bias may be slightly smaller. The high Bias for the smaller catchments is mainly caused by the discrepancy in area between the contributing area defined by the flow paths and the catchment area defined by the topographic boundary. At the 5, 2.5 and 1 km resolutions all Bias values apart from two are within ±10%, indicative of realistic simulation of the water balance, with all NS values greater than 0.60, again apart from two catchments. Contributing areas for the four grid resolutions are compared with the area defined by a high resolution topographic catchment boundary (Figure 4); catchments with an area greater than 2000 km^2 are adequately defined with the 10 km grid while catchments with an area less than 500 km^2 require a 1 km grid.

Table 3. Values of NS and Bias for 13 catchments, with catchments listed in order of increasing catchment area, for four model resolutions from CLASSIC-GB (best value for each catchment in bold) and for CLASSIC-catchment, for 1981–1983 (with 1980 as 1-year warm-up period)

Station Number	10 km		5 km		2.5 km		1 km		CLASSIC-Catchment	
	NS	Bias	NS	Bias	NS	Bias	NS	Bias	NS	Bias
39081	N/A	N/A	0.68	−33.5	0.78	−23.4	**0.80**	**−13.6**	0.79	5.7
27034	0.39	44.5	**0.78**	−5.5	0.72	**1.5**	**0.78**	2.8	0.59	2.8
47001	0.69	32.0	**0.83**	2.3	**0.83**	**−0.4**	0.82	0.6	0.80	0.3
71001	0.71	18.1	**0.79**	2.5	0.70	**0.0**	0.73	0.4	0.64	4.7
27041	**0.74**	−8.2	0.66	−7.1	0.66	**−4.9**	0.65	−5.3	0.93	3.4
43021	**0.68**	**−10.9**	0.63	−15.6	0.63	−16.8	0.60	−17.6	0.89	−10.1
12002	**0.64**	**2.2**	0.63	−6.7	0.57	−3.4	0.55	−3.4	0.74	-2.3
84013	**0.82**	7.1	0.79	**−1.3**	0.79	**−1.3**	0.80	−2.5	0.85	−1.3
27009	0.81	9.5	**0.83**	−4.7	0.76	4.2	0.80	**3.1**	0.92	8.6
54001	0.72	10.7	**0.77**	**−2.2**	0.66	5.8	0.66	4.6	0.90	11.3
15006	**0.50**	−7.7	**0.50**	−9.4	0.48	**−7.5**	0.46	−7.7	0.85	−7.4
54057	0.74	9.7	**0.81**	**−1.5**	0.78	2.8	0.78	2.1	0.85	6.0
39001	**0.84**	−7.4	0.82	**−4.6**	0.81	−5.9	0.81	−5.9	0.96	−0.9
Average*	0.69	8.3	**0.74**	−4.5	0.70	**−2.2**	0.70	−2.4	0.82	1.3

*excluding 39081.

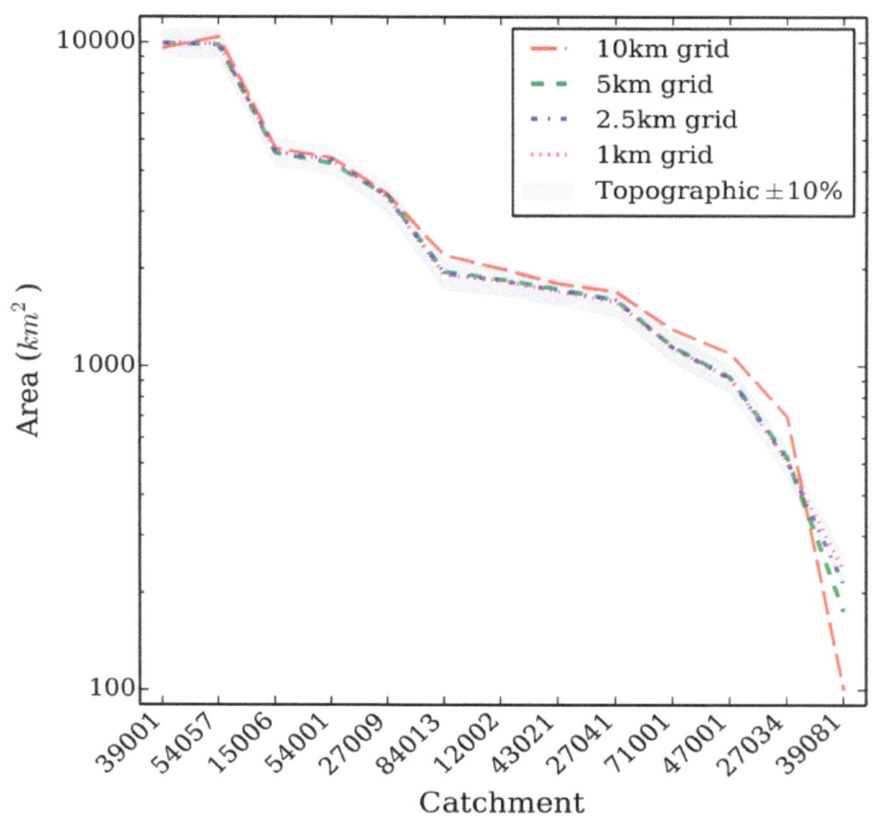

Figure 4.: Comparison of contributing area (km²) for the topographic boundary (used in CLASSIC-catchment) and the four spatial resolutions used in CLASSIC-GB.

Also of interest is how CLASSIC-GB performance for the 13 catchments compares with using CLASSIC-catchment (although datasets used with CLASSIC-catchment are not identical with those in HMF-GB). For most catchments CLASSIC-catchment gives higher NS values than CLASSIC-GB (Table 3), but for the four smallest catchments the position is reversed. Three of these catchments have a high proportion of peat soils (HOST classes 26–29) and/or the fast-responding HOST class 24 for which the smaller grid size (5 km or less in CLASSIC-GB compared with 10 km CLASSIC-catchment) and sub-daily modelling time-step may be an advantage.

Development of CLASSIC-GB

Catchments where performance is considerably worse using CLASSIC-GB are larger catchments with dominant HOST classes of 15 and 17 (15006, 54001, 54057) and those with a high groundwater component—HOST classes 1, 2 and 3 (27041, 39001, 39081 and 43021). For the first group of catchments visual inspection of modelled and observed hydrographs indicated that the modelled flows were insufficiently attenuated; that is the modelled flows have too high peaks, too fast recession and underestimate low flows.

For the second group of catchments differences relate, in part, to use of the same parameter response values for each of the three HOST classes, whereas regional variation was included in CLASSIC-catchment.

Although the routing module in HMF-GB allows parallel surface and sub-surface routing (see Section 2.1), initial CLASSIC-GB runs routed all generated runoff via surface pathways alone (to replicate more closely the channel routing in CLASSIC-catchment, where all runoff is routed together). To investigate whether performance could be improved using the greater attenuation provided by parallel surface and sub-surface routing, CLASSIC-GB was run with runoff split between surface and sub-surface pathways, with all runoff from HOST 1, 2 and 3 and the slow response from other HOST classes taking the sub-surface path. Generally, using sub-surface and return flow makes little difference to generated flows, but using the slower sub-surface wave speeds the hydrograph shape for the three catchments given above (15006, 54001, 54057) is slightly better although the timing of peaks is then often a day late so NS values are considerably lower. The decision was made to preserve the runoff-production scheme as in CLASSIC-catchment as much as possible, so some of the soil-drainage parameter values were reassessed to increase the delay and attenuation in these parameters, with all runoff routed as surface flow. Parameter values were tuned through a combination of visual assessment of daily hydrographs and assessment of measures of fit for ranges of likely parameter values for relevant HOST classes.

To improve the simulation of groundwater flow the regional variation in parameter values of HOST classes 1, 2 and 3, included in CLASSIC-catchment, was introduced into CLASSIC-GB, using a response-time index. The response-time index links a grid with sets of parameters for HOST classes 1, 2 and 3 which are used in the soil-drainage module. Most areas of Britain take default values for the three HOST classes but where there are major aquifers (see the Location Map in [41]) these have been divided into 12 regions allowing for different response-times within the three HOST classes. Each region is used to give a different value to one of the three HOST classes; the other two take the default values. The division is based on sensitivity of water level change to rainfall primarily using groundwater level records (Marsh, personal communication) but also hydrograph recession characteristics for catchments draining significant areas of HOST classes 1, 2 and 3. HOST class 1 (Chalk) is divided into five regions, HOST class 2 (limestone—Jurassic and Magnesian) into four regions and HOST class 3 (soft sandstones—main aquifers are in Permo-Triassic sandstones) into three regions. The other HOST classes use the same soil-drainage parameters for all grids. HOST class 4 is characterised by hard fissured bedrock, including Carboniferous limestone and fissured sandstone. It is therefore quite a diverse group, which combined with water flow through fissures (which may be very variable in size and alignment between catchments) makes it difficult to characterise the drainage response by one set of parameters. Both the spatial heterogeneity and wide geographical distribution mean it is not straightforward to divide HOST 4 into different regions.

Following the two sets of changes described above, NS values are mostly improved; particularly for catchments 15006, 27041, 54001 and 54057. Values for catchments 27034 and 47001 are slightly lower but these are smaller, responsive catchments which perform better with faster routing. Further small changes were made to some of the drainage parameter values so as not to bias the routing too much in favour of larger catchments.

Final Results

The enhanced version of CLASSIC-GB was assessed on a further 41 catchments, listed in the lower part of Table 1. Results for all 54 catchments, for four grid resolutions and three performance measures, are shown in 0, where the catchments are arranged in order of decreasing catchment area. Results are for a longer time period (1991–2000) than for the initial model runs (1981–1983) but the average pattern of performance is similar between the two time periods, indicating relatively stable simulation of flow characteristics. No results are given for the smaller catchments at 10 km resolution as they are too small to be realistically defined at this resolution. The shaded grey bands in 0 depict three levels of performance (see Table 2); the number of catchments within each band is given at the top of the figure. At least 50% of catchments tested achieve a Band 1 (highest) level of performance, with nearly 90% at least Band 2 level, for NS and Bias. The three catchments with consistently low NS values characterise three situations affecting comparison of observed and simulated flows; impact of water transfer and control of flow for power generation (04001), extensive urban development (28066) and simulation of flow in catchments with disparity in permeability between soils and substrata (33019). A more detailed investigation of the results is given in the following sections.

Examples of observed and simulated (5 km resolution) daily hydrographs for eight catchments are shown in Figure 6. The catchments are selected and arranged (left to right, top to bottom) to show the characteristic hydrograph shapes and flow ranges from different combinations of HOST classes with decreasing baseflow contribution. The slowest response, with recession curves dominating the annual hydrograph, is generated from areas underlain by chalk aquifers (HOST 1), while the fastest response, with little baseflow contribution, is generated from peat soils (HOST 26 to 29). The whole range of different observed runoff regimes is generally well replicated by the simulated flows from CLASSIC-GB.

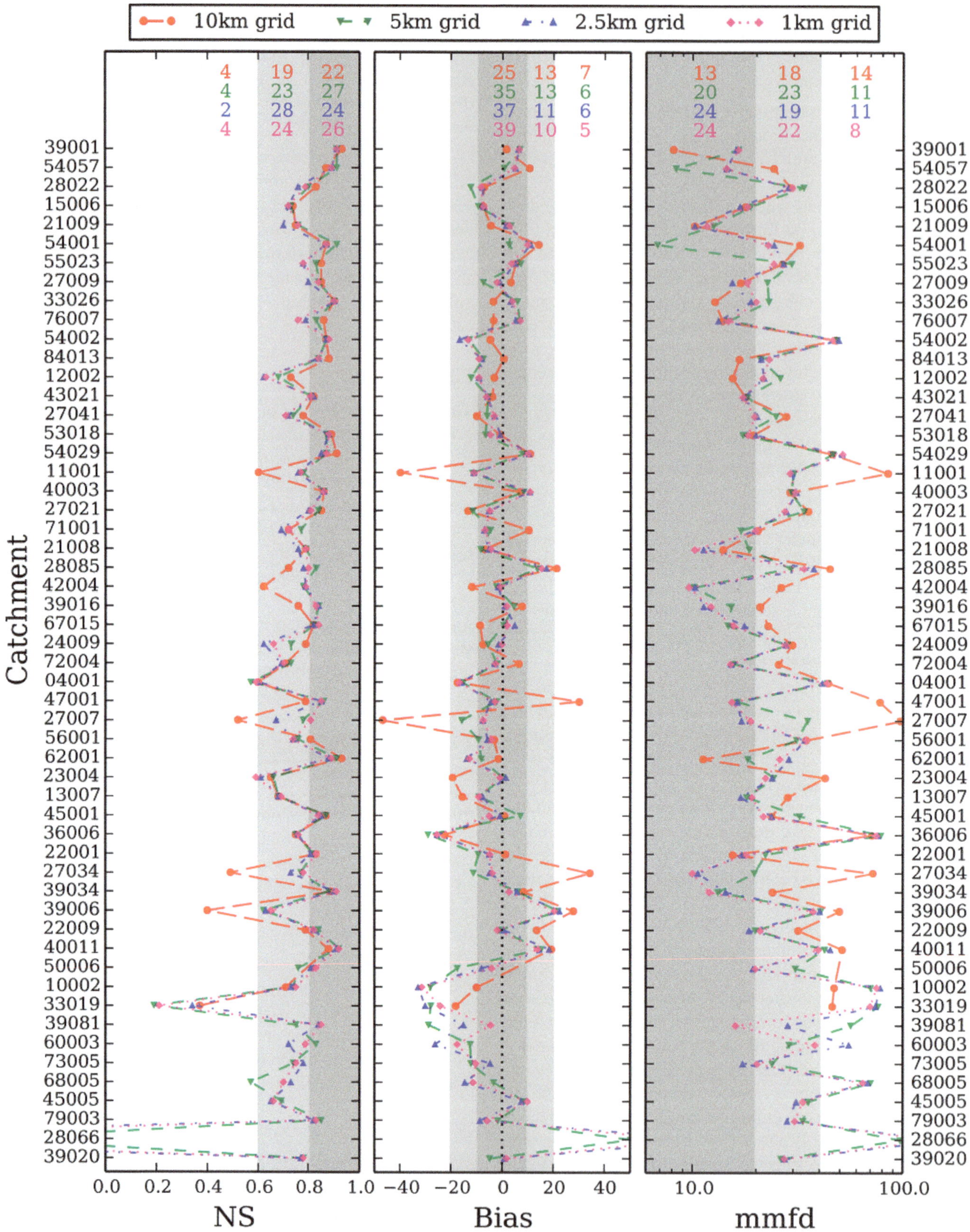

Figure 5: Values of NS, Bias and mmfd for 54 catchments, listed in order of decreasing catchment area, for four resolutions (10 km—red circles; 5 km—green down-triangles; 2.5 km—blue up-triangles; 1 km—magenta diamonds) for 1991–2000 (with 1990 as 1-year warm-up period). Performance bands are indicated by shading (Band 1—dark grey; Band 2—light grey; Band 3—white; see Table 2) with the number of catchments in each band, at each resolution, presented at the top.

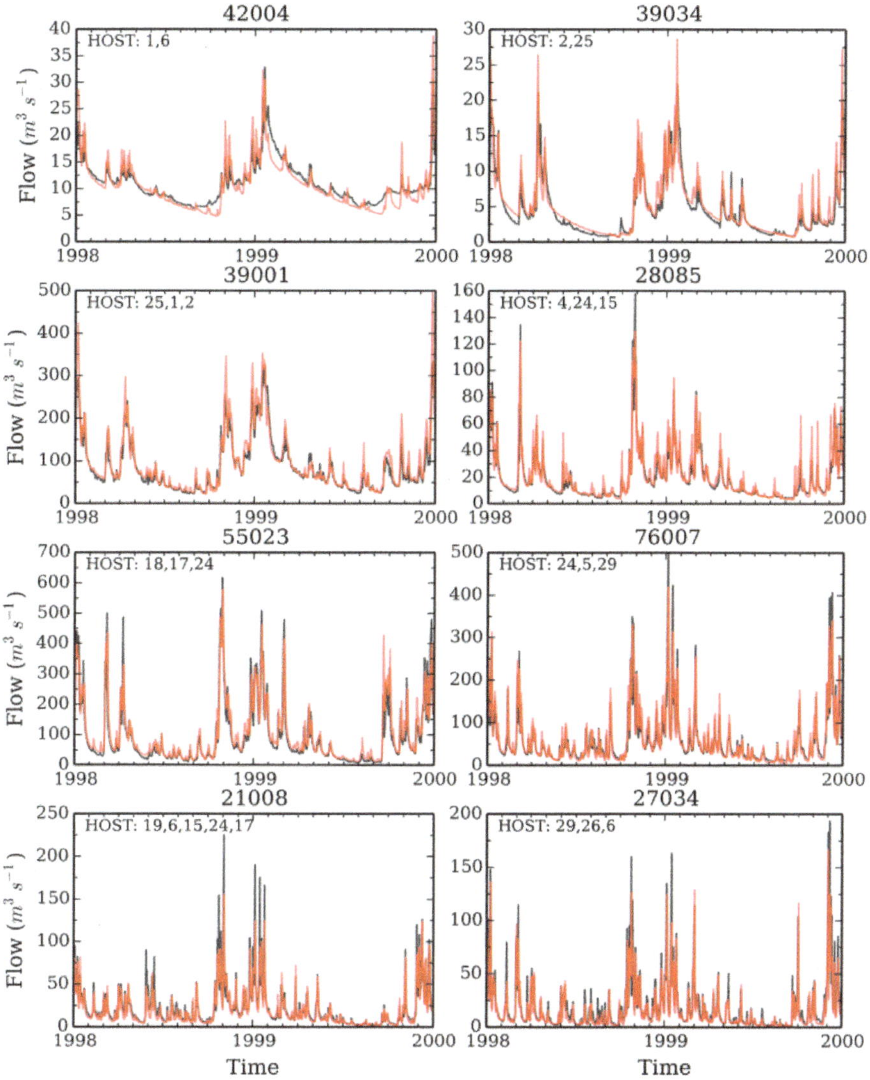

Figure 6: Example observed (black) and simulated (5 km resolution; red) hydrographs for eight catchments for the years 1998–1999. The main HOST classes for each catchment (from Table 1) are given in the top-left of each plot.

DISCUSSION

The results shown in Figure 5 have been averaged in a number of ways (Table 4) to show the impact of model resolution and to help in understanding how well the generalised parameter methodology for setting up CLASSIC-GB is able to reproduce the observed flow regime. The catchment characteristics used are catchment area, SAAR, BFI and maximum altitude. The average value of the three performance measures for all catchments is given at the top of Table 4 (45 catchments for 10 km resolution, 53 for 5, 2.5 and 1 km resolutions; omitting 28066). Comparing the average values across the resolutions shows that the four NS values are similar, the 10 km resolution provides the lowest average Bias and values of mmfd improve with finer resolution.

In terms of catchment area, Table 4 and Figure 5 clearly show that, for all four resolutions, model performance is better for larger catchments and generally improves with finer modelling resolution, though differences from 5 to 1 km are small for NS and Bias. Values of mmfd, though, are lowest at the finer resolutions, for the most part because the catchment area is better defined but also indicating that differences in hydrological processes which occur through the year (with changes in balance between rainfall and evaporation) are being better represented when modelling at a finer resolution. NS and Bias are not as sensitive to differences at the monthly/seasonal scale. For larger catchments (≥2000 km^2) performance is not improved by modelling at lower resolutions than 10 km. While better performance for larger catchment areas is not unexpected, as the setup methodology was developed for such catchments, it also indicates that the range of physical and climatic conditions across large catchments is being well represented through the gridded modelling framework. For smaller catchments, the effect of more detailed hydrological processes than are represented in the model may be evident and the soil-drainage response times and constant routing parameter values may still be slightly biased towards the timing of flow simulation in larger catchments. The noticeably higher Bias and mmfd for small catchments (<350 km^2) also partly reflects percentage differences being accentuated when observed discharge is small and non-natural. In large heterogeneous catchments using generalised parameters is likely to work better as differences in processes and responses across a catchment average out, but in small catchments it is more important to simulate the exact processes taking place, perhaps at a scale which is not represented in CLASSIC. This association between catchment area and model performance of CLASSIC-GB reflects concepts of spatial scale dependency and threshold behaviour [5].

Table 4: Average values of the three performance measures, for four resolutions, averaged overall and grouped by catchment area, $SAAR_{61-90}$, BFI and maximum altitude. The best value of each performance measure for each grouping method is highlighted in bold.

	Number of	10 km			5 km			2.5 km			1 km		
	Catchments	NS	Bias	mmfd	NS	Bias	mmfd	NS	Bias	mmfd	NS	Bias	mmfd
Overall		0.77	**−1.3**	33.3	**0.78**	−4.7	28.9	0.77	−3.6	27.7	**0.78**	−3.3	**26.8**
Area (km²)													
< 350	12	N/A	N/A	N/A	0.72	−9.5	42.7	0.74	−9.4	40.8	0.74	−7.1	37.9
350–999	14	0.71	−1.2	44.0	0.76	−5.5	29.1	0.74	−4.7	26.7	0.76	−5.2	26.3
1000–1999	15	0.78	−3.3	30.2	0.80	−2.8	23.8	0.77	−0.6	23.5	0.78	−0.9	22.9
≥ 2000	12	**0.85**	**0.3**	21.3	0.84	−1.5	**21.2**	0.82	−0.4	**21.2**	0.83	**−0.3**	21.5
$SAAR_{61-90}$ (mm)													
580–749	12	0.77	0.8	36.8	0.75	−4.2	43.9	0.78	−3.6	40.7	0.78	−2.4	38.0
750–949	16	0.79	−3.1	31.4	**0.82**	−5.4	24.7	0.80	−2.8	25.3	0.80	−3.0	25.0
950–1299	12	0.75	−3.2	36.0	0.78	**−0.1**	24.4	0.74	**−0.1**	**21.5**	0.75	−0.5	21.6
≥ 1300	13	0.74	1.7	29.4	0.78	−8.6	24.4	0.75	−7.9	24.5	0.77	−7.1	23.7
BFI													
0.27–0.49	17	0.75	**1.6**	35.9	0.79	−5.5	23.0	0.76	−3.1	**18.9**	0.78	−3.0	19.7
0.50–0.59	20	**0.82**	−3.4	29.6	0.79	−4.4	31.0	0.78	−4.8	32.8	0.79	−4.5	31.4
≥ 0.60	16	0.72	**−1.6**	35.3	0.76	−4.4	32.5	0.76	−2.8	30.8	0.76	−2.3	28.9
Max altitude (m)													
70–299	12	0.76	−2.8	34.5	0.75	−7.2	41.8	0.79	−7.3	39.4	0.77	−6.0	36.7
300–599	13	**0.82**	5.0	33.1	0.81	−1.3	28.1	0.79	−1.2	28.9	0.81	**−0.6**	27.0
600–799	14	0.75	2.1	33.8	0.80	−4.8	24.5	0.76	−2.8	**21.0**	0.79	−2.8	21.4
≥ 800	14	0.76	−7.9	32.1	0.76	−5.7	22.9	0.75	−3.6	23.5	0.75	−4.1	23.7

Looking at results grouped by SAAR suggests that, apart from drier catchments (SAAR < 750 mm), average rainfall has little impact on correspondence between observed and simulated flows. The implication of this is that soil-moisture accounting is being realistically modelled, as otherwise performance would relate to average rainfall. In wet catchments evaporation may be a small percentage of rainfall, and how soil-moisture is modelled may not be a determining factor in model performance. However, in catchments where evaporation is water-limited (PE higher than rainfall, [42]) or there is a finer balance between rainfall and evaporation, then the soil-moisture modelling is more critical. (Examples of how different combinations of rainfall and PE affect AE and runoff are given in [42] for three catchments in Britain; 72004—wet/energy-limited, 62001—energy/water-balanced and 39001—dry/water-limited). For drier catchments the average value of mmfd is higher than for the other rainfall bands, which is not evident for NS and Bias, but flow in such catchments is more likely to be affected by substantial water utilisation which results in high percentage differences at low flows. A similar pattern of results is found when altitude is considered; catchments with the lowest maximum altitude (< 300 m) have the lowest performance and, given the geography of Britain, these tend to correspond with the driest catchments. Performance for catchments with the highest altitude is slightly lower than those in the middle bracket (300–799 m) probably reflecting higher uncertainty in climatic data over uplands and modelling of snow-related processes. The average Bias becomes more negative and NS values decrease for the higher altitude groups, though these groups have the best values of mmfd. The results, showing that performance is predominantly independent of SAAR and altitude, are of importance when using the model for climate change studies (or with future climate data) where the balance between rainfall and evaporation in a catchment may change from wet/energy limited to water/energy balanced, for example.

The dominating factor controlling the time-response of flow is soils and substrata, with every catchment having a unique combination and spatial distribution of soil types and underlying substrate. Differences in runoff characteristics for different combinations of HOST classes are illustrated in Figure 6. The heterogeneity cannot be captured in a single characteristic, as noted in [11], but is partly represented by BFI, used as a measure of soil variation. There is little difference in NS and Bias with BFI for all resolutions, but average mmfd is smallest for catchments with low BFI (<0.50) and highest for those with mid-values (0.50–0.59). These results generally show that modelling using individual HOST soil types (and associated sets of parameter values) in the soil-drainage module is able to simulate the range of observed characteristics of rainfall-runoff in different catchments. Permeable catchments, with high values of BFI, can be the most difficult to simulate with generalised parameters (e.g., [12,43]). In CLASSIC-GB, as in CLASSIC-catchment, a number of features are included to specifically allow for differences between response times in permeable and semi-permeable catchments.

A further problem can occur in permeable catchments, notably chalk catchments, which have a permeable substrate with an overlying drift cover. In this case the response is modelled according to the HOST class of the drift soil but the observed flow response may be that of the permeable substrate. Areas where this occurs can be identified by sparse drainage networks typical of permeable regions in catchments with drift soils, normally associated with a denser channel network. In CLASSIC-catchment this problem was allowed for, where relevant, by manual adjustment of areas of appropriate HOST classes in the soil-drainage module. In CLASSIC-GB a generalisation has been included so that where both HOST

1 and 18 are present then the runoff response from the area of HOST 18 is modelled as HOST 1. For all catchments listed in Table 1where this conjunction occurs the NS values increased. For catchment 39001 the NS value increases from 0.82 to 0.91 (5 km resolution) when both the response-time index and allowance for drift soils are implemented. However, soils other than HOST 18 may have similar impacts, for which it is not appropriate to generalise. This appears to be the case for catchment 33019, which has low performance results for all resolutions. The catchment has a BFI of 0.78 and is entirely underlain by chalk, but has 70% drift cover of which only 14% is HOST 18. Other difficulties in simulating flow in permeable catchments include where the catchment area defined by a topographic boundary differs from that of the groundwater catchment, leading to bias in the water balance. Another HOST class where generalisation may result in poorer performance than expected is HOST 4. This class was noted in Section 3.2 as having a very variable response rate depending on the degree of fissuring and alignment of fissures in the bedrock. Catchment 54029, which has 25% HOST 4, has a Band 1 NS (0.87, 5 km resolution) but relatively poor simulation of mean monthly flow – overestimation of mean daily flows through the summer and underestimation of high flows in the winter resulting in a Band 3 mmfd (46.1, see Figure 5). In this case the response rate is faster than modelled with the default parameter values for HOST 4, whereas these values give good response times for catchment 28085 (Figure 6, 45% HOST 4).

Additional Analyses

The catchment with the lowest performance values is 28066, which is the second smallest catchment selected, but the main reason for selection was substantial urbanisation (urban extent 40% [36]). Visual inspection of observed and simulated hydrographs for 28066 shows reasonable timing of flow simulation but flows are greatly overestimated. The structure of CLASSIC includes a separate modelling pathway for urban areas (no soil-moisture store, urban drainage parameter), with the area determined from the land cover database using that designated as urban plus a proportion of suburban areas. Assessment of this urban pathway with CLASSIC-catchment for catchments with a range of urban percentages has shown generally good simulation of small summer peaks, the main time when contribution from urban areas is evident. Poor performance for a small urban catchment, therefore, suggests either that the method of modelling runoff from urban areas in CLASSIC is not appropriate for such catchments or that the specific characteristics of this catchment are not well simulated with a generalised approach. Fully allowing for urban land cover involves integration between the land cover and soils databases, which is not currently undertaken; allowance for the urban areas has been achieved by subtracting the urban percentage from the areas of the different soil types in a grid square by area weighting. This method may result in inappropriate simulation of flow response from the remaining soil areas, which will be more apparent in a small catchment. CLASSIC-GB was re-run taking the urban area as just urban without adding any proportion of suburban (all suburban areas modelled as grass), which improves the Bias for 28066 (from 51% to 18%) but still has a negative NS value. However, results for five other catchments with urban extent greater than 4.0% indicate that performance is generally better with the original method of estimating impermeable urban areas. But there clearly may be a difference in how catchments dominated by urban development should be modelled compared with catchments which are predominantly rural but contain scattered areas of urban development. Further modelling for other substantially urbanised catchments is required to determine if it is appropriate to use generalised methods for simulation of flow in such catchments.

Another catchment selected to see how well generalised methods perform where there is substantial alteration to the natural flow is 04001 where 'extensive volumes of surface storage [are] controlled for [Hydro Electric] power generation' [36]. Performance measures are Band 2 or 3 but visual inspection of gauged and simulated hydrographs shows that overall hydrological response is well modelled, while differences between the two flow series indicate when the natural flow is interrupted.

A simple example of how CLASSIC-GB can be used in an investigative way is using two methods for determining sub-daily temperatures in the snowmelt module. One (standard in HMF-GB) uses a sinusoidal variation between daily minimum and maximum temperature while the other uses a constant mean daily temperature (average of minimum and maximum). There was little difference in results between the two methods, but for 46% of the catchments tested a variable temperature gives slightly higher values of NS, for 39% the values are the same and for the remaining 15% a constant temperature gives slightly higher values.

CONCLUSIONS

The concept and practicality of a hydrological modelling framework (HMF-GB) has been demonstrated through the transfer of the runoff-production scheme from an existing catchment-based model (CLASSIC-catchment) to simulation of river flow at a national scale (CLASSIC-GB). The transfer is possible because of the generalised methodology for setting parameter values from physiographic properties developed for CLASSIC-catchment. The demonstration shows that the HMF could provide a suitable platform for broadening the application of other hydrological model runoff-production schemes.

Although parameter generalisation may generate lower model performance than individual catchment calibration, the assumed stationarity between catchment properties and hydrological processes ensures consistency of simulation between and within catchments and reduces the parameter uncertainty inherent in the calibration of individual catchments for defined time periods. It also allows simulation of flows at ungauged locations or in catchments where direct calibration is not appropriate due to the quality of the flow record. Generalisation can broaden understanding of hydrological processes (and inform model development) by showing where simulated flow responses consistently differ from those in the observed flow record; that is where a generalised approach does not give a good performance. The performance of CLASSIC-GB has been assessed by comparison of simulated flow with observed but the structure of the framework allows for output of intermediate hydrological variables, such as soil-moisture, which could be compared with appropriate field measurements [44]. Such comparison would also contribute to understanding of relationships between physical catchment properties and model representation and parameter values.

The main adjustment required in the transfer between the two versions of CLASSIC was reassessment of some of the soil-drainage parameter values to allow for the change in the routing procedure from catchment-specific routing parameter values to constant values of land and river wave speeds for connecting all flow paths. Results from the enhanced version of CLASSIC-GB show that performance measures, on average, are comparable with those from CLASSIC-catchment, and results from testing over a large group of catchments show that good model performance can be achieved without direct calibration of model parameters. At least 50% of catchments tested achieve a Band 1 level of performance, with nearly 90% a Band 2 level for Nash-Sutcliffe efficiency and water balance bias.

The ability to run CLASSIC at different grid resolutions within HMF-GB demonstrates the effect of averaging hydrological processes. Large catchment areas (>2000 km^2) can be well simulated with a 10 km grid resolution, a 5 km resolution is adequate for catchments larger than 500 km^2, but a finer resolution is required for smaller areas, mainly to ensure that there is a good correspondence between topographic catchment area and contributing grid-square area. However, in small catchments specific attributes of the catchment, such as urbanisation, may not be well represented by generalised methods of setting parameter values at the spatial scale of hydrological processes currently in CLASSIC. Factors such as spatially uniform land and river routing parameter values and use of daily rainfall data, while modelling at a sub-daily time step, may also be implicated in poorer model performance in small catchments.

Integral to the success of the generalisation is the use of the HOST classification system to generate the rainfall-runoff response times of a grid square by cumulating the runoff from each soil type present in the square. A response-time index has been implemented to improve the simulation of flow in areas with permeable substrates, such as chalk and limestone. This has enabled similar levels of model performance to be obtained for all catchment permeabilities, as defined by BFI. Simulation of runoff regimes may be compromised where there is a disparity between response from the substrate, notably chalk, and overlying, but hydrologically connected, drift soils.

HMF-GB provides a valuable platform for easily testing the effect of alternative model formulations, at different spatial resolutions, over the range of climatic and physical properties found nationally and combined in infinite patterns of catchment heterogeneity. Alternative model approaches include, for example, simple differences as in method of calculation of mean daily temperature in the snow module, through variation in the method for calculating soil-moisture, to the changes required to fully implement the HMF routing scheme through use of parallel surface and sub-surface routing. The availability of a range of other databases, in particular a mapping of European soils onto HOST classes [45], potentially widens the scope of the HMF to continental or global application. Currently, the HMF is setup with natural flow paths and simulates natural river flow; additional modules and appropriate databases could allow for water usage and artificial lateral transfer of water between grid boxes (e.g., through pipes and canals). Databases of, for example, climate change scenarios or different land cover enable wide-scale assessment of such impacts on the flow regime. The flexible infrastructure of the framework enables consistent studies, from catchment-based or detailed fine-resolution modelling to less detailed coarse-resolution modelling using very long runs or applying large ensembles.

ACKNOWLEDGMENTS

This work was supported by the NERC-CEH Water programme.

REFERENCES

1. Todini, E. Hydrological catchment modelling: Past, present and future. *Hydrol. Earth Syst. Sci.* 2007, *11*, 468–482.
2. Brath, A.; Montanari, A.; Moretti, G. Assessing the effect on flood frequency of land use change via hydrological simulation. *J. Hydrol.* 2006, *324*, 141–153.
3. Bastola, S.; Murphy, C.; Sweeney, J. The sensitivity of fluvial flood risk in Irish catchments to the range of IPCC AR4 climate change scenarios. *Sci. Total Environ.* 2011, *409*, 5403–5415.
4. Prudhomme, C.; Wilby, R.L.; Crooks, S.; Kay, A.L.; Reynard, N.S. Scenario-neutral approach to climate change impact studies: Application to flood risk. *J. Hydrol.* 2010, *390*, 198–209.
5. Hrachowitz, M.; Savenije, H.H.G.; Blöschl, G.; McDonnell, J.J.; Sivapalan, M.; Pomeroy, J.W.; Arheimer, B.; Blume, T.; Clark, M.P.; Ehret, U.; *et al.* A decade of Predictions in Ungauged Basins (PUB)—A review. *Hydrol. Sci. J.* 2013, *58*, 1198–1255.
6. Mazvimavi, D.; Meijerink, A.M.J.; Stein, A. Prediction of base flows from basin characteristics: A case study from Zimbabwe. *Hydrol. Sci. J.* 2004, *49*, 703–715.
7. Skaugen, T.; Onof, C. A rainfall-runoff model parameterized from GIS and runoff data. *Hydrol. Processes* 2014, *28*, 4529–4542.
8. Kay, A.; Jones, D.A.; Crooks, S.M.; Calver, A.; Reynard, N.S. A comparison of three approaches to spatial generalization of rainfall-runoff models. *Hydrol. Processes* 2006, *20*, 3953–3973.
9. Young, A.R.; Grew, R.; Keller, V.; Stannett, J.; Allen, S. Estimation of river flow time-series to support water resources management: The CERF model. In Proceedings of Sustainable Hydrology for the 21st Century, 10th British Hydrological Society National Hydrology Symposium, Exeter, UK, 15–17 September 2008; pp. 100–106.
10. Wagener, T.; Wheater, H.S. Parameter estimation and regionalization for continuous rainfall–runoff models including uncertainty. *J. Hydrol.* 2006, *320*, 132–154.
11. Oudin, L.; Kay, A.; Andréassian, V.; Perrin, C. Are seemingly physically similar catchments truly hydrologically similar? *Water Resour. Res.* 2010, *46*, W11558.
12. Bell, V.A.; Kay, A.L.; Jones, R.G.; Moore, R.J.; Reynard, N.S. Use of soil data in a grid-based hydrological model to estimate spatial variation in changing flood risk across the UK. *J. Hydrol.* 2009, *377*, 335–350.

13. Gudmundsson, L.; Wagner, T.; Tallaksen, L.M; Engeland, K. Evaluation of nine large scale hydrological models with respect to the seasonal runoff climatology in Europe. *Water Resour. Res.* 2012, *48*, W11504.

14. Wagener, T.; Wheater, H.S.; Gupta, H.V. *Rainfall–Runoff Modelling in Gauged and Ungauged Catchments*; Imperial College Press: London, UK, 2004.

15. Clark, M.P.; Slater, A.G.; Rupp, D.E.; Woods, R.A.; Vrugt, J.A.; Gupta, H.V.; Wagener, T.; Hay, L.E. Framework for Understanding Structural Errors (FUSE): A modular framework to diagnose differences between hydrological models. *Water Resour. Res.* 2008, *44*, W00B02.

16. Crooks, S.M.; Naden, P.S. CLASSIC: A semi-distributed modelling system. *Hydrol. Earth Syst. Sci.* 2007, *11*, 516–531.

17. Davies, H.N.; Bell, V. Assessment of methods for extracting low resolution river networks from high resolution digital data. *Hydrol. Sci. J.* 2009, *54*, 17–28.

18. Bell, V.A.; Kay, A.L.; Jones, R.G.; Moore, R.J. Development of a high resolution grid-based river flow model for use with regional climate model output. *Hydrol. Earth Syst. Sci.* 2007, *11*, 532–549.

19. Bell, V.A.; Moore, R.J. An elevation-dependent snowmelt model for upland Britain. *Hydrol. Processes* 1999, *13*, 1887–1903.

20. Environment Agency. *Continuous Estimation of River Flows (CERF) Technical Report: Estimation of Precipitation Inputs*; Project SC030240; Environment Agency: Bristol, UK, 2008.

21. Hough, M.; Palmer, S.; Weir, A.; Lee, M.; Barrie, I.A. *The Meteorological Office Rainfall and Evaporation Calculation System: MORECS Version 2.0 (1995). An Update to Hydrological Memorandum 45*; The Met. Office: Bracknell, UK, 1997.

22. Jenkins, G.J.; Perry, M.C.; Prior, M.J.O. *The Climate of the United Kingdom and Recent Trends*; Met Office Hadley Centre: Exeter, UK, 2007.

23. Kay, A.L.; Crooks, S.M. An investigation of the effect of transient climate change on snowmelt, flood frequency and timing in northern Britain. *Int. J. Climatol.* 2014.

24. Boorman, D.B.; Hollis, J.M.; Lilly, A. *Hydrology of Soil Types: A Hydrologically Based Classification of Soils in the United Kingdom*; IH Report No. 126; Institute of Hydrology: Wallingford, UK, 1995.

25. Morton, D.; Rowland, C.; Wood, C.; Meek, L.; Marston, C.; Smith, G.; Wadsworth, R.; Simpson, I.C. *Final Report for LCM2007—The New UK Land Cover Map*; Countryside Survey Technical Report No 11/07; Centre for Ecology & Hydrology: Wallingford, UK, 2011.

26. Morris, D.G.; Flavin, R.W. A digital terrain model for hydrology. In Proceedings of the 4th International Symposium on Spatial Data Handling, Zurich, Switzerland, 23–27 July 1990; Volume 1, pp. 250–262.

27. Fuller, R.M. The land cover map of Great Britain. *Earth Sp. Rev.* 1993, *2*, 13–18.

28. Fuller, R.M.; Smith, G.M.; Sanderson, J.M.; Hill, R.A.; Thompson, A.G. The UK land-cover map 2000: Construction of a parcel-based vector map from satellite images. *Cartogr. J.* 2002, *39*, 15–25.

29. Dunn, S.M.; Lilly, A. Investigating the relationship between a soils classification and the spatial parameters of a conceptual catchment scale hydrological model. *J. Hydrol.* 2001, *252*, 157–173.

30. Marechal, D.; Holman, I.P. Development and application of a soil classification based conceptual catchment scale hydrological model. *J. Hydrol.* 2005, *312*, 277–293.

31. Crooks, S.; Davies, H. Assessment of land use change in the Thames catchment and its effect on the flood regime of the river. *Phys. Chem. Earth* 2001, *26*, 583–591.

32. Kay, A.L.; Crooks, S.M.; Pall, P.; Stone, D. Attribution of Autumn/Winter 2000 flood risk in England to anthropogenic climate change: A catchment-based study. *J. Hydrol.* 2011, *406*, 97–112.

33. Prudhomme, C.; Crooks, S.; Kay, A.L.; Reynard, N.S. Climate change and river flooding: Part 1 Classifying the sensitivity of British catchments. *Clim. Change* 2013, *119*, 933–948.

34. Kay, A.L.; Crooks, S.M.; Reynard, N.S. Using response surfaces to estimate impacts of climate change on flood peaks: assessment of uncertainty. *Hydrol. Processes* 2013.

35. Prudhomme, C.; Haxton, T.; Crooks, S.; Jackson, C.; Barkwith, A.; Williamson, J.; Kelvin, J.; Mackay, J.; Wang, L.; Young, A.; *et al.* Future Flows Hydrology: an ensemble of daily river flow and monthly groundwater levels for use for climate change impact assessment across Great Britain. *Earth Syst. Sci. Data* 2013, *5*, 101–107.

36. National River Flow Archive. Available online: http://www.ceh.ac.uk/data/nrfa/index.html. (accessed on 4 June 2014).

37. *UK Hydrometric Register*; Hydrological Data UK Series; Marsh, T.J., Hannford, J., Eds.; Centre for Ecology & Hydrology: Wallingford, UK, 2008.

38. Nash, J.E.; Sutcliffe, J.V. River flow forecasting through conceptual models part 1—A discussion of principles. *J. Hydrol.* 1970, *10*, 282–290.

39. Crooks, S.; Young, A.; Jackson, C. *Modelling Protocol*, Science Report SC090016/PN4. 2012, p. 33. Available online: http://www.ceh.ac.uk/data/nrfa/nhmp/monthly_hs.html (accessed on 4 June 2014).

40. Biondi, D.; Freni, G.; Iacobellis, V.; Mascaro, G.; Montanari, A. Validation of hydrological models: Conceptual basis, methodological approaches and a proposal for a code of practice. *Phys. Chem. Earth* 2012, *42–44*, 70–76.

41. Hydrological Summaries of the National Hydrological Monitoring Programme. Available online: http://www.ceh.ac.uk/data/nrfa/nhmp/monthly_hs.html (accessed on 4 June 2014).

42. Kay, A.L.; Bell, V.A.; Blyth, E.M.; Crooks, S.M.; Davies, H.N. A hydrological perspective on evaporation: Historical trends and future projections in Britain. *J. Water Clim. Change* 2013, *4*, 193–208.

43. Bell, V.A.; Kay, A.L.; Cole, S.J.; Jones, R.G.; Moore,

R.J.; Reynard, N.S. How might climate change affect river flows across the Thames Basin? An area-wide analysis using the UKCP09 Regional Climate Model ensemble. *J. Hydrol.* 2012,*442–443*, 89–104.

44. UK Soil Moisture Monitoring Network. Available online: http://www.ceh.ac.uk/cosmos/index.html (accessed on 4 June 2014).

45. Schneider, M.K.; Brunner, F.; Hollis, J.M.; Stamm, C. Towards a hydrological classification of European soils: Preliminary test of its predictive power for the base flow index using river discharge data. *Hydrol. Earth Syst. Sci.* 2007,*11*, 1501–1513.

CHAPTER 18

LARGE-SCALE HYDROLOGICAL MODELING AND DECISION-MAKING FOR AGRICULTURAL WATER CONSUMPTION AND ALLOCATION

Yang Yu, Markus Disse, Ruide Yu, Guoan Yu, Lingxiao Sun, Philipp Huttner and Christian Rumbaur

Keywords: MIKE HYDRO, water scarcity, agricultural water allocation, irrigation water demand, irrigation and land use scenarios, Tarim River basin

INTRODUCTION

Water scarcity is expected to increase in the dry regions of Asia [1]. Increasing population and farming practices will increase the stress on water resources in these arid regions. The needs of natural ecosystems have become another source of competition that increases pressure on water resources in the future [2]. The management of irrigation water use continues to pose increasing challenges to decision-makers and planners in developing countries, particularly those facing limited water resources. The Tarim River is one of the largest inland rivers in the world and located in the northwest of China, known worldwide for its extreme climatic situation and vulnerable ecosystem. With the high-intensity and large-scale exploitation of water resources at the upper reaches of the Tarim River, the water curtailment downstream has brought great negative effects for the ecosystem in the Tarim River basin [3,4].

The Tarim River originates at the confluence of the Aksu, Yarkand, Hotan and Kaidu-Konqi rivers, and flows into the Taitema Lake, east of Xinjiang. With a main stem of 1321 km, it is the longest inland river in China. The total area of the Tarim River basin is 1.02×10^6 km^2. The basin contains 42 counties and cities, with a total population of 8.26×10^6. The basin has become the most important source of cotton in China. Due to relatively low rainfall in the region, irrigation is the primary source of water for agriculture along the Tarim River [5]. Although the volume of flow from the river's headwaters is sensitive to precipitation [6], the main stem of the river is dominated by human activities such as agricultural irrigation [7]. Water allocations specify the amount of water that can be extracted each year by license holders. During the past six decades, as conflicts among the water users from upstream to downstream has continuously been severe, and intensive exploitation of water resources has caused serious environmental problems in the basin, a comprehensive spatial-temporal assessment of variability in flows and water allocation is imperative to work towards sustainable management of water resources on the Tarim River.

Previous studies show that, due to the effects of climate change, annual runoff in the river's headwaters has significantly increased since the 1950s. Average air temperature experienced a significant increase in the last few decades, and average annual precipitation increased by 6–8 mm per decade [6]. However, surface runoff in the main stem has dramatically decreased in the 1970s, 1980s and 1990s [7]. Since the 1970s, less than 1/4 of the discharge from the main stem of the Tarim River has been reaching the downstream areas in the average year. However, the lower reach of the river serves as the "Green Corridor" in protecting the natural environment from encroachment by the Taklamakan and Kuluk deserts. The Green Corridor is an essential ecological corridor for maintaining ecosystem balance in the lower reach. Its area has sharply reduced because of serious wind erosion,

Citation: Yu, Y.; Disse, M.; Yu, R.; Yu, G.; Sun, L.; Huttner, P.; Rumbaur, C. Large-Scale Hydrological Modeling and Decision-Making for Agricultural Water Consumption and Allocation. Water 2015, 7, 2821-2839. doi:10.3390/w7062821.

land desertification and degeneration of natural vegetation [4]. This indicates the challenge of maintaining sustainable water resources in the river oases. In this extremely arid region close to Taklamakan Desert, water is the key factor in maintaining local ecosystems. Increased farming and water use since the 1950s have led to significant hydrological and environmental degradation [8,9,10,11,12,13]. In this context, water saving and higher crop irrigation efficiencies are crucial to the future allocation of water resources.

Large-scale hydrological models have been used to assess the impacts of land use changes and other human activities on water resources [14,15,16]. These models represent a compromise between the limited insight obtained from a lumped model and the large amount of data and computational power required from a distributed model. Consequently, a more suitable approach would be a semi-distributed model wherein the overland flow and unsaturated zone are represented by their lumped-parameter semi-empirical counterparts to simplify the overall analysis [17]. MIKE HYDRO is an integrated, multipurpose, map-based decision support tool developed by the Danish Hydraulic Institute (DHI). The Basin model type is the MIKE HYDRO module used for a variety of model applications, covering integrated water resources management (IWRM), water resources assessment, water allocation, reservoir operation and other types of analysis [18]. It enables detailed simulations of water resources and land use in the catchment areas. The scale of these models may range from large river basins to smaller local projects.

This study aims to simulate the possible effects of irrigation and land use scenarios in water savings for the main stem of the Tarim River. Higher crop irrigation efficiencies and optimized water allocation strategies are discussed to seek practical approach of increasing ecological water for the restoration of downstream Green Corridor. The study is an important part of the five-year SuMaRiO (Sustainable Management of River Oases along the Tarim River/China, 2011–2016) project. Section 2describes the development of the large-scale hydrological model for the main stem of the Tarim River, including routing method, definition of the four sub-catchments, crop modules and the calibration of monthly data from three gauging stations. Section 3 describes the design of scenarios for irrigation and land use at the sub-catchment level. The results of modeling and analyses are presented in Section 4. Finally, key study findings and conclusions are presented in Section 5.

HYDROLOGICAL MODELING

Study Area

The study area of main stem Tarim River starts from Alar and ends at Taitema Lake (Figure 1). The area is characterized by an extremely arid desert climate with low precipitation and high evapotranspiration. The average annual temperature ranges from 10.6 to 11.5 °C and precipitation from 17.4 to 42.0 mm [19]. In this arid region, the annual reference evapotranspiration may reach 3000 mm and is relatively high in July and August. Because of the region's aridity, no overland flow is generated from rainfall at the alluvial plains of the main stem. Peak flows, caused by melting glacier water in the upper mountain areas of the Tarim River basin, occur in July, August and September. River discharge data is collected from four gauging stations: Alar, Xinqiman, Yingbaza and Qiala (Figure 1).

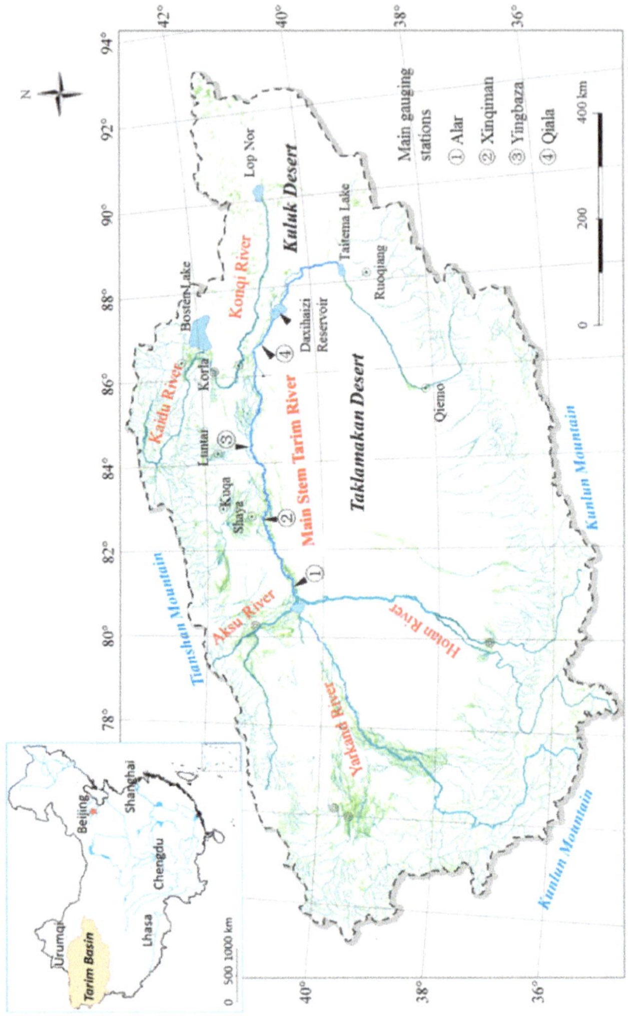

Figure 1: Tarim River basin and main stem river gauging stations.

Average annual discharge in the main stem during the last 60 years is approximately 4.5 × 10^9 m^3 (Figure 2). Discharge of upper and lower reaches were measured in Alar and Qiala gauging station, respectively. Discharge of the lower reach indicates a significant drop from the 1970s, due to overexploitation of water resources in the upper and middle reaches of the river basin. However, since the Tarim River basin comprehensive management project funded by the Chinese central government was initiated, this decline has reversed during the past decade. The increasing availability of water has had a positive effect on the recovery of the Green Corridor in the lower reach [20]. However, water scarcity still remains a critical problem in this arid region.

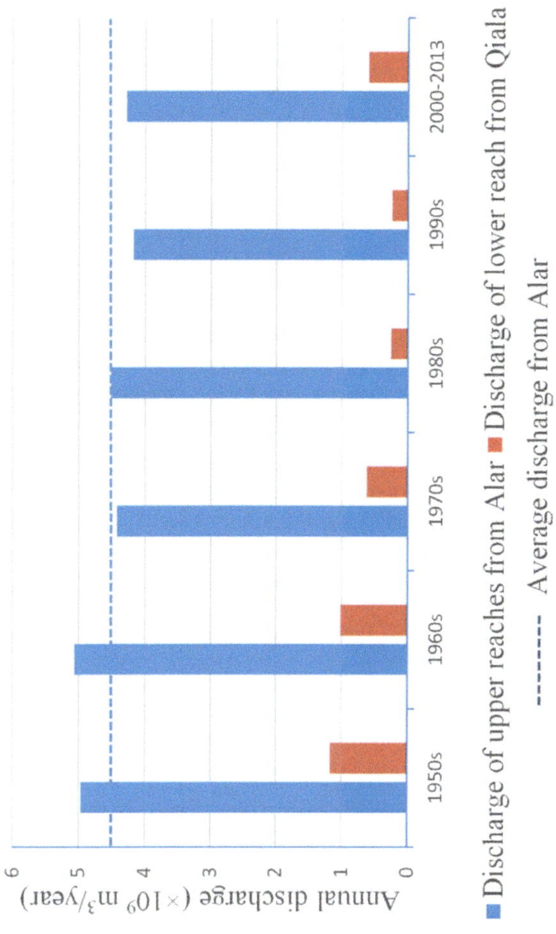

Figure 2: Annual discharge of the main stem Tarim River in the past six decades.

MIKE HYDRO Model Setup

MIKE HYDRO is a comprehensive deterministic, semi-distributed and physically-based modeling system for the simulation of water flow, water consumption, water quality, and sediment transport. It has an integrated modular structure with basic computational modules for hydrology and hydrodynamics. One basic MIKE HYDRO module is MIKE HYDRO basin, containing modules for evapotranspiration, snow melt, overland and channel flow, crop irrigation, and exchange between aquifer and rivers. A map layer coordinates the parallel running of the process components. Rivers and catchments are depicted in the map layer. Parameters and data requirements are outlined in a tabular layer. Simulation results are obtained from a result layer.

Sub-Catchments and Data

In this study, the study area was divided into four sub-regions, namely Alar-Xinqiman (A), Xinqiman-Yingbaza (B), Yingbaza-Qiala (C), and Qiala-Taitema Lake (D). The sub-regions were named "sub-catchments" in the model (Figure 3) and their borders were automatic delineation from MIKE HYDRO, rather than real catchment borders. Because the hydrological parameters within each sub-catchment were aggregated, the region borders did not affect the simulation. The sub-catchments were separated by the gauging stations shown in Figure 1. In each sub-catchment, the hydrologic parameters were aggregated in an effort to simplify the overall analysis. In the main stem Tarim River, the upper, middle and lower reaches extend from Alar to Yingbaza, Yingbaza to Qiala and Qiala to Taitema Lake, respectively. The borders of sub-catchments are automatic catchment borders from MIKE HYDRO. Because the features in each sub-catchment are lumped together, the shapes of sub-catchment borders does not effect on the simulation results.

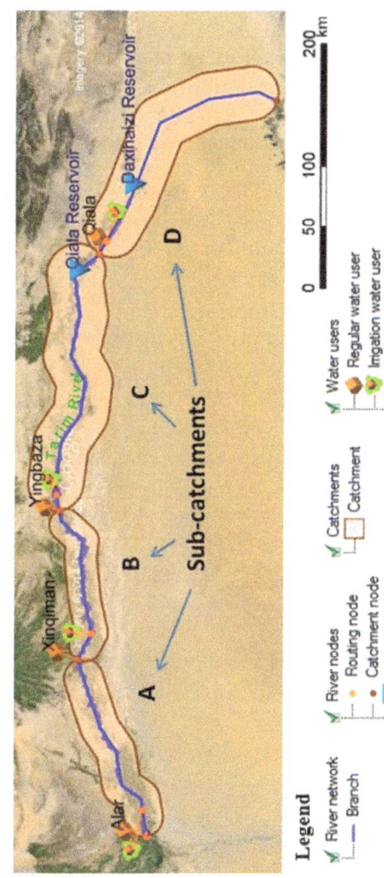

Figure 3: Sub-catchments in the main stem of Tarim River in MIKE HYDRO.

River nodes, branches, catchments, reservoirs, water users and their connections were all specified in a network model (Figure 3). The model design comprised two types of water users considered by MIKE HYDRO: "irrigation water users" (e.g., cotton, wheat and tomatoes) and "regular water users" (non-irrigation water users). Each irrigation water user was assigned with one irrigation field wherein different irrigation methods, crop modules and soil types were specified. On each irrigated field, a crop sequence was characterized by a crop type, sowing date and reference to the irrigation method used to irrigate the crop. A crop sequence lasted until the end of the previous growth stage [18]. For each regular water user, a water demand time series was specified to represent the total amount of water required to fulfill domestic and ecological water demand in each sub-catchment. In the middle and lower reaches, the Qiala and Daxihaizi reservoirs were included to simulate the operations using operating rule curves. Since both reservoirs had significant surface areas, evaporation and infiltration of reservoirs were factors important to water balance calculations. Data for parameterization in the study were acquired and sorted by Xinjiang Institute of Ecology and Geography, Chinese Academy of Sciences (CAS).

Muskingum Routing

The Muskingum method is a commonly used hydrologic routing method in situations requiring a variable storage-discharge relationship [21]. The Muskingum method models the storage volume of flooding in a river channel. The key parameters in Muskingum routing are travel time (K) and weighting coefficient (X). The value of X depends on the shape of the wedge storage, and ranges from 0 for reservoir-type storage to 0.5 for a full wedge.

The Muskingum method assumes that water surface in the reach is a uniform continuous surface profile between upstream and downstream ends of the section. It also assumes that K and X are constant through the range of flows [22]. The Muskingum parameters K and X are best derived from stream flow measurements. In natural rivers, X has a value between 0 and 0.3 with a mean value of 0.2. Greater accuracy in determining X may not be necessary because the results are relatively insensitive to the value of this parameter. In this MIKE HYDRO model of the Tarim River, X was chosen to be 0.2. The measured K values are shown in Table 1.

Table 1: Muskingum routing parameter K for sub-catchments

Muskingum Routing	Upper Reaches		Middle Reach	Lower Reach
Sub-Catchment	A	B	C	D
River Length (km)	189	258	398	428
K (h)	51	86	158	198

Crop Factors and Growth Stages

The FAO-56 Dual Crop Coefficient method applied in the model concerns the parameters to be assigned to each crop. Those parameters included the share of the total irrigated area devoted to crops, number of sowing days for each crop, root depth (RD), maximum height (MH), basal crop coefficient (K_{cb}) and length (days) of the growing stage for each crop. Growing stages were categorized as follows: initial stage (INI), development stage (DEV), mid-season (MID), and late season (LAT). Seven crop modules were established in MIKE HYDRO based on field surveys and statistical data (Table 2).

Table 2: Crop factors and growth stages in study area. Data on the share (%) and sowing day were collected from the CAS. Length (days), RD (mm), MH (m), and K_{cb}values were based on a FAO publication [23], with some adjustment based on field surveys of the study area

Crops	Share (%)	Sowing Day	Length (Days)				RD (mm)	MH (m)	K_{cb}		
			INI	DEV	MID	LAT			INI	MID	LAT
Wheat	2.7	03.21	15	25	40	20	1500	1	0.4	1.2	0.5
Maize	1.6	04.16	20	25	60	15	1700	2	0.4	1.2	0.7
Sugarbeet	5.1	03.26	25	35	60	45	1200	0.5	0.5	1.2	0.8
Bean	1.7	04.21	20	30	30	10	700	0.4	0.4	1.1	0.9
Melon	2.1	04.01	25	35	40	20	1500	0.4	0.5	1	0.8
Cotton	82.6	04.21	25	45	50	40	1700	1.5	0.5	1.2	0.8
Tomato	2.1	04.11	35	40	50	25	1500	0.6	0.5	1.2	0.8

Parameterization was based on the average values of the irrigated fields in the sub-catchments. K_{cb} was defined as the ratio of crop evapotranspiration over the reference evapotranspiration. In the initial MID and LAT stage, K_{cb} was assumed to be constant and follow a linear variation between INI and MID. The relationship of reference evapotranspiration (ET_0), crop evapotranspiration (ET_c) and actual evapotranspiration (ET_a) is given in Equations (1) and (2) as follows.

$$ET_c=K_{cb}\times ET_0 \quad (1)$$

$$ET_a=K_s \times ET_c \quad (2)$$

where K_s is the water stress coefficient that describes the effect of water stress on crop transpiration and was determined in MIKE HYDRO by considering soil water availability in the irrigated field. Due to the higher performance of the FAO-56 Penman-Monteith method, ET_0 was computed from meteorological data. ET_c was calculated by the crop module, which was influenced by the determination of crop factors [23]. ET_a was calculated by MIKE HYDRO, which identified the actual rate of crop evapotranspiration under the effects of soil water stress.

The sowing days for the crops differs each year due to weather conditions in the study area. Wheat refers to spring wheat in the crop module. Watermelon and muskmelon were considered in the same category as melons. Cotton is the dominant crop in the sub-catchments, representing 82.6% of the total irrigated area. Maximum RDs of cotton depend largely on soil temperature and moisture condition. Soil salinity affects crop development, crop transpiration and, hence, biomass production and harvestable yield. Water in the root zone becomes less available for root extraction when salts build up in the soil profile [24]. In some cases, the cotton root of cotton may reach a depth of 2.2 m to obtain water. After a few tests in the MIKE HYDRO model, the simulation results were not sensitive to the variance of RD around the average level. Therefore, it was regarded as a reliable method for determining average values in the module. Chinese jujube is also widely cultivated in the study area. However, it does not fit into a standard crop module. Therefore, in the model, jujube was considered to be a regular water user rather than an irrigation water user.

Discharge and Calibration

The MIKE HYDRO model for the main stem Tarim River was calibrated manually for stream flow using monthly data from 2006 to 2008 (Figure 4). The aim of the calibration was to input discharge data to investigate water allocation scenarios, rather than input climate data to investigate river discharge. Therefore, rainfall-runoff model was not essential in the calibration scheme and there were no future predictions.

Calibration was mainly conducted on water losses. Evaporation, seepage losses and soil porosity are optimized values in the model calibration. All these three parameters did not possess very large influence on the stream flow. Evaporation data was computed from meteorological data by FAO-56 Penman-Monteith method. Since the climate data was only acquired from Alar meteorological station, the evaporation data along the main stem Tarim River was given a range of 95%–105% for calibration. The calibrated values of evaporation varied slightly different from year to year, and from sub-catchment to sub-catchment. Based on the information collected from the study area, seepage losses were within a range from 0.0007 to 0.0011 by the fraction of seepage losses from stream flow. The calibration results were 0.0007 for all the sub-catchments. Soil porosity was another modified parameter during the calibration. Different types of soil (sand, loam, clay, *etc.*) possessed different ranges of porosities, and the calibrated values varied slightly in different sub-catchments.

NAM automatic calibration was also conducted in the model (Table 3), but it possessed almost no effect on the stream flow. This phenomenon corroborated the theory that because of the region's aridity, no overland flow was generated from rainfall in the study area.

Table 3: Important parameters for NAM automatic calibration

NAM Parameters	Parameter Descriptions	Units	Value Ranges	Calibrated Values
Umax	Maximum water content in surface storage	mm	10–20	17.79
Lmax	Maximum water content in root zone storage	mm	100–300	166.25
CQOF	Overland flow runoff coefficient	-	0.1–1	0.51
CKIF	Time constant for routing interflow	h	200–1000	533.28
CK1	Time constant 1 for routing overland flow	h	10–50	22.98
CK2	Time constant 2 for routing overland flow	h	10–50	10
TOF	Root zone threshold value for overland flow	-	0–0.99	0.56
TIF	Root zone threshold value for interflow	-	0–0.99	0.53
TG	Root zone threshold value for groundwater recharge	-	0–0.99	0.03
CKBF	Time constant for routing base flow	h	1000–4000	2179.01
CQLOW	Lower base flow, recharge to lower reservoir	percentage	0–100	0
CKLOW	Time constant for routing lower base flow	h	1000–30,000	10,000

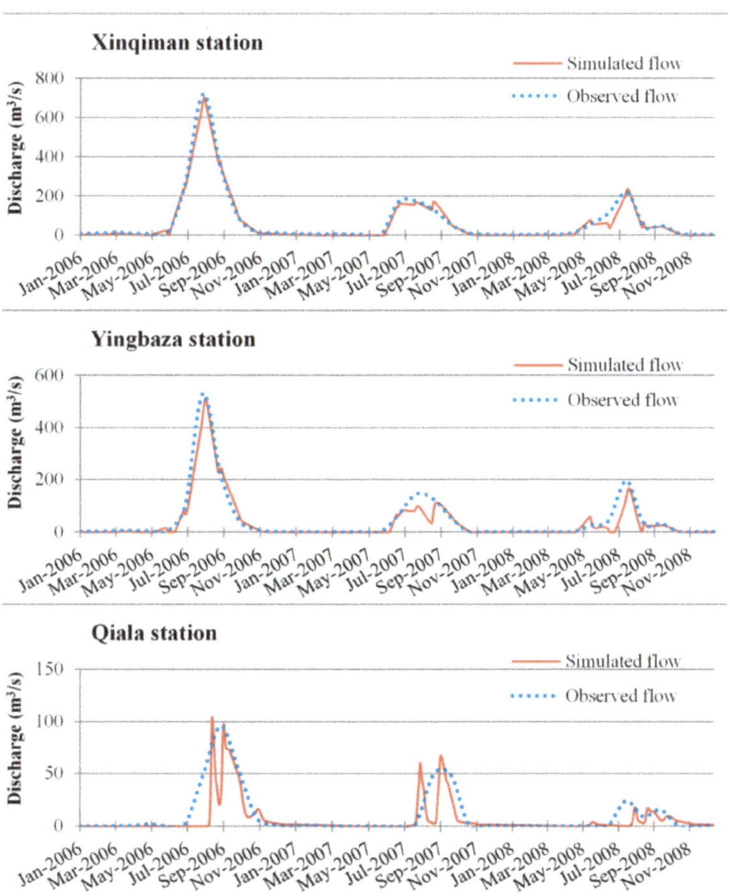

Figure 4: Observed and simulated discharge from three gauging stations: 2006–2008.

The simulation period was not enough to cover the extreme climate situations. Because river discharges of the main stem were around the annual average level in 2006–2008, the simulation results did not take into account the extreme arid years (e.g., in 2009, discharge in the main stem was less than half of the annual average discharge). Since the simulation results were not mainly dependent on climate conditions, as long as it was not an extremely dry year, the study would be suitable for applications in other cases.

The model adequately reproduced the patterns of observed discharges and their magnitudes, although the simulated flow was underestimated. The reason of this underestimation likely resulted from the absence of groundwater in the model. Groundwater is pumped for irrigation in the study area. Since the 1980s, the water table has annually declined at a rate of 20 cm, attributable to increases in water resources development and utilization [9]. In the model, groundwater use was not considered and irrigation was supplied entirely by the surface water. Consequently, the curve for simulated flows would experience an obvious reduction during the crop-growing season compared with that for the observed flow. The simulated peak flow at the Qiala station was lower than the observed flow. Presumably, it was caused by the pumping water from Boston Lake, which increases flows in the lower reach. The influence of this water transfer was evaluated by Xu *et al.* [20], who concluded that the conveyance of water to the lower reach of the Tarim River has a positive effect on local agricultural development and the river ecosystem.

To quantitatively evaluate the performance of the model calibration, four error indices were considered: (a) the Nash-Sutcliffe efficiency (NSE); (b) the root mean square error (RMSE); (c) the RMSE-observations standard deviation ratio (RSR); and (d) the % Bias. NSE is defined in Equation (3) and compares the relative magnitude of squared residuals to the variance of the observed flow [25]. RMSE in Equation (4) is used to examine the square root of the mean squared difference between the observed and simulated flows. RSR standardizes RMSE using the standard deviation in the observations [26] in Equation (5). The % Bias is given in Equation (6) to assess the values of residuals to the observed flow from the gauging stations.

$$NSE = 1 - \frac{\sum_{t=1}^{t=n}[(Q_{\text{sim}})_t - (Q_{\text{obs}})_t]^2}{\sum_{t=1}^{t=n}\left[(Q_{\text{obs}})_t - \left(\overline{Q_{\text{obs}}}\right)_t\right]^2} \tag{3}$$

$$RMSE = \sqrt{\frac{1}{n}\sum_{t=1}^{t=n}[(Q_{\text{sim}})_t - (Q_{\text{obs}})_t]^2} \tag{4}$$

$$RSR = \frac{RMSE}{\sqrt{\frac{1}{n}\sum_{t=1}^{t=n}\left[(Q_{\text{obs}})_t - \left(\overline{Q_{\text{obs}}}\right)_t\right]^2}} \tag{5}$$

$$\%\,Bias = \frac{\sum_{t=1}^{t=n}[(Q_{\text{sim}})_t - (Q_{\text{obs}})_t]}{\sum_{t=1}^{t=n}(Q_{\text{obs}})_t} \times 100 \tag{6}$$

where Q_{sim} is the simulated discharge (m^3/s), Qobs is the observed discharge (m^3/s), $\overline{Q_{obs}}$ is the mean observed discharge (m^3/s) and n is the number of observations, with time intervals t.

The statistical performances from Table 4 could be considered satisfactory compared to the model evaluation guidelines by Moriasi *et al.* [27]. The simulated discharges were very similar to the observed discharges from the gauging stations, especially the Xinqiman and Yingbaza stations in the upper and middle reaches. Because the efficiency coefficient was sensitive to extreme values, the high values of the NSE (0.88, 0.86 and 0.92) indicated a good match of simulated and observed discharges during flood seasons. RMSE values at the Xinqiman and Yingbaza stations were acceptable in light of the high discharges at both stations. RSR standardizes RMSE by including the standard deviation of the observations, and both values represented good ratings. The increasing tendency of the % Bias from upstream to downstream could be attributed to the absence of groundwater discharge and the pumping of water near the Qiala station. If daily values rather than monthly values are used, it is likely that the calibration would have a lower performance. However, daily time series for the calibration of river flows are not available.

Table 4: Evaluation of calibration performance for three gauging stations: 2006–2008

Gauging Stations	NSE	RMSE (m^3/s)	RSR	% Bias
Xinqiman	0.88	14.7	0.11	−2.41
Yingbaza	0.86	11.53	0.12	−3.42
Qiala	0.92	3.58	0.10	−8.24

The model was lacking of validation mainly because of data limitation. This problem would have influence on the accuracy of the model, especially on the river discharge, evaporation and seepage losses. Consequently, the model could not provide reliable quantitative time series data on water losses, but as long as the qualitative effects and the mutual influences of the four sub-catchments were still valid, simulation results on the water consumption and allocation would be reliable. Furthermore, since river discharge data in the model was not generated from rainfall-runoff model, and no future predictions were made by the MIKE HYDRO model, the accuracy of the simulation results would not be largely affected by the absence of validation. Other authors have also experienced similar difficulties when monitoring data were limited [28,29].

DESIGN OF SCENARIOS

Based on recent trends in agricultural practices, different scenarios were developed. The proposed scenarios were designed to provide stakeholders and decision makers with improved insight into water scarcity and solutions for agricultural water allocation.

Irrigation Scenarios

Total Available Water (TAW) Scenarios

When calculating the soil-water balance, the amount of water stored in the root zone can be expressed as an equivalent water depth (W_r) or as root zone depletion (D_r). TAW is the amount of water that a crop can extract from its root zone. Its magnitude depends on the soil type and rooting depth. The water content above field capacity cannot be held against the forces of gravity and will drain out because the water content below the wilting point cannot be extracted by plant roots. Accordingly, TAW in the root zone is the difference between the water content at field capacity and the wilting point. At field capacity, D_r is zero and at the permanent wilting point D_r is equal to TAW [23].

The design of the TAW scenarios assumed that irrigation starts when soil moisture content reaches the specified fraction of TAW. From TAW = 0.7 to TAW = 0.1, seven scenarios were investigated to show the yield performance by different crops. As crop yield declined with the decrease of TAW, the primary purpose of the scenarios was to find the suitable fraction of TAW for each crop at which less water is consumed while maintaining a relatively high crop yield.

Irrigation is essential for stabilizing and increasing crop yields. Stewart [30] derived the relationship between the decrease in relative yield and the relative evapotranspiration deficit. The water use-yield relationship was determined using Equation (7).

$$(1-\frac{Y_a}{Y_m}) = K_y(1-\frac{ET_a}{ET_c}) \tag{7}$$

where Y_a is the actual harvest yield (t/ha), Y_m is the maximum harvest yield (t/ha) and K_y is the yield response factor representing the effect of reduction in evapotranspiration on lost yield. K_y values were obtained from FAO Irrigation and Drainage Paper No. 33 [31].

Water-Saving Irrigation Scenarios

In the study area, the implementation of water-saving irrigation is developing rapidly by the method of drip irrigation under mulch (DIUM). By applying DIUM method in the irrigation fields, spray loss (SL) and wetting fraction (WF) will be significantly reduced compared with sprinkler irrigation. SL corresponds to the fraction of irrigation water that is evaporated before reaching the soil surface. WF determines the fraction of field surface being wetted during irrigation. SL and WF values are calculated from Equations (8) and (9):

$$SL=SL_S \times\%SL_S+SL_D \times\%SL_D \tag{8}$$

$$WF=WF_S \times\%WF_S+WF_D \times\%WF_D \tag{9}$$

where SL_S is the SL for sprinkler irrigation, SL_D is the SL for DIUM, WF_S is the WF for sprinkler irrigation, WFD is the WF for DIUM. Each of these parameters is multiplied by its percentage applied in the irrigated fields. Only the sprinkler irrigation and DIUM methods were considered in this scenario. Consequently, the sum of $\%SL_S$ and $\%SL_D$ as well as the sum of $\%WF_S$ and $\%WF_D$ were assumed to be 1.

DIUM is mainly applied in the cotton fields where the ground surface was approximately 80% covered with transparent polythene film as mulch [32]. SL is set to be 0.5 for sprinkler irrigation and 0.1 for DIUM in the model. WF is close to 1 in sprinkler irrigation and reduced to 0.1 for DIUM in MIKE HYDRO irrigation module [18].

Five water-saving scenarios were designed based on the percentage of DIUM applied in the total irrigated area, 10%, 30%, 50%, 70% and 100%, respectively. The primary goal of these scenarios was to investigate how much water could be saved by applying DIUM to the irrigated fields.

Land Use Scenarios

Three land-use scenarios were designed based on the information collected from farmers, surveys, interviews with decision makers in the region, and the Tarim River basin water resources management ordinance in Xinjiang Uygur Autonomous Region. The ordinance was created by the local government for the sustainable management of water resources. The impetus for these scenarios included eco-system protection, high prices of agricultural products and government regulations restricting agricultural water use in the Tarim River basin. In each sub-catchment the hydrological features were lumped together, so the spatial distribution of land use within each sub-catchment was not considered. The land use scenarios were considered among the sub-catchments. The three scenarios included the agricultural land use decrease scenario (LUD scenario), the land use increase scenario (LUI scenario) and crop type change scenario (CTC scenario). These scenarios were considered independently based on practical situations and local policies.

The LUD scenario reflects the government policy to restrict agricultural water use in the upstream, thus increasing free-flowing water for downstream areas. Under this scenario, approximately 20% of the agricultural land would be abandoned in sub-catchments A and B.

The LUI scenario assumes more available water in downstream areas resulting from new water conveyance projects in the lower reach [20]. Under this scenario, the amount of new agricultural farmland in sub-catchment D would be increased by 20%.

The CTC scenario reflects the proposal for using less irrigation water by substituting different types of crops. Apocynum grows in the arid climate of Central Asia and provides an income for local people. It withstands higher levels of soil salinity and consumes less water than cotton [33]. Apocynum is used as a medicinal plant and a fiber crop. The leaves provide raw material for tea and the stems provide raw material for the textile and paper industries. The CTC scenario assumed that 20% of the cotton would be replaced by apocynum in all the sub-catchments.

RESULTS AND DISCUSSION

Actual Crop Evapotranspiration (ET_a) and Deep Percolation (DP)

While ET_a refers to the upward loss of water by the vegetated surface, deep percolation (DP) is defined as the downward movement of water through the soil profile beyond the root zone. The results of simulations for ET_a and DP in the four sub-catchments are shown in Figure 5. ET_a primarily occurs from June to September, while DP primarily appears in a shorter period from July to August. Compared with ET_a, the shorter period of DP may be due to low soil moisture, which restricts crop growth in this arid region [34]. The rapid increase and decrease of DP at the beginning and end period indicates that DP mainly occurs when soil moisture reach a certain content. ET_a conformed to DP in a slowly increasing trend from sub-catchment A to B and from B to C. In sub-catchment D, both factors showed a dramatic drop, reflecting the water shortage in this lower reach [5]. The total ET_a for 2006 in sub-catchments A–D was 666, 709, 864 and 190 mm, respectively. This reflected high water stress in the lower reach of sub-catchment D. The highest levels of ET_a and DP both occurred in sub-catchment C during July. Irrigation water consumption was relatively high in this district. In the non-growing season, there was no ET_a or DP. This finding was consistent with Yuan *et al.* [35], who found that in the early and late stage of the dormancy period of vegetation, ET_a was approximately zero, whereas the level of ET_c was still large.

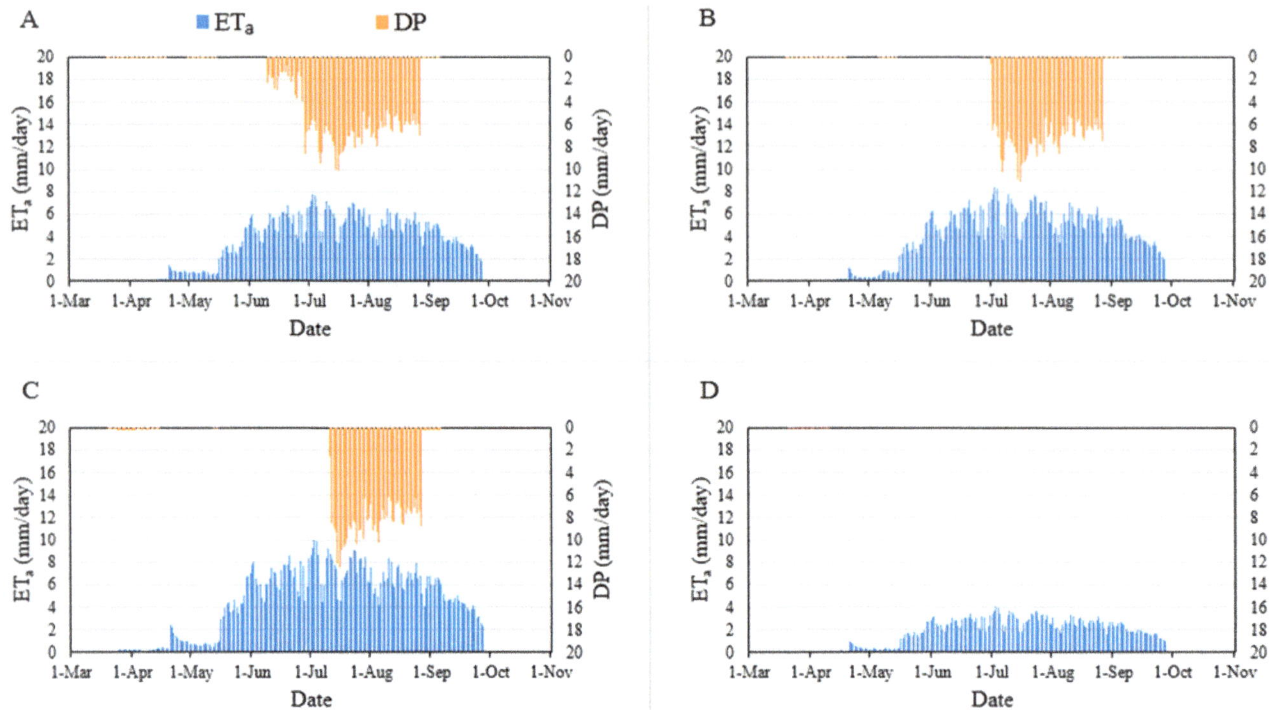

Figure 5: Actual crop evapotranspiration (ET_a) and deep percolation (DP) in sub-catchments A–D during the crop-growing season.

Effects of Scenarios

Irrigation Scenarios

Total Available Water (TAW) Scenarios

Optimal crop production depends greatly on available soil water [36]. Analyzing crop yields for every crop type under the same conditions of water scarcity represents a sensitivity analysis. It shows the highly differing susceptibility of crop types to water stress.

Figure 6 demonstrates robustness in the performance of cotton yield with reductions in the fraction of TAW. The yield of ginned cotton dropped 6%, from TAW of 0.7 to TAW of 0.4. The yields of wheat and bean also showed robust performances when irrigation water was decreased. Crop yields in response to TAW were investigated beginning at 0.7, where none of the crops suffered a reduction in yield. At a TAW of 0.6, most crops maintained a relatively high yield without a rapid decrease, except for tomatoes. Nearly half of the tomato production was stopped due to lack of water at TAW of 0.4. This demonstrated that tomatoes should be considered only in areas where the supply of irrigation water is ensured. At the same level of TAW, the yield of maize, sugarbeet and melon suffered a significant decline. However, because cotton was the most important crop in the Tarim River basin [37], its cultivation should be prioritized in areas where optimum conditions can be maintained. To maintain cotton production at relatively high levels, irrigation at a TAW level of 0.4 is recommended.

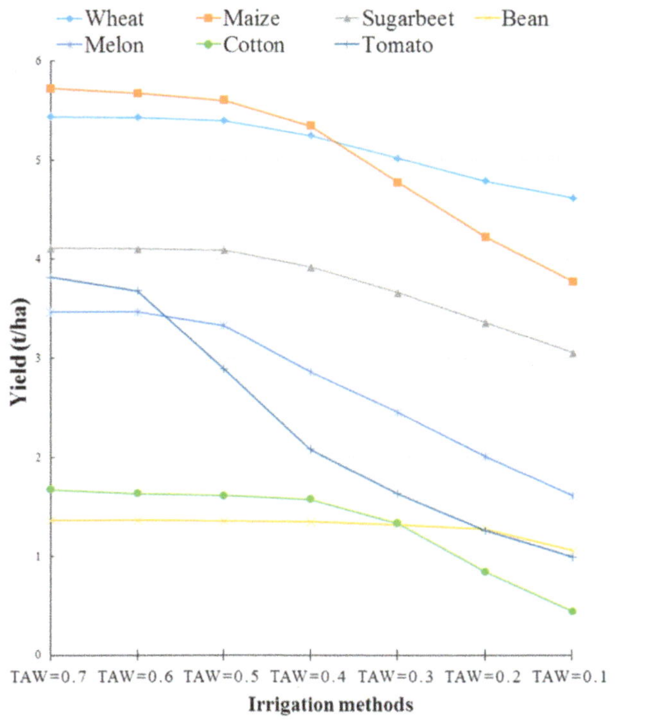

Figure 6: Irrigation scenarios based on fraction of total available water (TAW). Average yield performance of crops showing in the whole irrigation field within three-year simulation period.

Water-Saving Irrigation Scenarios

Spray loss (SL) and wetting fraction (WF) showed marked reductions with an increase in the percentage of DIUM applied in irrigated fields (Table 5), whereas water saving (WS) and reduction of water demand deficit (RWDD) showed positive effects. WS was only 6% and RWDD was only 5% when DIUM represented 10% of total irrigation. However, WS and RWDD increased to 40% and 30%, respectively, when the level of DIUM applied to all irrigated fields was 100%. This indicates the potential value of DIUM for saving water and reducing water deficits. Even when DIUM is at 50%, WS and RWDD reflected notable changes at 25% and 16%, respectively. When DIUM increased from 10% to 100%, SL decreased from 46% to 10% and WF reduced from 91% to 10%. This reflects the substantial benefits to water savings resulting from the application of these technologies.

Table 5: Summary of five DIUM scenarios

% DIUM	% SL	% WF	% WS	% RWDD
10	46	91	6	5
30	38	73	17	12
50	30	55	25	16
70	22	37	32	22
100	10	10	40	30

Results of modeling reveal that a considerable amount of water could be saved in with the introduction of water-saving irrigation. Considering the effort of applying DIUM on the irrigated fields, it was an arduous task to immediately raise the DIUM to 100%. However, even at DIUM of 70%, significant potential for saving water and reducing the water demand deficit exists. RWDD was slightly less affected than WS by the increase of DIUM. However, this abatement could result in a crucial change in water scarcity, with the highest value of 30% on the irrigated fields. Similar results were reported by Hu *et al.* [38], who concluded that water-saving economical irrigation can greatly increase water use efficiency.

Land Use Scenarios

The effects of the scenarios on irrigation water demand and water deficits at the sub-catchment level are shown in Figure 7.

In the LUD scenario, the reduction of farmland in sub-catchments A and B indicates a positive effect on the reduction of irrigation water demand. A substantial decline of 100 mm irrigation water demand could be achieved in the upper reaches. Moreover, LUD possessed a broader potential benefit on % irrigation water deficit reduction. All the sub-catchments showed notable decrease on water deficit. A similar land use investigation on the headwater tributary of Tarim River was conducted by Huang *et al.* [13], who found that the expansion of irrigation area ultimately leads to the increase in water consumption and reduces water availability.

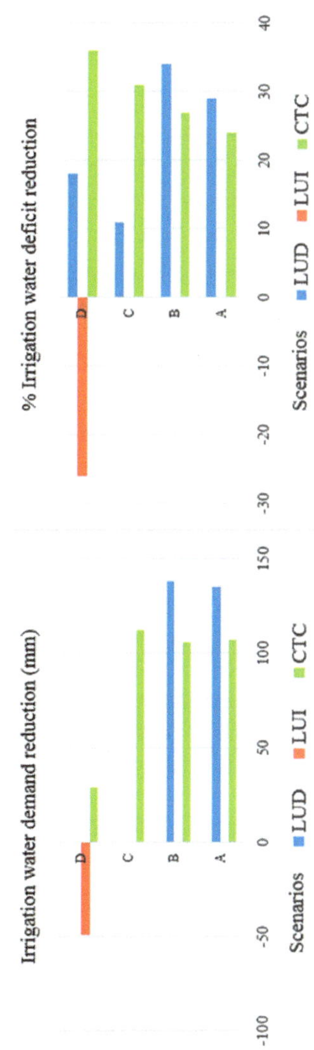

Figure 7: Effects of land use scenarios LUD (land use decrease), LUI (land use increase) and CTC (crop type change), with irrigation water demand reduction and % irrigation water deficit reduction as indicators in sub-catchments A–D.

In the LUI scenario, irrigation water demand was raised by nearly 50 mm with the increase of farmland in the lower reach. % irrigation water deficit was also increased by over 25% in the region. Since the increase of farmland in the upper and middle reaches had already been strictly restricted by the local government policies, the increase of farmland in the lower reach was also found to be harmful to the water saving purpose, and therefore should not be recommended.

In the CTC scenario, both irrigation water demand and % irrigation water deficit were largely reduced in all sub-catchments. Irrigation water demand dropped over 100 mm in the upper and middle reaches. % Irrigation water deficits were also significantly decreased, with a maximum reduction of 36% in the lower reach. Rotation of crop types can be an effective approach to reduce irrigation water demand and deficit. Therefore, scenario CTC is recommended for future land use developments.

CONCLUSIONS

The MIKE HYDRO model adequately represented the river discharge and irrigation water allocation in the main stem Tarim River. By adapting the K_{cb} values, the actual evapotranspiration could also be reliably simulated. In general, the model performed satisfactorily.

In 2006, the total ET_a in sub-catchments A–D was 666, 709, 864 and 190 mm, respectively, which illustrates the condition of high water stress for crops in the lower reach of sub-catchment D. Meanwhile, the highest ET_a and DP both occurred in sub-catchment C. Irrigation water consumption was relatively high in this district.

Optimal crop production depends greatly on available soil water. Cotton demonstrates a robust yield performance with a reduction in the fraction of TAW. The yield of ginned cotton dropped 6%, from TAW of 0.7 to TAW of 0.4. To maintain cotton production at a relatively high level, while reducing total water use, a TAW of 0.4 is recommended. Wheat, maize, sugar beets, beans, melons and cotton all showed a higher compatibility with water stress compared with tomatoes. Tomatoes should be grown only in areas where the supply of irrigation water is ensured.

WS and RWDD increased to 40% and 30%, respectively, with DIUM at 100% in all irrigated fields. This indicates the potential value of DIUM in saving water and reducing water deficits. DIUM could largely ameliorate the water shortage in the lower reach, where the natural Tugai vegetation is functioning as an important ecosystem by restraining dust. This water-saving technique should be applied whenever possible.

A considerable reduction in the demand for irrigation water and water deficits can be achieved by changing land use. Over 100 mm of irrigation water demand was saved by reducing the amount of farmland in sub-catchments A and B. In both sub-catchments, irrigation water deficit decreased by more than 25%. Reduced areas of farmland in the upper reaches could have considerable benefits for all the sub-catchments. The substitution of apocynum for cotton could be an effective approach in reducing irrigation water demand and water deficit. Therefore, scenario CTC is recommended for future land use developments in the region. The most important feature of the MIKE HYDRO model is the clarification of the mutual interference of the sub-catchments along the main stem Tarim River. Issues of agricultural water consumption and allocation cannot be effectively managed in separate regions. Clear evidence was presented to provide stakeholders and decision makers with relevant information about the anticipated effects of each scenario. Because the hydrological model has already been established in the study area, future simulations of water availability and allocation can be undertaken. Groundwater and the spatial distribution of water users in the sub-catchments will be addressed in the next phase of our research. The results of MIKE HYDRO simulations can contribute to decision-making for sustainable management of water and land use in the Tarim River basin.

ACKNOWLEDGMENTS

This study was supported by German-Sino bilateral collaboration research project SuMaRiO funded by the German Federal Ministry of Education and Research. Within SuMaRiO research group, we are grateful to Thevs Niels and Shamaila Zia-Khan's work, which helped the progress of our research. We thank our partners from the Xinjiang Institute of Ecology and Geography, Chinese Academy of Sciences who helped with our fieldwork and provided the basic data and information in the cooperated project (NSFC-UNEP, No. 41361140361). Suggestions and comments by the two reviewers on this paper are greatly appreciated. This work was supported by the German Research Foundation (DFG) and the Technische Universität München within the Open Access Publishing Funding Programme.

REFERENCES

1. Schmitz, C.; Lotze-Campen, H.; Gerten, D.; Dietrich, J.P.; Bodirsky, B.; Biewald, A.; Popp, A. Blue water scarcity and the economic impacts of future agricultural trade and demand. *Water Resour. Res.* 2013, *49*, 3601–3617. [Google Scholar] [CrossRef]
2. Cai, X.M.; Rosegrant, M.W. Optional water development strategies for the Yellow River Basin: Balancing agricultural and ecological water demands. *Water Resour. Res.* 2004, *40*. [Google Scholar] [CrossRef]
3. Chen, Y.; Xu, Z. Plausible impact of global climate change on water resources in the Tarim River Basin. *Sci. China Ser. D Earth Sci.* 2005, *48*, 65–73. [Google Scholar] [CrossRef]
4. Chen, Y.; Zhang, X.; Zhu, X.; Li, W.; Zhang, Y.; Xu, H. Analysis on the ecological benefits of the stream water conveyance to the dried-up river of the lower reaches of Tarim River, China. *Sci. China Ser. D Earth Sci.* 2004, *47*, 1053–1064. [Google Scholar] [CrossRef]
5. Thevs, N. Water scarcity and allocation in the Tarim Basin: Decision structures and adaptations on the local level. *J. Curr. Chin. Aff.* 2011, *40*, 113–137. [Google Scholar]
6. Chen, Y.N.; Takeuchi, K.; Xu, C.; Chen, Y.P.; Xu, Z.X. Regional climate change and its effects on river runoff in the Tarim Basin, China. *Hydrol. Process.* 2006, *20*, 2207–2216. [Google Scholar] [CrossRef]
7. Hao, X.M.; Chen, Y.; Xu, C.; Li, W. Impacts of climate change and human activities on the surface runoff in

the Tarim River Basin over the last fifty years. *Water Resour. Manag.* 2008, *22*, 1159–1171. [Google Scholar] [CrossRef]

8. Ling, H.B.; Xu, H.L.; Fu, J.Y. High- and low-flow variations in annual runoff and their response to climate change in the headstreams of the Tarim River, Xinjiang, China. *Hydrol. Process.* 2013, *27*, 975–988. [Google Scholar] [CrossRef]
9. Feng, Q.; Endo, K.N.; Cheng, G.D. Towards sustainable development of the environmentally degraded arid rivers of China-a case study from Tarim River. *Environ. Geol.* 2001, *41*, 229–238. [Google Scholar] [CrossRef]
10. Qi, F.; Wei, L.; Jianhua, S.; Yonghong, S.; Yewu, Z.; Zongqiang, C.; Haiyang, X. Environmental effects of water resource development and use in the Tarim River basin of northwestern China. *Environ. Geol.* 2005, *48*, 202–210. [Google Scholar] [CrossRef]
11. Hao, X.; Chen, Y.; Li, W. The driving forces of environmental change during the last 50 years in the Tarim River Basin.*Acta Geogr. Sin.* 2006, *61*, 262–272. [Google Scholar]
12. Chen, Y.; Ye, Z.; Shen, Y. Desiccation of the Tarim River, Xinjiang, China, and mitigation strategy. *Quat. Int.* 2011, *244*, 264–271. [Google Scholar] [CrossRef]
13. Huang, S.; Krysanova, V.; Zhai, J.; Su, B. Impact of intensive irrigation activities on river discharge under agricultural scenarios in the semi-arid Aksu River basin, northwest China. *Water Resour. Manag.* 2015, *29*, 945–959. [Google Scholar] [CrossRef]
14. Arnold, J.G.; Srinivasan, R.; Muttiah, R.S.; Williams, J.R. Large area hydrologic modeling and assessment part I: Model development. *J. Am. Water Resour. Assoc.* 1998, *34*, 73–89. [Google Scholar] [CrossRef]
15. Biancamaria, S.; Bates, P.D.; Boone, A.; Mognard, N.M. Large-scale coupled hydrologic and hydraulic modeling of the Ob river in Siberia. *J. Hydrol.* 2009, *379*, 136–150. [Google Scholar] [CrossRef]
16. Paz, A.R.; Bravo, J.M.; Allasia, D.; Collischonn, W.; Tucci, C.E.M. Large-Scale hydrodynamic modeling of a complex river network and floodplains. *J. Hydrol. Eng.* 2010, *15*, 152–165. [Google Scholar] [CrossRef]
17. Gunduz, O. Coupled Flow and Contaminant Transport Modeling in Large Watersheds. Ph.D. Thesis, Georgia Institute of Technology, Atlanta, GA, USA, 2004. [Google Scholar]
18. Danish Hydraulic Institute (DHI). *MIKE HYDRO User Manual*; DHI: Hoersholm, Denmark, 2014. [Google Scholar]
19. Chen, Y.N.; Xu, C.C.; Hao, X.M.; Li, W.H.; Chen, Y.P.; Zhu, C.G. Fifty-Year climate change and its effect on annual runoff in the Tarim River Basin, China. *Quat. Int.* 2009, *208*, 53–61. [Google Scholar]
20. Xu, H.; Ye, M.; Li, J. The water transfer effects on agricultural development in the lower Tarim River, Xinjiang of China. *Agric. Water Manag.* 2008, *95*, 59–68. [Google Scholar] [CrossRef]
21. Chow, V.T.; Maidment, D.R.; Mays, L.W. *Applied Hydrology*; Tata McGraw-Hill Education: New York, NY, USA, 1988; p. 572. [Google Scholar]
22. Veissman, W.; Lewis, G.L. *Introduction to Hydrology*, 5th ed.; Prentice Hall: Upper Saddle River, NJ, USA, 2003; p. 612. [Google Scholar]
23. Allen, R.G.; Pereira, L.S.; Raes, D.; Smith, M. *Crop Evapotranspiration—Guidelines for Computing Crop Water Requirements*; FAO Irrigation and Drainage Paper 56; FAO: Rome, Italy, 1998; Volume 300. [Google Scholar]
24. Steduto, P.; Hsiao, T.C.; Fereres, E.; Raes, D. *Crop Yield Response to Water*; FAO: Rome, Italy, 2012; pp. 17–40. [Google Scholar]
25. Nash, J.; Sutcliffe, V. River flow forecasting through conceptual models, part I—A discussion of principles. *J. Hydrol.*1970, *10*, 282–290. [Google Scholar] [CrossRef]
26. Legates, D.R.; McCabe, G.J. Evaluating the use of "goodness-of-fit" measures in hydrologic and hydroclimatic model validation. *Water Resour. Res.* 1999, *35*, 233–241. [Google Scholar] [CrossRef]
27. Moriasi, D.N.; Arnold, J.G.; van Liew, M.W.; Bingner, R.L.; Harmel, R.D.; Veith, T.L. Model evaluation guidelines for systematic quantification of accuracy in watershed simulations. *Trans. ASABE* 2007, *50*, 885–900. [Google Scholar] [CrossRef]
28. Girolamo, D.; Porto, L. Land use scenario development as a tool for watershed management within the Rio Mannu Basin. *Land Use Policy* 2012, *29*, 691–701. [Google Scholar] [CrossRef]
29. Molina-Navarro, E.; Trolle, D.; Martínez-Pérez, S.; Sastre-Merlín, A.; Jeppesen, E. Hydrological and water quality impact assessment of a Mediterranean limno-reservoir under climate change and land use management scenarios. *J. Hydrol.* 2014, *509*, 354–366. [Google Scholar] [CrossRef]
30. Stewart, J.I.; Hagan, R.M.; Pruitt, W.O.; Danielson, R.E.; Franklin, W.T.; Hanks, R.J.; Riley, J.P.; Jackson, E.B. *Optimizing Crop Production through Control of Water and Salinity Levels in the Soil*; Utah State University: Logan, UT, USA, 1977; p. 67. [Google Scholar]
31. Doorenbos, J.; Kassam, A.H. *Yield Response to Water*; Irrigation and Drainage Paper; FAO: Rome, Italy, 1979; Volume 33, p. 257. [Google Scholar]
32. Zia-Khan, S.; Spreer, W.; Pengnian, Y.; Zhao, X.; Othmanli, H.; He, X.; Müller, J. Effect of dust deposition on stomatal conductance and leaf temperature of cotton in northwest China. *Water* 2014, *7*, 116–131. [Google Scholar] [CrossRef]
33. Thevs, N.; Zerbe, S.; Kyosev, Y.; Rozi, A.; Tang, B.; Abdusalih, N.; Novitskiy, Z. *Apocynum venetum* L. and *Apocynum pictum* Schrenk (Apocynaceae) as multi-functional and multi-service plant species in Central Asia: A review on biology, ecology, and utilization. *J. Appl. Bot. Food Qual.* 2012, *85*, 159–167. [Google Scholar]
34. Ma, X.; Chen, Y.; Zhu, C.; Li, W. The variation in soil moisture and the appropriate groundwater table for desert riparian forest along the Lower Tarim River. *J. Geogr.*

Sci. 2011, *21*, 150–162. [Google Scholar] [CrossRef]

35. Yuan, G.; Zhang, P.; Shao, M.A.; Luo, Y.; Zhu, X. Energy and water exchanges over a riparian *Tamarix* spp. stand in the lower Tarim River basin under a hyper-arid climate. *Agric. For. Meteorol.* 2014, *194*, 144–154. [Google Scholar] [CrossRef]
36. Stričevič, R.; Čaki, E. Relationships between available soil water and indicators of plant water status of sweet sorghum to be applied in irrigation scheduling. *Irrig. Sci.* 1997, *18*, 17–21. [Google Scholar]
37. Liu, Y.; Chen, Y. Impact of population growth and land-use change on water resources and ecosystems of the arid Tarim River basin in western China. *Int. J. Sustain. Dev. World Ecol.* 2006, *13*, 295–305. [Google Scholar] [CrossRef]
38. Hu, S.; Song, Y.; Zhou, H.; Tian, C. Experimental study on water use efficiency of cotton in the Tarim River Basin.*Agric. Res. Arid Areas* 2001, *20*, 66–70.

INDEX

F

G

H

I

J

L

M

N

O

P

R

S

T

U

V

W

Y